T0182106

Undergraduate Lecture Notes in Physics

Series editors

Neil Ashby, University of Colorado, Boulder, CO, USA

William Brantley, Department of Physics, Furman University, Greenville, SC, USA

Matthew Deady, Physics Program, Bard College, Annandale-on-Hudson, NY, USA

Michael Fowler, Department of Physics, University of Virginia, Charlottesville, VA, USA

Morten Hjorth-Jensen, Department of Physics, University of Oslo, Oslo, Norway

Undergraduate Lecture Notes in Physics (ULNP) publishes authoritative texts covering topics throughout pure and applied physics. Each title in the series is suitable as a basis for undergraduate instruction, typically containing practice problems, worked examples, chapter summaries, and suggestions for further reading.

ULNP titles must provide at least one of the following:

- An exceptionally clear and concise treatment of a standard undergraduate subject.
- A solid undergraduate-level introduction to a graduate, advanced, or non-standard subject.
- A novel perspective or an unusual approach to teaching a subject.

ULNP especially encourages new, original, and idiosyncratic approaches to physics teaching at the undergraduate level.

The purpose of ULNP is to provide intriguing, absorbing books that will continue to be the reader's preferred reference throughout their academic career.

More information about this series at http://www.springer.com/series/8917

Jochen Pade

Quantum Mechanics for Pedestrians 1

Fundamentals

Second Edition

 Springer

Jochen Pade
Institut für Physik
Universität Oldenburg
Oldenburg, Germany

ISSN 2192-4791   ISSN 2192-4805   (electronic)
Undergraduate Lecture Notes in Physics
ISBN 978-3-030-00463-7   ISBN 978-3-030-00464-4   (eBook)
https://doi.org/10.1007/978-3-030-00464-4

Library of Congress Control Number: 2018954852

Originally published with the title: *Original Quantum Mechanics for Pedestrians 1: Fundamentals*
1st edition: © Springer International Publishing Switzerland 2014
2nd edition: © Springer Nature Switzerland AG 2018
This work is subject to copyright. All rights are reserved by the Publisher, whether the whole or part of the material is concerned, specifically the rights of translation, reprinting, reuse of illustrations, recitation, broadcasting, reproduction on microfilms or in any other physical way, and transmission or information storage and retrieval, electronic adaptation, computer software, or by similar or dissimilar methodology now known or hereafter developed.
The use of general descriptive names, registered names, trademarks, service marks, etc. in this publication does not imply, even in the absence of a specific statement, that such names are exempt from the relevant protective laws and regulations and therefore free for general use.
The publisher, the authors and the editors are safe to assume that the advice and information in this book are believed to be true and accurate at the date of publication. Neither the publisher nor the authors or the editors give a warranty, express or implied, with respect to the material contained herein or for any errors or omissions that may have been made. The publisher remains neutral with regard to jurisdictional claims in published maps and institutional affiliations.

This Springer imprint is published by the registered company Springer Nature Switzerland AG
The registered company address is: Gewerbestrasse 11, 6330 Cham, Switzerland

Preface to the Second Edition, Volume 1

The first edition of 'Physics for Pedestrians' was very well received. Repeatedly, I was asked to extend the considerations to relativistic phenomena. This has now been done in this second edition. Volume 1 contains elements of relativistic quantum mechanics, and Volume 2 contains elements of quantum field theory.

These extensions are placed in the Appendix. They are not comprehensive and complete presentations of the topics, but rather concise accounts of some essential ideas of relativistic quantum physics.

Furthermore, for the sake of completeness and to guarantee a consistent notation, there are outlines of relevant topics such as special relativity, classical field theory, and electrodynamics.

In addition, a few minor bugs have been fixed and some information has been updated.

I gratefully thank Svend-Age Biehs, Heinz Helmers, Stefanie Hoppe, Friedhelm Kuypers and Lutz Polley who have helped me in one way or another to prepare this second edition.

Oldenburg, Germany   Jochen Pade
February 2018

Preface to the First Edition, Volume 1

There are so many textbooks on quantum mechanics—do we really need another one?

Certainly, there may be different answers to this question. After all, quantum mechanics is such a broad field that a single textbook cannot cover all the relevant topics. A selection or prioritization of subjects is necessary *per se*, and moreover, the physical and mathematical foreknowledge of the readers has to be taken into account in an adequate manner. Hence, there is undoubtedly not only a certain leeway, but also a definite need for a wide variety of presentations.

Quantum Mechanics for Pedestrians has a thematic blend that distinguishes it from other introductions to quantum mechanics (at least those of which I am aware). It is not just about the conceptual and formal foundations of quantum mechanics, but from the beginning and in some detail it also discusses both current topics as well as advanced applications and basic problems as well as epistemological questions. Thus, this book is aimed especially at those who want to learn not only the appropriate formalism in a suitable manner, but also those other aspects of quantum mechanics addressed here. This is particularly interesting for students who want to teach quantum mechanics themselves, whether at the school level or elsewhere. The current topics and epistemological issues are especially suited to generate interest and motivation among students.

Like many introductions to quantum mechanics, this book consists of lecture notes which have been extended and complemented. The course which I have given for several years is aimed at teacher candidates and graduate students in the master's program, but is also attended by students from other degree programs. The course includes lectures (two sessions/four hours per week) and problem sessions (two hours per week). It runs for 14 weeks, which is reflected in the 28 chapters of the lecture notes.

Due to the usual interruptions such as public holidays, it will not always be possible to treat all 28 chapters in 14 weeks. On the other hand, the later chapters in particular are essentially independent of each other. Therefore, one can make a selection based on personal taste without losing coherence. Since the book consists of extended lecture notes, most of the chapters naturally offer more material than will fit into a two-hour lecture. But the 'main material' can readily be presented within this time; in addition, some further topics may be treated using the exercises.

Before attending the quantum mechanics course, the students have had among others an introduction to atomic physics: Relevant phenomena, experiments, and simple calculations should therefore be familiar to them. Nevertheless, experience has shown that at the start of the lectures, some students do not have enough substantial and available knowledge at their disposal. This applies less to physical and more to the necessary mathematical knowledge, and there are certainly several reasons for this. One of them may be that for teacher training; not only the quasi-traditional combination physics/mathematics is allowed, but also others such as physics/sports, where it is obviously more difficult to acquire the necessary mathematical background and, especially, to actively practice its use.

To allow for this, I have included some chapters with basic mathematical knowledge in the Appendix, so that students can use them to overcome any remaining individual knowledge gaps. Moreover, the mathematical level is quite simple, especially in the early chapters; this course is not just about practicing specifically elaborated formal methods, but rather we aim at a compact and easily accessible introduction to key aspects of quantum mechanics.

As remarked above, there are a number of excellent textbooks on quantum mechanics, not to mention many useful Internet sites. It goes without saying that in writing the lecture notes, I have consulted some of these, have been inspired by them and have adopted appropriate ideas, exercises, etc., without citing them in detail. These books and Internet sites are all listed in the bibliography and some are referred to directly in the text.

A note on the title *Quantum Mechanics for Pedestrians*: It does not mean 'quantum mechanics light' in the sense of a painless transmission of knowledge à la Nuremberg funnel. Instead, 'for pedestrians' is meant here in the sense of autonomous and active movement—step by step, not necessarily fast, from time to time (i.e., along the more difficult stretches) somewhat strenuous, depending on the level of understanding of each walker—which will, by the way, become steadily better while walking on.

Speaking metaphorically, it is about discovering on foot the landscape of quantum mechanics; it is about improving one's knowledge of each locale (if necessary, by taking detours); and it is perhaps even about finding your own way.

By the way, it is always amazing not only how far one can walk with some perseverance, but also how fast it goes—and how sustainable it is. 'Only where you have visited on foot, have you really been.' (Johann Wolfgang von Goethe).

Klaus Schlupmann, Heinz Helmers, Edith Bakenhus, Regina Richter, and my sons, Jan Philipp and Jonas have critically read several chapters. Sabrina Milke assisted me in making the index. I enjoyed enlightening discussions with Lutz Polley, while Martin Holthaus provided helpful support and William Brewer made useful suggestions. I gratefully thank them and all the others who have helped me in some way or other in the realization of this book.

Contents

Contents of Volume 2

Introduction

Quantum mechanics is probably the most accurately verified physical theory existing today. To date, there has been no contradiction from any experiments; the applications of quantum mechanics have changed our world right up to aspects of our everyday life. There is no doubt that quantum mechanics 'functions'—it is indeed extremely successful. On a formal level, it is clearly unambiguous and consistent and (certainly not unimportant)—as a theory—it is both aesthetically satisfying and convincing.

The question in dispute is the 'real' meaning of quantum mechanics. What does the wavefunction stand for, and what is the role of chance? Do we actually have to throw overboard our classical and familiar conceptions of reality? Despite the nearly century-long history of quantum mechanics, fundamental questions of this kind are still unresolved and are currently being discussed in a lively and controversial manner. There are two contrasting positions (along with many intermediate views): Some see quantum mechanics simply as the precursor stage of the 'true' theory (although eminently functional); others see it as a valid, fundamental theory itself.

This book aims to introduce its readers to both sides of quantum mechanics, the established side and the side that is still under discussion. We develop here both the conceptual and formal foundations of quantum mechanics, and we discuss some of its 'problem areas.' In addition, this book includes applications—oriented fundamental topics, some 'modern' ones—for example, issues in quantum information— and 'traditional' ones such as the hydrogen and the helium atoms. We restrict ourselves to the field of nonrelativistic physics, although many of the ideas can be extended to the relativistic case.[1] Moreover, we consider only time-independent interactions.

In introductory courses on quantum mechanics, the practice of formal skills often takes priority (this is subsumed under the slogan 'shut up and calculate'). In accordance with our objectives here, we will also give appropriate space to the discussion of fundamental questions. This special blend of basic discussion and

---

[1]In the second edition, some essentials of relativistic quantum mechanics are added; see the Appendix.

modern practice is in itself very well suited to evoke interest and motivation in students. This is, in addition, enhanced by the fact that some important fundamental ideas can be discussed using very simple model systems as examples. It is not coincidental that some of the topics and phenomena addressed here are treated in various simplified forms in high-school textbooks.

In mathematical terms, there are two main approaches used in introductions to quantum mechanics. The first one relies on differential equations (i.e., analysis) and the other one on vector spaces (i.e., linear algebra); of course, the 'finished' quantum mechanics is independent of the route of access chosen. Each approach (they also may be called the Schrodinger and the Heisenberg routes) has its own advantages and disadvantages; the two are used in this book on an equal footing.

The roadmap of the book is as follows:

The foundations and structure of quantum mechanics are worked out step by step in the first part (Volume 1, Chaps. 1–14), alternatively from an analytical approach (odd chapters) and from an algebraic approach (even chapters). In this way, we avoid limiting ourselves to only one of the two formulations. In addition, the two approaches reinforce each other in the development of important concepts. The merging of the two threads starts in Chap. 12. In Chap. 14, the conclusions thus far reached are summarized in the form of quite general postulates for quantum mechanics.

Especially in the algebraic chapters, we take up current problems early on (interaction-free quantum measurements, the neutrino problem, quantum cryptography). This is possible since these topics can be treated using very simple mathematics. Thus, this type of access is also of great interest for high-school level courses. In the analytical approach, we use as elementary physical model systems the infinite potential well and free particle motion.

In the second part (Volume 2, Chaps. 15–28), applications and extensions of the formalism are considered. The discussion of the conceptual difficulties (measurement problem, locality and reality, etc.) again constitutes a central theme, as in the first volume. In addition to some more traditionally oriented topics (angular momentum, simple potentials, perturbation theory, symmetries, identical particles, scattering), we begin in Chap. 20 with the consideration of whether quantum mechanics is a local realistic theory. In Chap. 22, we introduce the density operator in order to consider in Chap. 24 the phenomenon of decoherence and its relevance to the measurement process. In Chap. 27, we continue the realism debate and explore the question as to what extent quantum mechanics can be regarded as a complete theory. Modern applications in the field of quantum information can be found in Chap. 26.

Finally, we outline in Chap. 28 the most common interpretations of quantum mechanics. Apart from this chapter, a general statement applies: While it is still a controversial issue as to which (if indeed any) of the current interpretations is the 'correct' one, an introduction to quantum mechanics must take a concrete position and has to present the material in a coherent form. In this book, we choose the version commonly known as the 'standard interpretation.'

A few words about the role of mathematics:

In describing objects that—due to their small size—are beyond our everyday experience, quantum mechanics cannot be formulated completely in terms of everyday life and must therefore remain to some extent abstract. A deeper understanding of quantum mechanics cannot be achieved on a purely linguistic level; we definitely need mathematical descriptions.[2] Of course, one can use analogies and simplified models, but that works only to a certain degree and also makes sense only if one is aware of the underlying mathematical apparatus, at least in broad terms.[3]

It is due to this interaction of the need for mathematical formulations and the lack of intuitive access that quantum mechanics is often regarded as 'difficult.' But that is only part of the truth; to be sure, there are highly formalized and demanding aspects. Many wider and interesting issues, however, are characterized by very simple principles that can be described using only a basic formalism.

Nevertheless, beginners in particular perceive the role of mathematics in quantum mechanics as discouraging. Three steps serve to counter this impression or, in the optimum case, to avoid it altogether:

First, we keep the mathematical level as simple as possible and share the usual quite nonchalant attitude of physicists toward mathematics. In particular, the first chapters go step by step, so that the initially diverse mathematics skills of the readers are gradually brought up to a common level.

In addition, we use very simple models, toy models so to speak, especially in the first part of the book, in order to treat the main physical ideas without becoming involved in complicated mathematical questions. Of course, these models are only rough descriptions of actual physical situations. But they manage with relatively simple mathematics, do not require approximation methods or numerics, and yet still permit essential insights into the fundamentals of quantum mechanics.[4] Only in Volume 2, more realistic models are applied, and this is reflected occasionally in a somewhat more demanding formal effort.

The third measure involves exercises and some support from the Appendix. At the end of almost every chapter, there is a variety of exercises, some of them dealing with advanced topics. They invite the reader to work with the material in

---

[2]This applies at least to physicists; for as Einstein remarked: 'But there is another reason for the high repute of mathematics: it is mathematics that offers the exact natural sciences a certain measure of security which, without mathematics, they could not attain.' To give a layman without mathematical training an understanding of quantum mechanics, one will (or must) rely instead on math-free approaches.

[3]Without appropriate formal considerations, it is impossible to understand, for example, how to motivate the replacement of a physical measurement variable by a Hermitian operator.

[4]We could instead also make use of the large reservoir of historically important experiments. But their mathematical formulation is in general more complex, and since in the frame of our considerations they do not lead to further-reaching conclusions than our 'toy models,' we restrict ourselves to the latter for clarity and brevity.

order to better assimilate and more clearly grasp it, as well as of course to train the necessary formal skills.[5]

The learning aids in the Appendix include chapters with some basic mathematical and physical background information; this allows the reader to refresh 'passive' knowledge without the need to refer to other sources or to become involved with new notations.

Moreover, the no doubt unusually extensive Appendix contains the solutions to many of the exercises and, in addition, some chapters in which further-reaching questions and issues are discussed; although these are very interesting in themselves, their treatment would far exceed the framework of a lecture course.

The footnotes with a more associative character can be skipped on a first reading.

A note on the term 'particle': Its meaning is rather vague in physics. On the one hand, it denotes 'something solid, not wavelike'; on the other hand 'something small', ranging from the elementary particles as structureless building blocks of matter, to objects which themselves are composed of constituent 'particles' like the α particle and other atomic nuclei or even macroscopic particles like sand grains. In quantum mechanics, where indeed it is often not even clear whether a particular object has mainly particle or mainly wave character, the careless use of the term may cause confusion and communication problems.

Accordingly, several terms which go beyond 'wave' or 'particle' have been suggested, such as quantal particle, wavical, wavicle, quantum object, quanton. Throughout this book, we will use the term 'quantum object,' unless there are traditionally established terms such as 'identical particles' or 'elementary particles.' The consistent use of 'quantum object' instead of 'particle' may perhaps seem somewhat pedantic, but we hope that it will help to ensure that fewer false images stick in the minds of readers; it is for this reason that this term is also found in many high-school textbooks.

Quantum mechanics is a fundamental theory of physics, which has given rise to countless applications. But it also extends deep into areas such as philosophy and epistemology and leads to thinking about 'what holds the world together at its core'; in short, it is also an intellectual adventure. The fascinating thing is that the more one becomes acquainted with quantum mechanics, the more one realizes how simple many of its central ideas really are.[6] It would be pleasing if *Quantum Mechanics for Pedestrians* could help to reveal this truth.

---

[5]'It is a great support to studying, at least for me, to grasp everything that one reads so clearly that one can apply it oneself, or even make additions to it. One is then inclined to believe in the end that one could have invented everything himself, and that is encouraging.' Georg Christoph Lichtenberg, *Scrap Books,* Vol. J (1855).

[6]'The less we know about something, the more complicated it is, and the more we know about it, the easier it is. This is the simple truth about all the complexities.' Egon Friedell, in *Kulturgeschichte der Neuzeit; Kulturgeschichte Agyptens und des alten Orients (Cultural history of modern times; the cultural history of Egypt and the ancient Near East).*

Let us close with a remark by Richard Feynman which holds true not only for physics in general, but even more for quantum mechanics: 'Physics is like sex: Sure, it may give some practical results, but that's not why we do it.'

Overview of Volume 1

In the following 14 chapters, we want to work out the fundamental structure of quantum mechanics, *videlicet* on the basis of a few simple models. The use of these 'toy systems' has two advantages.

First, their simplicity allows us to identify the essential mechanisms of quantum mechanics without getting lost in complex mathematical considerations. These mechanisms, which we summarize in Chap. 14 in the form of postulates, can nevertheless be formulated in a rather general manner.

Second, we can emphasize the essential ideas very quickly in this manner, so that we can treat and understand current topics quite soon along the trail.

Part I
Fundamentals

Chapter 1
Towards the Schrödinger Equation

We construct an equation that is valid for matter in the nonrelativistic domain, but also allows for wave-like solutions. This is the Schrödinger equation; it describes the dynamics of a quantum system by means of the time evolution of the wavefunction.

Many different paths lead to the goal of this chapter, the *Schrödinger equation* (SEq). We choose a traditional one, in which wave properties and the relationship between energy and momentum are the defining elements. Another approach (quantum hopping) can be found in Appendix J, Vol. 1. Certainly that approach is more unconventional, but on the other hand, it makes the basic physical principles more clearly manifest. Of course, the two approaches both lead to the same result.

After a few words about the construction of new theories, we consider solutions of the classical wave equation. It will turn out that the wave equation is not suitable for describing quantum-mechanical phenomena. But we learn in this way how to construct the 'right' equation, i.e. the Schrödinger equation. We restrict our considerations to sufficiently low velocities so that we can ignore relativistic effects.[1]

1.1 How to Find a New Theory

Classical mechanics cannot explain a goodly number of experimental results, such as the interference of particles (two-slit experiment with electrons), or the quantization of angular momentum, energy etc. (Stern-Gerlach experiment, atomic energy levels). A new theory is needed—but how to construct it, how do we find the adequate new physical concepts and the appropriate mathematical formalism?

---

[1]Relativistic effects are explicitly considered in Appendix U, Vol. 1 (relativistic quantum mechanics) and Appendix W, Vol. 2 (quantum field theory).

© Springer Nature Switzerland AG 2018
J. Pade, *Quantum Mechanics for Pedestrians 1*, Undergraduate Lecture
Notes in Physics, https://doi.org/10.1007/978-3-030-00464-4_1

The answer is: There is no clearly-prescribed recipe, no deductive or inductive 'royal road'. To formulate a new theory requires creativity or, in simpler terms, something like 'intelligent guessing'.[2] Of course, there are experimental and theoretical frameworks that limit the arbitrariness of guessing and identify certain directions. Despite this, however, it is always necessary to think of something new which does not exist in the old system—or rather, cannot and must not exist in it. The transition from Newtonian to relativistic dynamics requires as a new element the hypothesis that the speed of light must have the same value in all inertial frames. This element does not exist in the old theory—on the contrary, it contradicts it and hence cannot be inferred from it.

In the case of quantum mechanics (QM), there is the aggravating circumstance that we have no sensory experience of the microscopic world[3] which is the actual regime of quantum mechanics. More than in other areas of physics, which are closer to everyday life and thus more intuitive,[4] we need to rely on physical or formal analogies,[5] we have to trust the models and mathematical considerations, as long as they correctly describe the outcome of experiments, even if they are not in accord with our everyday experience. This is often neither easy nor familiar[6]—in quantum mechanics particularly, because the meaning of some terms is not entirely clear. In fact, quantum mechanics leads us to the roots of our knowledge and understanding of the world, and this is why in relation to certain questions it is sometimes called 'experimental philosophy'.

In short, we cannot derive quantum mechanics strictly from classical mechanics or any other classical theory[7]; new formulations must be found and stand the test of experiments. With all these caveats and preliminary remarks, we will now start along our path to quantum mechanics.[8]

---

[2]How difficult this can be is shown e.g. by the discussion about quantum gravity. For dozens of years, there have been attempts to merge the two basic realms of *quantum theory* and *general relativity theory*—so far (2018) without tangible results.

[3]Evolution has made us (more or less) fit for the demands of our everyday life—and microscopic phenomena simply do not belong to that everyday world. This fact, among others, complicates the teaching of quantum mechanics considerably.

[4]Insofar as e.g. electrodynamics or thermodynamics are intuitive...

[5]In the case of quantum mechanics, a physical analogy would be for example the transition from geometrical optics (= classical mechanics) to wave optics (= quantum mechanics). If one prefers to proceed abstractly, one can for instance replace the Poisson brackets of classical mechanics by commutators of corresponding operators—whereby at this point it naturally remains unclear without further information why one should entertain such an idea.

[6]This also applies e.g. to the special theory of relativity with its 'paradoxes', which contradict our everyday experience.

[7]This holds true in a similar way for all fundamental theories. For instance, Newtonian mechanics cannot be inferred strictly from an older theory; say, Aristotle's theory of motion. In the frame of classical mechanics, Newton's axioms are principles which are not derivable, but rather are postulated without proof.

[8]"A journey of a thousand miles begins with a single step". (Lao Tzu).

1.2   The Classical Wave Equation and the Schrödinger Equation

This approach to the Schrödinger equation is based on analogies, where the mathematical formulation in terms of differential equations[9] plays a central role. In particular, we take as a basis the physical principle of *linearity* and in addition the non-relativistic relation between energy and momentum, i.e.

$$E = \frac{p^2}{2m}. \tag{1.1}$$

By means of the *de Broglie relations*[10]

$$E = \hbar\omega \text{ and } p = \hbar k, \tag{1.2}$$

Equation (1.1) may be rewritten and yields the *dispersion relation*[11]:

$$\omega = \frac{\hbar^2}{2m}k^2. \tag{1.3}$$

In the following, we will examine special solutions of differential equations, namely plane waves, and check whether their wavenumber k and frequency ω satisfy the dispersion relation (1.3).

A remark on the constants k and ω: They are related to the wavelength λ and the frequency ν by $k = 2\pi/\lambda$ and $\omega = 2\pi\nu$. In quantum mechanics (and in other areas of physics), one hardly ever has to deal with λ and ν, but almost *exclusively* with k and ω. This may be the reason that in physics, ω is usually called 'the frequency' (and not the *angular* frequency).

1.2.1   From the Wave Equation to the Dispersion Relation

As a result of interference phenomena, the double-slit experiment and other experiments suggest that the electron, to put it rather vaguely, is 'somehow a kind of wave'.

---

[9]Some basic facts about differential equations can be found in Appendix E, Vol. 1.

[10]The symbol h was introduced by Planck in 1900 as an auxiliary variable ('Hilfsvariable' in German, hence the letter h) in the context of his work on the black body spectrum. The abbreviation $\hbar$ for $\frac{h}{2\pi}$ was probably used for the first time in 1926 by P.A.M. Dirac. In terms of frequency ν/ wavelength λ, the de Broglie relations are $E = h\nu$ and $p = \frac{h}{\lambda}$. In general, the symmetrical form (1.2) is preferred.

[11]The term 'dispersion relation' means in general the relationship between ω and k or between E and p (it is therefore also called the energy-momentum relation). Dispersion denotes the dependence of the velocity of propagation of a wave on its wavelength or frequency, which generally leads to the fact that a wave packet made up of different wavelengths diverges (disperses) over time.

Now we have learned in mechanics and electrodynamics that the classical wave equation

$$\frac{\partial^2 \Psi\,(\mathbf{r},t)}{\partial t^2} = c^2 \left(\frac{\partial^2 \Psi\,(\mathbf{r},t)}{\partial x^2} + \frac{\partial^2 \Psi\,(\mathbf{r},t)}{\partial y^2} + \frac{\partial^2 \Psi\,(\mathbf{r},t)}{\partial z^2} \right) = c^2 \nabla^2 \Psi\,(\mathbf{r},t) \quad (1.4)$$

describes many kinds of waves (acoustic, elastic, light waves, etc.). Ψ contains the amplitude and phase of the wave; c is its velocity of propagation, assumed to be constant.[12] It seems obvious to first check this equation to see if it can explain phenomena such as particle interference and so on. To keep the argument as simple as possible, we start from the one-dimensional equation:

$$\frac{\partial^2 \Psi\,(x,t)}{\partial t^2} = c^2 \frac{\partial^2}{\partial x^2} \Psi\,(x,t)\,. \quad (1.5)$$

The results thus obtained can be readily generalized to three dimensions.

In the following, we will examine the question of whether or not the wave equation can describe the behavior of electrons. Though the answer will be 'no', we will describe the path to this answer in a quite detailed way because it shows, despite the negative result, how one can guess or construct the 'right' equation, namely, the Schrödinger equation.

But first, we would like to point out an important property of the wave equation: it is *linear*—the unknown function Ψ occurs only to the first power and not with other exponents such as Ψ^2 or $\Psi^{1/2}$. From this, it follows that when we know two solutions, Ψ_1 and Ψ_2, any arbitrary linear combination $\alpha\Psi_1 + \beta\Psi_2$ is also a solution. In other words: The *superposition principle* holds.

1.2.1.1   Separation of Variables

Equation (1.5) has a solution, for example as the function

$$\Psi\,(x,t) = \Psi_0 e^{i(kx - \omega t)} \quad (1.6)$$

with the wave number k and frequency ω. How can we find such solutions? An important constructive approach is the so-called *separation of variables* which can be used for all linear partial differential equations. This *ansatz*, for obvious reasons also called *product ansatz*, reads

$$\Psi\,(x,t) = f\,(t) \cdot g\,(x) \quad (1.7)$$

with yet-to-be-determined functions $f\,(t)$ and $g\,(x)$. Substitution into (1.5) leads, with the usual shorthand notation $\dot{f} \equiv \frac{df}{dt}$ and $g' \equiv \frac{dg}{dx}$, to

---

[12]The *Laplacian* $\frac{\partial^2}{\partial x^2} + \frac{\partial^2}{\partial y^2} + \frac{\partial^2}{\partial z^2}$ is written as ∇^2, since it is the divergence $(\nabla\cdot)$ of the gradient, i.e. $\nabla\,(\nabla f) = \nabla^2 f$ (see Appendix D, Vol. 1).

$$\ddot{f}(t) \cdot g(x) = c^2 f(t) \cdot g''(x); \tag{1.8}$$

or, after division by $f(t) \cdot g(x)$, to

$$\frac{\ddot{f}(t)}{f(t)} = c^2 \frac{g''(x)}{g(x)}. \tag{1.9}$$

At this point we can argue as follows: x and t each appear on only *one* side of the equation, respectively (i.e. they are separated). Since they are *independent* variables we can, for example, fix x and vary t independently of x. Then the equality in (1.9) can be satisfied for all x and t only if both sides are *constant*. To save extra typing, we call this constant α^2 instead of simply α. It follows that:

$$\frac{\ddot{f}(t)}{f(t)} = \alpha^2; \quad \frac{g''(x)}{g(x)} = \frac{1}{c^2}\alpha^2; \quad \alpha \in \mathbb{C}. \tag{1.10}$$

Solutions of these differential equations are the exponential functions

$$f(t) \sim e^{\pm \alpha t}; \quad g(x) \sim e^{\pm \frac{1}{c}\alpha x}. \tag{1.11}$$

The range of values of the yet undetermined constant α can be limited by the requirement that physically meaningful solutions must remain *bounded* for all values of the variables.[13] It follows that α cannot be real, because then we would have unlimited solutions for t or $x \to +\infty$ or $-\infty$. Exactly the same is true if α is a complex number[14] with a non-vanishing real part. In other words: α must be purely *imaginary*,

$$\alpha \in \mathbb{I} \to \alpha = i\omega; \quad \omega \in \mathbb{R}. \tag{1.12}$$

Since the term $\frac{\alpha}{c}$ occurs in (1.11), we introduce the following abbreviation:

$$k = \frac{\omega}{c}. \tag{1.13}$$

Thus we obtain for the functions f and g

$$f(t) \sim e^{\pm i\omega t}; g(x) \sim e^{\pm ikx}; \quad \omega \in \mathbb{R} \tag{1.14}$$

where, unless noted otherwise, we assume without loss of generality that $k > 0$, $\omega > 0$ (in general, it follows that $\omega^2 = c^2 k^2$ from (1.9)). All combinations of the functions f and g, such as $e^{i\omega t}e^{-ikx}$, $e^{-i\omega t}e^{ikx}$, etc., each multiplied by an arbitrary constant, are also solutions of the wave equation.

---

[13]This is one of the advantages of physics as compared to mathematics: under certain circumstances, we can exclude mathematically correct solutions due to physical requirements (see also Appendix E, Vol. 1).

[14]Some remarks on the subject of complex numbers are to be found in Appendix C, Vol. 1.

1.2.1.2 Solutions of the Wave Equation; Dispersion Relation

To summarize: the separation *ansatz* has provided us with solutions of the wave equation. Typically, they read for $k > 0$, $\omega > 0$:

$$\Psi_1(x,t) = \Psi_{01}e^{i\omega t}e^{ikx}; \quad \Psi_2(x,t) = \Psi_{02}e^{-i\omega t}e^{ikx}$$
$$\Psi_3(x,t) = \Psi_{03}e^{i\omega t}e^{-ikx}; \quad \Psi_4(x,t) = \Psi_{04}e^{-i\omega t}e^{-ikx}. \quad (1.15)$$

The constants Ψ_{0i} are arbitrary, since due to the linearity of the wave equation, a multiple of a solution is also a solution.

Which physical situations are described by these solutions? Take, for example:

$$\Psi_2(x,t) = \Psi_{02}e^{-i\omega t}e^{ikx} = \Psi_{02}e^{i(kx-\omega t)}. \quad (1.16)$$

Due to $k > 0$, $\omega > 0$, this is a *plane wave* moving to the right, just as are Ψ_2^*, Ψ_3 and Ψ_3^* (* means the complex conjugate). By contrast, Ψ_1 and Ψ_4 and their complex conjugates are plane waves moving to the left.[15] For a clear and intuitive argumentation, see the exercises at the end of this chapter.

Although a plane wave is quite a common construct in physics,[16] the waves found here cannot describe the behavior of electrons. To see this, we use the de Broglie relations

$$E = \hbar\omega \quad \text{and} \quad p = \hbar k. \quad (1.17)$$

From (1.13), it follows that:

$$\omega = kc, \quad (1.18)$$

and this gives with (1.17):

$$\frac{E}{\hbar} = c\frac{p}{\hbar} \quad \text{or} \quad E = p \cdot c. \quad (1.19)$$

This relationship between energy and momentum cannot apply to our electron. We have restricted ourselves to the nonrelativistic domain, where according to

---

[15]To determine whether a plane wave moves to the left or to the right, one can set the exponent equal to zero. For $k > 0$ and $\omega > 0$, one obtains for example for Ψ_1 or Ψ_4:

$$v = \frac{x}{t} = -\frac{\omega}{k} < 0.$$

Due to $v < 0$, this is a left-moving plane wave. In contrast, for $k < 0$ and $\omega > 0$, we have right-moving plane waves.

[16]Actually it is 'unphysical', because it extends to infinity and on the average is equal everywhere, and therefore it is localized neither in time nor in space. But since the wave equation is linear, one can superimpose plane waves (partial solutions), e.g. in the form $\int c(k) e^{i(kx-\omega t)}dk$. The resulting *wave packets* can be quite well localized, as will be seen in Chap. 15, Vol. 2.

$E = p^2/2m$, a doubling of the momentum increases the energy by a factor of 4, while according to (1.19), only a factor of 2 is found. Apart from that, it is not clear what the value of the constant propagation velocity c of the waves should be.[17] In short, with (1.18), we have deduced the wrong dispersion relation, namely $\omega = kc$ and not the non-relativistic relation $\omega = \frac{\hbar}{2m} k^2$ which we formulated in (1.3). This means that the classical wave equation is not suitable for describing electrons—we must look for a different approach.

A remark about the three-dimensional wave equation (1.4): Its solutions are plane waves of the form

$$\Psi(\mathbf{r}, t) = \Psi_0 e^{i(\mathbf{kr}-\omega t)}; \quad \mathbf{k} = (k_x, k_y, k_z), k_i \in \mathbb{R} \tag{1.20}$$

with $\mathbf{kr} = k_x x + k_y y + k_z z$ and $\omega^2 = c^2 \mathbf{k}^2 = c^2 |k|^2 = c^2 k^2$. The *wave vector* $\mathbf{k}$ indicates the direction of wave propagation. In contrast to the one-dimensional wave, the components of $\mathbf{k}$ usually have arbitrary signs, so that the double sign $\pm$ in (1.14) does not appear here.

1.2.2 From the Dispersion Relation to the Schrödinger Equation

We now take the opposite approach: We start with the desired dispersion relation and deduce from it a differential equation under the assumptions that plane waves are indeed solutions and that the differential equation is linear, i.e. schematically:

Wave equation $\underset{\text{plane waves, linear}}{\Longrightarrow}$ 'wrong' relation $E = cp$

Schrödinger equation $\underset{\text{plane waves, linear}}{\Longleftarrow}$ 'right' relation $E = \dfrac{p^2}{2m}$.

The energy of a classical force-free particle is given by

$$E = \frac{p^2}{2m}. \tag{1.21}$$

With the de Broglie relations, we obtain the dispersion relation

$$\omega = \frac{\hbar k^2}{2m}. \tag{1.22}$$

Now we look for an equation whose solutions are plane waves, e.g. of the form $\Psi = \Psi_0 e^{i(kx-\omega t)}$, with the dispersion relation (1.22). To achieve this, we

---

[17]Likewise, (1.19) does not apply to an electron in the relativistic domain, since $E \sim p$ holds only for objects with zero rest mass.

differentiate the plane wave once with respect to t and twice with respect to x (we use the abbreviations $\partial_x := \frac{\partial}{\partial x}$, $\partial_{xx} = \partial_x^2 = \frac{\partial^2}{\partial x^2}$, etc.):

$$\partial_t \Psi = -i\omega \Psi_0 e^{i(kx-\omega t)}$$
$$\partial_x^2 \Psi = -k^2 \Psi_0 e^{i(kx-\omega t)}. \tag{1.23}$$

We insert these terms into (1.22) and find:

$$\omega = \frac{1}{-i\Psi}\partial_t \Psi = \frac{\hbar}{2m}k^2 = \frac{\hbar}{2m}\left(-\frac{1}{\Psi}\partial_x^2 \Psi\right)$$
$$\rightarrow i\partial_t \Psi = -\frac{\hbar}{2m}\partial_x^2 \Psi. \tag{1.24}$$

Conventionally, one multiplies by $\hbar$ to finally obtain:

$$i\hbar\partial_t \Psi = -\frac{\hbar^2}{2m}\partial_x^2 \Psi, \tag{1.25}$$

or, in the three-dimensional case,

$$i\hbar\frac{\partial}{\partial t}\Psi = -\frac{\hbar^2}{2m}\nabla^2 \Psi. \tag{1.26}$$

This is the *free time-dependent Schrödinger equation*. As the name suggests, it applies to an interaction-free quantum object.[18] For motions in a field with potential energy V, we have (in analogy to the classical energy $E = \frac{p^2}{2m} + V$) the *(general) time-dependent Schrödinger equation*

$$i\hbar\frac{\partial}{\partial t}\Psi = -\frac{\hbar^2}{2m}\nabla^2 \Psi + V\Psi. \tag{1.27}$$

Written out in full detail, it reads:

$$i\hbar\frac{\partial}{\partial t}\Psi(\mathbf{r}, t) = -\frac{\hbar^2}{2m}\nabla^2 \Psi(\mathbf{r}, t) + V(\mathbf{r}, t)\Psi(\mathbf{r}, t). \tag{1.28}$$

It is far from self-evident that the potential[19] V should be introduced into the equation in this manner and not in some other way. It is rather, like the whole 'derivation' of (1.28), a reasonable attempt or a bold step which still has to prove itself, as described above.

---

[18] We repeat a remark from the Introduction: For the sake of greater clarity we will use in quantum mechanics the term 'quantum object' instead of 'particle', unless there are traditionally preferred terms such as 'identical particles'.

[19] Although it is the potential energy V, this term is commonly referred to as the *potential*. One should note that the two concepts differ by a factor (e.g. in electrostatics, by the electric charge).

We note that the SEq is *linear* in Ψ: From two solutions, Ψ_1 and Ψ_2, any linear combination $\alpha_1\Psi_1 + \alpha_2\Psi_2$ with $\alpha_i \in \mathbb{C}$ is also a solution (see exercises). This is a crucial property for quantum mechanics, as we shall later see again and again.

Two remarks concerning $\Psi(\mathbf{r}, t)$, which is called the *wavefunction*,[20] state function or, especially in older texts, the psi function (Ψ function), are in order: The first is rather technical and almost self-evident. In general, only $\mathbf{r}$ and t are given as arguments of the wave function. But since these two variables have the physical units meter and second (we use the International System of Units, SI), they do not occur alone in the wave function, but always in combination with quantities having the inverse units. In the solutions, we always use $\mathbf{kr}$ and ωt, where $\mathbf{k}$ has the unit m^{-1} and ω the unit s^{-1}.

The second point is more substantial and far less self-evident. While the solution of the classical wave equation (1.5) has a direct and very clear physical meaning, namely the description of the properties of the observed wave (amplitude, phase, etc.), this is not the case for the wavefunction. Its magnitude $|\Psi(\mathbf{r}, t)|$ specifies an amplitude—but an amplitude of what? What is it that here makes up the 'waves' (remember that we are discussing electrons)? This was never referred to concretely in the derivation—it was never necessary to do so. It is indeed the case that the wavefunction has no direct physical meaning (at least not in everyday terms).[21] Perhaps it can best be understood as a complex-valued field of possibilities. In fact, one can extract from the wavefunction the relevant physical data, with an often impressive accuracy, without the need of a clear idea of what it specifically means. This situation (one operates with something, not really knowing what it is) produces unpleasant doubts, uncertainties and sometimes learning difficulties, particularly on first contact with quantum mechanics. But it is the state of our knowledge—the wavefunction as a key component of quantum mechanics has *no direct physical meaning*—that is how things are.[22]

---

[20] Despite its name, the wavefunction is a solution of the Schrödinger equation and not of the wave equation.

[21] This is one of the major problems in the teaching of quantum mechanics e.g. in high schools.

[22] Notwithstanding its somewhat enigmatic character (or perhaps because of it?), the wavefunction appears even in thrillers. An example: Harry smiled. "Good. In classical physics, an electron can be said to have a certain position. But in quantum mechanics, no. The wavefunction defines an area, say, of *probability*. An analogy might be that if a highly contagious disease turns up in a segment of the population, the disease control center gets right on it and tries to work out the probability of its recurrence in certain areas. The wavefunction isn't an entity, it's nothing in itself, it describes probability." Harry leaned closer as if he were divulging a sexy secret and went on: "So what we've got, then, is the probability of an electron's being in a certain place at a certain moment. Only when we're *measuring* it can we know not only where it is but *if* it is. So the cat..." Martha Grimes, in *The Old Wine Shades*.

1.3 Exercises

1. Consider the relativistic energy-momentum relation

$$E^2 = m_0^2 c^4 + p^2 c^2. \tag{1.29}$$

   Show that in the nonrelativistic limit $v \ll c$, it gives approximately (up to an additive positive constant)

$$E = \frac{p^2}{2m_0}. \tag{1.30}$$

2. Show that the relation $E = p \cdot c$ (c is the speed of light) holds only for objects with zero rest mass.
3. A (relativistic) object has zero rest mass. Show that in this case the dispersion relation reads $\omega^2 = c^2 k^2$.
4. Let $k < 0$, $\omega > 0$. Is $e^{i(kx - \omega t)}$ a right- or left-moving plane wave?
5. Solve the three-dimensional wave equation

$$\frac{\partial^2 \Psi(\mathbf{r}, t)}{\partial t^2} = c^2 \nabla^2 \Psi(\mathbf{r}, t) \tag{1.31}$$

   explicitly by using the separation of variables.
6. Given the three-dimensional wave equation for a vector field $\mathbf{A}(\mathbf{r}, t)$,

$$\frac{\partial^2 \mathbf{A}(\mathbf{r}, t)}{\partial t^2} = c^2 \nabla^2 \mathbf{A}(\mathbf{r}, t). \tag{1.32}$$

   (a) What is a solution in the form of a plane wave?
   (b) Which condition must $\mathbf{A}_0$ satisfy if $\mathbf{A}$ is (a) a longitudinal, (b) a transverse wave?

7. Given are the SEq

$$i\hbar \frac{\partial}{\partial t} \Psi(\mathbf{r}, t) = -\frac{\hbar^2}{2m} \nabla^2 \Psi(\mathbf{r}, t) + V(\mathbf{r}, t) \Psi(\mathbf{r}, t) \tag{1.33}$$

   and two solutions $\psi_1(\mathbf{r}, t)$ and $\psi_2(\mathbf{r}, t)$. Show explicitly that any linear combination of these solutions is also a solution.
8. The wavefunction of a quantum object of mass m is given by

$$\psi(x, t) = \psi_0 \exp\left(-\frac{x^2}{2b^2} - i\frac{\hbar}{2mb^2} t\right), \tag{1.34}$$

   where b is a fixed length. Determine the potential energy $V(x)$ of the quantum object.

9. Given the plane waves

$$\Phi_1 (x, t) = \Phi_{01} e^{\pm i (kx - \omega t)}; \quad \Phi_2 (x, t) = \Phi_{02} e^{\pm i (kx + \omega t)}; \quad k, \omega > 0; \quad \Phi_{0i} \in \mathbb{R}. \tag{1.35}$$

Explain in a visual way that $\Phi_1 (x, t)$ is a right- and $\Phi_2 (x, t)$ is a left-moving plane wave.

Chapter 2
Polarization

In this chapter, we make the transition from classical mechanics to quantum mechanics by considering light polarization. This leads us directly to two key concepts of quantum mechanics, vector space and probability. For the first time, we encounter the problem of measurement in quantum mechanics.

The approach to quantum mechanics in the preceding chapter is based on the description of the time evolution of a state by means of a differential equation. In this chapter, we choose a different approach. We consider (for now) not the Schrödinger equation or another description of the space-time behavior, but instead the emphasis is now on how we can define *states* (for the moment, time-independent states).

Again we start from classical formulations which we 'pep up' quantum mechanically. For this purpose we first show that under certain circumstances, we can treat electromagnetic waves as if they propagate in a two-dimensional *complex vector space*.[1] As is known from optics, we can express the intensities of light waves as the absolute squares of their amplitudes. After reviewing classical formulations, we extend these ideas to the quantum-mechanical case by means of a reinterpretation which is, while not mandatory, very plausible. In this interpretation, the amplitudes do not lead to intensities, but instead to *probabilities*. At this point we will see for the first time that the concept of *measurement* in quantum mechanics is not as trivial as it is in classical mechanics.

---

[1]If this term is not familiar (or forgotten): The basic concepts are summarized in Appendix G, Vol. 1. In addition, we will return to this topic in Chap. 4. For the moment, it is enough to know that e.g. the set of all vectors $\begin{pmatrix} a_1 \\ a_2 \end{pmatrix}$ with $a_i \in \mathbb{C}$ forms a two-dimensional complex vector space. An important property is that any linear combination of two vectors is itself a valid vector in this space.

© Springer Nature Switzerland AG 2018
J. Pade, *Quantum Mechanics for Pedestrians 1*, Undergraduate Lecture
Notes in Physics, https://doi.org/10.1007/978-3-030-00464-4_2

We will base our discussion on the *polarization* of light, which should be familiar from lectures and lab courses.[2]

2.1 Light as Waves

We first derive the 'minimal description' of a classical electromagnetic wave. Using the common definition of linear and circular polarization, we see that we can describe these 'typical waves' in a two-dimensional complex vector space.

2.1.1 The Typical Shape of an Electromagnetic Wave

We start with the description of an electromagnetic plane wave as[3]

$$\mathbf{E}\left(\mathbf{r}, t\right) = \mathbf{E}_0 e^{i(\mathbf{kr}-\omega t)}; \;\; \mathbf{B} = \frac{\mathbf{k} \times \mathbf{E}}{c} \tag{2.1}$$

with $\mathbf{k} \cdot \mathbf{E}_0 = 0$ (a transverse wave, as follows from the first Maxwell equation. In a charge-free region of space, it states that $\mathbf{\nabla E} = 0$); $\omega^2 = c^2 \mathbf{k}^2$ (dispersion relation for zero rest mass), with $\mathbf{E}_0 \in \mathbb{C}^3$. In the following, we restrict our considerations to the electric field[4] $\mathbf{E}$; the magnetic field can be calculated from $\mathbf{E}$ by using (2.1).

It holds quite generally that the description of a plane wave can be made considerably simpler and more transparent by means of a suitable choice of the coordinate system, without losing any of its physical significance. We choose the new z axis to point in the direction of propagation of the wave, i.e. the $\mathbf{k}$ direction—in other words, $\mathbf{k} = (0, 0, k)$—and obtain

$$\mathbf{E}\left(\mathbf{r}, t\right) = \left(E_{0x}, E_{0y}, 0\right) e^{i(kz - \omega t)}. \tag{2.2}$$

The z component disappears due to the transverse nature of the wave (see exercises at the end of this chapter).

---

[2]From the theory of electromagnetism, we know that light is a transverse wave, i.e. that its electric field oscillates perpendicular to its direction of propagation. The *polarization* describes the orientation of this oscillation.

Polarization is often regarded as an esoteric and specialized topic, possibly because we cannot see directly whether light is polarized. However, it is a ubiquitous phenomenon in our environment—natural light is almost always polarized, at least partially. Many animals, such as bees or other insects, take advantage of this; they can detect and analyze light polarization. In our daily life, polarization is used e.g. in polarizing filters for cameras or some sunglasses. Moreover, the fundamentals of the formal treatment of polarization are also very simple, as we shall see below.

[3]We note that a real light wave is only approximately described by a plane wave, since that would have the intensity at all points and all times. However, this approximate description is common for several reasons, and suffices for our purposes here.

[4]In this connection also called the *light vector*.

The amplitude can be written quite generally as

$$E_{0x} = e^{i\alpha} |E_{0x}|; \quad E_{0y} = e^{i\beta} |E_{0y}|; \quad \alpha, \beta \in \mathbb{R}. \tag{2.3}$$

It follows that

$$\mathbf{E}(\mathbf{r}, t) = \left(|E_{0x}|, |E_{0y}| e^{i(\beta - \alpha)}, 0\right) e^{i(kz - \omega t + \alpha)}. \tag{2.4}$$

We can now put $\alpha = 0$ without loss of generality, since the last equation shows that any value of α can be compensated by a suitable choice of the zero of time. In order to avoid confusion, we rename β as δ. Then the typical form of an electromagnetic wave is given by[5]:

$$\mathbf{E}(\mathbf{r}, t) = \left(|E_{0x}|, |E_{0y}| e^{i\delta}, 0\right) e^{i(kz - \omega t)}. \tag{2.5}$$

2.1.2 Linear and Circular Polarization

For purposes of illustration (we will include some figures in the following), we consider in this subsection only the real part of the wave function (the imaginary part alone would be just as suitable):

$$E_x(\mathbf{r}, t) = |E_{0x}| \cos(kz - \omega t); \quad E_y(\mathbf{r}, t) = |E_{0y}| \cos(kz - \omega t + \delta). \tag{2.6}$$

$\delta \in \mathbb{R}$ can assume all possible values. One can, however, single out two basic cases, namely $\delta = 0$ (*linear polarization*) and $\delta = \pm\pi/2$ (*elliptical or circular polarization*).

2.1.2.1 Linear Polarization

With the choice $\delta = 0$, we have

$$E_x(\mathbf{r}, t) = |E_{0x}| \cos(kz - \omega t); \quad E_y(\mathbf{r}, t) = |E_{0y}| \cos(kz - \omega t). \tag{2.7}$$

It follows immediately that

$$E_y = \frac{|E_{0y}|}{|E_{0x}|} E_x. \tag{2.8}$$

This is a straight line on which the light vector oscillates back and forth—hence the name *linear* polarization, see Fig. 2.1.

Basic types of this polarization are obtained by setting one component equal to zero:

---

[5]The relative phase could of course be associated with the y component instead of the x component.

Fig. 2.1 Linear polarization. The z-axis points out of the image plane

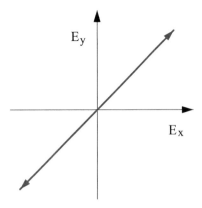

$$\text{horizontally polarized: } E_x\,(\mathbf{r},t) = |E_{0x}|\cos{(kz - \omega t)}\,; \quad E_y = \left|E_{0y}\right| = 0$$
$$\text{vertically polarized: } \quad E_x = |E_{0x}| = 0\,; \quad E_y\,(\mathbf{r},t) = \left|E_{0y}\right|\cos{(kz - \omega t)}\,. \tag{2.9}$$

The names are self-explanatory. Due to the vector character of the electric field, it follows readily that any linearly-polarized wave can be written as a superposition of horizontally- and vertically-polarized waves.

2.1.2.2 Elliptical and Circular Polarization

In this case, we choose $\delta = \pm\pi/2$ and this means that

$$E_x\,(\mathbf{r},t) = |E_{0x}|\cos{(kz - \omega t)}$$
$$E_y\,(\mathbf{r},t) = \left|E_{0y}\right|\cos{(kz - \omega t \pm \pi/2)} = \mp\left|E_{0y}\right|\sin{(kz - \omega t)}\,. \tag{2.10}$$

It follows from this that

$$\left(\frac{E_x}{|E_{0x}|}\right)^2 + \left(\frac{E_y}{\left|E_{0y}\right|}\right)^2 = 1. \tag{2.11}$$

The arrowhead of the light vector thus moves on an ellipse with semiaxes $|E_{0x}|$ and $\left|E_{0y}\right|$—hence the name *elliptical* polarization; see Fig. 2.2. The direction of rotation can be determined by using

$$\tan\vartheta = \frac{E_x}{E_y} = \mp\frac{\left|E_{0y}\right|}{|E_{0x}|}\tan{(kz - \omega t)} = \frac{\left|E_{0y}\right|}{|E_{0x}|}\tan{(\pm\omega t \mp kz)} \tag{2.12}$$

(most easily seen for fixed z).

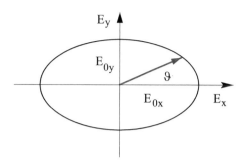

Fig. 2.2 Elliptical polarization. The z-axis is directed out of the image plane

In particular, for $|E_{0x}| = |E_{0y}|$, the ellipse becomes a circle and we have circularly-polarized light, i.e. right circularly-polarized light with the upper sign, left circularly-polarized light with the lower sign.[6]

2.1.3 From Polarization to the Space of States

In summary, we have in the complex representation (remember $e^{\pm i\pi/2} = \pm i$) for the linearly (horizontal h/vertical v) and circularly (right r/left l) polarized waves:

$$
\begin{aligned}
\mathbf{E}_h &= (|A_{0x}|, 0, 0)\, e^{i(kz-\omega t)} \\
\mathbf{E}_v &= (0, |B_{0x}|, 0)\, e^{i(kz-\omega t)} \\
\mathbf{E}_r &= (|C_{0x}|, i\,|C_{0x}|, 0)\, e^{i(kz-\omega t)} \\
\mathbf{E}_l &= (|C_{0x}|, -i\,|C_{0x}|, 0)\, e^{i(kz-\omega t)}.
\end{aligned}
\tag{2.13}
$$

So far, we have just repeated material that should be known from previous semesters. Now we turn to something that is (quite possibly) new. We begin by noting that the representation (2.13) is redundant and we can simplify it further.

2.1.3.1 Simplifying the Notation

To achieve this simplification, we have to restrict our world: it will consist exclusively of the waves given by (2.13). In particular, there is e.g. no other direction of propagation and no other wave number k. Then we can simplify as follows:

1. The factor $e^{i(kz-\omega t)}$ occurs everywhere, so we can omit it.
2. Since the third component is always zero, we suppress it. In other words, our little world is *two dimensional*.

---

[6]In physical optics, right and left circular polarization is usually defined the other way around (optics convention).

3. The notation as a row vector was chosen for typographical convenience; the correct notation is as a column vector.[7]
4. We fix the undetermined quantities $|A_{0x}|$ etc. in such a way that the respective vector has length 1, and thus represents a unit vector. We can then build up a general vector by taking appropriate linear combinations of these unit vectors.

In summary:

$$(|A_{0x}|, 0, 0)\, e^{i(kz-\omega t)} \xrightarrow{1.} (|A_{0x}|, 0, 0)$$
$$\xrightarrow{2.} (|A_{0x}|, 0) \xrightarrow{3.} \begin{pmatrix} |A_{0x}| \\ 0 \end{pmatrix} \xrightarrow{4.} \begin{pmatrix} 1 \\ 0 \end{pmatrix} \tag{2.14}$$

and[8]

$$(|C_{0x}|, i\,|C_{0x}|, 0)\, e^{i(kz-\omega t)} \xrightarrow{1.} (|C_{0x}|, i\,|C_{0x}|, 0)$$
$$\xrightarrow{2.} (|C_{0x}|, i\,|C_{0x}|) \xrightarrow{3.} \begin{pmatrix} |C_{0x}| \\ i\,|C_{0x}| \end{pmatrix} \xrightarrow{4.} \frac{1}{\sqrt{2}}\begin{pmatrix} 1 \\ i \end{pmatrix}. \tag{2.15}$$

In short, we go from a three-dimensional to a two-dimensional *complex vector space* which we call the *state space*. In this space, the states associated with the vectors (2.13) are written as *two-component vectors* which depend neither on position nor on time. For a convenient shorthand notation comparable to $\mathbf{E}_h$ in (2.13), we introduce the notation $|h\rangle$, $|v\rangle$, $|r\rangle$ and $|l\rangle$ for light in the linear horizontal, linear vertical, right circular and left circular polarized states, respectively. For these states we have the *representation* (denoted by the symbol $\cong$)

$$|h\rangle \cong \begin{pmatrix} 1 \\ 0 \end{pmatrix}; \quad |v\rangle \cong \begin{pmatrix} 0 \\ 1 \end{pmatrix}$$
$$|r\rangle \cong \frac{1}{\sqrt{2}}\begin{pmatrix} 1 \\ i \end{pmatrix}; \quad |l\rangle \cong \frac{1}{\sqrt{2}}\begin{pmatrix} 1 \\ -i \end{pmatrix}. \tag{2.16}$$

We emphasize that in our 'small world' this representation is completely equivalent to the representation in (2.13).

A note concerning the symbol $\cong$: actually, the representation (2.16) is just one of infinitely many possible ones, based on the fact that in (2.13), we identify the horizontal direction with the x axis. Of course, this is not to be taken for granted, since the y axis might just as well play the same role, given a corresponding orientation. Generally, one can start from any representation $|h\rangle \cong \frac{1}{\sqrt{a^2+b^2}}\begin{pmatrix} a \\ b \end{pmatrix}$ and

---

[7]In the following, we want to multiply vectors by *matrices*. In the usual notation, a matrix acts on a vector from the the left, which therefore—according to the usual rules of matrix multiplication—must be a column vector. See also Appendix F, Vol. 1, on linear algebra.

[8]The length of the vector $\begin{pmatrix} 1 \\ i \end{pmatrix}$ is given by $\sqrt{2}$; we explain the reasoning for this in Chap. 4.

$|v\rangle \cong \frac{1}{\sqrt{c^2+d^2}} \begin{pmatrix} c \\ d \end{pmatrix}$ with $a^*c + b^*d = 0$. We therefore identify specific representations by the special symbol $\cong$ and do not simply use the equals sign.[9]

The question may arise as to whether the representation (2.16) is not oversimplified. In this context, we recall the following: In physics, the objective of the formal description is not to describe 'nature' (whatever is meant by this term) *directly* but rather to find a *model* for a part of nature and to describe this model as accurately as possible. This is also reflected in the much-quoted 'accuracy' of the natural sciences. It is not the description of nature which is exact, but at most the formal treatment of the model (if at all).[10] Kepler's laws, for example, do not describe the conditions in the solar system exactly, as is well known: the planets influence each other, they are not point masses, there are moons and the solar wind, etc. Kepler's laws, however, are exact within the framework of the model 'point mass earth moves around point mass sun', and this model is correct and sufficiently precise for many applications.[11]

In this sense, the description by models is not unique, but depends on the particular question being considered. The general rule is: as easy as possible, as elaborate as necessary.[12] This is easily said, but of course it is not clear from the outset in all cases what it means in detail. In fact, it is *the* art in science to carve out meaningful, workable models from the 'jumble of reality', neither oversimplified nor overcomplicated.

For the following considerations, our quite modest, simplistic representation (2.16) will be sufficient: We need no direction of propagation, no plane waves, no explicit time behavior and so on.

2.1.3.2   Two Basic Systems

With $|h\rangle$, $|v\rangle$ and $|r\rangle$, $|l\rangle$ we have two pairs of linearly-independent vectors and therefore two basis systems for our two-dimensional vector space. They can be transformed into each other by

---

[9]Different symbols are in use to denote representations; Fließbach writes :=, for example. Apart from that, many authors denote representations not by a special symbol, but by simply writing =.

[10]Also, the general mathematical modelling uses concepts that are implemented only approximately in reality. A time-honored example is Euclidean geometry with its points and lines, which strictly speaking do not exist anywhere in our real world. Yet no one doubts that Euclidean geometry is extremely useful for practical calculations. "Although this may be seen as a paradox, all exact science is dominated by the idea of approximation" (Bertrand Russell).

[11]Most theoretical results are based on approximations or numerical calculations and are in this sense not strictly precise. This naturally applies a fortiori to experimental results. Even though there are high-precision measurements with small relative errors of less than a part per billion, it has to be noted that *each* measurement is inaccurate. Nevertheless, one can estimate this inaccuracy in general quite precisely; keyword 'theory of errors'.

[12]If several theories describe the same facts, one should prefer the simplest of them (this is the principle of parsimony in science, also called Occam's razor: "*entia non sunt multiplicanda praeter necessitatem*").

$$|r\rangle = \frac{|h\rangle + i|v\rangle}{\sqrt{2}}$$
$$|l\rangle = \frac{|h\rangle - i|v\rangle}{\sqrt{2}} \tag{2.17}$$

and

$$|h\rangle = \frac{|r\rangle + |l\rangle}{\sqrt{2}}$$
$$|v\rangle = \frac{|r\rangle - |l\rangle}{i\sqrt{2}} \tag{2.18}$$

These relations hold *independently* of the representation, and this is why we write $=$ here, and not $\cong$. Mathematically, (2.17) and (2.18) are basis transformations; physically, these equations mean that we can consider linearly-polarized light as a superposition of right and left circularly-polarized light—and of course vice versa.

2.1.3.3 Intensity and the Absolute Square Amplitude

If we send right circularly-polarized light through an analyzer (linearly horizontal/vertical), then the relative intensity of horizontally and vertically polarized light is 1/2, respectively. Where can we find this factor 1/2 in the expression $|r\rangle = \frac{|h\rangle + i|v\rangle}{\sqrt{2}}$? Clearly, we obtain the intensities (as usual) by calculating the squared sum of the coefficients (amplitudes), i.e.

$$\frac{1}{2} = \left(\frac{1}{\sqrt{2}}\right)^2 = \left|\frac{i}{\sqrt{2}}\right|^2. \tag{2.19}$$

Next, we consider light whose polarization plane is rotated by ϑ. The rotation matrix, which is known to be given by $\begin{pmatrix} \cos\vartheta & -\sin\vartheta \\ \sin\vartheta & \cos\vartheta \end{pmatrix}$, transforms the state $|h\rangle \cong \begin{pmatrix} 1 \\ 0 \end{pmatrix}$ into the rotated state $|\vartheta\rangle \cong \begin{pmatrix} \cos\vartheta \\ \sin\vartheta \end{pmatrix}$.[13] Hence we have $|\vartheta\rangle = \cos\vartheta\,|h\rangle + \sin\vartheta\,|v\rangle$. The absolute square of the coefficient of $|h\rangle$ is $\cos^2\vartheta$. This is known as the *Law of Malus*[14] and gives the relative intensity of the horizontally-polarized light.

Thus we have recovered the known relationship between intensity and absolute squared amplitude: For $|A\rangle = c_1|h\rangle + c_2|v\rangle$, the (relative) intensity of e.g. $|h\rangle$ is given by $|c_1|^2$, whereby $|c_1|^2 + |c_2|^2 = 1$ has to hold. In other words: the state $|A\rangle$ must be *normalized*.

---

[13]The active rotation (rotation of the vector by ϑ counterclockwise) is given by $\begin{pmatrix} \cos\vartheta & -\sin\vartheta \\ \sin\vartheta & \cos\vartheta \end{pmatrix}$; the passive rotation (rotation of the coordinate system) by $\begin{pmatrix} \cos\vartheta & \sin\vartheta \\ -\sin\vartheta & \cos\vartheta \end{pmatrix}$.

[14]Perhaps familiar from school or undergraduate laboratory courses?

2.2 Light as Photons

The above considerations are independent of the intensity of the light—they apply to an intense laser beam as well as to the dimmest glow. But if we can turn down the intensity of a light source sufficiently far, we eventually encounter a situation where the light consists of a stream of *single photons*.[15] Even then—and that is the crucial point—we assume that the above formulation remains valid. This is the above-mentioned jump from classical physics to quantum mechanics, which is not strictly derivable logically, but requires additional assumptions; there are in fact two of them. First, the existence of photons is assumed, which we take as an experimentally assured fact. Secondly, there is the assumption that expressions such as (2.17) and (2.18) retain their validity even for single photons.

Though it is not absolutely mandatory, as mentioned before, the second assumption is without any apparent alternatives—provided that light consists of a stream of photons—because we cannot draw the conclusion from the above considerations that these equations apply only above a certain number of photons. An additional degree of plausibility can be found in the fact that the wave character of light, for example in (2.17) and (2.18), never enters the arguments explicitly. And, finally, such an assumption—independently of plausibility—has to be proven by experiment, which of course has long since been done.

2.2.1 Single Photons and Polarization

We see that polarization is a property of *individual photons*. This fact is new and is by no means self-evident; thus, for individual photons we have e.g.

$$|r\rangle = \frac{|h\rangle + i|v\rangle}{\sqrt{2}}$$
$$|l\rangle = \frac{|h\rangle - i|v\rangle}{\sqrt{2}}. \tag{2.20}$$

However, the interpretation *must* be *different* from the case of 'classical' light, since a photon in state $|r\rangle$ whose linear polarization is measured (i.e. with respect to $|h\rangle$ or $|v\rangle$) cannot split up into two linearly-polarized photons (how would the energy

---

[15]Single-photon experiments are standard technology these days. In 1952, Schrödinger declared: "We never experiment with just one electron or atom or (small) molecule. In thought-experiments we sometimes assume that we do; this invariably entails ridiculous consequences." Times have changed: Precision experiments using a single photon or a single atom are the basis of e.g. today's time standard, and modern quantum-mechanical developments such as the quantum computer rest on those 'ridiculous consequences'. We recall that photons (as far as we know) have immeasurably small dimensions and are in this sense referred to as point objects (or point particles). Although they represent light of a specific wavelength, they do not have a spatial extension on the order of the wavelength of the light.

$E = \hbar\omega$ be divided up in that case?).[16] We must assume that we can infer the *probabilities* P of finding a photon, initially in the state $|r\rangle$, in a state $|h\rangle$ or $|v\rangle$ after its passage through e.g. a linear polarizing filter, from (2.20); namely

$$P(h) = \left|1/\sqrt{2}\right|^2 = \frac{1}{2} \text{ and } P(v) = \left|i/\sqrt{2}\right|^2 = \frac{1}{2}. \tag{2.21}$$

Therefore, one must beware of interpreting the expressions (2.20) incorrectly—it is not at all the case that an $|r\rangle$ photon consists of half a horizontally- and half a vertically-polarized object. Rather, (2.20) tells us that an $|r\rangle$ photon contains two *possibilities* to present itself in a measurement as either $|h\rangle$ or $|v\rangle$—but only one of these is realized in any given measurement. Before the measurement, however, the photon is in a *superposition* of the two states. This is a very common trait of quantum mechanical systems: states can be superposed.

This superposition principle is valid for *all* states or objects described by quantum mechanics, whether we attribute to them more wave- or more particle-like character. In the macroscopic domain, the superposition of states would lead to very unusual effects—for example, in a system with the two states |cow in barn⟩ and |cow in field⟩, or in the famous example of Schrödinger's cat, namely |dead cat⟩ and |live cat⟩. Our direct daily experience does not include such superposed states, and so certain quantum-mechanical phenomena are in conflict with 'common sense' (whatever that may be, exactly). But as mentioned above, our sensory apparatus was trained by evolution under macroscopic conditions[17] and our understanding of the world is based on corresponding model concepts. No one will seriously argue that therefore, the whole of nature should operate according to these 'daily life rules' which literally permeate our flesh and blood.

So when speaking of the paradoxes of quantum mechanics, we should recognize that the real paradox is simply that the rules of quantum mechanics (which we can indeed recognize, identify and formulate) proceed according to a different pattern from our familiar daily-life rules (i.e. 'common sense'). But quantum mechanics works, and indeed it works verifiably, consistently, reproducibly, and with an amazingly high degree of accuracy—in short, according to all scientific standards, it is a successful theory. Quantum mechanics is one of the best if not *the* best-validated basic theory in physics. Of course, in spite of this the question remains as to why there are only microscopic superposition states and apparently no macroscopic ones. This is a central problem of quantum mechanics, which we will address several times in various chapters in the following.

---

[16]In a vacuum, photons are indivisible, and that holds also for most interactions with matter. One has to work hard to 'cut' photons. This can be achieved for example in the interaction with certain nonlinear crystals, where a single photon breaks up into two photons of lower energy (parametric fluorescence, see Appendix I, Vol. 2). Devices for polarization measurement are of course manufactured in such a way that they leave the photons unsplit.

[17]Furthermore, in 'slow' conditions—the effects of the theory of relativity are beyond our daily life experience, as well.

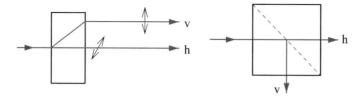

Fig. 2.3 Birefringence (*left*) and polarization beam splitter (*right*)

2.2.2 *Measuring the Polarization of Single Photons*

Back to our single photons: We produce one of them with a certain polarization and send it through the usual polarization filter (analyzer), which absorbs photons with the 'wrong' polarization,[18] or through an analyzer with two outputs, such as a birefringent crystal or a polarizing beam splitter (PBS); see Fig. 2.3. We assume an angle φ between the polarization direction and the analyzer axis. Whether a photon passes the analyzer or not, or alternatively where it leaves the PBS, can be predicted with certainty only if $\varphi = 0$ (the photon passes, or exits to the right); or $\varphi = \pi/2$ (the photon is absorbed, or exits downwards).[19] In all other cases, we can specify only the *probability* $P(\varphi)$ that the photon passes the absorbing analyzer or the PBS with the same polarization; it is given by $P(\varphi) = \cos^2 \varphi$.

We can interpret these facts as follows: Before a measurement, one cannot objectively determine whether the photon will pass through the analyzer or not. This is revealed only by the process of measurement. Which one of the two possibilities will be realized cannot be said before the measurement, but one can specify the respective probabilities of observing each of the two outcomes. We can summarize the general result: For a state such as $|z\rangle = c\,|x\rangle + d\,|y\rangle$, the term $P = |c|^2$ represents the probability of measuring the state $|x\rangle$ (assuming a normalized state $|z\rangle$, i.e. with $|c|^2 + |d|^2 = 1$). In symbolic shorthand notation,

$$\text{measuring probability} = |\text{coefficient}|^2. \tag{2.22}$$

This is a similar finding as in classical physics—with the very significant difference that the statement holds there for *intensities* , but here for *probabilities*.

We note that while it is possible to measure with sufficient accuracy the polarization which a classical wave had before the measurement, this is *in principle impossible* for a single photon of unknown polarization. Quantum objects do not always possess well-defined values of all physical quantities—a linearly-polarized photon 'has', for example, no well-defined circular polarization. If we send a

---

[18]We point out that this is not an exotic quantum-mechanical procedure—the eyes of every bee, or suitable sunglasses, perform precisely this kind of 'measurement process'.

[19]These cases can be produced by inserting a further analyzer whose orientation is $\varphi + 0$ or $\varphi + \pi/2$, for example.

horizontally linear-polarized photon through a linear analyzer, rotated through the angle φ, we find horizontally- and vertically-polarized light with the probabilities $\cos^2 \varphi$ and $\sin^2 \varphi$, respectively—in principle, with no ifs, ors, or buts.

2.2.2.1 The Ensemble

How can we verify experimentally that the calculated probabilities are correct? Obviously not in a *single* experiment. Because if we send e.g. a circularly-polarized photon through a PBS, it emerges on the other side as either horizontally or vertically polarized, and we have no information about the probabilities. So we need to repeat the measurement more than once. Now the term *ensemble* comes into play. In quantum mechanics, this term refers to a set of (strictly speaking) infinitely many identically-prepared copies of a system.[20] It is a fictitious set which has no counterpart in physical reality, but serves only for conceptual clarification. The 'strictly speaking' in brackets refers to the fact that often (and for practical reasons), N identical copies of an system are called an ensemble, if N is sufficiently large (even though not infinite).

The concept 'ensemble' allows us to calculate the probabilities for the occurrence of certain measurement values, and thus to predict them—they are given simply by the fraction of subsets of the ensemble which are characterized by the presence of those values. In the example given above, this is $P(\varphi) = \cos^2 \varphi$ for passing the analyzer or $\sin^2 \varphi$ for not passing through it.

We emphasize that the use of the word *ensemble* does *not* imply that the relevant physical quantities (here the polarization) have well-defined values which are distributed in some unknown way among the members of the ensemble. In the example, the ensemble consists of horizontally-polarized photons whose polarization properties are *not* defined with respect to an analyzer that is rotated by φ.

In practice, one can of course measure only finitely many systems; often one has to make do even with a *single* system (in the example, a single photon). But the predictions arising from the concept of ensembles are generally valid and apply (in terms of probabilities) to the particular case considered.[21]

So we can imagine that we prepare N systems in an identical manner and always measure the same variable, in our example the rate of observation of vertically- or horizontally-polarized photons behind the PBS.[22] For $N \to \infty$, the relative frequencies of occurrence of the different measurement results become the probabilities of the ensemble; in the above example $\cos^2 \varphi$ and $\sin^2 \varphi$.

---

[20]The systems need not be in the same state, but the preparation process must be the same.

[21]Just as the interference pattern in the double slit experiment builds up gradually from scattered spots over time.

[22]Another example of an ensemble are electrons which are prepared by a Stern–Gerlach apparatus and a velocity filter so that their spins are pointing upwards and their speeds are confined to a particular interval $(v - \Delta v, v + \Delta v)$. A further example is a set of hydrogen atoms in a particular excited state, whereby here the preparation refers to the energy, but not to the angular momentum of the state.

2.2.2.2 Ensemble or Single Object?

As we can see, the experimental verifiability of the theory is not guaranteed for a single quantum object, but rather requires an ensemble. Hence, one can argue that the formalism developed above (regardless of our derivation) is essentially a mathematical rule which applies only to an ensemble (so-called ensemble interpretation). Another position asserts however that the formalism applies also to an individual quantum object, as we have assumed to be the case. Both interpretations lead to the same results, but they are based on different concepts of 'reality'.

We encounter here for the first time a situation typical of quantum mechanics: The formalism and the verification of its predictions by measurements are uncontroversial (if we accept certain basic assumptions); the controversial issue concerns what quantum mechanics 'really' means. This debate is as old as quantum mechanics itself, and is still very much alive; there are a dozen or more different explanations (interpretations). We will discuss these questions often and will give an overview of current interpretations in Chap. 28, Vol. 2.

2.2.2.3 Do We Really Need Probabilities?

Finally, a remark about the concept of probability. In classical physics, probabilities reflect the fact that we do not know (or do not wish to know) enough about some of the properties of a system in order to calculate them explicitly. For instance, in the kinetic theory of gases, one is not interested in the behavior of a single molecule; a well-known example from a quite different field are the opinion polls before an election, where the behavior of individual voters is not of interest. Analogously, one could assume here that the occurrence of probabilities indicates that below the level of our discussion, there are some *hidden variables*, and if we were able to know them, we could formulate the whole process exactly without resorting to probabilities. This is an obvious idea which was brought up very soon after the emergence of quantum mechanics. It took nearly 40 years until a criterion was found to resolve this question in principle, and a few more years to disprove the idea of hidden variables experimentally based on that criterion—at least this holds for the major classes of hidden variables. More on this topic in Chaps. 20 and 27, Vol. 2.

According to our current knowledge, we cannot avoid the term 'probability' in quantum mechanics. It is, so to speak, a structural element of quantum mechanics, the sign that quantum mechanics deals first of all with possibilities, one of which is realized by a measurement, with a certain probability for different outcomes.

2.3 Exercises

1. Given an electromagnetic wave $\mathbf{E}(\mathbf{r}, t) = \mathbf{E}_0 e^{i(\mathbf{kr} - \omega t)}$ in a charge-free region of space (we consider the electric field only); show that the wave is transverse, i.e. that $\mathbf{k} \cdot \mathbf{E}_0 = 0$ holds (Hint: cf. the Maxwell equation $\nabla \mathbf{E} = 0$). Specialize to $\mathbf{k} = (0, 0, k)$.

2. Linear combinations

   (a) Express $|r\rangle$ as a linear combination of $|h\rangle$ and $|v\rangle$. Do the same for $|l\rangle$.
   (b) Express $|h\rangle$ as a linear combination of $|r\rangle$ and $|l\rangle$. Do the same for $|v\rangle$.

3. A phase shift of 90° is described by $e^{i\pi/2} = i$. What follows for a phase shift of 180°?

4. Elliptical polarization: Given the state $|z\rangle = \alpha \, |h\rangle + \beta \, |v\rangle$, with $|\alpha|^2 + |\beta|^2 = 1$; express $|z\rangle$ as a superposition of $|r\rangle$ and $|l\rangle$.

Chapter 3
More on the Schrödinger Equation

We first examine some general properties of the Schrödinger equation. Among other topics, the concept of *vector space* emerges—the solutions of the Schrödinger equation form such a space. In the Schrödinger equation, operators occur. We see that the order of the operators plays a role, provided that they do not commute.

In Chap. 1, we introduced the Schrödinger equation (SEq)

$$i\hbar\frac{\partial}{\partial t}\Psi\left(\mathbf{r},t\right) = -\frac{\hbar^2}{2m}\nabla^2\Psi\left(\mathbf{r},t\right) + V\left(\mathbf{r},t\right)\Psi\left(\mathbf{r},t\right). \tag{3.1}$$

For our considerations it is *the* basic differential equation of quantum mechanics. In view of of its central role, we want to examine in the following which properties the SEq has and which consequences follow from those properties. By separating out the time, one can obtain from (3.1) the *stationary Schrödinger equation* (also known as the time-independent Schrödinger equation), which for us is the workhorse of quantum mechanics. Finally, we make a few preliminary comments on operators, which in quantum mechanics are identified with measurable quantities.

3.1 Properties of the Schrödinger Equation

The SEq has several immediately recognizable features which model important physical properties and imply certain consequences. For example, one sees immediately that $\Psi\left(\mathbf{r},t\right)$ *must* be complex if the potential V is real (we will restrict ourselves to this case; the reason will be discussed later). Certain features are treated here only provisionally, while more detailed treatments follow in subsequent chapters. Some basic facts about differential equations are summarized in Appendix E, Vol. 1.

1. The SEq is *linear* in Ψ. If one has found two solutions Ψ_1 and Ψ_2, then any linear combination $c\Psi_1 + d\Psi_2$ is also a solution (with $c, d \in \mathbb{C}$). This means that one can superimpose the solutions - the *superposition principle* holds. This

© Springer Nature Switzerland AG 2018
J. Pade, *Quantum Mechanics for Pedestrians 1*, Undergraduate Lecture
Notes in Physics, https://doi.org/10.1007/978-3-030-00464-4_3

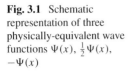

Fig. 3.1 Schematic representation of three physically-equivalent wave functions $\Psi(x)$, $\frac{1}{2}\Psi(x)$, $-\Psi(x)$

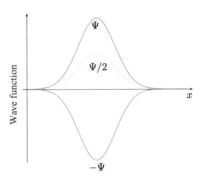

principle, known e.g. from the description of classical waves, has far-reaching consequences in quantum mechanics. Properties of the microscopic world which for our everyday understanding are very 'bizarre' are largely due to the seemingly trivial fact of linearity. Incidentally, due to this linearity, the total wave functions Ψ and $c\Psi$ are physically equivalent, or to be more precise, they *must* be physically equivalent; see Fig. 3.1.

Being a linear equation, the SEq always has the solution $\Psi \equiv 0$, the so-called *trivial solution*. This solution does not describe a physical state, as one can for example add arbitrary multiples of it to any other state without changing anything physically. In other words, if it turns out that the state of a physical system is described by the trivial solution, then we know that this state does not exist.

2. The SEq is a differential equation of *first order in time*. This means that for a given *initial condition* $\Psi(\mathbf{r}, t = 0)$, the wave function $\Psi(\mathbf{r}, t)$ is determined for *all* times (greater and less than zero). In other words, in the time evolution of $\Psi(\mathbf{r}, t)$, there are no stochastic or random elements — by specifying $\Psi(\mathbf{r}, t = 0)$, one uniquely defines the wave function for all past and future times.

3. The SEq is a differential equation of *second order in space*. To describe a specific given physical situation, the solution must satisfy certain *boundary conditions*.

4. The SEq determines, as we shall see below, which results are generally possible, but not which result will be realized in an actual measurement. This information must therefore come from somewhere else.[1]

The ability to form superpositions is a fundamental property of all elements of a *vector space* $\mathcal{V}$. In fact, it can easily be shown that the solutions of the SEq span a vector space over the complex numbers—see the definition in Appendix G, Vol. 1. Thus, we have a similar situation as for the polarization: The states of a system are described by elements of a vector space, in which the superposition principle applies. The dimensions of the spaces may be different—it is 2 in the case of polarization, while the dimension of the solution space of the SEq is unknown to us yet. But at least we have found with $\mathcal{V}$ a structure which is common to both the approaches of Chaps. 1 and 2.

---

[1]One can summarize the difference between classical mechanics and quantum mechanics in a bold and simple way as follows: Classical mechanics describes the time evolution of the *factual*, quantum mechanics (i.e. the SEq) describes the time evolution of the *possible*.

Things are different for the pair of concepts 'determinism—probability'. In this regard, our two approaches to quantum mechanics (still) do not match. Probabilities which we *had* to introduce in the algebraic approach to the transition from classical mechanics → quantum mechanics do not appear in the SEq. On the contrary, the SEq is a deterministic equation whose solutions are uniquely determined for all times, given the initial conditions. Hence, the apparent randomness of quantum mechanics (e.g. in radioactive decay) is *not* hidden in the SEq.

As we will see in the next chapters, chance comes into play through the wavefunction. We emphasize once again that the wavefunction as a solution of the SEq has *no direct, intuitive meaning* in the 'everyday world'. In this respect, the question of what Ψ 'actually is' cannot be answered in everyday terms. Perhaps the idea mentioned in Chap. 1, of a complex-valued field of possibilities, is the most appropriate.

3.2 The Time-Independent Schrödinger Equation

In Chap. 1, we found that the solutions of the *free* Schrödinger equation ((3.1) with $V \equiv 0$) are plane waves with the dispersion relation $\hbar\omega = \hbar^2 k^2/2m$. But what are the solutions for a non-vanishing potential V? The answer is: There are virtually no closed or analytical solutions in this case. Apart from just a handful of special potentials, one always has to deal with approximations or numerical results.

Nevertheless, the approach to the Schrödinger equation may be facilitated by separating out the variable t. This leads to the so-called *stationary* or time-independent Schrödinger equation which depends on space variables only. The prerequisite is, however, that the potential V must not depend on time:

$$V(\mathbf{r}, t) = V(\mathbf{r}). \tag{3.2}$$

Of course there are also physically reasonable potentials which do depend on time, but we will restrict ourselves to *time-independent potentials* in the following.

The method of choice is again the separation of variables. We insert the *ansatz*

$$\Psi(\mathbf{r}, t) = f(t) \cdot \varphi(\mathbf{r}) \tag{3.3}$$

into the Schrödinger Equation (3.1) and obtain

$$i\hbar \frac{\dot{f}}{f} = -\frac{\hbar^2}{2m}\frac{1}{\varphi}\nabla^2\varphi + V. \tag{3.4}$$

The right- and the left-hand sides must be constant (because the independent variables 'space' and 'time' are separated):

$$i\hbar \frac{\dot{f}}{f} = \text{const.} = E = \hbar\omega. \tag{3.5}$$

E and ω are a yet undetermined energy (this follows from its physical units) and frequency. The sign (i.e. E and not $-E$) is chosen in such a way that it agrees with the usual definition of the energy. A solution of this last equation is

$$f(t) = e^{-iEt/\hbar} = e^{-i\omega t} \tag{3.6}$$

or

$$\Psi(\mathbf{r}, t) = e^{-i\omega t} \varphi(\mathbf{r}). \tag{3.7}$$

E must be a real number, because otherwise the solutions would be unphysical, since they would tend for $t \to \infty$ or $t \to -\infty$ towards infinity and would not be bounded.

Inserting the wavefunction (3.7) into the *time-dependent* Schrödinger equation,

$$i\hbar \frac{\partial}{\partial t} \Psi(\mathbf{r}, t) = -\frac{\hbar^2}{2m} \nabla^2 \Psi(\mathbf{r}, t) + V(\mathbf{r}) \Psi(\mathbf{r}, t) \tag{3.8}$$

leads to the *time-independent* (= *stationary*) Schrödinger equation:

$$E\varphi(\mathbf{r}) = -\frac{\hbar^2}{2m} \nabla^2 \varphi(\mathbf{r}) + V(\mathbf{r}) \varphi(\mathbf{r}). \tag{3.9}$$

At this point, the possible values of E are not explicitly defined. We take up this issue again in Chap. 5.

In the last two equations, the expression $-\frac{\hbar^2}{2m} \nabla^2 + V(\mathbf{r})$ occurs. It is called the *Hamiltonian operator H* (or simply *Hamiltonian* for short) and is a central term in quantum mechanics:

$$H = -\frac{\hbar^2}{2m} \nabla^2 + V(\mathbf{r}). \tag{3.10}$$

With this, the time-dependent Schrödinger equation is written as

$$i\hbar \frac{\partial}{\partial t} \Psi = H\Psi. \tag{3.11}$$

Note that the expression (3.10) is just *one* possible form of the Hamiltonian, and indeed a particularly simple one. Other formulations (which are considered in the Appendix) contain vector potentials or describe relativistic situations.

The properties of the Schrödinger equation which we listed above hold true for *all* SEq (3.11), independently of the special form of the Hamiltonian H. This applies also to the method used to derive the time-independent SEq from the time-dependent SEq, as long as the otherwise arbitrary operator H does not depend on time. In that case, the separation *ansatz* $\psi(\mathbf{r}, t) = e^{-i\omega t} \varphi(\mathbf{r})$ *always* leads to the stationary SEq

$$H\varphi = E\varphi. \tag{3.12}$$

In a certain sense, this *is* quantum mechanics in short form.

3.3 Operators

Mathematically, the stationary SEq (3.12) is none other than an eigenvalue problem. You perhaps remember such problems from school days in the following form: Given a matrix A and a vector x, for which numbers $\lambda \neq 0$ do there exist solutions x of the equation $Ax = \lambda x$? The answer is that the allowed values of λ are given by the solutions of the *secular equation* $\det(A - \lambda) = 0$.

In the SEq (3.12), the Hamiltonian operator H appears on the left side instead of the matrix A. The concept of an *operator*[2] plays an essential role in quantum mechanics. While in the following chapters we will return repeatedly to this topic, here we give just a brief heuristic consideration or motivation. The term 'operator' can be best illustrated by 'manipulation' or 'tool'. To apply an operator A to a function means to manipulate this function in a prescribed manner.

For example, the operator $A = \frac{\partial}{\partial x}$ differentiates a function partially with respect to x. The operator $B = \frac{\partial}{\partial x} x$ multiplies a function by x and then differentiates the product. Products of operators are performed from right to left; ABf means that we first apply B to f and then A to Bf. In the following, we always take for granted that the functions have the properties which are required for the application of the operator under consideration. For instance, the functions on which $A = \frac{\partial}{\partial x}$ acts must be differentiable with respect to x.

The eigenvalue problem can be formulated in a general way. Consider a general operator A (which can be a matrix or a differential operator, for example). If the equation

$$Af = \alpha f \tag{3.13}$$

can be solved for certain numbers $\alpha \in \mathbb{C}$ (which means that there are solutions f), then α is called an *eigenvalue* of the operator A and f is called the associated *eigenfunction*. If one wants to emphasize that the function f is an element of a vector space, then f is called an *eigenvector* instead of an eigenfunction. The set of all eigenvalues is termed the *spectrum*; the spectrum can contain finitely or infinitely many elements. The eigenvalues may be countable (*discrete spectrum*) or uncountable (*continuous spectrum*); spectra can contain both discrete and continuous components.

If there are two or more (e.g. n) linearly-independent eigenfunctions corresponding to the same eigenvalue, one speaks of *degeneracy*. The eigenvalue is called n-fold degenerate, where n is the *degree of degeneracy*. Degeneracy is the consequence of a symmetry which is intrinsic to the problem; it can in principle be avoided by an arbitrary small, suitable 'perturbation operator'.

---

[2] A mapping between two vector spaces (whose elements can be functions, for example) is usually called an *operator*; a mapping from one vector space to its scalar field a *functional*. Integral transforms such as the Fourier or the Laplace transform can be viewed as integral operators.

In the interest of a unique terminology, we fix the difference between operator and function as follows: The domain of definition and the range of operators are vector spaces, while for functions, they are sets of numbers.

Here are two simple examples of eigenvalue problems:

1. Given the operator $\frac{\partial}{\partial x}$, the eigenvalue problem reads

$$\frac{\partial}{\partial x} f(x) = \gamma f(x); \gamma \in \mathbb{C}. \tag{3.14}$$

Obviously, we can solve this equation for all γ. The solution is

$$f(x) = f_0 e^{\gamma x}. \tag{3.15}$$

The spectrum is continuous and not degenerate.

2. Given the operator $\frac{\partial^2}{\partial x^2}$, the eigenvalue problem

$$\frac{\partial^2}{\partial x^2} f = \delta^2 f; \delta \in \mathbb{C}. \tag{3.16}$$

is clearly invariant under the exchange $x \to -x$, and its solutions are

$$f = f_{0+} e^{+\delta x} \text{ and } f = f_{0-} e^{-\delta x}. \tag{3.17}$$

The spectrum is continuous and doubly degenerate (for one value of δ^2 there exist the two linearly-independent eigenfunctions $e^{+\delta x}$ and $e^{-\delta x}$).

The limitation of the range of allowed functions in these two examples (for instance due to boundary conditions) can lead to a discrete spectrum; examples are found in the exercises. A (classical) example is the vibration of a violin string. The fundamental vibrational mode has the wavelength $\lambda = 2L$, where L is the length of the string (i.e. the position variable x along the string is bounded, $0 \le x \le L$). Other allowed solutions are harmonics of the fundamental mode, i.e. their frequencies are whole-number multiples of the fundamental frequency. The (countable) eigenvalues are these integer multiples, giving a discrete spectrum.

3.3.1 Classical Numbers and Quantum-Mechanical Operators

The SEq (3.1) formally resembles the expression for the classical energy

$$E = \frac{\mathbf{p}^2}{2m} + V. \tag{3.18}$$

Indeed, one can transform the *numerical* (3.18) into an *operator* equation and *vice versa*, if one identifies[3]:

$$x \leftrightarrow x \quad \text{or} \quad \mathbf{r} \leftrightarrow \mathbf{r}$$

$$p_x \leftrightarrow \frac{\hbar}{i} \frac{\partial}{\partial x} \quad \text{or} \quad \mathbf{p} \leftrightarrow \frac{\hbar}{i} \nabla. \tag{3.19}$$

$$E \leftrightarrow i\hbar \frac{\partial}{\partial t}$$

In this way, the expression (3.18) leads to the SEq (3.1) in its representation as an operator equation:

$$i\hbar \frac{\partial}{\partial t} = -\frac{\hbar^2}{2m} \nabla^2 + V(\mathbf{r}, t) = H. \tag{3.20}$$

We can motivate these 'translations' from classical to quantum-mechanical quantities as follows: We differentiate a plane wave

$$f = e^{i(kx - \omega t)} \tag{3.21}$$

with respect to x:

$$\frac{\partial f}{\partial x} = ike^{i(kx - \omega t)} = ikf. \tag{3.22}$$

In order to find the momentum, we multiply both sides with $\hbar/i$ and obtain with $p = \hbar k$

$$\frac{\hbar}{i} \frac{\partial f}{\partial x} = \hbar k f = pf \quad \text{or} \quad \frac{\hbar}{i} \frac{\partial}{\partial x} f = pf \quad \text{or} \quad \left(\frac{\hbar}{i} \frac{\partial}{\partial x} - p \right) f = 0. \tag{3.23}$$

The bracket in the last equation does not depend on the particular wave number k. Because of its linearity, this equation applies to all functions which we can generate by a superposition of plane waves (i.e. all 'sufficiently reasonable' functions), if we understand p to represent not the momentum of a single wave, but that of the whole new function. It is quite natural to define an *operator p* (*momentum operator*, usually just called p, sometimes also p_{op} or $\hat{p}$):

$$p = \frac{\hbar}{i} \frac{\partial}{\partial x}. \tag{3.24}$$

In this context, x is also called the *position operator*. This formulation may appear unnecessarily complicated at this point, since the application of the position operator simply means multiplication by x. But later on we will encounter other contexts where this is no longer the case. Here, we can at least motivate the terminology by the following parallel

---

[3] This small table is sometimes (rather jokingly) referred to as the 'dictionary of quantum mechanics.'

$$\text{application of the } \begin{array}{c} \text{momentum} \\ \text{position} \end{array} \text{ operator to } e^{i(kx-\omega t)} \text{ yields } \begin{array}{c} pe^{i(kx-\omega t)} \\ xe^{i(kx-\omega t)}. \end{array} \qquad (3.25)$$

The crucial point of the translations table (3.19), which in more sophisticated language is referred to as the *correspondence principle*,[4] is that it allows the translation of classical expressions into those of quantum mechanics. Some examples: The classical expression $E = \frac{p_x^2}{2m}$ becomes in quantum mechanics $i\hbar \frac{\partial}{\partial t} = -\frac{\hbar^2}{2m}\frac{\partial^2}{\partial x^2}$, and $E = \frac{\mathbf{p}^2}{2m}$ becomes $i\hbar\frac{\partial}{\partial t} = -\frac{\hbar^2}{2m}\nabla^2$. The classical angular momentum $\mathbf{l} = \mathbf{r} \times \mathbf{p}$ leads to the quantum-mechanical angular momentum operator $\mathbf{l} = \frac{\hbar}{i}\mathbf{r} \times \nabla$, and from the relativistic energy-momentum relation $E^2 = m_0^2 c^4 + p^2 c^2$, we obtain $-\hbar^2\frac{\partial^2}{\partial t^2} = m_0^2 c^4 - c^2\hbar^2\nabla^2$. This last expression is the so-called *Klein-Gordon equation* which describes free relativistic quantum objects with zero spin.

3.3.2 Commutation of Operators; Commutators

In this process of translation, however, problems can arise if we translate products of two or more variables. These are due to the fact that numbers commute, but in general operators do not.[5] As an illustrative example, we consider the classical expression xp_x which obviously equals $p_x x$. But this no longer applies to its quantum-mechanical replacement by operators

$$xp_x = x\frac{\hbar}{i}\frac{\partial}{\partial x} \neq p_x x = \frac{\hbar}{i}\frac{\partial}{\partial x}x = \frac{\hbar}{i}\left(1 + x\frac{\partial}{\partial x}\right). \qquad (3.26)$$

Anyone who is not sure about such considerations should transform the operator equations into 'usual' equations by applying the operators to a function (the function need not be specified in detail here, but must of course meet the necessary technical requirements). Then, for example, we have for the operator $\frac{\partial}{\partial x}x$ due to the product rule

$$\frac{\partial}{\partial x}xf = f + x\frac{\partial f}{\partial x} = \frac{\partial}{\partial x}xf = \left(1 + x\frac{\partial}{\partial x}\right)f \qquad (3.27)$$

or briefly, in operator notation,

$$\frac{\partial}{\partial x}x = 1 + x\frac{\partial}{\partial x}. \qquad (3.28)$$

---

[4]In the old quantum theory, the (Bohr) correspondence principle denoted an approximate agreement of quantum-mechanical and classical calculations for large quantum numbers. In modern quantum mechanics, correspondence refers to the assignment of classical observables to corresponding operators. This assignment, however, has mainly a heuristic value and must always be verified or confirmed experimentally. A more consistent procedure is for example the introduction of position and momentum operators by means of symmetry transformations (see Chap. 21 Vol. 2).

[5]It is known for example that for two square matrices A and B (= operators acting on vectors), in general $AB \neq BA$ holds.

The importance of the topic of 'operators' in quantum mechanics is based, among other things, on the fact that *measurable variables* (such as the momentum p_x) are represented by operators (such as $-i\hbar\partial_x$). If, as in (3.26), the *order* of the operators matters because of $x\frac{\hbar}{i}\frac{\partial}{\partial x} \neq \frac{\hbar}{i}\frac{\partial}{\partial x}x$, then this holds true also for the corresponding measurement variables. In other words, it makes a difference in quantum mechanics whether we measure first the position x and then the momentum p_x, or *vice versa*.

For the corresponding operators, the equality

$$(xp_x - p_x x) f = \frac{\hbar}{i}\left(x\frac{\partial f}{\partial x} - x\frac{\partial f}{\partial x} - f\right) = i\hbar f \tag{3.29}$$

holds, or

$$xp_x - p_x x = i\hbar. \tag{3.30}$$

Because differences of this kind play a key role in quantum mechanics, there is a special notation, namely a square bracket, called the *commutator*:

$$[x, p_x] = xp_x - p_x x = i\hbar. \tag{3.31}$$

For two operators A and B, the commutator[6] is defined as

$$[A, B] = AB - BA. \tag{3.32}$$

If it is equal to zero, A and B are called *commuting operators*.[7]

We repeat our remark that the *order is crucial* (of operators as well as of measurements). Of course there are commuting operators, such as for instance p_x and y or p_x and z, and so on. Position and momentum commute if and only if they do not refer to the same coordinate.

---

[6]The *anticommutator* is defined as
$$\{A, B\} = AB + BA$$
(despite the use of the same curly brackets, it is of course quite different from the Poisson brackets of classical mechanics).

[7]There is an interesting connection with classical mechanics which we have already mentioned briefly in a footnote in Chap. 1: In classical mechanics, the Poisson bracket for two variables U and V is defined as
$$\{U, V\}_{\text{Poisson}} = \sum_i\left(\frac{\partial U}{\partial q_i}\frac{\partial V}{\partial p_i} - \frac{\partial U}{\partial p_i}\frac{\partial V}{\partial q_i}\right),$$
where q_i and p_i are the positions and (generalized) momenta of n particles, $i = 1, 2, \ldots, 3n$. In order to avoid confusion with the anticommutator, we have added the (otherwise uncommon) index *Poisson*. If U and V are defined as quantum-mechanical operators, their commutator is obtained by setting $[U, V] = i\hbar\{U, V\}_{\text{Poisson}}$. Example: In classical mechanics, we choose $U = q_1 \equiv x$ and $V = p_1 \equiv p_x$. Then it follows that $\{q_1, p_1\}_{\text{Poisson}} = 1$, and we find the quantum-mechanical result $[q_1, p_1] = [x, p_x] = i\hbar$. This method, called 'canonical quantization', is considered in more detail in the relativistic sections in the Appendix.

A short remark concerning the problem of translation of 'ambiguous' terms such as xp_x: The problem can be resolved by symmetrization. The reason will be discussed in Chap. 13; here, it suffices to say that in this way one gets the correct quantum-mechanical expression. With the two possibilities xp_x and $p_x x$, we construct the symmetrized expression

$$A_{QM} = \frac{xp_x + p_x x}{2} = \frac{\hbar}{2i}\left(x\frac{\partial}{\partial x} + \frac{\partial}{\partial x}x\right) = \frac{\hbar}{2i}\left(1 + 2x\frac{\partial}{\partial x}\right). \tag{3.33}$$

However, this trick is hereafter hardly ever needed — quantum mechanics is very good-natured in a certain sense.[8] Consider, for example, the angular momentum $\mathbf{l} = \mathbf{r} \times \mathbf{p}$. Must it be symmetrized, i.e. $\mathbf{l} = \frac{\mathbf{r}\times\mathbf{p}}{2} - \frac{\mathbf{p}\times\mathbf{r}}{2}$, for the translation into quantum mechanics? The answer is 'no', because for it we have

$$l_x = (\mathbf{r} \times \mathbf{p})_x = yp_z - zp_y = \frac{\hbar}{i}\left(y\frac{\partial}{\partial z} - z\frac{\partial}{\partial y}\right), \tag{3.34}$$

and we see that we need not symmetrize, since position and momentum commute if they belong to different coordinates:

$$\partial_z yf(y,z) = y\partial_z f(y,z) \tag{3.35}$$

or

$$[x, p_x] = i\hbar; \quad [x, p_y] = [x, p_z] = 0; \quad \text{analogously for } y, z. \tag{3.36}$$

Actually, for the 'standard' operators, one can do without symmetrization.

One of the few counterexamples is the radial momentum $\mathbf{pr}/r$ which occurs e.g. in the formulation of the kinetic energy in spherical coordinates (see exercises). Another example is the *Lenz vector* $\mathbf{\Lambda}$. If a particle with mass m moves in a potential $U = -\frac{\alpha}{r}$, then the vector $\mathbf{\Lambda}$, defined by

$$\mathbf{\Lambda} = \frac{1}{m\alpha}(\mathbf{l} \times \mathbf{p}) + \frac{\mathbf{r}}{r}, \tag{3.37}$$

is a conserved quantity. For the translation into quantum mechanics, the term $\mathbf{l} \times \mathbf{p}$ must be symmetrized. For more on the Lenz vector, see Appendix G, Vol. 2.

---

[8] Actually that is good news, because this symmetrization is not without problems. Take for example $x^2 p$—is the symmetrized expression xpx, $\frac{1}{2}\left(x^2 p + px^2\right)$, $\frac{1}{3}\left(x^2 p + xpx + px^2\right)$, $\frac{1}{4}\left(x^2 p + 2xpx + px^2\right)$ or a completely different term? Or does everything lead to the same quantum-mechanical expression (as is indeed the case in this example)?

3.4 Exercises

1. Show explicitly that the solutions of the Schrödinger (3.1) span a vector space.
2. Calculate $\left[x, \frac{\partial^2}{\partial x^2} \right]$.
3. Given the relativistic energy-momentum relation $E^2 = m_0^2 c^4 + c^2 p^2$; from this dispersion relation, deduce a differential equation.
4. Separation: Deduce the time-independent Schrödinger equation from the time-dependent Schrödinger equation by means of the separation of variables.
5. Given the eigenvalue problem

$$\frac{\partial}{\partial x} f(x) = \gamma f(x); \quad \gamma \in \mathbb{C} \tag{3.38}$$

with $f(x)$ satisfying the boundary conditions $f(0) = 1$ and $f(1) = 2$, calculate the eigenfunction and eigenvalue.
6. Given the eigenvalue problem

$$\frac{\partial^2}{\partial x^2} f = \delta^2 f; \quad \delta \in \mathbb{C} \tag{3.39}$$

with $f(x)$ satisfying the boundary conditions $f(0) = f(L) = 0$ and $L \neq 0$, $\delta \neq 0$, calculate eigenfunctions and eigenvalues.
7. Given the nonlinear differential equation

$$y'(x) = \frac{dy(x)}{dx} = y^2(x). \tag{3.40}$$

$y_1(x)$ and $y_2(x)$ are two different nontrivial solutions of (3.40), i.e. $y_1 \neq const \cdot y_2$ and $y_1 y_2 \neq 0$.

(a) Show that a multiple of a solution, i.e. $f(x) = c y_1(x)$ with $c \neq 0, c \neq 1$, is not a solution of (3.40).
(b) Show that a linear combination of two solutions, i.e. $g(x) = a y_1(x) + b y_2(x)$ with $ab \neq 0$, but otherwise arbitrary, is not a solution of (3.40).
(c) Find the general solution of (3.40).

8. Radial momentum

(a) Show that the classical momentum $\mathbf{p}$ obeys

$$\mathbf{p}^2 = \left(\mathbf{p}\hat{\mathbf{r}} \right)^2 + \left(\mathbf{p} \times \hat{\mathbf{r}} \right)^2. \tag{3.41}$$

(b) Deduce the quantum-mechanical expression p_r for the classical radial momentum $\hat{\mathbf{r}}\mathbf{p} \, (= \mathbf{p}\hat{\mathbf{r}})$.

9. Show explicitly that the classical expression $\mathbf{l} = \mathbf{r} \times \mathbf{p}$ need not be symmetrized for the translation into quantum mechanics.

10. Given the operators $A = x\frac{d}{dx}$, $B = \frac{d}{dx}x$ and $C = \frac{d}{dx}$:

   (a) Calculate $Af_i(x)$ for the functions $f_1(x) = x^2$, $f_2(x) = e^{ikx}$ and $f_3(x) = \ln x$.

   (b) Determine $A^2 f(x)$ for arbitrary $f(x)$.

   (c) Calculate the commutators $[A, B]$ and $[B, C]$.

   (d) Compute $e^{iC}x^2 - (x + i)^2$. Prove the equation $e^{iC}e^{ikx} = e^{-k}e^{ikx}$.

Chapter 4
Complex Vector Spaces and Quantum Mechanics

In our complex vector space, we can define a scalar product. The properties of orthogonality and completeness lead to the important concept of a complete orthonormal system. The measurement process can be formulated by means of suitable projection operators.

Up to now, we have occasionally used the terms 'vector space' or 'state space'. In this chapter, we will address this concept in more detail. For reasons of simplicity, we will rely heavily on the example of *polarization*, where the basic formulations are of course independent of the specific realization and are valid for all two-dimensional state spaces (such as polarization states, electron spin states, a double-well potential, the ammonia molecule, etc.). Moreover, the concepts introduced here retain their meaning in higher-dimensional state spaces, as well. Therefore, we can introduce and discuss many topics by using the example of the simple two-dimensional state space. From the technical point of view, this chapter is about the discussion of some of the elementary facts of complex vector spaces. The basic definitions are given in Appendix G, Vol. 1.[1]

In Chap. 2, we introduced the polarization states which also apply to single photons[2]

$$
|h\rangle \cong \begin{pmatrix} 1 \\ 0 \end{pmatrix}; \quad |v\rangle \cong \begin{pmatrix} 0 \\ 1 \end{pmatrix}
$$
$$
|r\rangle \cong \frac{1}{\sqrt{2}} \begin{pmatrix} 1 \\ i \end{pmatrix}; \quad |l\rangle \cong \frac{1}{\sqrt{2}} \begin{pmatrix} 1 \\ -i \end{pmatrix}. \tag{4.1}
$$

These vectors are obviously elements of a two-dimensional complex vector space $\mathcal{V}$. In fact, one can convince oneself that all the axioms which apply to a vector space are satisfied; see Appendix G, Vol. 1. To put it simply, these axioms state in the end that one can perform all operations as usual—one can add vectors

---

[1]Of course, we treat these technical aspects not as an end in themselves, but because they are of fundamental importance for the physical description of natural phenomena in the context of quantum mechanics.

[2]For the notation $\cong$, see Chap. 2.

© Springer Nature Switzerland AG 2018
J. Pade, *Quantum Mechanics for Pedestrians 1*, Undergraduate Lecture
Notes in Physics, https://doi.org/10.1007/978-3-030-00464-4_4

and multiply them by a number, subject to the familiar rules such as the distributive law, etc. We note in this context that products of numbers and vectors commute, so that $c \cdot |z\rangle = |z\rangle \cdot c$ holds. Although the notation $|z\rangle \cdot c$ is perhaps unfamiliar, it is nevertheless absolutely correct.

Especially important is the fact that the elements of a vector space can be *superposed*—if $|x\rangle$ and $|y\rangle$ are elements of the vector space, then so is $\lambda |x\rangle + \mu |y\rangle$ with $\lambda, \mu \in \mathbb{C}$. In our example of polarization, this means that *each* vector (except the zero vector) represents a viable physical state.[3] This superposition principle[4] is anything but self-evident—just think for example of the state space which consists of all positions that are reachable in a chess game beginning from the starting position. Obviously, here the superposition principle does not hold, since the multiplication of such a state with a number or the addition or linear combination of states is simply not meaningful. Another example is the phase space of classical mechanics, in which the states are denoted by points—the addition of these points or states is not defined.

We will repeatedly come across the central importance of the superposition principle in quantum mechanics in the following sections and chapters.

4.1 Norm, Bra-Ket Notation

The familiar visual space $\mathbb{R}^3$ has the pleasant property that one can calculate the length of a vector and the angle between two vectors, namely by means of the scalar product. We want to implement these concepts also in the complex vector space, at least to some extent.

Following the familiar formula, the length L of the vector $\begin{pmatrix} a \\ b \end{pmatrix}$ would be $L^2 = a^2 + b^2$. But this is wrong, since the vector space is *complex*. Accordingly, due to $1 + i^2 = 0$, the vector $\begin{pmatrix} 1 \\ i \end{pmatrix}$ would have zero length, which evidently makes no sense.[5] Instead, the correct formula reads

$$L = |a|^2 + |b|^2 = aa^* + bb^*. \tag{4.2}$$

Making use of the usual rules of matrix multiplication, we can write this as the product of a row vector with a column vector[6]:

---

[3]Later on, we will meet vector spaces where this is no longer the case; keyword 'identical particles' or 'superselection rules'.

[4]We note that the superposition principle contains three pieces of information: (1) The multiplication of a state by a scalar is meaningful. (2) The addition of two states is meaningful. (3) Every linear combination of two states is again an element in the vector space.

[5]As we know, only the zero vector has length zero.

[6]We recall that * means complex conjugation.

$$L^2 = \begin{pmatrix} a^* & b^* \end{pmatrix} \begin{pmatrix} a \\ b \end{pmatrix}. \tag{4.3}$$

The space of the row vectors is called the dual space to $\mathcal{V}$. One obtains the vector $\begin{pmatrix} a^* & b^* \end{pmatrix}$ from the corresponding column vector by complex conjugation and inverting the roles of column and row (= transposing, symbol ^{T}). By this process, we obtain the *adjoint*,[7] which is denoted by a kind of superscripted cross

$$\begin{pmatrix} a^* & b^* \end{pmatrix} = \begin{pmatrix} a \\ b \end{pmatrix}^{*T} = \begin{pmatrix} a \\ b \end{pmatrix}^{\dagger}. \tag{4.4}$$

The operation is analogously defined for general $n \times m$-matrices: the adjoint is always obtained by complex conjugation and transposition. We note that the adjoint is a *very* important term in quantum mechanics.

We have denoted the elements of the vector space using the short-hand notation $|\,\rangle$. Analogously, we choose for the elements of the dual space the notation $\langle\,|$. The symbols are defined as follows:

$$|\,\rangle \quad \text{is called a } ket \tag{4.5}$$
$$\langle\,| \quad \text{is called a } bra.$$

This is the so-called *bra-ket notation* (from bracket = bra-(c)-ket), or *Dirac notation*, named after P.A.M. Dirac who first introduced it.[8] We have for example

$$|h\rangle^{\dagger} = \langle h| \text{ or } \begin{pmatrix} 1 \\ 0 \end{pmatrix}^{\dagger} = \begin{pmatrix} 1 & 0 \end{pmatrix}$$
$$\langle r|^{\dagger} = |r\rangle \text{ or } \tfrac{1}{\sqrt{2}} \begin{pmatrix} 1 & -i \end{pmatrix}^{\dagger} = \tfrac{1}{\sqrt{2}} \begin{pmatrix} 1 \\ i \end{pmatrix}. \tag{4.6}$$

With these concepts we can now define the length L of a vector $|z\rangle$ as $L^2 = \langle z|\,z\rangle$ (actually, one would expect to write $\langle z|\,|z\rangle$, but the double bar is omitted). Instead of length, the term *norm* is generally used. The designations are $\|\ \|$ or equivalently $|\ |$. For example, we have

$$\||h\rangle\|^2 = \langle h|\,h\rangle = \begin{pmatrix} 1 & 0 \end{pmatrix} \begin{pmatrix} 1 \\ 0 \end{pmatrix} = 1 \cdot 1 + 0 \cdot 0 = 1 \tag{4.7}$$

and correspondingly for $|r\rangle$

---

[7] Strictly speaking, there are two adjoints. The one considered here is called *Hermitian adjoint*; it applies so to say in non-relativistic considerations. In the relativistic case, there is another kind, called *Dirac adjoint* which is defined differently. The bulk of the book is devoted to non-relativistic considerations; here adjoint means always Hermitian adjoint.

[8] In the bra-ket notation, one cannot identify the dimension of the corresponding vector space (the same holds true for the familiar vector notations $\mathbf{v}$ or $\vec{v}$, by the way). If necessary, this information must be given separately.

$$\||r\rangle\|^2 = \langle r|\, r\rangle = \frac{1}{2}\left(1 \ -i\,\right)\binom{1}{i} = \frac{1}{2}(1\cdot 1 - i\cdot i) = 1. \qquad (4.8)$$

Both vectors have the length 1. Such vectors are called *unit vectors*; they are *normalized*. The term $\langle z|\, z\rangle$ is a *scalar product* (also called inner product or dot product); more about this topic is to be found in Chap. 11 and in Appendix G, Vol. 1. We remark that we can use an equals sign in (4.7) and (4.8) instead of $\cong$, since scalar products are *independent of the representation*.

A comment on the nomenclature: A complex vector space in which a scalar product is defined is called a *unitary space*.

4.2   Orthogonality, Orthonormality

Now that we know how to calculate the length of a vector, the question of the angle between two vectors still remains open. First, we note that we can also form inner products of different vectors, for example

$$\langle v|\, r\rangle = \frac{1}{\sqrt{2}}\left(0\ 1\right)\binom{1}{i} = \frac{i}{\sqrt{2}} \qquad (4.9)$$

Note: As with any scalar product, $\langle a|\, b\rangle$ is a (generally complex) *number*. For the adjoint of an inner product, for example, we have

$$(\langle v|\, r\rangle)^\dagger = \langle r|\, v\rangle = \frac{1}{\sqrt{2}}\left(1\ -i\,\right)\binom{0}{1} = -\frac{i}{\sqrt{2}}. \qquad (4.10)$$

So to form the adjoint of an expression we follow the procedure: (1) A number is replaced by its complex conjugate, $c^\dagger = c^*$; (2) A ket is replaced by the corresponding bra and *vice versa*; (3) The order of terms is reversed, for example $\langle a|\, b\rangle^\dagger = \langle a|\, b\rangle^* = \langle b|\, a\rangle$.

Regarding the question of the angle, in the following only one particular angle plays a role (apart from the angle zero), namely the right angle. One says that two vectors are *orthogonal* if their scalar product vanishes (this convention of terminology is valid also for non-intuitive higher-dimensional complex vector spaces). An example:

$$\langle v|\, h\rangle = \left(0\ 1\right)\binom{1}{0} = 0. \qquad (4.11)$$

More generally and in short form: $\langle a|\, b\rangle = 0 \leftrightarrow |a\rangle \perp |b\rangle$.

Note that the zero vector is orthogonal to itself and to all other vectors. Just as in the trivial solution of the SEq, it does not describe a physical state and is therefore in general not taken into account in considerations concerning orthogonality, etc.

Systems of vectors, all of which are normalized and pairwise orthogonal, play a special role.[9] Such a system of vectors is called an *orthonormal system (ONS)*. Two-dimensional examples are the systems $\{|h\rangle, |v\rangle\}$ and $\{|r\rangle, |l\rangle\}$; an example from three-dimensional visual space $\mathbb{R}^3$ are the three unit vectors lying on the coordinate axes. The general formulation reads: $\{|\varphi_n\rangle, n = 1, 2, \ldots\}$ is an ONS if and only if

$$\langle \varphi_i | \varphi_j \rangle = \delta_{ij} \tag{4.12}$$

where the *Kronecker delta* (Kronecker symbol) is defined as usual by

$$\delta_{ij} = \begin{cases} 1 \\ 0 \end{cases} \quad \text{for} \quad \begin{matrix} i = j \\ i \neq j \end{matrix}. \tag{4.13}$$

4.3 Completeness

We can write any vector $|z\rangle$ from our two-dimensional complex vector space as

$$|z\rangle = a\,|h\rangle + b\,|v\rangle \tag{4.14}$$

where $|h\rangle$ and $|v\rangle$ are orthonormal. Due to $\langle h| z\rangle = a\,\langle h| h\rangle + b\,\langle h| v\rangle = a \cdot 1 + b \cdot 0 = a$ (analogously for $\langle v| z\rangle$), this property leads to

$$\langle h| z\rangle = a \text{ and } \langle v| z\rangle = b. \tag{4.15}$$

We insert this and find[10]

$$\begin{aligned} |z\rangle &= \langle h| z\rangle\,|h\rangle + \langle v| z\rangle\,|v\rangle \\ &= |h\rangle\,\langle h| z\rangle + |v\rangle\,\langle v| z\rangle \\ &= \{|h\rangle\,\langle h| + |v\rangle\,\langle v|\}\,|z\rangle, \end{aligned} \tag{4.16}$$

or in other words (by comparing the left and right sides)[11]:

---

[9]In the two-dimensional vector space that we are currently addressing, such a system consists of course of two vectors; as stated above, the zero vector is excluded *a priori* from consideration.

[10]We repeat the remark that for products of numbers and vectors, it holds that $c \cdot |z\rangle = |z\rangle \cdot c$. Because $\langle h| z\rangle$ is a number, we can therefore write $\langle h| z\rangle\,|h\rangle$ as $|h\rangle\,\langle h| z\rangle$.

[11]In equations such as (4.17), the 1 on the right side is not necessarily the number 1, but is generally something that works like a multiplication by 1, i.e. a *unit operator*. For instance, this is the unit matrix when working with vectors. The notation 1 for the unit operator (which implies writing simply 1 instead of $\begin{pmatrix} 1 & 0 \\ 0 & 1 \end{pmatrix}$, for instance) is of course quite lax. On the other hand, as said before, the effect of multiplication by the unit operator and by 1 is identical, so that the small inaccuracy is generally accepted in view of the economy of notation. If necessary, 'one knows' that 1 means the unit operator. But there are also special notations for it, such as $\mathbb{E}$, I_n (where n indicates the

$$|h\rangle \langle h| + |v\rangle \langle v| = 1. \tag{4.17}$$

A term like $|x\rangle \langle y|$ is called a *dyadic product*. To get an idea of the meaning of such products, we choose the representation of row and column vectors. With

$$|h\rangle \cong \begin{pmatrix} 1 \\ 0 \end{pmatrix}; \quad \langle h| \cong \begin{pmatrix} 1 & 0 \end{pmatrix} \tag{4.18}$$

we have, according to the rules of matrix multiplication,

$$|h\rangle \langle h| \cong \begin{pmatrix} 1 \\ 0 \end{pmatrix} \begin{pmatrix} 1 & 0 \end{pmatrix} = \begin{pmatrix} 1 & 0 \\ 0 & 0 \end{pmatrix}. \tag{4.19}$$

As can be seen, dyadic products are matrices or, more generally, operators which can be applied to states and usually change them.
With

$$|v\rangle \langle v| \cong \begin{pmatrix} 0 \\ 1 \end{pmatrix} \begin{pmatrix} 0 & 1 \end{pmatrix} = \begin{pmatrix} 0 & 0 \\ 0 & 1 \end{pmatrix} \tag{4.20}$$

it follows that

$$|h\rangle \langle h| + |v\rangle \langle v| \cong \begin{pmatrix} 1 & 0 \\ 0 & 1 \end{pmatrix} \text{ (unit matrix)}. \tag{4.21}$$

This equation, or (4.17), indicates that the ONS $\{|h\rangle, |v\rangle\}$ is *complete*, i.e. it spans the whole space. Consequently, $\{|h\rangle, |v\rangle\}$ is a *complete orthonormal system* (*CONS*). Another one is, for example, $\{|r\rangle, |l\rangle\}$ (see exercises). The terminology is transferred readily to n-dimensional vector spaces: A CONS consists of states $\{|\varphi_n\rangle, n = 1, 2, \ldots\}$ which are normalized and pairwise orthogonal (orthonormality), and which span the whole space (completeness)[12]:

$$\langle \varphi_n| \varphi_m\rangle = \delta_{nm} \text{ (orthonormality)}$$
$$\sum_n |\varphi_n\rangle \langle \varphi_n| = 1 \text{ (completeness)} \tag{4.22}$$

With the methods developed so far, we can easily calculate the fractions of vertically—and horizontally-polarized light which are found e.g. in right circularly-polarized light. Of course, the example is simple enough to read off the answer directly from (4.1). But here, we are concerned with setting up a procedure that

---

dimension) and others. An analogous remark applies to the zero operator. By the way, we recall that in the case of vectors we write quite naturally $\vec{a} = 0$ and not $\vec{a} = \vec{0}$.

[12]For the summation we use almost exclusively the abbreviation $\sum_n$ (instead of $\sum_{n=1}^{\infty}$ or $\sum_{n=1}^{N}$ etc.). In the shorthand notation, the range of values of n must follow from the context of the problem at hand, if necessary.

works in any space. Basically, it is a multiplication by 1—but with 1 in a special notation. We have:

$$|r\rangle = 1 \cdot |r\rangle \overset{4.17}{=} (|h\rangle \langle h| + |v\rangle \langle v|) \cdot |r\rangle$$

$$= |h\rangle \langle h| r\rangle + |v\rangle \langle v| r\rangle = \frac{1}{\sqrt{2}} |h\rangle + \frac{i}{\sqrt{2}} |v\rangle, \tag{4.23}$$

where we have used $\langle h| r\rangle = \frac{1}{\sqrt{2}}$ and $\langle v| r\rangle = \frac{i}{\sqrt{2}}$ in the final step.

With (4.23), we have formulated the state $|r\rangle$ in the basis $\{|h\rangle, |v\rangle\}$. This being the case, the coefficients $1/\sqrt{2}$ and $i/\sqrt{2}$ are none other than the coordinates of $|r\rangle$ with respect to $|h\rangle$ and $|v\rangle$. However, the term *coordinate* is used quite rarely in quantum mechanics; instead, one speaks of *projection*,[13] which is perhaps an even more descriptive term.

For higher dimensions, the following applies: Given a vector space $\mathcal{V}$ and a CONS $\{|\varphi_n\rangle\}, n = 1, 2, \ldots\} \in \mathcal{V}$, any vector $|\psi\rangle \in \mathcal{V}$ can be represented as

$$|\psi\rangle = 1 \cdot |\psi\rangle = \sum_n |\varphi_n\rangle \langle \varphi_n |\psi\rangle = \sum_n c_n |\varphi_n\rangle ; \quad c_n = \langle \varphi_n |\psi\rangle \in \mathbb{C}. \tag{4.24}$$

The coefficients (coordinates) c_n are the projections of $|\psi\rangle$ onto the basis vectors $|\varphi_n\rangle$.

4.4 Projection Operators, Measurement

4.4.1 *Projection Operators*

As mentioned above, expressions like $|h\rangle \langle h|$ or $|\varphi_n\rangle \langle \varphi_n|$ act on states and are therefore *operators*. They are different from those that we met up with in the analytical approach of Chap. 3 (e.g. the derivative $\frac{\partial}{\partial x}$), but this is actually not surprising, since the states defined in the algebraic and the analytical approaches are quite different. We note, however, that there is a structure common to both approaches: in each case the states are elements of a vector space, and changes of these states are produced by operators.

The term $|h\rangle \langle h|$ is a particularly simple example of a *projection operator* (or projector). If P is a projection operator, we have[14]

$$P^2 = P. \tag{4.25}$$

---

[13] For the connection between inner product and projection, see Appendix F, Vol. 1.

[14] As we shall see in Chap. 13, a projection operator in quantum mechanics must meet a further condition (self-adjointness).

In fact, in the specific example $P = |h\rangle \langle h|$, we have, due to the normalization $\langle h | h\rangle = 1$, the equality

$$P^2 = |h\rangle \langle h | h\rangle \langle h| = |h\rangle \langle h| = P. \tag{4.26}$$

A further example of a projection operator is $|h\rangle \langle h| + |v\rangle \langle v|$, namely the projection onto the total space (because of $|h\rangle \langle h| + |v\rangle \langle v| = 1$).

The property $P^2 = P$ is actually very intuitive: if one filters out (= projects) a component of a total state by means of P, then a second projection does not change this component. In the matrix representation (4.19) with $P \cong \begin{pmatrix} 1 & 0 \\ 0 & 0 \end{pmatrix}$, this reads as follows:

$$\begin{pmatrix} 1 & 0 \\ 0 & 0 \end{pmatrix} \begin{pmatrix} a \\ b \end{pmatrix} = \begin{pmatrix} a \\ 0 \end{pmatrix}; \quad \begin{pmatrix} 1 & 0 \\ 0 & 0 \end{pmatrix} \begin{pmatrix} 1 & 0 \\ 0 & 0 \end{pmatrix} \begin{pmatrix} a \\ b \end{pmatrix} = \begin{pmatrix} a \\ 0 \end{pmatrix}. \tag{4.27}$$

4.4.1.1 Projection Operators and Measurement

Projection operators gain special importance from the fact that they can be used for the modelling of the measurement process. To see this, we start with a simple example, namely a right circularly-polarized state. Using (4.1), we write it as a superposition of linearly-polarized states:

$$|r\rangle = \frac{|h\rangle + i\,|v\rangle}{\sqrt{2}}. \tag{4.28}$$

We send this state $|r\rangle$ through an analyzer which can detect linearly-polarized states, e.g. a polarizing beam splitter (PBS). *Before* the measurement, we cannot say with certainty which one of the two linearly-polarized states we will measure. According to the considerations of Chap. 2, we can specify only the probabilities of measuring one of the states—in our example they are $\left|\frac{1}{\sqrt{2}}\right|^2 = \frac{1}{2}$ and $\left|\frac{i}{\sqrt{2}}\right|^2 = \frac{1}{2}$. We extend this idea to the more general state

$$|z\rangle = a\,|h\rangle + b\,|v\rangle; \quad |a|^2 + |b|^2 = 1, \tag{4.29}$$

for which the probabilities of obtaining the vertically or horizontally polarized state are given by $|b|^2$ or $|a|^2$.

After the measurement, we have a different state from before the measurement, namely either $|h\rangle$ or $|v\rangle$.[15] Since states can be changed only by the action of operators, we have to model this transition by an operator. This modelling should be as simple and universal as possible, in order to be independent of the specific experimental details. Let us assume that we have the state $|h\rangle$ after the measurement.

---

[15]In other words, due to the process of measuring, a superposition such as $|z\rangle = a\,|h\rangle + b\,|v\rangle$ 'collapses' e.g. into the state $|h\rangle$.

Then we can describe this process by applying $|h\rangle \langle h|$ to $|z\rangle$, i.e. the projection of $|z\rangle$ onto $|h\rangle$, which leads to $|h\rangle \langle h | z\rangle = a |h\rangle$ (with an analogous formulation for $|v\rangle$). As a result of this 'operation', we obtain the desired state $|h\rangle$, but multiplied by a factor a, the absolute square of which gives the probability of obtaining that state in a measurement.

Hence, we can model the measurement process $|z\rangle \rightarrow |h\rangle$ as follows:

$$\underset{\text{before measurement}}{|z\rangle} = a |h\rangle + b |v\rangle \overset{\text{projection}}{\rightarrow} |h\rangle \langle h| (a |h\rangle + b |v\rangle)$$

$$= a |h\rangle \overset{\text{normalization}}{\rightarrow} \underset{\text{after measurement}}{\frac{|a|}{a} |h\rangle} , \qquad (4.30)$$

where we obtain the final result with probability $|a|^2$. Occasionally, it is assumed that one can set the normalization factor equal to 1 after the measurement, which formally means $\frac{|a|}{a} = 1$. As we said above, an analogous formulation applies to the measurement result $|v\rangle$.

4.4.1.2 Extension to Higher Dimensions

The generalization to dimensions $N > 2$ is straightforward. Before the measurement, the state is a superposition of different states, i.e. $|\psi\rangle = \sum c_n |\varphi_n\rangle$, where $\{|\varphi_n\rangle\}, n = 1, \ldots\}$ is a CONS. After the measurement, we have just one of the states, e.g. $|\varphi_i\rangle$. The measurement process is modelled by the projection operator $P_i = |\varphi_i\rangle \langle \varphi_i|$. With a slightly different notation, we have

$$|\psi\rangle_{\text{before}} = \sum_n c_n |\varphi_n\rangle \rightarrow |\varphi_i\rangle \langle \varphi_i | \psi\rangle_{\text{before}} = |\varphi_i\rangle \langle \varphi_i| \sum_n c_n |\varphi_n\rangle = c_i |\varphi_i\rangle .$$

$$(4.31)$$

The probability of measuring this state is thus given by $|c_i|^2 = |\langle \varphi_i | \psi\rangle|^2$. After the measurement, we have again a normalized state, namely

$$|\psi\rangle_{\text{after, normalized}} = \frac{P_i |\psi\rangle}{|P_i |\psi\rangle|} = \frac{c_i |\varphi_i\rangle}{|c_i|} . \qquad (4.32)$$

We emphasize that the measurement process itself is *not* modelled, but only the situation immediately before and after the measurement. As an example, the situation is sketched in Fig. 4.1.

Fig. 4.1 Example sketch of
the coefficients c_n in (4.31)
and (4.32) before the
measurement (blue) and after
the measurement (red)

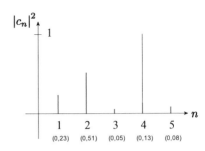

4.4.1.3 The Measurement Problem

Of course one may ask at this point, which mechanism picks out precisely the state $|\varphi_i\rangle$ from the superposition $\sum_n c_n |\varphi_n\rangle$, and not some other state. There is still no satisfactory answer to this question in spite of the advanced age of quantum mechanics. In fact, it is still an open problem, called the *measurement problem*. It is perhaps *the* conceptual problem of quantum mechanics. We shall meet it repeatedly in the following chapters. The different interpretations of quantum mechanics, at which we look closer in the last chapter of Vol. 2, are in some sense simply different ways of dealing with the measurement problem.

We note that the measurement problem has nothing to do with the extension to arbitrary dimensions, but applies even to the simplest systems. An example already treated in Chap. 2 and above is the right circularly-polarized photon, which we examine with respect to its possible linear polarization states. If we send

$$|r\rangle = \frac{|h\rangle + i\,|v\rangle}{\sqrt{2}} \tag{4.33}$$

through e.g. a PBS, we find either a horizontal or a vertical linearly-polarized photon, with probabilities $\frac{1}{2}$ in each case. Before the measurement, we cannot say which polarization we will obtain.

The key question is whether there is *in principle* such a selection mechanism. We have two alternatives. The first one: Yes, there is such a mechanism, although we do not know either the process (at present?) or the variables which it acts upon, the so-called *hidden variables*. If we knew these, we could describe the selection process that occurs during the measurement without any use of probabilities. The other alternative: No, there is no such mechanism. The selection of states in the course of the measurement is purely random—one speaks of *objective chance*.

The choice of the alternative in question must be decided experimentally. We have already noted in Chap. 2 that, all in all, relevant experiments do not support the existence of hidden variables. Therefore, we hereafter assume the existence of objective chance, but we will take up the measurement problem again in later chapters.

Fig. 4.2 Measurement of
the linear polarization of a
photon

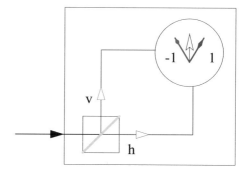

4.4.2   Measurement and Eigenvalues

To arrive at a more compact description of the measurement, we imagine that, after
the PBS, we have a detector which is connected to a display. For vertical polarization,
a pointer shows '−1', for horizontal polarization '+1'; see Fig. 4.2. This means that
after the measurement on the state $|z\rangle = a|h\rangle + b|v\rangle$, the value '−1' or '+1' is
displayed with the probabilities $|b|^2$ or $|a|^2$.

We now want to describe the measured physical quantity 'horizontal/vertical
polarization', encoded by ± 1. To this end, we choose a linear combination of the pro-
jection operators $|h\rangle\langle h|$ and $|v\rangle\langle v|$. The simplest non-trivial combination is clearly
the polarization operator P_L

$$P_L = |h\rangle\langle h| - |v\rangle\langle v| \cong \begin{pmatrix} 1 & 0 \\ 0 & -1 \end{pmatrix} = \sigma_z, \tag{4.34}$$

where σ_z is one of the three *Pauli matrices* (more on the Pauli matrices in the exercises). We note that the Pauli matrices, and thus P_L, are not projection operators;
here the P stands for 'polarization'.

The properties which are relevant for the measurement follow now by consid-
ering the *eigenvalue problem* $P_L|z\rangle = \mu|z\rangle$ (where we have now introduced the
eigenvalue problem, treated in the analytical approach already in Chap. 3, into the
algebraic approach considered here). As is easily verified (see exercises), P_L has
the eigenvalues $\mu = +1$ and $\mu = -1$ and the eigenvectors $|z_1\rangle = |h\rangle$ and $|z_{-1}\rangle = |v\rangle$.
This means that the eigenvectors describe the possible states and the eigenvalues the
possible pointer positions (measurement results) after the measurement—the pointer
position $+1$ tells us, for example, that after the measurement we have the state $|h\rangle$.

Similarly, we can imagine a measuring apparatus for circular polarization, in
which the physical quantity 'right/left circular polarization' is encoded by ± 1. We
describe this by

$$P_C = |r\rangle\langle r| - |l\rangle\langle l| \cong \begin{pmatrix} 0 & -i \\ i & 0 \end{pmatrix} = \sigma_y. \tag{4.35}$$

The eigenvalues of P_C are also ± 1, where the eigenvalue $+1$ belongs to the eigenvector $|r\rangle$ and the eigenvalue -1 to the eigenvector $|l\rangle$.

Finally, we treat a linear polarization state rotated by $45°$. For the rotated state, we have (see exercises):

$$|h'\rangle = \frac{|h\rangle + |v\rangle}{\sqrt{2}} \; ; \; |v'\rangle = \frac{-|h\rangle + |v\rangle}{\sqrt{2}}. \tag{4.36}$$

We describe the corresponding measuring apparatus by the operator

$$P_{L'} = |h'\rangle\langle h'| - |v'\rangle\langle v'| \cong \begin{pmatrix} 0 & 1 \\ 1 & 0 \end{pmatrix} = \sigma_x. \tag{4.37}$$

This operator has the eigenvalues ± 1, also. (For the determination of the eigenvalues and eigenvectors of the three Pauli matrices, see the exercises.)

We learn from these three examples that the information about possible measurement results lies in the eigenvalues of certain operators. For this purpose we have constructed three examples that yield information about certain polarization states. The question of how to extend these findings and how to represent general physically-measurable variables will be treated in the following chapters.

4.4.3 Summary

We summarize with the help of the example $|z\rangle = a\,|h\rangle + b\,|v\rangle$, with $|a|^2 + |b|^2 = 1$ and $ab \neq 0$. *Before* the measurement, we can say only that: (1) The pointer will display position '+1' with probability $|a|^2 = |\langle h\,|z\rangle|^2$, and position '−1' with probability $|b|^2$; and (2) Just one of the eigenvalues of $P_L = |h\rangle\langle h| - |v\rangle\langle v|$ will be realized, with the corresponding probability. *After* the measurement, one of the two eigenvalues is realized (the pointer displays one of the two possible values), and the photon is in the corresponding state (the associated normalized eigenstate of P_L), e.g. $\frac{\langle h\,|z\rangle}{|\langle h\,|z\rangle|}\,|h\rangle$. We cannot discern which mechanism leads to this choice, but can only specify the probabilities for the possible results. The process is irreversible—the initial superposition no longer exists, and it cannot be reconstructed from the measurement results from a single photon.[16] This is possible at most by measuring an ensemble of photons in the state $a\,|h\rangle + b\,|v\rangle$ many times. From the relative frequencies of occurrence of the pointer values ± 1, we can infer the quantities $|a|^2$ and $|b|^2$.

---

[16]In order to make it clear once more: If, for example, we measure an arbitrarily-polarized state $|z\rangle = a\,|h\rangle + b\,|v\rangle$ with $|a|^2 + |b|^2 = 1$ and $ab \neq 0$, we find with probability $|a|^2$ a horizontal linearly-polarized photon. This does *not* permit the conclusion that the photon was in that state before the measurement. It simply makes no sense in this case to speak of a of a well-defined value of the linear polarization ($+1$ or -1) before the measurement.

4.5 Exercises

1. Find examples for state spaces which

    (a) have the structure of a vector space,
    (b) do not have the structure of a vector space.

2. Polarization: Determine the length of the vector $\frac{1}{\sqrt{2}}\begin{pmatrix} 1 \\ i \end{pmatrix}$.

3. Given $\langle y| = i\begin{pmatrix} 1 & -2 \end{pmatrix}$ and $\langle z| = \begin{pmatrix} 2 & i \end{pmatrix}$, determine $\langle y| z \rangle$.

4. The Pauli matrices are

$$\sigma_x = \begin{pmatrix} 0 & 1 \\ 1 & 0 \end{pmatrix}; \quad \sigma_y = \begin{pmatrix} 0 & -i \\ i & 0 \end{pmatrix}; \quad \sigma_z = \begin{pmatrix} 1 & 0 \\ 0 & -1 \end{pmatrix} \tag{4.38}$$

    In addition to $\sigma_x, \sigma_y, \sigma_z$, the notation $\sigma_1, \sigma_2, \sigma_3$ is also common.

    (a) Show that $\sigma_i^2 = 1, i = x, y, z$.

    (b) Determine the commutators $[\sigma_i, \sigma_j] = \sigma_i\sigma_j - \sigma_j\sigma_i$ and the anticommuta-
    tors $\{\sigma_i, \sigma_j\} = \sigma_i\sigma_j + \sigma_j\sigma_i$ ($i \neq j$).

    (c) Calculate the eigenvalues and eigenvectors for each Pauli matrix.

5. Determine the eigenvalues and eigenvectors of the matrix

$$M = \begin{pmatrix} 1 & 4 \\ 2 & -1 \end{pmatrix}. \tag{4.39}$$

    Normalize the eigenvectors. Are they orthogonal?

6. Given the CONS $\{|a_1\rangle, |a_2\rangle\}$, determine the eigenvalues and eigenvectors of the operator

$$M = |a_1\rangle \langle a_1| - |a_2\rangle \langle a_2|. \tag{4.40}$$

7. Given a CONS $\{|\varphi_n\rangle\}$ and a state $|\psi\rangle = \sum_n c_n |\varphi_n\rangle$, $c_n \in \mathbb{C}$, calculate the coefficients c_n.

8. Show in bra-ket notation: The system $\{|r\rangle, |l\rangle\}$ is a CONS. Use the fact that $\{|h\rangle, |v\rangle\}$ is a CONS.

9. Given the operator $|h\rangle \langle r|$:

    (a) Is it a projection operator?
    (b) How does the operator appear in the representation (4.1)?
    (c) Given the state $|z\rangle$ with the representation $|z\rangle \cong \begin{pmatrix} z_1 \\ z_2 \end{pmatrix}$, apply the operator $|h\rangle \langle r|$ to this state (calculation making use of the representation).

(d) Use the concrete representation to prove the equality

$$(|h\rangle \langle r| z\rangle)^{\dagger} = \langle z| r\rangle \langle h| . \tag{4.41}$$

10. We choose the following representation for the states $|h\rangle$ and $|v\rangle$:

$$|h\rangle \cong \frac{1}{\sqrt{2}} \begin{pmatrix} i \\ 1 \end{pmatrix}; \quad |v\rangle \cong \frac{a}{\sqrt{2}\,|a|} \begin{pmatrix} 1 \\ i \end{pmatrix}. \tag{4.42}$$

(a) Show that the representing vectors form a CONS.
(b) Determine $|r\rangle$ and $|l\rangle$ in this representation. Specialize to the cases of $a = 1$, $-1, i, -i$.

11. Show that the three vectors

$$\mathbf{a} = \frac{1}{\sqrt{2}} \begin{pmatrix} 1 \\ i \\ 0 \end{pmatrix}; \quad \mathbf{b} = \begin{pmatrix} 0 \\ 0 \\ 1 \end{pmatrix}; \quad \mathbf{c} = -\frac{1}{\sqrt{2}} \begin{pmatrix} 1 \\ -i \\ 0 \end{pmatrix} \tag{4.43}$$

form a CONS. Do the same for

$$\mathbf{a} = \frac{1}{\sqrt{2}} \begin{pmatrix} 1 \\ 0 \\ -1 \end{pmatrix}; \quad \mathbf{b} = \frac{1}{2} \begin{pmatrix} 1 \\ \sqrt{2} \\ 1 \end{pmatrix}; \quad \mathbf{c} = \frac{1}{2} \begin{pmatrix} 1 \\ -\sqrt{2} \\ 1 \end{pmatrix}. \tag{4.44}$$

12. A three-dimensional problem: Given the CONS $\{|u\rangle, |v\rangle, |w\rangle\}$ and the operator[17]

$$L = |v\rangle \langle u| + (|u\rangle + |w\rangle) \langle v| + |v\rangle \langle w|. \tag{4.45}$$

(a) Determine the eigenvalues and eigenvectors of L.
(b) Show that the three eigenvectors form a CONS.

---

[17]Essentially, this operator is the x component of the orbital angular momentum operator for the angular momentum 1; see Chap. 16 Vol. 2.

Chapter 5
Two Simple Solutions of the Schrödinger Equation

The infinite potential well is the simplest model case for a discrete energy spectrum. We see that the eigenfunctions form a complete orthonormal system. Free motion is the simplest model case for a continuous spectrum. In both cases, we solve the initial-value problem. We make our first contact (within the analytical approach) with the interpretation of probability and measurements.

This chapter deals with the solutions of the SEq for two simple but important one-dimensional systems. First, we consider the *infinite potential well* as a simple model of a bounded system, then *force-free unlimited motion* as a simple model of an unbounded system. Here, 'bounded motion' means basically that the system is confined to a finite region, in contrast to unlimited motion.

The two examples in this chapter are of interest not only in view of our current state of knowledge in this course, but also because they provide further information. At the same time, they are mathematically so simple that they are treated as specific cases even at the school level. Among other things, we will see below that the striking differences between the two solutions can be attributed to 'just' their different boundary conditions.[1]

5.1 The Infinite Potential Well

We imagine a ping-pong ball which bounces back and forth between two fixed, infinitely rigid walls, whereby friction and gravity are switched off. We can represent the two walls by infinitely high potential barriers at $x = 0$ and $x = a$; for $0 < x < a$, the potential energy is zero. Classically, the ping-pong ball can have any speed or kinetic energy (it has no potential energy). This means that in Fig. 5.1, the ball can fly at any height (the height in the figure corresponds to the ball's kinetic energy, not to its position!).

---

[1]See also the exercises for Chap. 3.

© Springer Nature Switzerland AG 2018
J. Pade, *Quantum Mechanics for Pedestrians 1*, Undergraduate Lecture Notes in Physics, https://doi.org/10.1007/978-3-030-00464-4_5

Fig. 5.1 Infinite potential
well. In classical mechanics
(*left*), all energies are
allowed. In quantum
mechanics, only discrete
energy levels are allowed

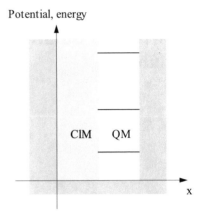

In contrast, the quantum-mechanical ping-pong ball can occupy only certain 'energy levels', as we will see below.[2] In other words, its energy is quantized. This system, which represents the prototype of a bounded problem in quantum mechanics, is called the *infinite potential well*:

$$V = \begin{cases} 0 \text{ for } 0 < x < a \\ \infty \text{ otherwise} . \end{cases} \qquad (5.1)$$

5.1.1 Solution of the Schrödinger Equation, Energy Quantization

The stationary SEq is given for $0 < x < a$ by:

$$E\varphi(x) = -\frac{\hbar^2}{2m}\varphi''(x) \qquad (5.2)$$

Outside the infinite potential well and at its edges (walls), the wavefunction vanishes identically

$$\varphi(x) \equiv 0 \text{ for } x \leq 0 \text{ and } a \leq x. \qquad (5.3)$$

Hence, the problem is described by (5.2) with the *boundary conditions*[3]

---

[2] Indeed, the quantum-mechanical ping-pong ball is quite a peculiar ball, namely an object described by a standing wave.

[3] A conclusive argument for these boundary conditions is given in Chap. 15, Vol. 2. For now, one might think (in an intuitive analogy to the wavefunction) of a rope which is clamped at both ends (although the question remains open as to what a rope has to do with this quantum-mechanical situation). Alternatively, one might consider a continuity requirement for the wavefunction at the walls to be plausible.

$$\varphi(0) = 0; \quad \varphi(a) = 0. \tag{5.4}$$

We write (5.2) in the form

$$\varphi'' = -\frac{2mE}{\hbar^2}\varphi. \tag{5.5}$$

In order to arrive at a more compact form, we make use of the de Broglie relation $p = \hbar k$ and obtain

$$E = \frac{p^2}{2m} = \frac{\hbar^2}{2m}k^2; \tag{5.6}$$

it then follows that

$$\varphi'' = -k^2\varphi. \tag{5.7}$$

This is the familiar differential equation for the classical harmonic oscillator, with the solutions

$$\varphi = Ae^{ikx} + Be^{-ikx}; \quad 0 < x < a; (A, B) \neq (0, 0), \tag{5.8}$$

where we assume without loss of generality that $k > 0$.[4] At this point, the energy E (and hence k) is not yet determined; we will find them in the next step.

The solution (5.8) contains the three free variables A, B and k, two of which can be fixed by the boundary conditions:

$$0 = \varphi(0) = A + B$$
$$0 = \varphi(a) = Ae^{ika} + Be^{-ika}. \tag{5.9}$$

This is a homogeneous system of equations for A and B. It follows that:

$$A = -B$$
$$0 = Ae^{ika} - Ae^{-ika}. \tag{5.10}$$

This yields for $A \neq 0$[5]:

$$e^{ika} - e^{-ika} = 0, \tag{5.11}$$

or, equivalently[6]:

$$\sin ka = 0. \tag{5.12}$$

Only when this condition is met does the system (5.9) have a nontrivial (i.e. physical) solution. Equation (5.12) can be satisfied only for certain values of k, namely $ka = n\pi, n \in \mathbb{N}$. Thus, there exist only discrete values for k:

[4]We have $k \neq 0$, since for $k = 0$, only the trivial solution is obtained.
[5]For $A = 0$, we would obtain the trivial solution.
[6]We recall that $\sin x = \frac{e^{ix}-e^{-ix}}{2i}$.

$$k = \left\{ \frac{\pi}{a}, \frac{2\pi}{a}, \frac{3\pi}{a}, \frac{4\pi}{a} \ldots \right\} = \{k_n\}; \; k_n = \frac{n\pi}{a}; \; n \in \mathbb{N}. \tag{5.13}$$

Accordingly, there are countably infinitely many solutions (= eigenfunctions) of the SEq, namely

$$\varphi_n(x) = 2i \, A \sin k_n x. \tag{5.14}$$

Due to the linearity of the SEq, one can choose the amplitude freely. With the choice[7]

$$2i \, A = \sqrt{\frac{2}{a}}, \tag{5.15}$$

we arrive at

$$\varphi_n(x) = \sqrt{\frac{2}{a}} \sin k_n x. \tag{5.16}$$

Because of the relation $E = \frac{\hbar^2 k^2}{2m}$, the energy can also assume only discrete values. These energy eigenvalues are given by:

$$E_n = \frac{\hbar^2}{2m} k_n^2 = \frac{\hbar^2}{2m} \frac{\pi^2}{a^2} n^2. \tag{5.17}$$

Since the SEq (5.2) has solutions only for certain eigenfunctions φ_n or energy levels E_n, one often writes the eigenvalue problem from the outset as

$$E_n \varphi_n(x) = -\frac{\hbar^2}{2m} \varphi_n''(x). \tag{5.18}$$

Thus, we have a *discrete* energy spectrum, which occurs whenever the quantum object is bounded or localized.

As is well known, the quantization of energy means the following: If we look at the energy of a quantum object in the infinite potential well, we always detect one of these eigenvalues, but never any intermediate values. In other words, the possible measured (energy) values are the eigenvalues of the (energy) operator, i.e. the Hamiltonian. We have already encountered the same situation in the algebraic approach in Chap. 4, where we saw that the possible measured polarization values are determined by the eigenvalues of the corresponding polarization operators. In fact, this is a general aspect of quantum mechanics: Physical quantities are represented by operators, and the eigenvalues of those operators are the experimentally measurable quantities.

Two more comments on the eigenfunctions:

1. The amplitude in (5.16) is chosen to give the greatest simplicity of the result. In principle, also the form

---

[7]This special choice will be justified below.

$$\varphi_n\,(x) = \sqrt{\frac{2}{a}}e^{i\delta_n}\,\sin k_n x \tag{5.19}$$

is possible, where $\delta_n \in \mathbb{R}$ is a phase shift. In order to avoid unnecessary restrictions, we will use the eigenfunctions in their complex form (5.19) in the following exemplary considerations, wherever appropriate.

2. If we take into account the time dependence (see below), a state of definite energy E_n is given by $\varphi_n\,(x)\,e^{-i\omega_n t} \sim \sin k_n x \cdot e^{-i\omega_n t}$, i.e. it is a *standing wave*.

5.1.2   Solution of the Time-Dependent Schrödinger Equation

How would a total solution for the wavefunction Ψ look? In Chap. 3, we started with the separation *ansatz*:

$$\Psi(x,t) = \varphi(x)e^{-i\omega t} \quad \text{with} \quad E = \hbar\omega. \tag{5.20}$$

The eigenfunctions $\varphi_n\,(x)$ are solutions of the stationary SEq with the eigenvalues E_n or ω_n. Therefore, *each* of the functions $\varphi_n\,(x)\,e^{-i\omega_n t}$ is a particular solution of the time-dependent SEq. Due to the linearity of the SEq, we obtain the general solution by superposition of *all* the particular solutions. It follows that[8]:

$$\Psi\,(x,t) = \sum_n c_n \varphi_n\,(x)\,e^{-i\omega_n t} \tag{5.21}$$

with

$$c_n \in \mathbb{C}\,;\ \ \varphi_n\,(x) = \sqrt{\frac{2}{a}}e^{i\delta_n}\,\sin k_n x;\ \ \omega_n = \frac{E_n}{\hbar} = \frac{\hbar k_n^2}{2m}. \tag{5.22}$$

Thus, we have integrated the SEq in closed form. We note that this is one of the few examples where this is possible.[9]

The coefficients c_n in (5.22) are determined by the particular choice of the system. If all the c_n vanish except for one, the system is in a definite energy state; otherwise, it is in a superposition of several states. With the last equations, the problem 'infinite potential well' is completely determined—we know, in closed form, all the eigenvalues, the corresponding eigenfunctions and thus the general form of the time-dependent solution. From (5.21), we see explicitly that the solutions $\Psi\,(x,t)$, as discussed in Chap. 3, are elements of a vector space $\mathcal{V}$. It holds for example that with

---

[8]We recall that we are using a shorthand notation for the summation $\sum_n$. The range of values of n must be clear from the context. Here, it would be $n = 1, \ldots .\infty$ or $\sum_{n=1}^{\infty}$.

[9]The form (5.21) for the general solution applies just as well to other potentials besides the infinite potential well considered here, although of course the eigenfunctions are then not the same as those in (5.22).

$$\Psi(x,t) = \sum_n c_n \varphi_n(x) e^{-i\omega_n t} \quad \text{and} \quad \Phi(x,t) = \sum_n d_n \varphi_n(x) e^{-i\omega_n t}, \qquad (5.23)$$

every linear combination $\Theta = \alpha\Psi + \beta\Phi$ may be written as

$$\Theta(x,t) = \sum_n (\alpha c_n + \beta d_n) \varphi_n(x) e^{-i\omega_n t} = \sum_n b_n \varphi_n(x) e^{-i\omega_n t}, \qquad (5.24)$$

and thus is also a solution.

However, one can still learn a lot more from this example. This is due to special properties of the eigenfunctions, whereby—and this is the salient point—these relationships are valid in general and not only for the infinite potential well. Thanks to these properties, the inclusion of the initial-value problem (and thus the proof that the solution of the SEq is determinate) is relatively easy, as we shall see in a moment.

5.1.3 Properties of the Eigenfunctions and Their Consequences

An essential property of the eigenfunctions (5.19) is their so-called *orthonormality*. As one can show,[10] the functions are *normalized*:

$$\int_0^a \varphi_n^*(x) \varphi_n(x)\, dx = 1 \qquad (5.25)$$

and *orthogonal*:

$$\int_0^a \varphi_m^*(x) \varphi_n(x)\, dx = 0; \quad m \neq n. \qquad (5.26)$$

Written compactly, they are *orthonormal*[11]:

$$\int_0^a \varphi_m^*(x) \varphi_n(x)\, dx = \delta_{nm}. \qquad (5.27)$$

Here, we integrate the product $\varphi_m^* \varphi_n$ and not $\varphi_m \varphi_n$, so that the expression is independent of the phase which occurs in (5.19). Because the wavefunctions vanish outside of the interval $[0, a]$, the integration can extend from $-\infty$ to ∞. We thus obtain the general formulation:

---

[10] See the exercises for this chapter.
[11] This explains also the choice which we made in (5.15) or (5.19).

Fig. 5.2 Example sketch of
two functions $f(x)$ and $g(x)$
which are orthogonal in the
sense of (5.28)

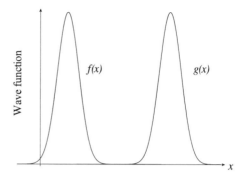

$$\int\limits_{-\infty}^{\infty} \varphi_m^* (x)\, \varphi_n (x)\, \mathrm{d}x = \delta_{nm}. \tag{5.28}$$

In fact, the eigenfunctions of *all* the Hamiltonians we consider possess this important property, provided the corresponding eigenvalues are discrete. Here, it is of
course assumed that the integrals exist, which means that the functions are *square-
integrable*.[12]

So far, we have used the term *orthonormal* in connection with 'usual' vectors, such
as column or row vectors or kets and bras, e.g. in Chap. 4 in the form $\langle \varphi_i \, | \varphi_j \rangle = \delta_{ij}$.
That now also *functions* such as (5.21) are deemed orthonormal may seem surprising
at first. It is due to the fact that, as mentioned above, these functions are *also* elements
of the vector space $\mathcal{V}$ of the solutions of the SEq,[13] and as such (i.e. as *vectors*), they
can be orthogonal to each other. Indeed, the form on the left side of (5.28) is a scalar
product, as is shown explicitly in Chap. 11. Hence, one has to distinguish between two
aspects: On one hand, $\varphi_n (x)$ is a function of x; on the other hand and simultaneously,
it is an element of the vector space $\mathcal{V}$ and in this sense a vector.[14] The orthogonality
of two functions to each other does not mean that the graphs of these two functions
intersect only at right angles or something similar; but rather that they, as members
of $\mathcal{V}$, behave as described in (5.28). This may appear as shown in Fig. 5.2.

By the way, an even function is always orthogonal to an odd one (with symmetric
limits of integration).

In addition to their orthonormality, the eigenfunctions (5.19) have the property
of *completeness*. Intuitively, this means that it is possible to formulate any solution
of the SEq for the infinite potential well as a superposition of these eigenfunctions,
as we have already noted in (5.21). Concerning their orthonormality, we use very

---

[12]Square-integrable (or quadratically integrable) over the interval $[a, b]$ are those functions $f(x)$
for which $\int_a^b |f(x)|^2 \, \mathrm{d}x < \infty$ holds. The short notation reads $f(x) \in L^2 [a, b]$. For $a = -\infty$ and
$b = \infty$, the notation $L^2 [\mathbb{R}]$ is common.

[13]Hence the often undifferentiated use of the terms eigenfunction and eigenvector.

[14]This use of the term vector has of course nothing to do with arrows or with the properties of
transformation behavior (polar and axial vectors).

similar formulations in both the algebraic and the analytic approaches, with

$$\langle \varphi_n | \varphi_m \rangle = \delta_{nm} \quad \text{or} \quad \int_{-\infty}^{\infty} \varphi_m^*(x)\, \varphi_n(x)\, \mathrm{d}x = \delta_{nm}. \tag{5.29}$$

The question of an analogous comparison for the completeness, which reads $\sum_n |\varphi_n\rangle \langle \varphi_n| = 1$ in the algebraic approach, will be taken up again only in Chap. 11. But here, we can already state that the eigenfunctions of the infinitely-deep potential well form a complete orthonormal system, a CONS.

5.1.4 Determination of the Coefficients c_n

Back to the example of the infinite potential well: For the general solution of the time-dependent SEq (i.e. the total wavefunction), we found the expression

$$\Psi(x, t) = \sum_n c_n \varphi_n(x)\, e^{-i\omega_n t}. \tag{5.30}$$

The eigenfunctions and eigenvalues are defined by the physical problem (i.e. the shape of the potential), while the actual behavior in time is determined by the choice of the coefficients c_n. If we know all the coefficients, we have uniquely determined the time dependence of $\Psi(x, t)$. On the other hand, the SEq is a differential equation of first order in time, which means that the specification of the initial condition $\Psi(x, 0)$ determines the temporal behavior. In other words, knowledge of the initial condition $\Psi(x, 0)$ gives the same information as knowledge of all the coefficients c_n.[15] So it must be possible to calculate: (i) $\Psi(x, 0)$ from knowledge of all the c_n's, and (ii) all the coefficients c_n from knowledge of $\Psi(x, 0)$.

This is trivial in the first case, since we have $\Psi(x, 0) = \sum_n c_n \varphi_n(x)$. For the other direction, we use the orthonormality of the eigenfunctions (5.27). An additional technical note: We always assume that the functions considered here are sufficiently well-behaved, that all series converge, and that we can interchange any limiting processes such as derivatives, integrals, and infinite sums. Of course, this must be shown explicitly for particular cases, but we will save ourselves some trouble and leave this job to others, and accept their results. Some remarks on this are given in Appendix D, Vol. 1.[16]

---

[15] At first sight, it may seem strange that one can compute *infinitely* many complex numbers c_n from *one* initial condition $\Psi(x, 0)$. But in fact, with $\Psi(x, 0)$ we have *uncountably* many values.

[16] "Physicists usually have a nonchalant attitude when the number of dimensions is extended to infinity. Optimism is the rule, and every infinite sequence is presumed to be convergent, unless proven guilty." A. Peres, *Quantum Theory*, p. 79.

We begin with

$$\Psi (x, 0) = \sum_n c_n \varphi_n(x). \tag{5.31}$$

Multiplying this equation from the left by $\varphi_m^*(x)$ and integrating yields:

$$\int_0^a \varphi_m^*(x) \Psi (x, 0) \, dx = \int_0^a \sum_n c_n \varphi_m^*(x) \varphi_n (x) \, dx. \tag{5.32}$$

Interchanging the integration and the summation and using the orthonormality (5.27) of the eigenfunctions leads to:

$$\sum_n c_n \int_0^a \varphi_m^*(x) \varphi_n (x) \, dx = \sum_n c_n \delta_{n,m} = c_m \tag{5.33}$$

or, compactly,

$$c_m = \int_0^a \varphi_m^*(x) \Psi (x, 0) \, dx. \tag{5.34}$$

Thus, the specification of the initial condition allows us to calculate uniquely all of the coefficients. It follows that

$$\Psi (x, t) = \sum_n \left(\int_0^a \varphi_n^*(x') \Psi \left(x', 0\right) \, dx' \right) \varphi_n (x) \, e^{-i\omega_n t} \tag{5.35}$$

gives an expression for the solution of the time-dependent SEq. We can read off from this equation directly that specifying the initial condition uniquely determines the time behavior of $\Psi (x, t)$ for all times.

5.2 Free Motion

As a second simple model system, we consider force-free unbounded motion. It is also described by the SEq[17]

$$i\hbar \dot{\Psi} (x, t) = -\frac{\hbar^2}{2m} \Psi''; \tag{5.36}$$

but here we assume that there are no limits on the motion. The quantum object is not localized and can move throughout all space.

---

[17]This equation is very similar to the heat equation $\dot{f} = \lambda \nabla^2 f$—apart from i in the SEq. As is well known, this 'small difference' is the mother of all worlds.

5.2.1 General Solution

As we know, special (particular) solutions of the problem are plane waves of the form

$$\Psi_{\text{part}}(x,t) = e^{i(kx-\omega t)}. \tag{5.37}$$

Since each $k \in \mathbb{R}$ is allowed, and thus also any energy $E = \frac{\hbar^2 k^2}{2m}$, we have a *continuous* energy spectrum. This case *always* occurs if the quantum object is not localized (i.e. is unbounded).

The general solution is the superposition of particular solutions, that is[18]

$$\Psi(x,t) = \int_{-\infty}^{\infty} c(k)e^{i(kx-\omega t)}\,dk. \tag{5.38}$$

At $t = 0$, we obtain

$$\Psi(x,0) = \int_{-\infty}^{\infty} c(k)e^{ikx}\,dk. \tag{5.39}$$

The specification of this initial condition determines the time evolution here, also, since the SEq is a differential equation of first order in time. Consequently, it must be possible to compute the coefficients $c(k)$ uniquely from $\Psi(x,0)$. This indeed works; by means of Fourier transformation,[19] we obtain immediately

$$c(k) = \frac{1}{2\pi} \int_{-\infty}^{\infty} \Psi(x,0)e^{-ikx}\,dx, \tag{5.40}$$

so that we can, in principle, determine the solution for any given initial distribution. Thus, we have again integrated the SEq in closed form. In a compact notation, the solution reads:

$$\Psi(x,t) = \frac{1}{2\pi} \int_{-\infty}^{\infty} \left(\int_{-\infty}^{\infty} \Psi(x',0)e^{-ikx'}\,dx' \right) e^{i(kx-\omega t)}\,dk \quad \text{with} \quad \omega = \frac{\hbar k^2}{2m}. \tag{5.41}$$

Again in this case, we see immediately that the wavefunction is determinate.

---

[18] Integral and not sum, because k is a continuous 'index'. The integration variable k may of course also assume negative values here.

[19] Some basics on Fourier transformation can be found in Appendix H, Vol. 1.

5.2.2 Example: Gaussian Distribution

A concrete standard example is based on the initial condition

$$\Psi(x, 0) = \frac{1}{(\pi b_0^2)^{\frac{1}{4}}} \exp\left(-\frac{x^2}{2b_0^2}\right) e^{iKx}, \qquad (5.42)$$

where we assume $K > 0$ without loss of generality.[20] Without the factor e^{iKx}, the center of the distribution would be stationary, i.e. it would remain at the same position. For the following discussion, we concentrate on the absolute square of the wavefunction. Initially, it is given by

$$\rho(x, 0) = |\Psi(x, 0)|^2 = \frac{1}{\sqrt{\pi} b_0} \exp\left(-\frac{x^2}{b_0^2}\right). \qquad (5.43)$$

This function has the form of a Gaussian bell curve with its maximum $\rho_{max} = (\sqrt{\pi} b_0)^{-1}$ at $x = 0$. The width of the curve is given by $2b_0$; it is measured between the points where the curve has the value $\rho = \rho_{max}/e$.[21]

In this example, one can determine $\Psi(x, t)$ exactly, but the calculation is tedious and will be omitted here.[22] One arrives eventually at

$$\rho(x, t) = |\Psi(x, t)|^2 = \frac{1}{\sqrt{\pi} b(t)} \exp\left(-\frac{\left(x - \frac{\hbar K}{m} t\right)^2}{b^2(t)}\right), \qquad (5.44)$$

where $b(t)$ is given by

$$b(t) = \sqrt{b_0^2 + \left(\frac{\hbar t}{b_0 m}\right)^2} \qquad (5.45)$$

with $b(0) = b_0$. Obviously, the function $b(t)$ increases monotonically with t and tends towards $\frac{\hbar t}{b_0 m}$ for $t \to \infty$.

Equation (5.44) again represents a Gaussian curve, with its maximum at $x = \frac{\hbar K}{m} t$ and width $2b(t)$. This means that the curve becomes wider and wider with increasing t, while its maximum moves with constant velocity $v = \frac{\hbar K}{m}$ to the right. Its height is given by $\rho_{max}(x, t) = (\sqrt{\pi} b(t))^{-1}$, i.e. it decreases continuously due to the monotonic form of $b(t)$. In short, the distribution $\rho(x, t)$ becomes steadily wider and flatter—it 'goes fuzzy' or becomes 'smeared out'; see Fig. 5.3.

This concludes our mathematical findings. However, the question remains as to what this 'smearing out' means physically. One thing is clear: It cannot mean that the

---

[20]The quite specific form of the coefficients is due to the normalization.

[21]Occasionally, this width is referred to as the halfwidth, although the function has dropped not to $1/2$, but to $1/e$ of its maximum value.

[22]A slightly more detailed analysis can be found in Appendix D, Vol. 2 (wave packets).

Fig. 5.3 Spreading of the
density distribution (5.44).
Arbitrary units; maximum at
$t = 0$, normalized to 1

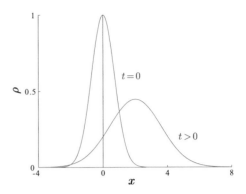

object (electron, etc.) is itself smeared out—an electron is, within the framework of
our considerations, always an (indivisible) point object. We will discuss this point in
more detail in Chap. 7. Here, we simply mention in anticipation that the question boils
down to the interpretation of $\rho(x, t)$ as a *probability density*. It allows us by means
of $\int_a^b \rho(x, t)$ to calculate the probability of finding the quantum object in the interval
$[a, b]$. Describing the spreading of the Gaussian curve means that the wavefunction
from which the probability is calculated spreads out (and not the quantum object
itself). In other words, the uncertainty with which we can determine the location of a
quantum object, $\Delta x \approx b(t)$, increases over time. With this interpretation of $\rho(x, t)$,
we have introduced the term 'probability' also into the analytical approach.

However, this concept makes sense only if the effects are noticeable (i) very
strongly for microscopic objects, and (ii) nearly not at all for macroscopic objects.
Everyday things around us do not have a spreading probability of being found at
a particular location, in contrast to objects in the microscopic world. In order to
arrive at a numerical estimate, we compute the time t_{2b_0} after which the width of a
bell-shaped curve, initially b_0, has doubled, that is $b\left(t_{2b_0}\right) = 2b_0$. It follows that:

$$\sqrt{b_0^2 + \left(\frac{\hbar t_{2b_0}}{b_0 m}\right)^2} = 2b_0 \text{ and thus } t_{2b_0} = \sqrt{3}\frac{m}{\hbar}b_0^2. \tag{5.46}$$

We calculate this doubling time t_{2b_0} for two examples ($\hbar \approx 10^{-34}$ kg m^2/s):

1.   A 'Grain of sand': $m = 1$ g, $b_0 = 1$ mm:

$$t_{2b_0,\text{grain}} = 1.7 \cdot \frac{10^{-3}}{10^{-34}} 10^{-6}\text{s} = 1.7 \cdot 10^{25}\text{s} \approx 5.4 \cdot 10^{17}\text{years}; \tag{5.47}$$

2.   An 'Electron', $m = 10^{-30}$ kg, $b_0 = 10^{-10}$ m:

$$t_{2b_0,\text{electron}} = 1.7 \cdot \frac{10^{-30}}{10^{-34}} 10^{-20}\text{s} \approx 1.7 \cdot 10^{-16}\text{s}. \tag{5.48}$$

Fig. 5.4 Characterization of the energy spectrum for an arbitrary potential, depending on the localizability of the quantum object

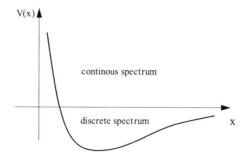

We see clearly the difference between a macroscopic and a microscopic object. We note that this calculation is only about orders of magnitude, not 'exact' values, and that the results apply only if the objects are completely isolated during the time t_{2b_0} (i.e. they do not interact with anything else in the universe). And of course we know that the 'grain of sand' with $m = 1$ g is a many-particle system with internal interactions.

5.3 General Potentials

A few words about the nature of the energy spectrum are in order. We have found that the energy spectrum of the infinite potential well is discrete, whereas it is continuous for unlimited motions. That does not mean that every system has either a discrete *or* a continuous energy spectrum. Consider, for example, the hydrogen atom, i.e., proton plus electron. When the electron is in a bound state, we have discrete energies. If we ionize the atom, thus separating the electron from the nucleus so that it can move freely and without limit, then it can move with *any* kinetic energy—we have a continuous energy spectrum in this range. The situation is shown schematically in Fig. 5.4. In short, there are many systems whose energy spectrum has both a discrete and a continuous part.[23]

We will take up the formal treatment of this question later, where we will also see that continuous systems are mathematically more difficult than discrete ones. To circumvent these problems, one can resort to a 'trick', which helps to ensure that the *entire* spectrum is discrete.

We outline the basic idea: For this we start from an arbitrary (sufficiently well-behaved) potential $V(x)$ that vanishes at infinity. Now let us imagine that we put the system under consideration *in addition* into a potential well with infinitely high potential walls on all sides, see Fig. 5.5. The walls should be so far away that we can assume that their existence has no measurable influence on the 'local' physics

---

[23] In fact, it may also be the case that discrete and continuous spectra overlap, or that discrete levels are embedded in the continuum, as we shall see using the example of the helium atom in Chap. 23, Vol. 2.

Fig. 5.5 At sufficiently high resolution, the apparent continuum of energy eigenvalues proves to consist of closely-spaced individual levels

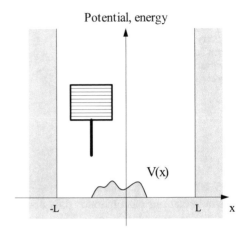

(i.e. in our lab). In particular, the potential V is negligible (zero) at the location of the walls. The stationary SEq $E\varphi(x) = -\frac{\hbar^2}{2m}\varphi''(x) + V(x)\varphi(x)$ is a *second-order* differential equation with respect to x and accordingly has *two* linearly-independent fundamental solutions, $\varphi_1(kx)$ and $\varphi_2(kx)$, with $k^2 = 2mE/\hbar^2$. For the following argument, it does not matter exactly what form these functions take, it is sufficient that they exist. Each solution of the stationary SEq for the energy E can be expressed as a linear combination $\varphi(x) = A\varphi_1(kx) + B\varphi_2(kx)$. If we now imagine infinitely high potential walls at $x = \pm L$, then $\varphi(x)$ must vanish there. It follows that

$$A\varphi_1(-kL) + B\varphi_2(-kL) = 0 \tag{5.49}$$
$$A\varphi_1(kL) + B\varphi_2(kL) = 0.$$

This is a homogeneous system of equations for the quantities A and B. This system is solvable[24] if

$$\varphi_1(-kL)\varphi_2(kL) - \varphi_2(-kL)\varphi_1(kL) = 0 \tag{5.50}$$

applies. This equation can be satisfied only for certain values of kL. For a given L, this is therefore a determining equation for k, with countably infinitely many solutions k_n. Hence the energy is discrete.

---

[24]Moreover, it follows from (5.49) for instance that

$$B = -\frac{\varphi_1(kL)}{\varphi_2(kL)}A$$

and so

$$\Phi(x) = A\left[\varphi_1(kx) - \frac{\varphi_1(kL)}{\varphi_2(kL)}\varphi_2(kx)\right],$$

leaving only *one* remaining free constant (and one *must* remain because of the linearity of the SEq).

The larger L is, the closer the energy levels lie together. We can visualize this by the fact that for sufficiently large n, the influence of the potential $V(x)$ is small (i.e. the main influence arises from the infinite potential well) and the energy levels are given approximately by

$$E_n \approx \frac{\hbar^2 k_n^2}{2m} = \frac{\hbar^2 \pi^2}{2m} \frac{n^2}{L^2}. \tag{5.51}$$

The difference between these energy levels is

$$E_n - E_{n-1} \approx \frac{\hbar^2 \pi^2}{2m} \frac{2n-1}{L^2}. \tag{5.52}$$

For sufficiently large L, one can reduce this difference to below any measurable value. In other words, we have in this case discrete energy eigenvalues, but they are so dense that they look to us like a (quasi-)continuum; cf. Fig. 5.5.

A numerical example: If the potential walls were a light year apart, then the differences between two neighboring energy levels for an electron are of the order of 10^{-50} eV (see exercises).

Finally, we note that another 'trick' for the discretization of the spectrum is the introduction of *periodic boundary conditions* of the form $\varphi(x+L) = \varphi(x)$. In this way, one can model solids, or also motions on a cylinder or a torus. Two examples can be found in the exercises.

5.4 Exercises

1. Given the free stationary SEq

$$E\Phi(x) = -\frac{\hbar^2}{2m} \Phi''(x), \tag{5.53}$$

   formulate the corresponding equation for the Fourier transform of Φ.
2. Given the stationary SEq

$$E\Phi(x) = -\frac{\hbar^2}{2m} \Phi''(x) + V(x)\Phi(x), \tag{5.54}$$

   formulate the corresponding equation for the Fourier transform of Φ.
3. The Hamiltonian has discrete nondegenerate eigenvalues $E_n, n = 1, 2, \ldots$. What is the general solution of the time-dependent SEq?
4. Infinite potential well: Show that the eigenfunctions in the form $\varphi_n(x) = \sqrt{\frac{2}{a}} e^{i\delta_n} \sin(k_n x)$ constitute an orthonormal system of functions ($\int_0^a \varphi_m^*(x)\varphi_n(x) = \delta_{mn}$). Hint: The integrals can be calculated for example by means of

$\sin x \sin y = \frac{\cos(x-y)-\cos(x+y)}{2}$ or the exponential representation of the sine functions.

5. Infinite potential well: Formulate the general solution of the time-dependent SEq and verify that specification of the initial condition determines the wave function. Concretize the considerations to the special cases ($C \in \mathbb{C}$ is an arbitrary complex constant):

   (a) $\Psi(x, t = 0) = C\delta(x - \frac{a}{2})$;
   (b) $\Psi(x, t = 0) = C$;
   (c) $\Psi(x, t = 0) = Ce^{iKx}$.

6. Given the three-dimensional SEq $E\psi(\mathbf{r}) = -\frac{\hbar^2}{2m}\nabla^2\psi(\mathbf{r})$, which energy eigenvalues are allowed if one imposes the following periodic boundary conditions: $\psi(x, y, z) = \psi(x + L_x, y, z) = \psi(x, y + L_y, z) = \psi(x, y, z + L_z)$?

7. An electron is located between the two walls of an infinite potential well, which are one light year apart. Calculate roughly the magnitude of the difference between two adjacent energy levels.

8. Find examples for functions which

   (a) are integrable, but not square-integrable;
   (b) are square-integrable, but not integrable.

9. Given the stationary SEq

$$E\varphi(x) = -\frac{\hbar^2}{2m}\varphi''(x) + V(x)\varphi(x), \tag{5.55}$$

rewrite this equation for a dimensionless independent variable.

10. A short outlook into string theory (compactified or rolled-up dimensions): String theory assumes that the elementary building blocks of nature are not point objects, but rather one-dimensional objects (strings) with a certain energy—comparable to an object in a one-dimensional potential well. Strings have a spatial extension of order of the Planck length and live in higher-dimensional spaces (e.g. dim = 10 or dim = 26), where only four dimensions are not rolled up (compactified)—quite similar to our following simple example.[25]
For the formal treatment, we take the two-dimensional SEq

---

[25]When a writer like Terry Pratchett couples the idea of rolled-up dimensions with other physical paradigms, it reads like this: "..and people stopped patiently building their little houses of rational sticks in the chaos of the universe and started getting interested in the chaos itself—partly because it was a lot easier to be an expert on chaos, but mostly because it made really good patterns that you could put on a T-shirt.

And instead of getting on with proper science, scientists suddenly went around saying how impossible it was to know anything, and that there wasn't really anything you could call reality to know anything about, and how all this was tremendously exciting, and incidentally did you know there were possibly all these little universes all over the place but no-one can see them because they are all curved in on themselves? Incidentally, don't you think this is a rather good T-shirt?" Terry Pratchett, in *Witches Abroad*, A Discworld Novel.

Fig. 5.6 The 'cylinder
world' of our toy string

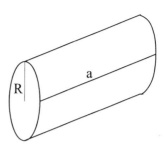

$$-\frac{\hbar^2}{2m}\left(\frac{\partial^2\psi}{\partial x^2}+\frac{\partial^2\psi}{\partial y^2}\right)=E\psi \tag{5.56}$$

as starting point. In the x direction, we have an infinite potential well

$$V=\begin{cases}0 & \text{for } 0<x<a\\ \infty & \text{otherwise}\end{cases} \tag{5.57}$$

and for the y coordinate we postulate

$$\psi(x,y)=\psi(x,y+2\pi R). \tag{5.58}$$

So we have a combination of two different boundary conditions: In the x direction, $\psi(0,y)=\psi(a,y)=0$ applies, while in the y direction the periodic boundary condition $\psi(x,y)=\psi(x,y+2\pi R)$ is valid. In other words, the quantum object 'lives' on the surface of a cylinder of length a and of radius R, see Fig. 5.6. The problem is now to calculate the possible energy levels. Discuss in particular the situation when $R\ll a$.

11. Given the free one-dimensional SEq (5.36) and the function $\Phi(x)$, show that

$$\Psi(x,t)=A\frac{1}{\sqrt{t}}\int_{-\infty}^{\infty}e^{\frac{im}{2\hbar}\frac{(x-y)^2}{t}}\Phi(y)\,dy \tag{5.59}$$

is a solution (A is a normalization constant).

Chapter 6
Interaction-Free Measurement

We discuss an experiment which provides an example of the unusual effects that may result
from the superposition of states, and of the peculiarities that may be associated with the
quantum-mechanical measurement process. In addition, we make the acquaintance of unitary
operators.

Self-interference, i.e. the interference of a quantum object with itself, is a fascinating phenomenon of quantum mechanics, which we discuss below in terms of the
interaction-free quantum measurement. The experiment is based on the principle of
the Mach–Zehnder interferometer (MZI). It shows the existence of quantum superpositions as clearly as the famous double-slit experiment, but it is by comparison
formally and experimentally much 'handier', so that it is increasingly finding its way
into textbooks. At the same time, it also allows for the treatment of further-reaching
questions. That is why we meet the MZI not only in many modern basic experiments,
but also for example in the field of quantum information, where we can realize basic
functions of the quantum computer by means of the MZI and its components (see
the closing remarks to this chapter).

6.1 Experimental Results

6.1.1 Classical Light Rays and Particles in the
Mach–Zehnder Interferometer

6.1.1.1 Light Rays

The experimental setup consists of a Mach–Zehnder interferometer and two pho-
todetectors, which respond to incident light; see Fig. 6.1. Coherent light enters the
apparatus at the lower left and is split by a *beam splitter* (or half-silvered mirror)

© Springer Nature Switzerland AG 2018
J. Pade, *Quantum Mechanics for Pedestrians 1*, Undergraduate Lecture
Notes in Physics, https://doi.org/10.1007/978-3-030-00464-4_6

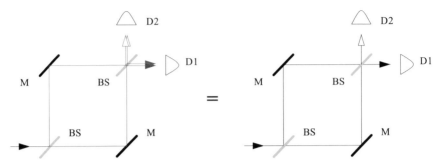

Fig. 6.1 Schematic of a MZI. BS = beam splitter, M = mirror, D = detector. Left: At the first beam splitter, the light is split into two sub-beams (blue and red do not signify the colors of the light beams, but serve only for better visualization), and these two sub-beams are split again at the second beam splitter, resulting in four sub-beams. The figure at the right is a compact description of these facts, which we will use in the following

into two beams.[1] These beams impinge, after reflection by a mirror, on a second beam splitter, so as to produce a total of four sub-beams, two each of which meet at one of the two detectors. The experimental finding is now that the upper detector D2 *never* responds and the lower detector D1 *always* responds. In other words, the relative intensity I on D1 is given by $I_1 = 1$, and on D2 by $I_2 = 0$. Here, we assume on the whole ideal conditions: the optical paths 'above' and 'below' have exactly the same length, there is no absorption by the mirrors, the efficiency of the detectors is 100%, and so on.

This different behavior of the two detectors may perhaps be surprising, since the experimental setup appears to be completely symmetrical at first glance. But in fact, its symmetry is broken, as long as the light enters the first beam splitter only in the horizontal and not also in the vertical direction (and with the same intensity).

The following consideration shows why the two detectors react differently: That part of the lower beam which after the second beam splitter enters D2 or D1 undergoes a reflection (1 × mirror) or two reflections (1 × mirror, 1 × beam splitter), while the part of the upper beam which after the second beam splitter enters D2 or D1 undergoes three reflections (1 × mirror, 2 × beam splitter) or two reflections (1 × mirror, 1 × beam splitter). In other words, the detector D1 sees two light beams with the same history (i.e. the same phase), which consequently interfere constructively. In contrast, the detector D2 sees two beams with different histories. We will show immediately that this indeed gives destructive interference.

In a variant of the experimental setup, we use a blocker that absorbs the upper beam or scatters it out of the MZI; see Fig. 6.2. Obviously, the upper and lower sub-beams now cannot interfere and the intensities at the detectors are given by $I_1 = I_2 = 1/4$.

---

[1] The two beams can in principle be separated quite far apart. In this way, the non-classical effects of certain quantum-mechanical setups can be demonstrated more impressively than in the double-slit experiment.

Fig. 6.2 MZI with a blocker
in the upper beam

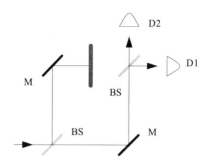

6.1.1.2 Particles in the MZI

What happens if we send *particles* ($m \neq 0$) instead of light waves through the
apparatus? Of course we have to replace the beam splitters by devices which let the
particles pass or reflect them with probabilities of 1/2, but otherwise the experimental
setup remains the same. If we now interpret the number of particles per detector as
intensities, it follows directly that for the case without a blocker, $I_1 = I_2 = 1/2$
holds, and for the case with a blocker, $I_1 = I_2 = 1/4$.

6.1.1.3 Comparison: Light–Particles

It follows that with a blocker, the intensities are given by $I_1 = I_2 = 1/4$, regardless
of whether we use waves or particles. For the case without a blocker, however, there
is a distinctive difference, since for waves we have $I_1 = 1$ and $I_2 = 0$, while for
particles, $I_1 = I_2 = 1/2$. So we can conclude that if we perform an experiment (in
the sense of a black-box setup) without the blocker and measure $I_2 = 0$, then we
know that a wave and not a particle has passed through the apparatus. These results
are summarized in the Table 6.1.

6.1.2 Photons in the Mach–Zehnder Interferometer

6.1.2.1 Single-Photon Experiments (MZI Without Blocker)

We let light enter the MZI and reduce its intensity (similar to the polarization experiments of Chap. 2). Since our previous considerations do not rely on the intensity of
the incident light, they should also apply to the limit of vanishing light intensity. This

Table 6.1 Intensities at the
two detectors

	Without blocker	With blocker
Wave	$I_1 = 1; \quad I_2 = 0$	$I_1 = \frac{1}{4}; \quad I_2 = \frac{1}{4}$
Particle	$I_1 = \frac{1}{2}; \quad I_2 = \frac{1}{2}$	$I_1 = \frac{1}{4}; \quad I_2 = \frac{1}{4}$

means that eventually there is only one photon in the MZI at a given time. In fact, the experimental findings are: even if we operate with single photons, only detector D1 responds, while D2 remains silent, or $I_1 = 1$ and $I_2 = 0$.

So we must conclude that a single photon is a wave and not a particle. On the other hand, a photon is a point object as far as we know. Our everyday understanding perceives the situation as contradictory: An object can be both point-like and wave-like. But our cognitive abilities are, as we have mentioned before, formed and trained by evolution in our macro-physical environment and not under quantum-mechanical conditions.

In addition, we have to conclude that due to the interference effect, the photon 'somehow' interacts with itself. It is not intuitively obvious how this takes place. Certainly, it is not the case that the photon splits into two smaller fragments. We have here the same problem as in the double slit experiment—if there are two possibilities which can be realized by a quantum-mechanical system, then certain interference phenomena will appear which have no classical analogues (self-interference).

As we said previously, it is perhaps best to imagine a quantum-mechanical *possibility landscape*, in which the quantum object (photon, electron, ...) moves. A superposition of possibilities yields a new landscape with new features, in which the object moves differently than in the landscape of only one possibility.

6.1.2.2 Interaction-Free Measurement (MZI With Blocker)

With a beam blocker, we have $I_1 = I_2 = 1/4$, and that means that in 25% of all cases, detector 2 responds. This in turn implies that we know in these cases that there is a blocker in the apparatus without the photon having interacted directly with the blocker (otherwise it would have disappeared from the apparatus and could not be detected in either detector).[2] This situation is called an *interaction-free quantum measurement*. Below, we make some critical remarks about this terminology.

The whole issue can be formulated more sensationally[3] by choosing a bomb[4] as the blocker. The bomb is so sensitive that a single photon is enough to cause it to detonate[5]—so to speak, just seeing the bomb means that it explodes.[6] We can use

---

[2]Thus, there are apparently physical effects influenced by *potential but unrealized* events, that is, events that *could* have happened, but did not actually occur. Such events are called *counterfactual* (not corresponding to the facts).

[3]A.C. Elitzur and L. Vaidman, "Quantum Mechanical Interaction-Free Measurements", Foundations of Physics 23, 987 (1993).

[4]In order to avoid the militaristic note, some textbooks use 'cracker test' instead of 'bomb test', but this sounds a bit whimsical.

[5]"A physical experiment which makes a bang is always worth more than a quiet one. Therefore a man cannot strongly enough ask of Heaven: If it wants to let him discover something, may it be something that makes a bang. It will resound into eternity." Georg Christoph Lichtenberg, *Scrap Books*, Vol. F (1147).

[6]This remark seems a bit exaggerated, but in fact the rods of the human eye can apparently react to even a single photon. The cones, responsible for color vision, need about 100 times stronger

this fact in the following setup: Assume that we have a black-box MZI, and we do not know whether it contains a bomb or not. The task now is to clarify this issue. It cannot be solved by means of classical physics. Quantum mechanics helps us— at least, we know in a quarter of the cases that a bomb is hidden in the apparatus without its blowing up in our faces. In fact, one can increase the 'efficiency' in a somewhat modified apparatus to virtually 100% by exploiting the so-called quantum Zeno effect. More about this in Appendix L, Vol. 1.

We have here again - as a purely quantum-mechanical effect - the *superposition of possibilities* (self-interference) that makes possible this surprising result. Of course, it is again not the case that the photon 'splits' up and, quasi by way of trial and error, passes at the same time through both arms of the MZI. The superposition of the possibilities provides precisely the different landscape mentioned above, in which the photon propagates in a different way. We can most easily describe this propagation by means of probabilities—if we let a photon start through the apparatus, it will end up with a probability of 1/4 in detector 2, and then we know that a bomb is in the beam path. But if we (in whatever way) know which arm the photon has passed through (*which-way information*), the landscape of possibilities or probabilities changes dramatically: in 50% of the cases the bomb explodes, in the other 50% nothing exciting happens. Formulated as a 'standard rule': If the path is known/unknown, then the probabilities/amplitudes are added:

$$\text{path is } \genfrac{}{}{0pt}{}{\text{known}}{\text{unknown}} \rightarrow \text{add } \genfrac{}{}{0pt}{}{\text{probabilities}}{\text{amplitudes}} . \tag{6.1}$$

Further comments on which-way experiments (or delayed-choice experiments) are found in Appendix M, Vol. 1.

In this context, the term *wave-particle duality* occasionally crops up. What is meant is this: Depending on the experimental situation, a quantum system shows either particle-like or wave-like features. We take as an example electrons in the double slit experiment. If we allow for interference, then the electrons show their 'wave nature'; if we want to see them as particles, e.g. by following their path through the slits, they show their 'particle nature'. Dualism in this context means that these properties are complementary—either particle or wave, but we can never measure both at the same time. We can state briefly and quite generally that asking for a particular property of an object leads to an answer that puts that property in the foreground and suppresses the other (complementary) property.

On closer inspection, the term wave-particle duality seems, however, to be redundant, or to favor misunderstandings, since it supports the widespread but erroneous notion that, before a measurement, a quantum object is actually a particle or a wave. That is a misinterpretation which can cloud the mind in the process of learning quantum mechanics. Indeed, before a measurement, quantum objects in general do *not* have well-defined properties. It is therefore understandable that one is often advised

---

excitation. See e.g. Davide Castelvecchi, People can sense single photons, *Nature*, https://doi.org/ 10.1038/nature.2016.20282 (Jul 2016).

to omit completely the terms 'wave-particle duality' or 'complementarity'; indeed, doing so does not cause a noticeable loss in understanding.[7]

A quantum object is simply something for which we have no detailed everyday terms, and depending on how we look at it, it seems to be more like a particle or more like a wave (but in fact it is neither)—it is simply a quantum object.[8] We could call it informally a 'quob,' but would it then be more familiar or intuitive?

6.2   Formal Description, Unitary Operators

To arrive at a simple, clear-cut description of states, we choose as the only distinctive criterion their direction of motion—either horizontal or vertical. Thus, we neglect polarization, beam profile, explicit time behavior and so on. We describe the conditions with and without a blocker under the assumption that we have two identical beam splitters.

6.2.1   First Approach

We divide the setup into four regions, as shown in Fig. 6.3. We denote the state in the region i by $|z_i\rangle$. With regard to the simplest possible description as just mentioned, we represent $|z_i\rangle$ as a superposition of horizontal $|H\rangle$ and vertical $|V\rangle$ propagation directions,[9] where these states constitute a CONS in a two-dimensional vector space. One can see that the propagation is horizontal in region 1 and both vertical and horizontal in region 2. Accordingly, we can write $|z_1\rangle = |H\rangle$ and $|z_2\rangle = c_1 |H\rangle + c_2 |V\rangle$. To determine the numbers c_1 and c_2, we take into account that (i) the relative phase shift is $90° \triangleq \frac{\pi}{2}$ (see Appendix K, Vol. 1), which corresponds to $e^{i\pi/2} = i$, and (ii) that the intensity in a *half-silvered mirror*[10] is equal for 'horizontal' and

---

[7]The Feynman Lectures on Physics, 5th Edition, 1970, Vol II, p. 37-1: "Newton thought that light was made up of particles, but then it was discovered that it behaves like a wave. Later, however (in the beginning of the twentieth century), it was found that light did indeed sometimes behaves like a particle. Historically, the electron, for example, was thought to behave like a particle, and then it was found that in many respects it behaved like a wave. So it really behaves like neither. Now we have given up. We say: 'It is like *neither*.'" Richard P. Feynman, S. Tomonaga and J. Schwinger were awarded the Nobel Prize in Physics 1965 for their fundamental work in quantum electrodynamics.

[8]We note at this point, more generally, that the practice of declaring all things perceived to simply 'exist' may be inadequate. Instead, one should first look at perception itself and examine its predictability. Therefore, in quantum mechanics we need advanced methods, because we cannot come to grips with the 'perceptions' (observations, measurements) by simply using intuitive, classical instruments. To obtain the information relevant to quantum mechanics, we have to think and act in a largely formal manner.

[9]Not to be confused with the polarization states $|h\rangle$ and $|v\rangle$.

[10]For asymmetrical beam splitters (reflectance $\neq$ transmittance), see the exercises.

Fig. 6.3 Division of the
MZI into four regions

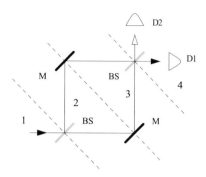

'vertical', i.e. $|c_1|^2 = |c_2|^2$. Thus, it follows that $|z_2\rangle = c[|H\rangle + i\,|V\rangle]$. We postpone
the determination of the constant c and summarize:

$$|H\rangle \underset{\text{beam splitter}}{\rightarrow} c[|H\rangle + i\,|V\rangle] \tag{6.2}$$

and analogously

$$|V\rangle \underset{\text{beam splitter}}{\rightarrow} c[|V\rangle + i\,|H\rangle]. \tag{6.3}$$

At a mirror, we have a phase shift of $180° \,\hat{=}\, \pi$ or $e^{i\pi} = -1$ and therefore

$$|H\rangle \underset{\text{mirror}}{\rightarrow} -|V\rangle; \quad |V\rangle \underset{\text{mirror}}{\rightarrow} -|H\rangle. \tag{6.4}$$

All in all, we have

$$
\begin{aligned}
|z_1\rangle &= |H\rangle \\
|z_2\rangle &= c[|H\rangle + i\,|V\rangle] \\
|z_3\rangle &= -c[|V\rangle + i\,|H\rangle] \\
|z_4\rangle &= -c^2[|V\rangle + i\,|H\rangle] - ic^2[|H\rangle + i\,|V\rangle] = -2ic^2\,|H\rangle.
\end{aligned} \tag{6.5}
$$

It follows immediately that only detector 1 responds, while detector 2 remains dark,
as is indeed observed experimentally.

We can define the constant c as follows: We assume that the setup operates without
losses—what goes in, comes out. This manifests itself in the fact that the norms are
equal, or more precisely must be equal, $\langle z_i|\, z_i\rangle = \langle z_j|\, z_j\rangle$. The simplest choice is
$-2ic^2 = 1$ or $c = \pm e^{i\pi/4}/\sqrt{2}$. We choose the upper sign and find $c = \frac{1+i}{2}$.

For the case with a blocker, we have analogously

$$
\begin{aligned}
|z_1\rangle &= |H\rangle \\
|z_2\rangle &= c[|H\rangle + i\,|V\rangle]
\end{aligned}
$$

$$|z_3\rangle = -c\,|V\rangle \tag{6.6}$$

$$|z_4\rangle = -c^2\,[|V\rangle + i\,|H\rangle] = \frac{1}{2i}\,[|V\rangle + i\,|H\rangle]\,.$$

We see that also in this case, the intensities measured by the two detectors are displayed correctly. We note that the transition $|z_2\rangle \rightarrow |z_3\rangle$ does not conserve the norm: $\langle z_2|\,z_2\rangle = 2\,|c|^2 \neq \langle z_3|\,z_3\rangle = |c|^2$. It is the absorbing effect of the blocker which leads to this inequality.

6.2.2 Second Approach (Operators)

We have just described the experiment with 'states and arrows'. A more compact approach is permitted by using operators. We can describe the effect of a beam splitter by an operator T, and the effect of a mirror without or with a blocker by S and S'. Without the blocker, this leads to:

$$
\begin{aligned}
|z_1\rangle &= \text{initial state} \\
|z_2\rangle &= T\,|z_1\rangle \\
|z_3\rangle &= S\,|z_2\rangle = ST\,|z_1\rangle \\
|z_4\rangle &= T\,|z_3\rangle = TS\,|z_2\rangle = TST\,|z_1\rangle = \text{final state}
\end{aligned}
\tag{6.7}
$$

and with the blocker, to:

$$|z_1\rangle = \text{initial state};\quad |z_4\rangle = TS'T\,|z_1\rangle = \text{final state}. \tag{6.8}$$

The operators are applied in sequence from right to left: $TST\,|z_1\rangle = T\,(S\,(T\,|z_1\rangle))$.

To obtain an explicit formulation for T, we consider the effect of this operator on the basis vectors. According to (6.2) and (6.3), we have

$$T\,|H\rangle = \frac{1+i}{2}\,[|H\rangle + i\,|V\rangle];\ \ T\,|V\rangle = \frac{1+i}{2}\,[i\,|H\rangle + |V\rangle]\,. \tag{6.9}$$

Using the completeness relation $|H\rangle\,\langle H| + |V\rangle\,\langle V| = 1$ leads to

$$T\,|H\rangle\,\langle H| + T\,|V\rangle\,\langle V| = T = \frac{1+i}{2}\,[|H\rangle + i\,|V\rangle]\,\langle H| + \frac{1+i}{2}\,[i\,|H\rangle + |V\rangle]\,\langle V|, \tag{6.10}$$

or compactly,

$$T = \frac{1+i}{2}\,[1 + i\,|H\rangle\,\langle V| + i\,|V\rangle\,\langle H|]\,. \tag{6.11}$$

Analogously, the 'mirror-operator' without a blocker is given by:

$$S = -|H\rangle \langle V| - |V\rangle \langle H| \tag{6.12}$$

and with the blocker by

$$S' = -|V\rangle \langle H|. \tag{6.13}$$

We learn from this that operators can generally be represented as linear combinations of dyadic products of the basis vectors.

It is easily verified that with (6.11)–(6.13), we have

$$TST = 1 \tag{6.14}$$

and

$$TS'T = \frac{1}{2}[1 + i|H\rangle \langle V| - i|V\rangle \langle H|], \tag{6.15}$$

so that we obtain again from the initial state $|z_1\rangle = |H\rangle$ the final states $|z_4\rangle = |H\rangle$ and $|z_4\rangle = \frac{1}{2}[|H\rangle - i|V\rangle]$ for the case with and without the blocker, respectively. For the explicit representation of the operators and their products as matrices, see the exercises.

The adjoint of the operator T is

$$T^\dagger = \frac{1-i}{2}[1 - i|H\rangle \langle V| - i|V\rangle \langle H|] \tag{6.16}$$

and it follows that

$$T^\dagger T = TT^\dagger = 1. \tag{6.17}$$

Analogously, the same holds true for S, but not for S', since here an (irreversible) absorption is included:

$$SS^\dagger = S^\dagger S = 1; \quad S'S'^\dagger = |V\rangle \langle V|; \quad S'^\dagger S' = |H\rangle \langle H|. \tag{6.18}$$

In fact, the operators T and S share an important property—they are *unitary*. As a generalization of (6.17), an operator (or matrix) U is unitary if

$$U^\dagger U = UU^\dagger = 1 \quad \text{or} \quad U^\dagger = U^{-1}. \tag{6.19}$$

The name 'unitary' stems from the fact that certain expressions are left unchanged under the transformation performed by the operator—in a way, it acts similarly to multiplication by 1. For example, the scalar product and thus also the norm are invariant. To show this, we start with two states $|\varphi\rangle$ and $|\psi\rangle$, and the unitary transformed states $|\varphi'\rangle = U|\varphi\rangle$ and $|\psi'\rangle = U|\psi\rangle$. Remember that a product of operators is reversed[11] in the adjoint, $(AB)^\dagger = B^\dagger A^\dagger$. This means that

---

[11] This is well known from linear algebra, e.g. when transposing or inverting matrices.

$$\left(\left|\psi'\right\rangle\right)^{\dagger} = (U\left|\psi\right\rangle)^{\dagger} \rightarrow \left\langle\psi'\right| = \left\langle\psi\right|U^{\dagger}. \tag{6.20}$$

It follows that

$$\left\langle\psi'\mid\varphi'\right\rangle = \left\langle\psi\right|U^{\dagger}U\left|\varphi\right\rangle = \left\langle\psi\mid\varphi\right\rangle, \tag{6.21}$$

i.e. the scalar product is conserved. Unitary transformations can always be understood in the end as a coordinate or basis transformation, even if the corresponding space is more elaborate than our two-dimensional space. These transformations conserve in particular scalar products, hence also lengths and angles, and they are reversible (because $U^{-1} = U^{\dagger}$ exists). Irreversible processes (e.g. measurements) can therefore not be represented by unitary transformations.

6.3 Concluding Remarks

As mentioned earlier in this chapter, the MZI is an essential tool for many modern fundamental experiments, both theoretically and experimentally. Due to lack of space, we can only outline some of them here, but a more detailed discussion is found in the Appendices (L and M, Vol.1; J, P and Q, Vol. 2). At the end of this chapter, we will take a closer look at the term 'interaction-free'.

6.3.1 Extensions

Out of a great number of applications of the MZI, we have selected those which are understandable with our present knowledge and which do not require additional concepts such as the Aharonov–Bohm effect.

6.3.2 Quantum Zeno Effect

There is an extension of the 'bomb test' which uses the *quantum Zeno effect*. This effect essentially implies that one can prevent the change of a system under appropriate circumstances by frequently-repeated measurements ('a watched pot never boils'). The experiment uses a modified MZI setup and is based on the observation of the polarization state of photons. In principle, one can achieve an efficiency of up to 100% (see Appendix L, Vol. 1).

Fig. 6.4 Delayed-choice experiment. The second beam splitter can be removed or inserted

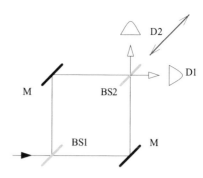

6.3.3 Delayed-Choice Experiments

Here, we use the familiar MZI setup, but the second beam splitter BS2 can be removed or inserted *after* the photon has passed the first beam splitter (hence the name 'delayed decision or choice'); see Fig. 6.4. The operation can be executed so quickly that a 'notification' of the photon would have to be superluminal.

Thus, the photon has to 'decide' whether it passes through the MZI as a coherent superposition (BS2 is inserted and only D1 responds) or whether it passes through only one of the two arms (BS2 is removed, the respective detector responds). The salient point is that the photon must take the decision after it has passed the first beam splitter (and possibly the mirror) but *before* we decide whether BS2 will be left in the path or not. That would mean (at least in a classical argument) that the photon must know before entering BS1 whether BS2 will be left or removed. In other words, the photon had to know about our future decision. Does that mean that delayed-choice experiments prove that certain events may have a retroactive action with respect to time?[12]

With a similar setup, one can produce a *quantum eraser*, with which one can subsequently delete ('erase') which-way information in certain experiments and thus restore their interference effects (see Appendix M, Vol. 1).

6.3.4 The Hadamard Transformation

The *Hadamard transformation* plays an important role in quantum information. It can be carried out by means of the MZI. Another method that can be experimentally realized uses the combination of a beam splitter and a phase shifter. Written as a

---

[12]Experiments are not confined to small distances. See e.g. F. Vedovato et al., Extending Wheeler's delayed-choice experiment to space, *Science Advances* Vol. 3, no. 10, https://doi.org/10.1126/sciadv. 1701180 (Oct 2017), where a delayed-choice experiment is reported with a propagation distance of up to 3500 km.

2×2 matrix, the Hadamard transformation H is (see Appendix P, Vol. 2 for the case $n \times n$):

$$H = \frac{1}{\sqrt{2}} \begin{pmatrix} 1 & 1 \\ 1 & -1 \end{pmatrix}. \tag{6.22}$$

6.3.5 From the MZI to the Quantum Computer

The MZI, with additional phase shifts, can be described as a network consisting of three simple *quantum logic gates*, namely as a combination of two Hadamard gates and a phase shifter. On this basis, other building blocks of quantum information such as the CNOT gate can be constructed (see Chap. 26, Vol. 2, and Appendix Q, Vol. 2).

6.3.6 Hardy's Experiment

This experiment combines an interaction-free measurement with quantum entanglement. This concept (which we will learn about in Chap. 20, Vol. 2) is another key aspect of quantum mechanics which has no classical counterpart. The experiment consists essentially of two superimposed MZI's (see Appendix J, Vol. 2).

6.3.7 How Interaction-Free is the 'Interaction-Free' Quantum Measurement?

Finally, a few words about the adjective 'interaction-free'. Indeed, we should always put it in quotation marks. This is due to the fact that, strictly speaking, this experiment can never be completely 'interaction-free'; there is an operator that describes the behavior of the photon in the interferometer, and this operator takes on a different form depending on whether a bomb is placed in the light path or not. Taking this into account, the term 'measurement with minimal interaction' is more correct and insofar preferable.

   This is because there is a fundamental limit to the attainable sensitivity of the detonator of the bomb, and the measurement can be called interaction-free at most within the limits of this sensitivity. The reason for this limitation is the *uncertainty principle* $\Delta x \Delta p \geq \hbar/2$. It is the basis for the following argument: If the bomb (the detonator) is located with an uncertainty Δx, then a given momentum uncertainty Δp results (for $\Delta x \to 0$, we would have $\Delta p = \infty$). To prevent the bomb from going off 'by itself', the detonator must not respond to momentum transfers smaller than Δp. In other words, the uncertainty principle necessarily requires that the bomb have an 'ignition threshold'. Under such circumstances one cannot speak of 'interaction-free'; a more appropriate term is *measurement with minimal interaction*. Along with a momentum transfer, there is also a possible energy transfer. The fact that this transfer can be very small in macroscopic objects ($\sim 1/M$) and vanishes in the limit $M \to \infty$ does not fundamentally alter the situation.

Conclusion: There is no 'interaction-free' quantum measurement, i.e. a measurement without interaction, but at most a measurement with minimal interaction. It is perhaps surprising that the term 'interaction-free' has established itself in the physics community (almost) without difficulty. On the other hand, one must admit that this term is very striking and much more effective in catching public attention than the more correct expressions (just as the term 'ozone hole' is in use rather than the more correct 'stratospheric region of low ozone concentration'). Thus, the interaction-free quantum measurement is another example of the fact that physics operates not only as pure science, but also through its perceptions by the larger society.

6.4 Exercises

1. Show that for all $|z_i\rangle$ in (6.5), $\||z_i\rangle\|^2 = 1$ holds.
2. Given a MZI with symmetrical beam splitters, calculate the final state with and without a blocker if the initial state is given by $\alpha|H\rangle + \beta|V\rangle$.
3. Given an operator A with

$$A|H\rangle = a|H\rangle; \; A|V\rangle = b|V\rangle, \tag{6.23}$$

   determine the explicit form of A.
4. Which eigenvalues can a unitary operator have?
5. Circularly- and linearly-polarized states are connected by $|r\rangle = \frac{1}{\sqrt{2}}|h\rangle + \frac{i}{\sqrt{2}}|v\rangle$ and $|l\rangle = \frac{1}{\sqrt{2}}|h\rangle - \frac{i}{\sqrt{2}}|v\rangle$. Show that this basis transformation is unitary (or that the transformation matrix is unitary).
6. Give the matrix representations of the operators T, S and S' from (6.11)–(6.13) and their combinations TST and $TS'T$.
7. Given the operator

$$U = a|H\rangle\langle H| + b|H\rangle\langle V| + c|V\rangle\langle H| + d|V\rangle\langle V| \cong \begin{pmatrix} a & b \\ c & d \end{pmatrix}; \tag{6.24}$$

   for which values of the coefficients is U a unitary operator? In other words: How is the general two-dimensional unitary transformation formulated?
8. Given a MZI without a blocker and with asymmetrical beam splitters (transmittance $\neq$ reflectance), determine the properties required of the beam splitters in order that a beam entering horizontally activates only detector 1, while detector 2 remains dark.

Chapter 7
Position Probability

We establish the concept of probability within the analytical approach to quantum mechanics
in the form of the position probability density and its associated probability current density

In the algebraic approach to quantum mechanics, we introduced early on the notion of
probability. Now we want to develop this concept in the analytical approach, as well,
and furthermore we aim at merging the two approaches gradually. The problem is as
follows: In the algebraic approach, probabilities appear rather naturally (due to the
plausible redefinition of intensity → probability). The SEq, however, is deterministic.
An initial state fixes the time evolution of the wavefunction for all times, and clearly
this leaves no room for chance.

Therefore, probabilities cannot come into play from the SEq itself, but only
through the wavefunction $\Psi(x, t)$. As we already briefly mentioned in Chap. 5,
the absolute square of $\Psi(x, t)$ can be regarded as the *position probability density*,[1]
usually denoted by the letter ρ:

$$\rho(x, t) = \Psi^*(x, t)\,\Psi(x, t) = |\Psi(x, t)|^2 \qquad (7.1)$$

The interpretation of ρ as a probability density is not at all obvious, and in the
early days of quantum mechanics it took some time until Max Born arrived at this
concept. At that point (and still for us at present), it was a hypothesis or conjecture
which had to prove itself by leading to consistent results and conclusions (which of
course it did).[2]

We will develop this concept in the following and will discuss its consequences.

---

[1] The probability w (probability of finding the quantum object in a given region of space) is obtained
by integrating the probability density ρ, as in $w = \int \rho dV$, over the spatial region of interest.
Analogously, the mass m is given as an integral over the mass density ρ as $m = \int \rho dV$.

[2] Especially when one is speaking to lay people about probabilities in quantum mechanics, one
should always keep in mind that this is a conceptually difficult notion. On the one hand, there is
the wavefunction with its abstractness, not understandable in everyday terms. On the other hand, it
is just this wavefunction which allows us to determine concrete values of probabilities. The How

© Springer Nature Switzerland AG 2018
J. Pade, *Quantum Mechanics for Pedestrians 1*, Undergraduate Lecture
Notes in Physics, https://doi.org/10.1007/978-3-030-00464-4_7

7.1 Position Probability and Measurements

7.1.1 Example: Infinite Potential Wall

This section is intended to serve primarily as a brief motivation.

We want to calculate the probability of finding an object with well-defined energy in the infinite potential well within the interval $0 < x_1 < x_2 < a$. Classically this is quite simple[3]; the probability is evidently given by

$$w^{cl}_{x_1,x_2} = \frac{x_2 - x_1}{a}. \tag{7.2}$$

For the quantum-mechanical analysis, we assume a state with the given energy E_n (see Chap. 5):

$$\Psi(x,t) = e^{-iE_n t/\hbar}\sqrt{\frac{2}{a}}\sin\frac{n\pi}{a}x; \; E_n = \frac{\hbar^2}{2m}\left(\frac{n\pi}{a}\right)^2; \; n = 1, 2, \ldots \tag{7.3}$$

We consider the expression

$$w^{qm}_{x_1,x_2} = \int_{x_1}^{x_2} \Psi^*(x,t)\,\Psi(x,t)\,dx. \tag{7.4}$$

As outlined in Chap. 5, we choose as the first factor under the integral not $\Psi(x,t)$, but rather the complex conjugate wavefunction $\Psi^*(x,t)$. This guarantees that we always obtain *positive* expressions for the probability, since $\Psi^*(x,t)\Psi(x,t) \geq 0$; this is as required of a probability. A simple calculation leads us to

$$w^{qm}_{x_1,x_2} = \frac{x_2 - x_1}{a} - \frac{\sin\left(n\pi\frac{x_2-x_1}{a}\right)\cos\left(n\pi\frac{x_2+x_1}{a}\right)}{n\pi}. \tag{7.5}$$

The comparison of (7.2) and (7.5) suggests the interpretation of $w^{qm}_{x_1,x_2}$ as the probability of finding the object in the interval $[x_1, x_2]$. This has the consequence that we can interpret $\Psi^*(x,t)\Psi(x,t) = |\Psi|^2$ as a probability *density*. We see (compare also Fig. 7.1) that the quantum-mechanical probability becomes increasingly similar to the classical one with increasing n, i.e. with increasing energy. This behavior is typical of many quantum-mechanical phenomena: The quantum character becomes all the more clearer, the lower the energies (low with respect to the energy scale of the system under consideration), and vice versa.

---

and Why are certainly not intuitively obvious and cannot be formulated convincingly with the aid of familiar everyday ideas.

[3]The velocity is constant between the turning points.

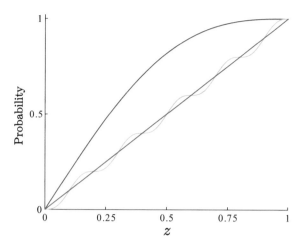

Fig. 7.1 Position probability (7.5) as a function of $z = \frac{x_2 - x_1}{a}$ for $x_2 = a\frac{1+z}{2}$ and $x_1 = a\frac{1-z}{2}$. The situation is shown for $n = 1$ (*red*), $n = 10$ (*green*) and $n = 1000$ (*blue*). The latter case is graphically indistinguishable from the classical straight line $w^{cl}_{x_1,x_2}$ given in (7.2)

7.1.2 Bound Systems

We start with the time-dependent SEq:

$$i\hbar \dot{\Psi}(x, t) = H\Psi(x, t). \tag{7.6}$$

Using the separation *ansatz*

$$\Psi(x, t) = e^{-i\frac{Et}{\hbar}} \varphi(x), \tag{7.7}$$

we obtain the time-independent SEq:

$$H\varphi(x) = E\varphi(x). \tag{7.8}$$

In this paragraph we assume that there are only discrete and no continuous eigenvalues, as discussed in Sect. 5.3. The eigenvalues and the eigenfunctions are given by $E_n = \hbar\omega_n$ and $\varphi_n(x)$, and the total solution reads

$$\Psi(x, t) = \sum_n c_n \varphi_n(x) e^{-i\frac{E_n t}{\hbar}}; \quad c_n \in \mathbb{C}, \tag{7.9}$$

with the initial state

$$\Psi(x, 0) = \sum_n c_n \varphi_n(x). \tag{7.10}$$

The *orthonormality* of the eigenfunctions, already mentioned in Chap. 5, is important for the following discussion (as we will show later on, this is a common feature of the eigenfunctions of *all* the Hamiltonians we will deal with in this book):

$$\int\limits_{-\infty}^{\infty} \varphi_n^* (x) \, \varphi_l (x) \, \mathrm{d}x = \delta_{nl}. \tag{7.11}$$

Since the total wavefunction Ψ is a solution of a linear differential equation, multiples of it are also solutions. We choose a multiple so that Ψ is normalized, i.e.

$$\int\limits_{-\infty}^{\infty} |\Psi(x,t)|^2 \, \mathrm{d}x = \int\limits_{-\infty}^{\infty} \rho(x,t)\mathrm{d}x = 1. \tag{7.12}$$

In short, we can always assume Ψ to be normalized.

We now interpret $|\Psi(x,t)|^2$ as a (position) probability density. Thus, the last equation implies that the quantum object is located with probability 1 (i.e. with certainty) *somewhere* in space, as it must be. The probability that the object is localized in a particular region, say $a \le x \le b$ at time t, is given by (as in (7.4)):

$$w(a \le x \le b, t) = \int\limits_a^b |\Psi(x,t)|^2 \, \mathrm{d}x. \tag{7.13}$$

Clearly, this probability is always positive definite, and the total probability $\int_{-\infty}^{\infty} |\Psi(x,t)|^2 \, \mathrm{d}x$ equals one.[4] Thus we have found for the wavefunction not an immediate, but at least an indirect physical significance,[5] in that its absolute square can be viewed as the position probability density.[6]

The extension of these considerations to three dimensions causes no problems.

7.1.2.1 Conclusions

What are the conclusions one can draw? We insert the total wavefunction (7.9) into (7.12) and obtain in the first step

$$1 \overset{!}{=} \int\limits_{-\infty}^{\infty} \sum_n c_n^* \varphi_n^* (x) \, e^{i\omega_n t} \sum_l c_l \varphi_l (x) \, e^{-i\omega_l t} \mathrm{d}x. \tag{7.14}$$

---

[4]The fact that we can actually interpret this as a probability is shown by the general definition. A probability measure μ on $\mathbb{R}$ is a mapping μ from the set of intervals (which are given here by the integration intervals) into the unit interval $[0, 1]$ which meets the following requirements: (i) $\mu(I) = 1 \ge 0$ for all intervals I (positive definite), (ii) $\mu(\mathbb{R}) = 1$ (normalized), (iii) $\mu(I_1 \cup I_2) = \mu(I_1) + \mu(I_2)$ for all pairwise disjoint intervals I_1 and I_2 (additivity property or σ additivity).

[5]The wavefunction itself is non-intuitive—it is just a complex-valued field of possibilities, as mentioned above.

[6]Since the concept is unique, one often omits the term 'position' and uses the more compact 'probability density'. We will do so also, for the most part.

Under the usual assumption that we can interchange the sum and the integral, and
with the notation of the two sums as a double sum, we obtain

$$
1 = \sum_{n,l} c_n^* e^{i\omega_n t} c_l e^{-i\omega_l t} \int_{-\infty}^{\infty} \varphi_n^*(x)\,\varphi_l(x)\,dx
$$

$$
= \sum_{n,l} c_n^* c_l e^{i(\omega_n - \omega_l)t} \delta_{n,l} = \sum_n c_n^* c_n = \sum_n |c_n|^2. \tag{7.15}
$$

In other words: the fact that Ψ is normalized is equivalent to

$$
\sum_n |c_n|^2 = 1. \tag{7.16}
$$

This equation is valid independently of time, so we can limit ourselves in our further
considerations to $t = 0$.

With (7.16), we have found the same relation as in the algebraic approach: The
absolute squares of the coefficients give the probabilities of finding the corresponding
states or quantum numbers. We illustrate this point by means of two concepts: mean
value and collapse.

Mean value. We consider an ensemble of identically-prepared systems (7.10),
where the measured quantity is the energy.[7] If we measure N members of the
ensemble, we observe the state $\varphi_n(x)$ (or the energy E_n) r_n times, where of course
$N = \sum_n r_n$. As usual, the *mean value* of the energy is found to be

$$
E_{\text{mean value}} = \sum_n h_n E_n \tag{7.17}
$$

with the relative frequencies of occurrence $h_n = r_n/N$. For $N \rightarrow \infty$, the relative
frequencies h_n become the probabilities $|c_n|^2$ of measuring the state $\varphi_n(x)$ or the
energy E_n, and we obtain the *expectation value*

$$
E_{\text{expectation value}} = \sum_n |c_n|^2 E_n. \tag{7.18}
$$

These concepts and the question of how they can be extended to continuous variables
will be discussed further in Chap. 9.

Collapse. We can apply the concept of probability to individual systems, with which
one mainly deals in practice. *Before* a single measurement, we can say that by measuring (7.10) we will obtain *one* of the states φ_n, say $\varphi_j(x)$, with the probability
$w_j = |c_j|^2$. *After* the measurement, the system is in a well-defined state, let us say
$\varphi_l(x)$. Thus we know immediately *after* the measurement the state of the system
with certainty:

---

[7]Those who wish may consider the infinite potential well as a concrete example.

$$c_n = 0 \text{ for } n \neq l \text{ directly after measurement} \tag{7.19}$$

or, formulated explicitly,[8]

$$\Psi_{\text{before}}(x, t) = \sum_n c_n \varphi_n(x) e^{-i\frac{E_n t}{\hbar}} \xrightarrow[\text{measurement}]{} \Psi_{\text{after}}(x, t) = \frac{c_l}{|c_l|} \varphi_l(x) e^{-i\frac{E_l t}{\hbar}}.$$
$$\tag{7.20}$$

We see that the measurement has forced the system into a unique state. We have already met up with this process of *state reduction* in the algebraic approach. That it now also occurs here is *not* due to the SEq. There is an additional element, namely, our interpretation of the wavefunction as a *probability amplitude* or a complex-valued field of possibilities.

The following picture emerges: The SEq describes the unperturbed time evolution of a quantum system. This evolution is interrupted by the measurement process, which changes the wavefunction. One also speaks of the *collapse of the wavefunction*. After the measurement, the system is again subject to the time evolution described by the SEq.[9]

As in the algebraic approach, there are open questions concerning the measurement. For example, if the measurement process is not included in the SEq, does this mean that measurement is not a quantum-mechanical process? Or is our description by means of the SEq plus measurement process insufficient? Or is it simply the best we can ever achieve, because nature is in reality not as simple as described by our theories? In short, what does 'to measure' actually mean in quantum mechanics?

7.1.3 Free Systems

In the case of free, unbounded systems, we have seen that an initial situation of the form (we limit ourselves to one dimension)

$$\rho(x, 0) = |\Psi(x, 0)|^2 = \frac{1}{\sqrt{\pi} b_0} \exp\left(-\frac{x^2}{b_0^2}\right) \tag{7.21}$$

evolves in the course of time into

$$\rho(x, t) = |\Psi(x, t)|^2 = \frac{1}{\sqrt{\pi} b(t)} \exp\left(-\frac{\left(x - \frac{\hbar K}{m} t\right)^2}{b^2(t)}\right), \tag{7.22}$$

with $b(t)$ given by

---

[8]The state must be normalized after the measurement; this is expressed by the factor $\frac{c_l}{|c_l|}$.

[9]We note again that the exact process of measurement itself is not described.

$$b(t) = \sqrt{b_0^2 + \left(\frac{\hbar t}{b_0 m}\right)^2}. \tag{7.23}$$

We repeat the following remark (cf. Sect. 5.2.2): We model here the quantum-like behavior of *material bodies* ($m \neq 0$) such as electrons. The spreading of $\rho(x, t)$ does not mean that the electron itself is 'smeared out' in space (in that case, the smearing would also apply to the electron's properties such as its mass and charge), like a mound of honey which flattens and spreads. It is the *wavefunction*, which determines the position probability, that disperses, rather than the object itself. In other words, the uncertainty with which we can determine the location of a quantum object increases in the course of time: $\Delta x \approx b(t)$.

Again the question arises: What happens when we perform a measurement? Let us assume that we have arrayed detectors along the entire x axis, each of length a. Now we release a free quantum object; the detectors are still switched off. We wait sufficiently long to be sure that $\Delta x \approx b(t) \gg a$ holds. Then we measure the position of the object by activating the detectors — one of them will respond. At that moment, the spatial uncertainty, which had grown steadily before our measurement, shrinks abruptly to a, and the wavefunction is correspondingly modified.[10] This means that we again observe the connection between the measurement process and the collapse of the wavefunction (or state reduction).

The considerations about the mean value which were outlined above for discrete eigenvalues cannot readily be applied to continuous measurements of quantities such as the position or the momentum. We will address this issue again in Chap. 9, and will formulate it so generally that the nature of the eigenvalue spectrum will no longer be relevant.

7.2 Real Potentials

The probability density ρ is positive definite, which follows directly from the definition (7.1). The probability of localizing the quantum object at time t in the interval $[x_1, x_2]$ is given by $W(x_1 < x < x_2; t) = \int_{x_1}^{x_2} \rho(x, t)\, dx$. In order to indeed interpret ρ as a probability density, the equation[11]

$$\int_{-\infty}^{\infty} \rho(x, t)\, dx \overset{!}{=} 1 \quad \forall t \tag{7.24}$$

must hold. In words: The quantum object must be located somewhere in space, and this must be true *at all times*. Therefore, two requirements must be met:

---

[10]We can regard this state as the initial condition for a new cycle of free propagation, in which case one refers to the measurement process as a (state-)preparation.

[11]We assume that there are neither creation nor annihilation processes.

1. The integral $\int_{-\infty}^{\infty} |\Psi (x, t)|^2 \, dx$ has to exist, at least at a certain time t. If it does,

   we can normalize the wave function so that at this time t, $\int_{-\infty}^{\infty} |\Psi (x, t)|^2 \, dx = 1$

   holds.
2. In addition, we must show that the normalization constant does not change, so that

   $\int_{-\infty}^{\infty} |\Psi (x, t)|^2 \, dx = 1$ is valid for *all times*.

Can we always satisfy these two requirements?

The first requirement means that $\Psi (x, t)$ must be square integrable in view of the interpretation of $|\Psi(x, t)|^2$ as a probability density. This is certainly the case over a finite interval of space when the wavefunction is sufficiently smooth or 'well-behaved' (i.e. does not have singularities, etc.), which we always assume in the following. In order that the integral from $-\infty$ to ∞ exists, the condition

$$\Psi \underset{|x|\to\infty}{\sim} |x|^{\alpha} \; ; \alpha < -\frac{1}{2} \tag{7.25}$$

must be fulfilled in addition, at least at some time t. One often describes this condition by saying that the wavefunction must *approach zero rapidly enough* at infinity.[12] We note in this context that there may be correct mathematical solutions of differential equations that must still be excluded for physical reasons. More about this issue is included in some of the following chapters and in Appendix E, Vol. 1.

As to the second requirement[13]: We have to show that $\int_{-\infty}^{\infty} \rho (x, t) \, dx = 1$ holds at all times. This means that

$$\frac{d}{dt} \int_{-\infty}^{\infty} \rho (x, t) \, dx \overset{!}{=} 0, \tag{7.26}$$

and it follows[14] that

$$0 \overset{!}{=} \int_{-\infty}^{\infty} \frac{\partial}{\partial t} \Psi^* \Psi dx = \int_{-\infty}^{\infty} \left(\dot{\Psi}^* \Psi + \Psi^* \dot{\Psi} \right) dx. \tag{7.27}$$

We replace the time derivatives making use of the SEq

---

[12]In three dimensions, the condition is slightly different. Because of $\int dV = \int r^2 dr \sin \vartheta d\vartheta d\varphi$, the wavefunction has to go to zero as r^{α} with $\alpha < -\frac{3}{2}$.
[13]With the conceptual framework derived in later chapters, the proof may be formulated in a considerably shorter way.
[14]As always, we assume the commutability of differentiation and integration. See Appendix D, Vol. 1.

$$i\hbar\dot{\Psi} = -\frac{\hbar^2}{2m}\Psi'' + V\Psi \tag{7.28}$$

and find, assuming a *real potential* $V \in \mathbb{R}$,[15]

$$0 \overset{!}{=} \int_{-\infty}^{\infty} \left[\left(\frac{\hbar}{2mi}\Psi^{*''} - \frac{V}{i\hbar}\Psi^* \right) \Psi + \Psi^* \left(-\frac{\hbar}{2mi}\Psi'' + \frac{V}{i\hbar}\Psi \right) \right] dx$$

$$= \frac{\hbar}{2mi} \int_{-\infty}^{\infty} \left(\Psi^{*''}\Psi - \Psi^*\Psi'' \right) dx. \tag{7.29}$$

We transform the second derivatives w.r.t. spatial coordinates by partial integration. It follows that

$$0 \overset{!}{=} \frac{\hbar}{2mi} \left[(\Psi^{*'}\Psi)_{-\infty}^{\infty} - \int_{-\infty}^{\infty} \Psi^{*'}\Psi' dx \right] - \frac{\hbar}{2mi} \left[(\Psi^*\Psi')_{-\infty}^{\infty} - \int_{-\infty}^{\infty} \Psi^{*'}\Psi' dx \right]. \tag{7.30}$$

The integrals cancel each other and we finally obtain

$$\left(\Psi^{*'}\Psi - \Psi^*\Psi' \right)\Big|_{-\infty}^{\infty} \overset{!}{=} 0. \tag{7.31}$$

This condition is fulfilled due to (7.25), since for $\alpha < -\frac{1}{2}$, we have $\Psi'\Psi \underset{|x|\to\infty}{\sim}$ $|x|^{2\alpha-1} \to 0$.

We see that the probability concept is inherently consistent, if the wavefunction vanishes at infinity rapidly enough and if the potential is real. These are very important properties, which we assume to be *always* fulfilled from now on.[16]

7.3 Probability Current Density

In the following, an expression for the (position) probability current density is derived. We rely on the *continuity equation*[17]

$$\frac{\partial\rho}{\partial t} + \nabla\mathbf{j} = \mathbf{0}. \tag{7.32}$$

---

[15] Here, the potential may depend on the time t.

[16] Complex potentials are required when one wants to describe e.g. absorption processes. These potentials are also called *optical potentials* (referring to the complex optical refractive index whose imaginary part describes absorption). An example is found in the exercises.

[17] The derivation of the continuity equation is given in Appendix N, Vol. 1.

This equation is a differential formulation of a global conservation law. It is valid not only for the mass density, but in fact applies to all densities (e.g. the charge density) for which integral conservation laws hold (e.g. global conservation of charge).

In particular, we assume the validity of the continuity equation for the probability density of quantum mechanics. Thus, we can calculate the probability current density $\mathbf{j}$. For the sake of simplicity, we consider only the one-dimensional problem and extend the result at the end to three dimensions.

In one dimension, the continuity equation reads

$$\dot{\rho}(x,t) + \frac{\partial}{\partial x} j(x,t) = 0. \tag{7.33}$$

To derive the relationship between j and Ψ, we insert $\rho = |\Psi|^2$. With $\dot{\rho} = \dot{\Psi}^*\Psi + \Psi^*\dot{\Psi}$ and the SEq

$$\dot{\Psi} = -\frac{\hbar}{2mi}\Psi'' + \frac{V\Psi}{i\hbar}, \tag{7.34}$$

we can rewrite the continuity equation as

$$\left(\frac{\hbar}{2mi}\Psi^{*\prime\prime} - \frac{V^*\Psi^*}{i\hbar}\right)\Psi + \Psi^*\left(-\frac{\hbar}{2mi}\Psi'' + \frac{V\Psi}{i\hbar}\right) + \frac{\partial}{\partial x}j = 0. \tag{7.35}$$

Since we assume $V \in \mathbb{R}$, the potential terms cancel. It follows that

$$\begin{aligned}
\frac{\partial}{\partial x}j &= \frac{\hbar}{2mi}\left(\Psi^*\Psi'' - \Psi^{*\prime\prime}\Psi\right) \\
&= \frac{\hbar}{2mi}\left(\Psi^*\Psi'' - \Psi^{*\prime\prime}\Psi + \Psi^{*\prime}\Psi' - \Psi^{*\prime}\Psi'\right) \\
&= \frac{\hbar}{2mi}\left(\frac{\partial}{\partial x}\Psi^*\Psi' - \frac{\partial}{\partial x}\Psi^{*\prime}\Psi\right).
\end{aligned} \tag{7.36}$$

Integration gives[18]:

$$j(x,t) = \frac{\hbar}{2mi}(\Psi^*\Psi' - \Psi^{*\prime}\Psi). \tag{7.37}$$

We have thus found an expression for the probability current density. We already know that it vanishes at infinity; see (7.31).

The extension of the probability current density to three dimensions yields in a straightforward manner:

$$\mathbf{j}(\mathbf{r},t) = \frac{\hbar}{2mi}(\Psi^*\nabla\Psi - \Psi\nabla\Psi^*). \tag{7.38}$$

---

[18] Actually, there could still be a constant of integration on the right-hand side, but it is set equal to zero due to the requirement $j = 0$ for $\Psi = 0$.

As an (unphysical, but familiar[19]) example, we consider a plane wave

$$\Psi(\mathbf{r}, t) = Ae^{i(\mathbf{kr}-\omega t)}. \tag{7.39}$$

With

$$\nabla\Psi(\mathbf{r}, t) = Ai\mathbf{k}e^{i(\mathbf{kr}-\omega t)}, \tag{7.40}$$

it follows that

$$\mathbf{j}(\mathbf{r}, t) = \frac{\hbar}{2mi}\left(i\mathbf{k}AA^* + i\mathbf{k}AA^*\right) = \frac{\hbar\mathbf{k}}{m}|A|^2. \tag{7.41}$$

Because of $\rho = \Psi^*\Psi = |A|^2$, we obtain the well-known relationship

$$\mathbf{j} = \frac{\hbar\mathbf{k}}{m}\rho = \frac{\mathbf{p}}{m}\rho := \mathbf{v}\rho \tag{7.42}$$

where $\mathbf{v}$ is the velocity of e.g. a maximum of the wave.[20]

We make some general remarks on the one-dimensional probability current density $j = \frac{\hbar}{2im}\left(\varphi^*\varphi' - \varphi^{*'}\varphi\right)$:

1. For $\varphi(x) = Ae^{\alpha x}$ ($\alpha \in \mathbb{R}$, $A \in \mathbb{C}$), we have

$$j = \frac{\hbar}{2im}\left(\alpha|A|^2 e^{2\alpha x} - \alpha|A|^2 e^{2\alpha x}\right) = 0. \tag{7.43}$$

With real exponents, j disappears. To put it graphically, this does not mean that nothing flows into the region or out of it, but rather that whatever flows in must also flow out again.

2. For $\varphi(x) = Ae^{i\gamma x}$ ($\gamma \in \mathbb{R}$, $A \in \mathbb{C}$), we have

$$j = \frac{\hbar}{2im}\left(i\gamma|A|^2 + i\gamma|A|^2\right) = \frac{\hbar}{m}\gamma|A|^2. \tag{7.44}$$

So there is a 'net flow', that is, something is actually transported.

---

[19]Unphysical, because the infinitely-extended plane wave, whose magnitude is one everywhere, does not represent a physical object. The fact that we can still make use of plane waves in quantum mechanics is due to the linearity of quantum mechanics, which allows us to construct wave packets with physically reasonable behavior by superposition of plane waves.

[20]We note that the 'velocity' of a quantum object is a seldom-used notion in quantum mechanics. The momentum is the central quantity. 'Velocity' will appear only in the context of Galilean transformations (relative motion of inertial frames, Chap. 21, Vol. 2) and in the Bohmian interpretation of quantum mechanics (see Chap. 28, Vol. 2), which is based on classical mechanics.

7.4 Exercises

1. Show for $\rho = |\psi(x, t)|^2$ that:

$$\int_{-\infty}^{\infty} \rho(x, t)\, dx = 1 \ \forall t. \tag{7.45}$$

Here, we assume that (i) the potential is real, and (ii) $\Psi \underset{x \to \infty}{\sim} x^a$, with $a < -\frac{1}{2}$.

2. Infinite potential well: Given the wave functions

   (a) $\Psi(x, t) = e^{-i\omega_n t} \sqrt{\frac{2}{a}} \sin \frac{n\pi}{a} x$

   (b) $\Psi(x, t) = c_n e^{-i\omega_n t} \sqrt{\frac{2}{a}} \sin \frac{n\pi}{a} x + c_m e^{-i\omega_m t} \sqrt{\frac{2}{a}} \sin \frac{m\pi}{a} x$,

   Calculate for both cases the probability of finding the quantum object in the interval (x_1, x_2)

$$w_{x_1, x_2}^{qm} = \int_{x_1}^{x_2} \Psi^*(x, t)\, \Psi(x, t)\, dx. \tag{7.46}$$

3. Given the SEq $i\hbar\dot\psi = H\psi$ with a real potential, derive from the continuity equation constructively (i.e. not just proving by insertion) that $\mathbf{j}$ is given by

$$\mathbf{j} = \frac{\hbar}{2mi} (\psi^* \nabla \psi - \psi \nabla \psi^*). \tag{7.47}$$

4. Calculate j (one-dimensional) for $\psi = A e^{\gamma x}$ and $\psi = A e^{i\gamma x}$, with $\gamma \in \mathbb{R}$ and $A \in \mathbb{C}$.

5. Calculate $\mathbf{j}(\mathbf{r}, t)$ for $\Psi(\mathbf{r}, t) = A e^{i(\mathbf{kr} - \omega t)}$.

6. Given a modification of the infinite potential well, namely the potential

$$V(x) = \begin{cases} iW \text{ for } 0 < x < a \\ \infty \ \text{ otherwise} \end{cases} ; \ W \in \mathbb{R}, \tag{7.48}$$

   calculate the energy spectrum and show that the norm of the (time-dependent) total wavefunction is independent of time only for $W = 0$.

Chapter 8
Neutrino Oscillations

Thus far, in the algebraic approach we have not considered the question of the time evolution of a system. We now want to tackle this topic on the basis of a problem of current interest. In addition, we meet up with Hermitian operators, and we address once again the problem of measurement.

8.1 The Neutrino Problem

As is well known, the *neutrino* ν was originally postulated by Wolfgang Pauli in order to 'save' the conservation of energy in beta decay. As it turned out later after careful examination. Each of the three elementary particles, the electron e, the muon μ and the tauon τ has its 'own' neutrino, i.e. ν_e, ν_μ, and ν_τ.[1] The rest mass of all three neutrinos seemed to be vanishingly small, and it was generally assumed to be zero.

Change of scene: We consider now the sun and the particles which it emits. Among them are the three neutrino species, and those in a certain ratio, which can be determined reasonably reliably on the basis of current solar models. But measurements on earth yielded a rather different value for this ratio. The question was: Are the solar models incorrect, or is something wrong with our description of neutrinos?

---

[1] Wolfgang Pauli in 1930 initially chose the name 'neutron.' The term 'neutrino' was introduced later by Enrico Fermi. In 1956, the electron neutrino was detected experimentally for the first time, and in 1962, the muon neutrino. The tauon was observed in 1975, but the corresponding neutrino only in 2000. There may be still other types of neutrinos. These (as yet hypothetical) *sterile neutrinos* interact only via gravity and not—like the other neutrinos—through the weak interaction (hence the adjective 'sterile'). See e.g. D. Castelvecchi, Icy telescope throws cold water on sterile neutrino theory, *Nature*, https://doi.org/10.1038/nature.2016.20382 (Aug 2016), and literature referenced there.

© Springer Nature Switzerland AG 2018
J. Pade, *Quantum Mechanics for Pedestrians 1*, Undergraduate Lecture
Notes in Physics, https://doi.org/10.1007/978-3-030-00464-4_8

There were good arguments to regard the solar models as correct. So something had to be changed in the description of the neutrinos. And it was this: If one assumes that the rest masses of the neutrinos are not exactly zero, the three neutrino species can change into each other over the course of time (*neutrino oscillations*); that is, on the way from the sun to the earth. In this way it could be explained that on earth, we measure a different relative abundance of the three neutrinos than is predicted by the solar models.

8.2 Modelling the Neutrino Oscillations[2]

We will now describe the process of neutrino oscillations, as simply as possible. In order to make clear the principle, we confine ourselves to a simpler model with only two neutrinos, since the computations for three neutrinos are more complicated. A few words about the three-dimensional case can be found at the end of this chapter.

In this chapter, we will for once visit the field of relativistic phenomena. We can do so because we need only the statement that there is a Hamiltonian (and in particular its energy eigenvalues) for the physical problem, without having to worry about its specific form or any details of the interaction.

8.2.1 States

We start with the production of neutrinos (e.g. in the sun or in an accelerator) as a superposition of two states $|\nu_1\rangle$ and $|\nu_2\rangle$ with *well-defined, different rest masses* m_{01} and m_{02}, called mass (eigen-)states. The momenta are equal, but the total energies E_1 and E_2 are therefore different.[3] Without loss of generality, we can set $\Delta m := m_{01} - m_{02} > 0$ and hence $\Delta E := E_1 - E_2 > 0$ or $\Delta \omega = \omega_1 - \omega_2$ (with $\omega = E/\hbar$). The states form a CONS; $\langle \nu_i | \nu_j \rangle = \delta_{ij}$ and $|\nu_1\rangle \langle \nu_1| + |\nu_2\rangle \langle \nu_2| = 1$.[4]

For certain reasons, one cannot measure the states $|\nu_1\rangle$ and $|\nu_2\rangle$ directly, but only superpositions of these states, which we call (referring to the actual situation) the electron neutrino and the muon neutrino, $|\nu_e\rangle$ and $|\nu_\mu\rangle$ (also termed *flavor states*). We have

$$|\nu_e\rangle = \cos\vartheta\,|\nu_1\rangle + \sin\vartheta\,|\nu_2\rangle$$
$$|\nu_\mu\rangle = -\sin\vartheta\,|\nu_1\rangle + \cos\vartheta\,|\nu_2\rangle. \tag{8.1}$$

---

[2]The importance of the issue can be seen e.g. from the fact that the Nobel Prize in Physics 2015 was awarded jointly to Takaaki Kajita (born 1959, Japanese physicist) and Arthur B. McDonald (born 1943, Canadian physicist) "for the discovery of neutrino oscillations, which shows that neutrinos have mass".

[3]We remind the reader: $E^2 = m_0^2 c^4 + p^2 c^2$.

[4]Call the Hamiltonian for free neutrino motion H. We have $H|\nu_1\rangle = E_1|\nu_1\rangle$ and $H|\nu_2\rangle = E_2|\nu_2\rangle$ with $\Delta E = E_1 - E_2 > 0$.

Here, ϑ is an (abstract) angle, called the mixing angle. The states $|\nu_e\rangle$ and $|\nu_\mu\rangle$ also form a CONS. Therefore, we can represent $|\nu_1\rangle$ and $|\nu_2\rangle$ as a superposition of $|\nu_e\rangle$ and $|\nu_\mu\rangle$. We then have

$$|\nu_1\rangle = \cos\vartheta \, |\nu_e\rangle - \sin\vartheta \, |\nu_\mu\rangle$$
$$|\nu_2\rangle = \sin\vartheta \, |\nu_e\rangle + \cos\vartheta \, |\nu_\mu\rangle. \tag{8.2}$$

In fact, these transformations are nothing more than rotations by the angle $\pm\vartheta$ within a two-dimensional space, or equivalently, a change of basis, which is described by the well-known transformation

$$\begin{pmatrix} \cos\vartheta & \pm\vartheta\sin\vartheta \\ \mp\sin\vartheta & \cos\vartheta \end{pmatrix}. \tag{8.3}$$

It represents a particularly simple example of a unitary matrix.

8.2.2 Time Evolution

Next, we want to investigate the time evolution of the states $|\nu_e\rangle$ and $|\nu_\mu\rangle$. To this end we use use the fact, found from the analytical approach, that the time evolution of a state with well-defined energy E is described by the factor $e^{-iEt/\hbar}$. Although this requirement suggests itself, it is not self-evident that it must be satisfied here. If we accept that it holds true (or regard it as an axiom for the moment), we find: If at time zero an initial state $|z(t=0)\rangle = |z(0)\rangle$ exists with the *well-defined energy* $E = \hbar\omega$, its time evolution is described by

$$|z(t)\rangle = |z(0)\rangle \, e^{-iEt/\hbar}. \tag{8.4}$$

This is a very important and universally valid fact in quantum mechanics. It follows that

$$i\hbar \frac{d}{dt} |z(t)\rangle = E \, | z(t)\rangle. \tag{8.5}$$

If we assume that E is an eigenvalue of an operator H, we have essentially 'recovered' the free SEq.[5]
We see that the time evolution (8.4) is a unitary process that conserves the norm:

$$\langle z(t)| \, z(t)\rangle = \langle z(0)| \, e^{i\omega t} e^{-i\omega t} \, | z(0)\rangle = \langle z(0)| \, z(0)\rangle. \tag{8.6}$$

We now take a muon-neutrino as the initial state $|\nu(0)\rangle$, i.e. $|\nu(0)\rangle = |\nu_\mu\rangle$. Then it follows with (8.1) for the time evolution:

---

[5]Here, H denotes a (still) unknown operator and *not* the well-known operator $-\frac{\hbar^2}{2m}\nabla^2 + V$. Double meanings of this type are quite common in quantum mechanics. We will learn the reason for this in later chapters.

$$|\nu(t)\rangle = -\sin\vartheta\,|\nu_1\rangle\,e^{-i\omega_1 t} + \cos\vartheta\,|\nu_2\rangle\,e^{-i\omega_2 t}. \tag{8.7}$$

Evidently, we have $|\langle\nu_1\,|\nu\rangle|^2 = \left|-\sin\vartheta e^{-i\omega_1 t}\right|^2 = \sin^2\vartheta$—this would be the probability of obtaining $|\nu_1\rangle$ in a measurement. But since we can measure only the states $|\nu_e\rangle$ and $|\nu_\mu\rangle$, we have to project the corresponding portions out of $|\nu(t)\rangle$, by means of the projection operators $|\nu_e\rangle\langle\nu_e|$ and $|\nu_\mu\rangle\langle\nu_\mu|$. With (8.2) we find, for example, $\langle\nu_e|\,\nu_1\rangle = \cos\vartheta$ and $\langle\nu_e|\,\nu_2\rangle = \sin\vartheta$. It follows for the electron neutrino that

$$|\nu_e\rangle\,\langle\nu_e\,|\nu(T)\rangle = \left[-\sin\vartheta\cos\vartheta e^{-i\omega_1 T} + \cos\vartheta\sin\vartheta e^{-i\omega_2 T}\right]|\nu_e\rangle. \tag{8.8}$$

It is seen that this term includes both frequencies ω_1 and ω_2 and thus displays a very different behavior from the mass states. We obtain the probability of measuring $|\nu_e\rangle$ by the usual application of the absolute square of the prefactor (see the exercises):

$$p_e\,(T) = \left|-\sin\vartheta\cos\vartheta e^{-i\omega_1 T} + \cos\vartheta\sin\vartheta e^{-i\omega_2 T}\right|^2 = \sin^2 2\vartheta\cdot\sin^2\left(\frac{\Delta\omega}{2}T\right). \tag{8.9}$$

8.2.3 Numerical Data

Equation (8.9) shows that the probability to find the neutrino in the state $|\nu_e\rangle$ depends periodically on time, where the period is $\tau = \frac{2\pi}{\Delta\omega}$. The neutrino oscillates between the states $|\nu_e\rangle$ and $|\nu_\mu\rangle$; see Fig. 8.1. This is quite similar to two coupled pendulums which show beats, in which the energy flows periodically from one pendulum to the other.

To get a feeling for the order of magnitudes, we perform a rough calculation. We can assume in good approximation that the neutrinos, due to their low mass, are moving with nearly the speed of light. In space, we have a period of length $L = c\tau = c\frac{2\pi}{\Delta\omega}$. We approximate the difference $\Delta\omega$ by (see the exercises)

$$\hbar\Delta\omega = \frac{c^4}{2pc}\left(m_1^2 - m_2^2\right) := \frac{c^4\Delta m^2}{2pc}. \tag{8.10}$$

It follows

$$L = c\frac{2\pi}{\Delta\omega} = \frac{4\pi\hbar}{c^2}\frac{p}{\Delta m^2}. \tag{8.11}$$

This term is most easily evaluated in the theoretical unit system in which $\hbar = c = 1$ and energies and masses are measured in eV, see Appendix B, vol. 1.[6] The

---

[6]Numerical examples: the electron in this system of units has a rest mass of about 0.5 MeV. The accelerator LHC operates with protons of energies of up to 7 TeV.

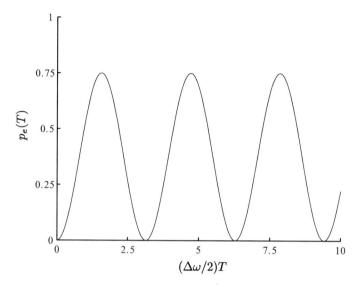

Fig. 8.1 $p_e(T)$ of (8.9) for $\vartheta = \pi/6$

mass difference between neutrinos[7] is about $\Delta m^2 \approx 10^{-3} (\text{eV})^2$, the momentum is $10\,\text{GeV} = 10^{10}\,\text{eV}$. We then find

$$L \mathrel{\hat{=}} 4\pi \frac{10^{10}}{10^{-3}\,\text{eV}} = 4\pi \frac{10^{19}}{\text{MeV}}, \tag{8.12}$$

and with the conversion of units of length $\frac{1}{\text{MeV}} \mathrel{\hat{=}} 0.1973 \times 10^{-12}\,\text{m}$ it follows finally

$$L = 4\pi \times 10^{19} \times 0.1973 \times 10^{-12}\,\text{m} \approx 25000\,\text{km}. \tag{8.13}$$

Of course we should not take the numerical value too seriously—we have considered only two instead of three neutrinos, and just the uncertainty regarding the mass difference leaves a wide margin for error. What is instead important is that we can at least qualitatively describe an effect such as neutrino oscillations, and this with only the simplest of formal means.

8.2.4 Three-Dimensional Neutrino Oscillations

The neutrino question remains an issue of ongoing research, since there are still some unresolved problems.[8] We will not go into this more deeply, but give only a very

---

[7]Of course, this is a key parameter—if it is $10^{-6}\,\text{eV}$ instead of $10^{-3}\,\text{eV}$, then the length increases correspondingly by a factor of 1000.

[8]A recent review which also contains the values cited in Table 8.1 is given by G.L. Fogli et al., 'Global analysis of neutrino masses, mixings and phases: entering the era of leptonic CP violation searches', http://arXiv.org/abs/1205.5254v3 (2012).

Table 8.1 Values of the mixing angles

$s_{12}^2 = 0.307$	or	$\theta_{12} \approx 34°$
$s_{23}^2 = 0.5$	or	$\theta_{23} \approx 45°$
$s_{13}^2 = 0.021$	or	$\theta_{13} \approx 8°$

brief comment on the three-dimensional problem. One assumes three flavor states and three mass states:

$$\begin{pmatrix} \nu_e \\ \nu_\mu \\ \nu_\tau \end{pmatrix} = U \cdot \begin{pmatrix} \nu_1 \\ \nu_2 \\ \nu_3 \end{pmatrix}. \tag{8.14}$$

With the abbreviations $s_{ij} = \sin\theta_{ij}$ and $c_{ij} = \cos\theta_{ij}$, the transformation matrix U is written as

$$U = \begin{pmatrix} 1 & 0 & 0 \\ 0 & c_{23} & s_{23} \\ 0 & -s_{23} & c_{23} \end{pmatrix} \cdot \begin{pmatrix} c_{13} & 0 & s_{13}e^{-i\delta} \\ 0 & 1 & 0 \\ -s_{13}e^{i\delta} & 0 & c_{13} \end{pmatrix} \cdot$$

$$\times \begin{pmatrix} c_{12} & s_{12} & 0 \\ -s_{12} & c_{12} & 0 \\ 0 & 0 & 1 \end{pmatrix} \cdot \begin{pmatrix} e^{i\alpha_1/2} & 0 & 0 \\ 0 & e^{i\alpha_2/2} & 0 \\ 0 & 0 & 1 \end{pmatrix}. \tag{8.15}$$

The first three of these four unitary matrices describe (from left to right) the changes $\nu_\mu \leftrightarrow \nu_\tau$, $\nu_e \leftrightarrow \nu_\tau$, and $\nu_e \leftrightarrow \nu_\mu$. The phases δ (Dirac phase) and α_i (Majorana phase) are introduced as a result of further considerations.[9] As a product of unitary matrices, the matrix U is itself again unitary (see the exercises).

For the angles, current values are given in Table 8.1.

For the mass differences, one finds $\delta m^2 = \Delta m_{21}^2 = m_2^2 - m_1^2 = 7.5 \times 10^{-5}\,\text{eV}^2$ and $\Delta m^2 = \Delta m_{23}^2 = m_3^2 - \frac{m_1^2 + m_2^2}{2} = 2.5 \times 10^{-3}\,\text{eV}^2$. The unit eV is defined in Appendix B, Vol. 1.[10]

For several reasons, new discoveries of the neutrino's properties are expected to change our understanding of the universe. Thus, neutrinos are a topic of ongoing research, see e.g. E. Gibney, Morphing neutrinos provide clue to antimatter mystery, *Nature* https://doi.org/10.1038/nature.2016.20405 (Aug 2016). A most important

---

[9]The first three matrices are (except for the phase shift δ) the rotation matrices $D_x(\theta_{23})\,D_y(\theta_{13})\,D_z(\theta_{12})$. The first matrix describes e.g. a rotation by the angle θ_{23} around the x axis.

[10]The values for the mixing angles and the mass differences are from Neutrino Mixing - Particle Data Group, pdg.lbl.gov/2017/listings/rpp2017-list-neutrino-mixing.pdf (30. 5. 2017). The precise determination of these angles is a current topic; see for instance Eugenie S. Reich, 'Neutrino oscillations measured with record precision', *Nature* 08 March 2012, where the measurement of the angle θ_{13} is discussed, or P. Adamson et al. (NOvA Collaboration), Measurement of the Neutrino Mixing Angle θ_{23} in NOvA, *Phys. Rev. Lett* 118, 151802 (10. 4. 2017).

open issue is the absolute mass scale of neutrinos. This question will be investi-
gated e.g. by the KATRIN experiment, launched in June 2018 (KArlsruhe TRItium
Neutrino, Karlsruhe, Germany).

8.3 Generalizations

8.3.1 Hermitian Operators

In this section, we want to generalize the findings obtained on the basis of the neutrino
problem. First, we extend the formulation (8.5) to a 'proper' SEq

$$i\hbar \frac{d}{dt} |\psi(t)\rangle = H |\psi(t)\rangle . \tag{8.16}$$

Apart from the mere analogy to the SEq in the analytical approach, the motivation
for this step is that we want to find a linear differential equation of first order in time
also for the algebraic approach. It is clear that we have at this point no information
about the operator H which appears in (8.16)—neither how it is constructed internally
(spatial derivatives as in the Laplace operator cannot occur here), nor about its relation
to the Hamiltonian used in the analytical approach. These points will be discussed
in later chapters.

Here we want to clarify which properties H must have in order that the evolution
of $|\psi(t)\rangle$ be unitary, which means that the scalar product $\langle\psi(t)|\psi(t)\rangle$ must be
constant for all times. With this in mind, we write (8.16) and the adjoint equation in
compact form

$$i\hbar \left|\dot{\psi}(t)\right\rangle = H |\psi(t)\rangle; \quad -i\hbar \left\langle\dot{\psi}(t)\right| = \langle\psi(t)| H^\dagger. \tag{8.17}$$

If $\langle\psi(t)|\psi(t)\rangle$ does not depend on time, it follows that

$$i\hbar \frac{d}{dt} \langle\psi(t)|\psi(t)\rangle = i\hbar \langle\dot{\psi}(t)|\psi(t)\rangle + i\hbar \langle\psi(t)|\dot{\psi}(t)\rangle = 0. \tag{8.18}$$

We insert (8.17) and obtain

$$- \langle\psi(t)| H^\dagger |\psi(t)\rangle + \langle\psi(t)| H |\psi(t)\rangle = \langle\psi(t)| H - H^\dagger |\psi(t)\rangle = 0. \tag{8.19}$$

Since this equation holds for every $|\psi(t)\rangle$, it follows that $H^\dagger = H$.

In general, an operator A is called *self-adjoint* or *Hermitian* if $A = A^\dagger$. The
importance of such operators in quantum mechanics lies in the fact that all physically-
measurable quantities are represented by self-adjoint operators. Indeed, Hermitian

operators have real eigenvalues[11] as we want to show now. Let A be a Hermitian operator, $A = A^\dagger$. Then the eigenvalue problem and its adjoint version read:

$$A\,|a_n\rangle = \lambda_n\,|a_n\rangle \quad \text{and} \quad \langle a_n|\,A^\dagger = \lambda_n^*\,\langle a_n|. \tag{8.20}$$

Multiplication of the first equation from the left by $\langle a_n|$ and of the second equation from the right by $|a_n\rangle$ leads, due to $A = A^\dagger$, to

$$\langle a_n|\,A\,|a_n\rangle = \lambda_n\,\langle a_n\,|a_n\rangle \quad \text{and} \quad \langle a_n|\,A^\dagger\,|a_n\rangle = \langle a_n|\,A\,|a_n\rangle = \lambda_n^*\,\langle a_n\,|a_n\rangle. \tag{8.21}$$

The comparison shows $\lambda_n = \lambda_n^*$, i.e. $\lambda_n \in \mathbb{R}$. Other properties of Hermitian operators are discussed in the following chapters.

In Chap. 4, we made the acquaintance of projection operators, and in Chap. 6 of unitary operators, and now Hermitian operators join in.[12] The good news is that the zoo of operators[13] of quantum mechanics is complete—we will be concerned *only* (to be exact, with *one* exception) with these three types of operators (or the corresponding matrices or other representations):

$$\begin{array}{ll} A = A^\dagger & \text{Hermitian operator} \\ AA^\dagger = A^\dagger A = 1 & \text{unitary operator} \\ A^2 = A & \text{projection operator.} \end{array} \tag{8.22}$$

The names are used also for the corresponding matrices and representations. We outline in brief form the applications of these operators: We can represent physically-measurable quantities by Hermitian operators; the unperturbed time evolution of a system is described by a unitary operator; and the measurement process can be modelled with the help of projection operators.

8.3.2   Time Evolution and Measurement

We denote the eigenvalues and eigenvectors of H in (8.16) by E_n and $|\varphi_n\rangle$. The general solution as a generalization of (8.7) is a superposition of the eigenvectors:

$$|\psi(t)\rangle = \sum_n c_n\,|\varphi_n\rangle\,e^{-iE_n t/\hbar} \tag{8.23}$$

where the integration constants c_i are determined by the initial conditions (see the exercises).

---

[11] Since measured values are real, we can interpret them as eigenvalues of Hermitian operators.

[12] These properties are not mutually exclusive: A unitary operator or a projection operator can also be e.g. Hermitian.

[13] Since these operators exhibit only a few species and are fairly well-behaved, one could also speak of a 'pet zoo'.

A measurement interrupts the time evolution of $|\psi(t)\rangle$ as described in (8.23). If we want to measure e.g. the state $|\chi\rangle$, then we can describe it by projecting onto $|\chi\rangle$; that is by the term $|\chi\rangle\langle\chi|\psi\rangle$ corresponding to (8.8). Here, $|\langle\chi|\psi\rangle|^2$ is the probability that we actually obtain $|\chi\rangle$ from a measurement:

$$|\langle\chi|\psi\rangle|^2 = \sum_n c_n \langle\chi|\varphi_n\rangle e^{-i E_n t/\hbar} \sum_m c_m^* \langle\varphi_m|\chi\rangle e^{i E_m t/\hbar}$$

$$= \sum_{n,m} c_n c_m^* \langle\chi|\varphi_n\rangle \langle\varphi_m|\chi\rangle e^{-i(E_n - E_m)t/\hbar}. \tag{8.24}$$

After or due to the measurement, we have the state $|\chi\rangle$ instead of $|\psi\rangle$.

We remark that all these considerations hold for systems of arbitrary dimensions.

8.4 Exercises

1. Given that $|\nu_1\rangle\langle\nu_1| + |\nu_2\rangle\langle\nu_2| = 1$, show: $|\nu_e\rangle\langle\nu_e| + |\nu_\mu\rangle\langle\nu_\mu| = 1$.

2. Show that the matrices $\begin{pmatrix} c & 0 & se^{-i\delta} \\ 0 & 1 & 0 \\ -se^{i\delta} & 0 & c \end{pmatrix}$ and $\begin{pmatrix} 1 & 0 & 0 \\ 0 & c & s \\ 0 & -s & c \end{pmatrix}$ with $\delta \in \mathbb{R}$ are unitary. The abbreviations s and c stand for $\sin\alpha$ and $\cos\alpha$.

3. Show that the product of two unitary matrices is also unitary.

4. Is the beam splitter operator T from Chap. 6,

$$T = \frac{1+i}{2}[1 + i|H\rangle\langle V| + i|V\rangle\langle H|], \tag{8.25}$$

a Hermitian, a unitary or a projection operator? $\{|H\rangle, |V\rangle\}$ is a CONS.

5. Given $A = \begin{pmatrix} 1 & i \\ -i & 1 \end{pmatrix}$:

   (a) Show that A is Hermitian, but not unitary.
   (b) Calculate e^{cA}.

6. Given the operators[14]

$$L_1 = \frac{|v\rangle(\langle u| + \langle w|) + (|u\rangle + |w\rangle)\langle v|}{\sqrt{2}}$$

$$L_2 = \frac{-|v\rangle(\langle u| - \langle w|) + (|u\rangle - |w\rangle)\langle v|}{i\sqrt{2}}$$

$$L_3 = |u\rangle\langle u| - |w\rangle\langle w|. \tag{8.26}$$

---

[14]These are essentially the three components of the orbital angular momentum operator for angular momentum 1; see Chap. 16, Vol. 2.

(a) Are these Hermitian, unitary or projection operators?
(b) Calculate $[L_1, L_2]$.

7. Show that the time evolution

$$|\nu(t)\rangle = -\sin\vartheta\, |\nu_1\rangle\, e^{-i\omega_1 t} + \cos\vartheta\, |\nu_2\rangle\, e^{-i\omega_2 t} \qquad (8.27)$$

is unitary.
8. Determine explicitly $\langle \nu_e | \nu(t)\rangle$ in (8.8), and $\langle \nu_\mu | \nu(t)\rangle$.
9. Determine explicitly p_e in (8.9), and p_μ.
10. Prove (8.10); find an approximation for ΔE in the case of very small rest masses.
11. Given the state

$$|\psi(t)\rangle = \sum_n c_n\, |\varphi_n\rangle\, e^{-iE_n t/\hbar} \qquad (8.28)$$

with the initial condition $|\psi(0)\rangle$. $\{|\varphi_n\rangle\}$ is a CONS. How are the constants c_n related to the initial conditions?
12. Given two CONS $\{|\varphi_i\rangle\}$ and $\{|\psi_i\rangle\}$. A quantum system is in the superposition $|z\rangle = \sum_i d_i\, |\psi_i\rangle$.

(a) Calculate the probability of measuring the quantum system in the state $|\varphi_k\rangle$.
(b) Show that $\sum_k p_k = 1$.

13. Given the model system

$$i\hbar \frac{d}{dt} |\psi(t)\rangle = H\, |\psi(t)\rangle \quad \text{with } H = 1 + A\sigma_y;\ A > 0, \qquad (8.29)$$

where σ_y is the y-Pauli matrix:

(a) Determine the eigenvalues and eigenvectors of H;
(b) How does the general expression $|\psi(t)\rangle$ read for a time-dependent state?
(c) How is $|\psi(t)\rangle$ expressed for the initial state $|\psi(t=0)\rangle = \begin{pmatrix} 1 \\ 0 \end{pmatrix}$?
(d) Assume that we measure $|\psi(t)\rangle$ from part c. With which probability will we find the state $|\chi\rangle = \begin{pmatrix} 1 \\ 0 \end{pmatrix}$ (i.e. the initial state)?

Chapter 9
Expectation Values, Mean Values, and Measured Values

The probability concept is further expanded. In addition, we look at Hermitian operators in more detail. The time behavior of mean values leads to the notion of conserved quantities.

We continue the discussion started in Chap. 7 on the calculation of the mean value of measured quantities, and generalize the formalism so that it is also applicable to the continuous case. As in the algebraic approach, the formulations lead us again to Hermitian operators, which are of particular importance in quantum mechanics. Furthermore, we address *conserved quantities* and establish a connection to classical mechanics.

9.1 Mean Values and Expectation Values

9.1.1 Mean Values of Classical Measurements

In classical physics, it is assumed that there is a 'true' value of each physical quantity which can be measured. Measuring this quantity several times, e.g. the location x, one will in general obtain different values x_i, where each value occurs with the frequency of occurrence n_i. The cause of the discrepancies between different values are inadequacies in the measuring apparatus (apart from the possibly varying skill of the experimenter). For l different readings, the total number of measurements amounts to $N = \sum_{i=1}^{l} n_i$. The *mean value* $\langle x \rangle$ (average) is defined as[1]

$$\langle x \rangle = \frac{\sum_{i=1}^{l} n_i x_i}{\sum_{i=1}^{l} n_i} = \frac{\sum_{i=1}^{l} n_i x_i}{N} = \sum_{i=1}^{l} \tilde{n}_i x_i \text{ with } \sum_{i=1}^{l} \tilde{n}_i = 1, \qquad (9.1)$$

---

[1] Instead of $\langle x \rangle$, the notation $\bar{x}$ is also common.

© Springer Nature Switzerland AG 2018
J. Pade, *Quantum Mechanics for Pedestrians 1*, Undergraduate Lecture
Notes in Physics, https://doi.org/10.1007/978-3-030-00464-4_9

where the $\tilde{n}_i = n_i/N$ are the relative frequencies which in the limit $l \to \infty$ become the probabilities w_i (Law of Large Numbers). In this limit, the average becomes the *expected value* or 'true' value:

$$\langle x \rangle = \sum_i p_i x_i \text{ with } \sum_i p_i = 1. \tag{9.2}$$

This averaging concept is also applicable to sets of *continuous* data. We perform the familiar transition, known from school,[2] of going from a sum $\sum$ to an integral $\int$, and obtain with the probability density[3] $\rho(x)$

$$\langle x \rangle = \int \rho(x) x \, dx \text{ with } \int \rho(x) \, dx = 1. \tag{9.3}$$

We generalize the last equation to three dimensions:

$$\langle x \rangle = \int \rho(\mathbf{x}) x \, dV \text{ with } \int \rho(x) \, dV = 1. \tag{9.4}$$

9.1.2 Expectation Value of the Position in Quantum Mechanics

We wish to transfer these ideas to quantum mechanics. $\Psi(\mathbf{r}, t)$ is the solution of the time-dependent SEq. With the probability density

$$\rho = |\Psi(\mathbf{r}, t)|^2 \tag{9.5}$$

(recall that Ψ must be normalized), we can, as described above, determine the probability w of finding the quantum object in a spatial region G as $w(G) = \int_G \rho \, dV$. If we now ask for its mean position in x-direction, we can formulate in analogy to (9.4)

$$\langle x \rangle = \int \Psi^*(\mathbf{r}, t) \Psi(\mathbf{r}, t) x \, dV. \tag{9.6}$$

For a discussion of this situation, we imagine an ensemble of N identically-prepared quantum objects, all of which we launch at time $t = 0$ from $x = 0$ (one dimensional, moving to the right). After a time T, we find the ensemble member i at the position x_i. Then the mean value of x is given by $\langle x \rangle = \sum_i x_i$, and this value agrees increasingly

---

[2] See also the chapter 'Discrete and continuous' in Appendix T, Vol. 1.
[3] For the special choice $\rho(x) = \sum_i p_i \delta(x - x_i)$, we obtain the expression (9.2) from (9.3). For the delta function $\delta(x)$, see Appendix H. Vol. 1.

better with the value given by (9.6) as N increases. In the limit $N \to \infty$, (9.6) is obtained exactly.[4]

In three dimensions, it follows that

$$\langle \mathbf{r} \rangle = \int \Psi^* (\mathbf{r}, t) \, \Psi (\mathbf{r}, t) \, \mathbf{r} \, dV. \tag{9.7}$$

9.1.3   Expectation Value of the Momentum in Quantum Mechanics

We now examine the momentum of the quantum object (the following calculation is one dimensional; the three-dimensional case is given below). We assume that the expectation value $\langle p \rangle$ obeys the equation

$$\frac{d}{dt} \langle x \rangle = \frac{1}{m} \langle p \rangle . \tag{9.8}$$

This is an *assumption* at this point,[5] which has to prove itself in the following (i.e. above all, it must lead to self-consistent results). It follows that:

$$\langle p \rangle = m \frac{d}{dt} \langle x \rangle = m \frac{d}{dt} \int_{-\infty}^{\infty} \Psi^* \Psi x \, dx$$

$$= m \int_{-\infty}^{\infty} \left(\dot{\Psi}^* \Psi + \Psi^* \dot{\Psi} \right) x \, dx = m \int_{-\infty}^{\infty} \dot{\Psi}^* \Psi x \, dx + c.c. \tag{9.9}$$

c.c. means the complex conjugate of the preceding term. We replace the time derivatives by space derivatives by means of the SEq $i\hbar\dot{\Psi} = -\frac{\hbar^2}{2m} \Psi'' + V\Psi$. We obtain (note: $V \in \mathbb{R}$):

$$\langle p \rangle = \frac{\hbar}{2i} \int_{-\infty}^{\infty} \Psi^{*''} \Psi x \, dx + c.c. \tag{9.10}$$

Partial integration yields

$$\langle p \rangle = \frac{\hbar}{2i} \left[\left(\Psi^{*'} \Psi x \right)_{-\infty}^{\infty} - \int_{-\infty}^{\infty} \Psi^{*'} \left(\Psi' x + \Psi \right) dx \right] + c.c. \tag{9.11}$$

---

[4]In fact, we cannot measure a point-like position x_i, but instead only an interval Δx_i which contains the quantum object. However, with the idea that we can make the interval arbitrarily small, we can accept the above argument as a limiting case. More will be said on this issue in Chap. 12.
[5]$\langle \frac{d}{dt} x \rangle = \frac{1}{m} \langle p \rangle$ would be better.

Since the wavefunction vanishes rapidly enough at infinity,[6] the integrated part vanishes. What remains is:

$$\langle p \rangle = \left\{ -\frac{\hbar}{2i} \int_{-\infty}^{\infty} \Psi^{*\prime} \Psi' x \, dx - \frac{\hbar}{2i} \int_{-\infty}^{\infty} \Psi^{*\prime} \Psi \, dx \right\} + c.c. \tag{9.12}$$

The first term cancels with its $c.c.$ It follows that

$$\langle p \rangle = -\frac{\hbar}{2i} \int_{-\infty}^{\infty} \Psi^{*\prime} \Psi \, dx + c.c.$$

$$= -\frac{\hbar}{2i} \int_{-\infty}^{\infty} \Psi^{*\prime} \Psi \, dx + \frac{\hbar}{2i} \int_{-\infty}^{\infty} \Psi^* \Psi' \, dx. \tag{9.13}$$

A further integration by parts of $-\frac{\hbar}{2i} \int_{-\infty}^{\infty} \Psi^{*\prime} \Psi \, dx$ leads to

$$\langle p \rangle = \frac{\hbar}{2i} \int_{-\infty}^{\infty} \Psi^* \Psi' \, dx - \frac{\hbar}{2i} \left\{ \Psi^* \Psi \Big|_{-\infty}^{\infty} - \int_{-\infty}^{\infty} \Psi^* \Psi' \, dx \right\}. \tag{9.14}$$

Again, the integrated term $\Psi^* \Psi \big|_{-\infty}^{\infty}$ vanishes and we obtain the result

$$\langle p \rangle = \frac{\hbar}{i} \int_{-\infty}^{\infty} \Psi^* \Psi' \, dx, \tag{9.15}$$

or, if the other term in (9.13) is integrated by parts:

$$\langle p \rangle = -\frac{\hbar}{i} \int_{-\infty}^{\infty} \Psi^{*\prime} \Psi \, dx. \tag{9.16}$$

As one can easily see, these terms can be written by using the momentum operator. With $p = \frac{\hbar}{i} \frac{d}{dx}$, we obtain[7]

$$\langle p \rangle = \int_{-\infty}^{\infty} \Psi^* \left(\frac{\hbar}{i} \frac{d}{dx} \Psi \right) dx = \int_{-\infty}^{\infty} \left(\frac{\hbar}{i} \frac{d}{dx} \Psi \right)^* \Psi \, dx$$

$$= \int_{-\infty}^{\infty} \Psi^* (p\Psi) \, dx = \int_{-\infty}^{\infty} (p\Psi)^* \Psi \, dx. \tag{9.17}$$

Two comments:

1. The equality $\int_{-\infty}^{\infty} \Psi^* (p\Psi) \, dx = \int_{-\infty}^{\infty} (p\Psi)^* \Psi \, dx$ plays an important role in quantum mechanics in a slightly different notation: it applies not only to the

---

[6]Recall that this behavior is necessary for the interpretation of $|\Psi|^2$ as a probability density. See Chap. 7.

[7]As usual we use the same symbol, p, for the physical quantity 'momentum' and the corresponding operator. What is meant in each case should be clear from context.

momentum, but equally to *all* measurable physical quantities. We will return to this point later.

2. The equality holds only if the wavefunction vanishes sufficiently rapidly at infinity.

In the three-dimensional case, we find accordingly[8]:

$$\langle \mathbf{p} \rangle = \int \Psi^* \left(\frac{\hbar}{i} \nabla \Psi \right) dV = \int \left(\frac{\hbar}{i} \nabla \Psi \right)^* \Psi dV$$

$$= \int \Psi^* (\mathbf{p}\Psi) \, dV = \int (\mathbf{p}\Psi)^* \, \Psi dV. \tag{9.18}$$

Similarly, one can derive the expectation value of the energy. It follows that

$$\langle E \rangle = \langle H \rangle = \int \Psi^* H \Psi dV = \int (H\Psi)^* \Psi dV. \tag{9.19}$$

9.1.4 General Definition of the Expectation Value

We summarize the results obtained so far:

$$\langle \mathbf{r} \rangle = \int \Psi^* \mathbf{r} \Psi \, dV = \int (\mathbf{r} \Psi)^* \Psi \, dV$$

$$\langle \mathbf{p} \rangle = \int \Psi^* \mathbf{p} \Psi dV = \int (\mathbf{p}\Psi)^* \Psi dV$$

$$\langle H \rangle = \int \Psi^* H \Psi dV = \int (H\Psi)^* \Psi dV. \tag{9.20}$$

We generalize to an arbitrary operator A representing a measurable variable [9] and define its expectation value $\langle A \rangle$ by

$$\langle A \rangle = \int \Psi^* A \Psi dV. \tag{9.21}$$

We do not require equality with $\int (A\Psi)^* \Psi dV$, as we did in (9.20). Actually, this does not apply to any operator, but only to a certain class of operators (Hermitian

---

[8] As mentioned in Chap. 4, we use the shorthand notation of summation $\sum_n$ instead of $\sum_{n=1}^{\infty}$. Similarly, we also abbreviate integrals: $\int \Psi dV$ is *not* an indefinite, but a definite integral, which is carried out over the whole domain of definition of Ψ, namely $\int \Psi dV \equiv \int_{\text{domain of definition}} \Psi dV$. The range of integration is explicitly specified only in exceptional cases.

[9] An example is the angular momentum, $\mathbf{l} = \mathbf{r} \times \mathbf{p}$.

operators); we come back to this point in a moment. We note that we have a tool with (9.21)[10] to connect quite generally operators of quantum mechanics with measurable quantities.[11]

We first want to recover the expressions found in Chap. 7. For this purpose we start from the eigenvalue problem

$$H\varphi_n(x) = E_n\varphi_n(x); \quad \int \varphi_n^*(x)\,\varphi_m(x)\,\mathrm{d}x = \delta_{nm}. \tag{9.22}$$

The total state is

$$\Psi(x,t) = \sum_n c_n \varphi_n(x)\, e^{-iE_n t/\hbar}. \tag{9.23}$$

Then it follows that

$$
\begin{aligned}
\langle H \rangle &= \int \Psi^* H \Psi \mathrm{d}x = \int \sum_{n,m} c_n^* \varphi_n^*(x)\, e^{iE_n t/\hbar} H c_m \varphi_m(x)\, e^{-iE_m t/\hbar} \mathrm{d}x \\
&= \int \sum_{n,m} c_n^* \varphi_n^*(x)\, e^{iE_n t/\hbar} E_m c_m \varphi_m(x)\, e^{-iE_m t/\hbar} \mathrm{d}x \\
&= \sum_{n,m} c_n^* E_m c_m\, e^{i(E_n - E_m)t/\hbar} \int \varphi_n^*(x)\, \varphi_m(x)\, \mathrm{d}x \\
&= \sum_{n,m} c_n^* E_m c_m\, e^{i(E_n - E_m)t/\hbar} \delta_{nm} = \sum_n |c_n|^2\, E_n.
\end{aligned}
\tag{9.24}
$$

So we have found again the familiar expression for the expectation value. As mentioned above, the definition (9.21) has the advantage that it is readily applicable to continuous variables such as the position (see exercises).

Some remarks are in order:

1. The expectation value depends on the state. If necessary, one can include this by using e.g. the notation $\langle A \rangle_\Psi = \int \Psi^* A \Psi \mathrm{d}V$.
2. In general, the expectation value is time dependent, but this is often not explicitly stated. For pure energy states (proportional to $e^{-i\omega t}$), however, the time dependence cancels out when averaging over e.g. x. In such cases, the expectation values are independent of time (see exercises).
3. A remark just to clarify our concepts: Strictly speaking, the mean value refers to a data set from the past, i.e. to a previously performed measurement, and it is formulated in terms of relative frequencies of occurrence. By contrast, the expectation value, as a conjecture about the future, is formulated with probabilities and is the theoretically-predicted mean. However, in quantum mechanics the notions

---

[10] Another one we have already discussed above (e.g. in Chap. 4), namely that only the eigenvalues of operators can occur as measured values.

[11] One can show that this type of averaging must follow under very general conditions (Gleason's theorem, see Appendix T, Vol. 2).

of expectation value and mean value are often used interchangeably due to a certain nonchalance of the physicists, and for a finite number of measurements, the term probability w_i is applied instead of relative frequency (as indeed 'ensemble' is used also for a finite set of identically prepared systems). A brief example in Appendix O, Vol. 1, illustrates the difference between the mean value and the expectation value.

4. As stated at the beginning of this chapter, a repeated measurement of a *classical* quantity yields a different value each time. With continued repetition, the mean value of all these measured values shows an increasingly better agreement with the true value. If the measuring instruments were *ideal*, we would obtain the same value every time.

   In contrast, in quantum mechanics, successive measurements of an identical ensemble can in general give *different* values (corresponding to different eigenvalues of the measured physical quantity) even with an ideal measurement apparatus.[12] We have already mentioned that in a single experiment we can obtain only *one eigenvalue* of the operator which corresponds to the measured physical quantity. Which one of the eigenvalues this will be cannot be predicted before the experiment (if the state is given by a superposition). In other words: Quantum-mechanical variables generally have *no* 'true' value.

   When we speak of the expectation *value* of a physical quantity A, this therefore does not imply that A *has* necessarily this value in the sense that classical quantities have 'true' values. Instead of the expectation value, it would therefore be more cautious and unbiased to speak of the *expected measured value* or the like. However, such a terminology is not often used.

9.1.5 Variance, Standard Deviation

A convenient measure of the deviation from the mean value of a classical variable A (no doubt familiar from introductory laboratory courses) is the *mean square deviation* or *variance* $(\Delta A)^2$:

$$(\Delta A)^2 = \left\langle (A - \langle A \rangle)^2 \right\rangle = \left\langle A^2 \right\rangle - \langle A \rangle^2 . \tag{9.25}$$

To obtain the same physical unit as for A, one takes the root and obtains with ΔA the *standard deviation*, also called *dispersion* or *uncertainty*. A brief motivation for the form of this expression is found in Appendix O, Vol. 1.

   We adopt this concept also in quantum mechanics. Since the uncertainty ΔA depends in general on the state Ψ of the system, there is also the notation $\Delta_\Psi A$ or

---

[12]If we send e.g. a horizontally linearly-polarized photon through a linear analyzer, rotated by the angle φ, we obtain, as discussed in Chap. 2, different measurement results (horizontal or vertical polarization) with the probabilities $\cos^2 \varphi$ and $\sin^2 \varphi$. This is true *in principle* and is not due to shortcomings of the measurement apparatus.

the like. An example is given in the exercises, treating the uncertainty of position and momentum in the infinite potential well.

We repeat a note on the significance of the standard deviation in quantum mechanics (cf. remark 4, above): In classical physics, the standard deviation ΔA is a measure of the dispersion of the measured values which arises due to instrumental imperfections. In quantum mechanics, the meaning is quite different; ΔA is not due to instrumental errors, but is an unavoidable genuine quantum effect. Successive measurements can yield different values even for ideal measuring equipment.[13] If and only if $\Delta_\Psi A = 0$ is the quantum object in an eigenstate of the measured operator A and all members of an ensemble have the same value of the physical quantity A. An exercise illustrates that statement by considering the energy in the infinite potential well.

In this sense, the quantum-mechanical dispersion may be seen as a measure of the extent to which a system 'has' a value for A ($\Delta A = 0$) or 'has not' ($\Delta A > 0$). Thus, if in the infinite potential well (of width a), the energy eigenstate φ_n has the position uncertainty $\Delta x = \frac{a}{2}\sqrt{\frac{n^2\pi^2-6}{3n^2\pi^2}} \sim 0.3a$ (see the exercises), this does not mean that each (single) position measurement always has an error of this magnitude, but rather that the quantum object simply does not have a position in the classical sense. In other words, the concept of 'exact location' is not appropriate to this quantum-mechanical problem. More on this issue will be given in later chapters.

9.2 Hermitian Operators

An essential property of measurement results is that they are *real*. If the expectation values represent measurable quantities, they must also be real. Therefore, it must hold that

$$\langle A \rangle = \langle A \rangle^* \tag{9.26}$$

or

$$\int \Psi^* (A\Psi)\,dV = \int (A\Psi)^* \Psi dV. \tag{9.27}$$

All operators in the small table (9.20) share this property. One can furthermore show for these operators that for two arbitrary functions[14] Ψ_1 and Ψ_2, the equation

$$\int \Psi_1^* A\Psi_2 dV = \int (A\Psi_1)^* \Psi_2 dV \tag{9.28}$$

---

[13] In view of this, some other expression than 'standard deviation' would be more appropriate in quantum mechanics to indicate the spread of measurement results, but the mathematical simplicity of this expression has led to its widespread use.

[14] Arbitrary only insofar as the two functions have to satisfy the necessary technical requirements and the integrals have to exist. We note that the Hermiticity of operators may depend on the functions on which they act. This point is addressed explicitly in the exercises for this chapter.

holds. Generally, an operator A satisfying (9.28) is called *Hermitian*. We have already met up with Hermitian operators (and their representation as Hermitian matrices) in the algebraic approach, but these operators do not seem to have much to do with (9.28). But, contrary to that appearance, they in fact amount to the same thing, as we will see in more detail in Chap. 11 under the topic 'matrix mechanics'.

As just pointed out, the expectation values of Hermitian operators are real. This makes sense, since in quantum mechanics *all* measurable quantities are represented by Hermitian operators. In addition, Hermitian operators in general have other very practical features: they have only *real eigenvalues* (which represent the possible individual measured values), and, in the case of a nondegenerate spectrum, their *eigenfunctions* are *pairwise orthogonal* to each other (as we have already seen in the example of the infinite potential well). We now want to prove these two properties.

9.2.1 Hermitian Operators Have Real Eigenvalues

We consider the eigenvalue equation

$$Af_n = a_n f_n; \ n = 1, 2, \ldots \tag{9.29}$$

where the operator A is Hermitian:

$$\int f_m^* A f_n \mathrm{d}V = \int (Af_m)^* f_n \mathrm{d}V. \tag{9.30}$$

We want to show now that its eigenvalues are real, i.e. $a_n = a_n^*$.

To this end, we write the two equations:

$$Af_n = a_n f_n; \ (Af_n)^* = a_n^* f_n^*. \tag{9.31}$$

We multiply the left equation by f_n^* and the right one by f_n:

$$f_n^* A f_n = f_n^* a_n f_n; \ (Af_n)^* f_n = a_n^* f_n^* f_n. \tag{9.32}$$

Integration over all space yields

$$\int f_n^* A f_n \mathrm{d}V = a_n \int f_n^* f_n \mathrm{d}V; \ \int (Af_n)^* f_n \mathrm{d}V = a_n^* \int f_n^* f_n \mathrm{d}V. \tag{9.33}$$

Because of the Hermitian property of A, the two left-hand sides of these equations are the same. Therefore also the right sides have to be equal:

$$a_n \int f_n^* f_n \mathrm{d}V = a_n^* \int f_n^* f_n \mathrm{d}V \leftrightarrow (a_n - a_n^*) \int f_n^* f_n \mathrm{d}V = 0. \tag{9.34}$$

Due to $\int f_n^* f_n \mathrm{d}V = 1 \neq 0$, it follows that

$$a_n = a_n^* \leftrightarrow a_n \in \mathbb{R}. \tag{9.35}$$

We see that the eigenvalues of a Hermitian operator are real. This also holds, as said above, for the expectation values. We note again that the result of measuring a physical quantity can only be one of the eigenvalues of the corresponding operator.

9.2.2 Eigenfunctions of Different Eigenvalues Are Orthogonal

Given a Hermitian operator A and the eigenvalue equation

$$A f_n = a_n f_n; \tag{9.36}$$

the spectrum is assumed to be nondegenerate. Then we have

$$\int f_m^* f_n \mathrm{d}V = 0 \text{ for } n \neq m. \tag{9.37}$$

In order to show this, we begin with

$$A f_n = a_n f_n; \quad (A f_m)^* = a_m f_m^*. \tag{9.38}$$

a_m is real, as we have just shown. We extend the equations

$$f_m^* A f_n = a_n f_m^* f_n; \quad (A f_m)^* f_n = a_m f_m^* f_n \tag{9.39}$$

and integrate:

$$\int f_m^* A f_n \mathrm{d}V = a_n \int f_m^* f_n \mathrm{d}V; \quad \int (A f_m)^* f_n \mathrm{d}V = a_m \int f_m^* f_n \mathrm{d}V. \tag{9.40}$$

Since A is Hermitian, the left sides are equal. It follows that:

$$(a_n - a_m) \int f_m^* f_n \mathrm{d}V = 0. \tag{9.41}$$

Since $n \neq m$ (and because there is no degeneracy), we have $a_n \neq a_m$. Then we conclude:

$$\int f_m^* f_n \mathrm{d}V = 0 \text{ for } n \neq m. \tag{9.42}$$

We can generalize this equation by including the case $m = n$. Since we always normalize the eigenfunctions, the result reads:

$$\int f_m^* f_n \mathrm{d}V = \delta_{nm}. \tag{9.43}$$

In other words, Hermitian operators always have real eigenvalues and their eigenfunctions constitute an ON system.

9.3 Time Behavior, Conserved Quantities

Examination of the time behavior of expectation values leads to the concept of *conserved quantities*. In addition, we can establish a connection to classical mechanics.

9.3.1 Time Behavior of Expectation Values

Since the wavefunction depends upon time, the expectation value of a physical quantity

$$\langle A \rangle = \int \Psi \, (\mathbf{r}, t)^* \, A \Psi (\mathbf{r}, t) \mathrm{d}V \tag{9.44}$$

will in general also be time dependent.

We consider the first time derivative of $\langle A \rangle$ and express the derivatives of the wavefunction using the SEq $i\hbar \dot{\Psi} = H\Psi$, while assuming that the potential V in $H = -\frac{\hbar^2}{2m}\nabla^2 + V$ is real. It follows that:

$$
\begin{aligned}
i\hbar \frac{\mathrm{d}}{\mathrm{d}t} \langle A \rangle &= i\hbar \int \dot{\Psi}^* A \Psi \mathrm{d}V + i\hbar \int \Psi^* \dot{A} \Psi \mathrm{d}V + i\hbar \int \Psi^* A \dot{\Psi} \mathrm{d}V \\
&= -\int (H\Psi)^* A \Psi \mathrm{d}V + i\hbar \int \Psi^* \dot{A} \Psi \mathrm{d}V + \int \Psi^* A H \Psi \mathrm{d}V \\
&\underset{H \text{ Hermitian}}{=} -\int \Psi^* H A \Psi \mathrm{d}V + i\hbar \int \Psi^* \dot{A} \Psi \mathrm{d}V + \int \Psi^* A H \Psi \mathrm{d}V \\
&= \int \Psi^* (AH - HA) \Psi \mathrm{d}V + i\hbar \left\langle \frac{\partial}{\partial t} A \right\rangle.
\end{aligned}
\tag{9.45}
$$

Here, we have used the Hermiticity of the Hamiltonian:

$$
\begin{aligned}
\int \Psi_1^* H \Psi_2 \mathrm{d}V = \int (H\Psi_1)^* \Psi_2 \mathrm{d}V \quad \text{or} \\
\int \Psi^* H A \Psi \mathrm{d}V = \int (H\Psi)^* A \Psi \mathrm{d}V.
\end{aligned}
\tag{9.46}
$$

$(AH - HA)$ is evidently the commutator of A and H. Then

$$i\hbar\frac{\mathrm{d}}{\mathrm{d}t}\langle A\rangle = \int \Psi^*[A, H]\Psi\mathrm{d}V + i\hbar\left\langle\frac{\partial}{\partial t}A\right\rangle, \qquad (9.47)$$

or in compact form,

$$i\hbar\frac{\mathrm{d}}{\mathrm{d}t}\langle A\rangle = \langle[A, H]\rangle + i\hbar\left\langle\frac{\partial}{\partial t}A\right\rangle. \qquad (9.48)$$

Practically all of the operators which we consider below do not depend explicitly on time.[15] In that case, $\frac{\partial}{\partial t}A$ is zero, and it thus follows that:

$$i\hbar\frac{\mathrm{d}}{\mathrm{d}t}\langle A\rangle = \langle[A, H]\rangle, \text{ if } A \text{ is not explicitly time dependent.} \qquad (9.49)$$

Although we will deal hereafter only with time-independent Hamiltonians, we nevertheless remark that the reasoning leading to (9.48) and (9.49) applies to both time-dependent and time-independent Hamiltonians. The key feature is the Hermiticity of H.

9.3.2 Conserved Quantities

Let us assume that we have an operator A which (i) is not explicitly time-dependent, i.e. $\frac{\partial A}{\partial t} = 0$, and which (ii) commutes with H, i.e. $[A, H] = 0$. Then it follows from (9.49) that:

$$i\hbar\frac{\mathrm{d}}{\mathrm{d}t}\langle A\rangle = 0 \text{, if } \frac{\partial}{\partial t}A = 0 \text{ and } [A, H] = 0. \qquad (9.50)$$

In other words, the expectation value $\langle A\rangle$ (and the associated physical quantity) remains constant over time, in which case one speaks of a *conserved quantity* or a *constant of the motion*.[16] As is well known, conserved quantities play a special role in physics: They (or the underlying symmetries) allow for a simpler description of a system.[17] For time-independent operators, the statements 'A commutes with H' and 'A is a conserved quantity' are equivalent. Thus, we have an effective instrument at our disposal for determining whether or not a given operator represents a conserved quantity.

---

[15]Examples are the momentum operator $\mathbf{p} = \frac{\hbar}{i}\nabla$ or the Hamiltonian, if the potential is time-independent.

[16]Or also of a 'good quantum number', if required.

[17]We will take a closer look at this question in Chap. 21, Vol. 2, 'Symmetries'.

9.3.3 Ehrenfest's Theorem

The question of whether position and momentum are conserved quantities leads to a connection with classical mechanics. It also provides a retrospective confirmation (in the sense of a self-consistent approach) of (9.8).

The physical problem is three-dimensional. We begin with the x component of the momentum. With $H = H_0 + V = \frac{p^2}{2m} + V$, we have

$$[p_x, H] = p_x H - H p_x = p_x H_0 + p_x V - H_0 p_x - V p_x$$
$$= \frac{\hbar}{i} \frac{\partial}{\partial x} V - V \frac{\hbar}{i} \frac{\partial}{\partial x} = \frac{\hbar}{i} \frac{\partial V}{\partial x} + \frac{\hbar}{i} V \frac{\partial}{\partial x} - \frac{\hbar}{i} V \frac{\partial}{\partial x} = \frac{\hbar}{i} \frac{\partial V}{\partial x}. \qquad (9.51)$$

For the time behavior of p_x, it follows that:

$$\frac{d}{dt} \langle p_x \rangle = - \left\langle \frac{\partial V}{\partial x} \right\rangle \text{ or } \frac{d}{dt} \langle \mathbf{p} \rangle = - \langle \nabla V \rangle. \qquad (9.52)$$

Next, we consider the position x. We have:

$$[x, H] = x H - H x = x H_0 + x V - H_0 x - V x$$
$$= x \frac{p^2}{2m} - \frac{p^2}{2m} x = x \frac{p_x^2}{2m} - \frac{p_x^2}{2m} x = -\frac{\hbar^2}{2m} \left(x \frac{\partial^2}{\partial x^2} - \frac{\partial^2}{\partial x^2} x \right) \qquad (9.53)$$

and it follows that

$$[x, H] = \frac{\hbar^2}{m} \frac{\partial}{\partial x} \text{ or } [\mathbf{r}, H] = \frac{\hbar^2}{m} \nabla. \qquad (9.54)$$

With $\mathbf{p} = \frac{\hbar}{i} \nabla$, this yields for the time behavior:

$$\frac{d}{dt} \langle \mathbf{r} \rangle = \frac{1}{m} \langle \mathbf{p} \rangle. \qquad (9.55)$$

Hence, we have recovered our starting point, (9.8), which means that we have obtained a confirmation of our *ansatz* in the sense of a self-consistent description.

We summarize the results of this section. For the expectation values of position and momentum, we have[18]:

$$\frac{d}{dt} \langle \mathbf{r} \rangle = \frac{1}{m} \langle \mathbf{p} \rangle \text{ and } \frac{d}{dt} \langle \mathbf{p} \rangle = - \langle \nabla V \rangle. \qquad (9.56)$$

The form of the equations is reminiscent of the classical Hamilton equations for a particle,

---

[18]So we find $m \frac{d^2}{dt^2} \langle \mathbf{r} \rangle = - \langle \nabla V \rangle = \langle \mathbf{F}(\mathbf{r}) \rangle$. In principle, one must still show that $\langle \mathbf{F}(\mathbf{r}) \rangle = \mathbf{F}(\langle \mathbf{r} \rangle)$ (or one defines the force accordingly).

$$\frac{d}{dt}\mathbf{r} = \frac{1}{m}\mathbf{p} \text{ and } \frac{d}{dt}\mathbf{p} = -\nabla V, \tag{9.57}$$

which can be written in this simple case as a Newtonian equation of motion:

$$\frac{d\mathbf{p}}{dt} = m\frac{d^2\mathbf{r}}{dt^2} = -\nabla V = \mathbf{F}. \tag{9.58}$$

In short: The quantum-mechanical expectation values obey the corresponding classical equations. This (and therefore also the (9.56)) is called Ehrenfest's theorem.[19]

9.4 Exercises

1. Given a Hermitian operator A and the eigenvalue problem $A\varphi_n = a_n\varphi_n$, $n = 1, 2, \ldots$, show that:

   (a) The eigenvalues are real.
   (b) The eigenfunctions are pairwise orthogonal. Here, it is assumed that the eigenvalues are nondegenerate.

2. Show that the expectation value of a Hermitian operator is real.
3. Show that

$$\int \Psi_1^* A\Psi_2 dV = \int (A\Psi_1)^* \Psi_2 dV \tag{9.59}$$

   holds for the operators $\mathbf{r}$, $\mathbf{p}$, H. Restrict the discussion to the one-dimensional case. What conditions must the wavefunctions satisfy?
4. Show that for the infinite potential well (between 0 and a), $\langle x \rangle = \frac{a}{2}$.
5. Given the infinite potential well with walls at $x = 0$ and $x = a$; we consider the state

$$\Psi(x, t) = \sqrt{\frac{2}{a}} \sin\left(\frac{n\pi}{a}x\right) e^{-i\omega_n t}. \tag{9.60}$$

   (a) Determine the position uncertainty Δx.
   (b) Determine the momentum uncertainty Δp.

6. In the infinite potential well, a normalized state is given by

$$\Psi(x, t) = c_n\varphi_n(x)e^{-i\omega_n t} + c_m\varphi_m(x)e^{-i\omega_m t}; \quad c_n, c_m \in \mathbb{C}; \quad n \neq m. \tag{9.61}$$

   Calculate $\langle x \rangle$.

---

[19]We note that also the general law (9.48) for the time dependence of mean values is sometimes called Ehrenfest's theorem.

7. Consider an infinite square well with potential limits at $x = 0$ and $x = a$. The initial value of the wavefunction is $\Psi (x, 0) = \Phi \in \mathbb{R}$ for $b - \varepsilon \leq x \leq b + \varepsilon$ and $\Psi (x, 0) = 0$ otherwise (of course, $0 \leq b - \varepsilon$ and $b + \varepsilon \leq a$). Remember that the eigenfunctions $\varphi_n (x) = \sqrt{\frac{2}{a}} \sin k_n x$ with $k_n = \frac{n\pi}{a}$ form a CONS.

(a) Normalize the initial state.
(b) Calculate $\Psi (x, t)$.
(c) Find the probability of measuring the system in the state n.

8. Show that for the expectation value of a physical quantity A,

$$i\hbar \frac{d}{dt} \langle A \rangle = \langle [A, H] \rangle + i\hbar \left\langle \frac{\partial}{\partial t} A \right\rangle \qquad (9.62)$$

holds. Show that for time-independent operators, the expectation value of the corresponding physical quantity is conserved, if A commutes with H.

9. Show that

$$\frac{d}{dt} \langle \mathbf{r} \rangle = \frac{1}{m} \langle \mathbf{p} \rangle \text{ and } \frac{d}{dt} \langle \mathbf{p} \rangle = -\langle \nabla V \rangle. \qquad (9.63)$$

10. Under which conditions is the orbital angular momentum $\mathbf{l} = \mathbf{r} \times \mathbf{p}$ a conserved quantity?

11. Given the Hamiltonian H with a discrete and nondegenerate spectrum E_n and eigenstates $\varphi_n (\mathbf{r})$, show that the energy uncertainty ΔH vanishes, iff the quantum object is in an eigenstate of the energy.

Chapter 10
Stopover; Then on to Quantum Cryptography

We first compare formulations of the analytic and algebraic approaches to quantum mechanics. In the second section, we see that the properties of the measurement process in quantum mechanics permit an encryption method which is in principle absolutely secure.

10.1 Outline

This chapter is exceptional insofar as the formalism is not developed further. Rather, it serves to collect our previously acquired knowledge, to compare and to check where there are open questions of form or content. In the second part of the chapter, we take up quantum cryptography. We will see that even allegedly abstract or theoretical peculiarities of quantum mechanics, such as those of the measurement process, can have immediate practical applications.

10.2 Summary and Open Questions

First, we collect the essential concepts and structures of quantum mechanics which we have developed up to now in the preceding chapters. Comparison of the analytical and algebraic approaches (ana and ala) shows, on the one hand, that there are many significant parallels, but also that a few components are missing in each case. To keep the text readable, we dispense with detailed remarks about which chapter introduced or treated the particular subject matter.

© Springer Nature Switzerland AG 2018
J. Pade, *Quantum Mechanics for Pedestrians 1*, Undergraduate Lecture
Notes in Physics, https://doi.org/10.1007/978-3-030-00464-4_10

10.2.1 Summary

10.2.1.1 States

We started with *states*, which we wrote in the analytical approach in terms of a position- and time-dependent wavefunction $\psi\,(\mathbf{r}, t)$, and in the algebraic approach as a ket $|\psi\,(t)\rangle$ or its representation as a column vector. It should be pointed out again that $|\psi\,(t)\rangle$ does *not* depend on position, but only on time. The states are in both cases elements of vector spaces.

10.2.1.2 Time-Dependent SEq, Hamiltonian

The time behavior of both approaches is described by the time-dependent SEq. It reads

$$i\hbar\frac{\partial}{\partial t}\psi\,(\mathbf{r}, t) = H\psi\,(\mathbf{r}, t)$$
$$i\hbar\frac{\mathrm{d}}{\mathrm{d}t}\,|\psi\,(t)\rangle = H\,|\psi\,(t)\rangle. \tag{10.1}$$

In the analytical approach, H is the Hamiltonian $-\frac{\hbar^2}{2m}\nabla^2 + V$, while in the algebraic approach it is an abstract operator, about which we know almost nothing so far— except that it can be represented as a matrix.[1] In any case, H is a Hermitian operator, where this property is defined in the ana by integrals, in the ala by scalar products:

$$\int \psi^* H\varphi\mathrm{d}V = \int (H\psi)^* \,\varphi\mathrm{d}V$$
$$\langle\psi|\,H\,|\varphi\rangle = \langle\psi|\,H^\dagger\,|\varphi\rangle. \tag{10.2}$$

10.2.1.3 Mean Value and Expectation Value

If the system is in the state $\psi\,(\mathbf{r}, t)$, we can obtain the expectation value $\langle A \rangle$ of an operator A (that is, the expectation value of the corresponding physical quantity) as

$$\langle A \rangle = \int \psi^* H\psi\mathrm{d}V. \tag{10.3}$$

A corresponding formulation in the algebraic approach is still pending.

We note that the expectation value of a time-independent operator A in general depends on time because of $\psi = \psi\,(\mathbf{r}, t)$. It is a conserved quantity (i.e. is independent of time) if the operator A commutes with H, i.e. $[A, H] = 0$.

---

[1] A comment on the notation: although the Hamiltonians of the two approaches in (10.1) are completely different mathematical objects, it is customary to denote them with the same symbol H. The same holds for the eigenfunctions and vectors.

10.2.1.4 Time-Independent SEq, Eigenvalues and Eigenvectors

The time evolution of the states can be developed by means of the eigenvalues and eigenvectors of H. We assume discrete, nondegenerate spectra with eigenvalues E_n. We denote the analytical eigenvectors (eigenfunctions) by $\varphi_n (\mathbf{r})$, the algebraic ones by $|\varphi_n\rangle$. They are the solutions of the eigenvalue problems (stationary SEq):

$$H\varphi_n (\mathbf{r}) = E_n \varphi_n (\mathbf{r})$$
$$H |\varphi_n\rangle = E_n |\varphi_n\rangle .$$

(10.4)

We note that the range of values of n can be finite or infinite.

Since H is a Hermitian operator in both of these approaches, its eigenfunctions $\{\varphi_n (\mathbf{r})\}$ or $\{|\varphi_n\rangle\}$ form an orthonormal system (ONS):

$$\int \varphi_m^* (\mathbf{r}) \, \varphi_n (\mathbf{r}) \, dV = \delta_{nm}$$
$$\langle \varphi_m | \varphi_n \rangle = \delta_{nm}.$$

(10.5)

In the ala, we described the completeness of an ONS by $\sum_n |\varphi_n\rangle \langle \varphi_n| = 1$. An analogous formulation in the ana is still pending.

10.2.1.5 Time-Dependent Solution

Using eigenvalues and eigenvectors, the time-dependent solution of the SEq can be written as

$$\psi (\mathbf{r}, t) = \sum_n c_n \varphi_n (\mathbf{r}) \, e^{-i E_n t/\hbar}$$
$$|\psi (t)\rangle = \sum_n c_n |\varphi_n\rangle \, e^{-i E_n t/\hbar}.$$

(10.6)

Being solutions of the deterministic SEq (10.1), these states are defined uniquely and for all times by specifying an initial condition $\psi (\mathbf{r}, 0)$ or $|\psi (0)\rangle$. We can see this by using (10.5) to obtain

$$\int \varphi_m^* (\mathbf{r}) \, \psi (\mathbf{r}, 0) \, dV = \sum_n c_n \int \varphi_m^* (\mathbf{r}) \, \varphi_n (\mathbf{r}) \, dV = \sum_n \delta_{nm} c_n = c_m$$
$$\langle \varphi_m | \psi (0)\rangle = \sum_n c_n \langle \varphi_m | \varphi_n \rangle = \sum_n \delta_{nm} c_n = c_m.$$

(10.7)

or, more compactly,

$$c_n = \int \varphi_n^* (\mathbf{r}) \, \psi (\mathbf{r}, 0) \, dV$$
$$c_n = \langle \varphi_n | \psi (0)\rangle .$$

(10.8)

Up to this point, the formalisms developed in the ana and ala are very similar, in spite of some differences (definition of a Hermitian operator, state, SEq as a differential equation or as a matrix equation). We conclude that there obviously *must* be a close connection. For example, the (10.8) suggest that the integral $\int \varphi_n^* (\mathbf{r}) \, \psi (\mathbf{r}, 0) \, dV$ of the ana corresponds to a scalar product in the ala. We take up this issue again in the next chapter.

10.2.1.6   Measurement, Probability

The formalism of quantum mechanics just outlined is strictly deterministic. A random element occurs only if we want to obtain information about the system by means of a measurement. We have seen in previous chapters that the coefficients of the form $|c_n|^2$ which appear in (10.6) give the probabilities of finding the system in the state $\varphi_n (\mathbf{r})$ or $|\varphi_n\rangle$. With quantized values (such as the energy or the state of a neutrino, i.e. muon neutrino or electron neutrino, etc.), one can always measure only *one* of the values of the spectrum. Other results are not possible.

In the ala, we formulated the measurement process by using projection operators. If we want to measure e.g. the state $|\chi\rangle$, we model this by applying the projection operator $P_\chi = |\chi\rangle \langle\chi|$ to the state $|\psi\rangle$:

$$|\chi\rangle \langle\chi |\psi\rangle = c |\chi\rangle . \tag{10.9}$$

Here, the term $|c|^2 = |\langle\chi |\psi\rangle|^2$ denotes the probability of in fact obtaining the state $|\chi\rangle$ by a measurement on $|\psi\rangle$. In the ana, we have not yet introduced projection operators. The parallelism of the descriptions in the ana and ala suggests, however, that there must be an equivalent in the ana.

10.2.1.7   Measurement, Collapse

Through the measurement, the system is transferred from the state $|\psi\rangle$ into $|\chi\rangle$ (provided $c = \langle\chi |\psi\rangle \neq 0$). In the formulation of the ala, this can be written as

$$|\psi\rangle \underset{\text{with probability } |c|^2}{\longrightarrow} \frac{P_\chi |\psi\rangle}{\left| P_\chi |\psi\rangle \right|} = \frac{\langle\chi |\psi\rangle}{|\langle\chi |\psi\rangle|} |\chi\rangle. \tag{10.10}$$

On measurement, a superposition of states generally breaks down[2] and the result is *one* single state. We have described this behavior in terms of state reduction or collapse of the state. After the measurement, we again have a normalized state, where a possibly remaining global phase is irrelevant,[3] since states are physically the same if they differ only by a phase (we will discuss this point later, in Chap. 14). The state

---

[2] In other words, if the initial and final states are not the same.
[3] Quantum mechanics is very well-behaved in this sense.

after the measurement can be interpreted as a new initial state[4] (at time T, we start our clock again), which evolves in a unitary manner until the next measurement.

We remark again that the actual measurement process itself is not modelled, but rather only the situation before and after the measurement.

10.2.2 Open Questions

The descriptions in the ana and ala outlined above leave open some questions which we now summarize briefly. These questions are in part of a more formal nature, and in part concern content (although this division is not necessarily clearcut). The answers to the open questions will be provided in the following chapters.

10.2.2.1 Formal Questions

As mentioned above, the great similarity of the expressions (10.1) and (10.8) suggests that there is a direct connection between the two approaches and the corresponding formulations. So it must be clarified, for example, which relationship exists between the description of states as kets and as wavefunctions. As a result, we will find among other things a representation of the projection operator in the ana, thus far defined only in the ala. In addition, this connection must explain the different formulations, as in (10.8); this is also true for the definitions of the Hamiltonians in the two approaches (as $-\frac{\hbar^2}{2m}\nabla^2 + V$ and as a matrix) which at first glance seem quite distinct.

Another topic still to be treated is that of degenerate as well as continuous spectra. This will be done in Chap. 12.

10.2.2.2 Questions of Content

A measurement, as described in (10.10), is generally (i.e. for $|c|^2 \neq 1$) not reversible, so it is not a unitary process. Assuming the validity of the projection principle for determining the measurement probabilities, we must explain how this state reduction comes about, i.e. the transition from a superposition such as $|\nu(t)\rangle$ to a single state such as $|\nu_e\rangle$. Meanwhile, it is accepted that this collapse of the state is a non-local, i.e. superluminal effect.[5] Some of following considerations answer part of the open questions, but another part is still poorly understood and still under discussion. We will come back to these topics in several chapters in volume 2.

---

[4] In this case one speaks of 'state preparation'.

[5] This makes it perhaps understandable that Einstein dismissed it as 'spooky action at a distance'. It can be shown that the effect is not suitable for the superluminal transmission of information—the validity of the theory of relativity thus remains unquestioned.

To avoid misunderstandings: Here, we have a problem at the level of the *interpretation* of quantum mechanics; that is, of its *comprehension*. On a formal level—technically, so to speak—quantum mechanics works extremely well; it is simply *fapp* (after a proverbial expression due to John S. Bell: 'Ordinary quantum mechanics is just *f*ine for *a*ll *p*ractical *p*urposes').[6]

We will resume the discussion of the issues of content in Chap. 14. In the rest of this chapter, we will examine a practical application of quantum mechanics—to some extent a case of *fapp*.

10.3 Quantum Cryptography

There are some popular misconceptions about quantum mechanics. The 'quantum jump' is symptomatic—what in quantum mechanics means the 'smallest possible change' has become in everyday language a metaphor for a giant leap, a radical change.[7]

Two other misconceptions are that quantum mechanics always requires an enormous mathematical apparatus,[8] and that the abstract peculiarities of quantum mechanics such as the measurement problem are at most of theoretical interest. That both assertions are wrong is shown by *quantum cryptography*.[9] In fact, it is based on a peculiarity of the quantum-mechanical measurement process and can, in its simplest formulation, be described *without any formula* at all,[10] as we will see shortly. Of course, one can describe the whole situation more formally, but here we have one of the admittedly very few examples where this is not necessarily required.

The procedure is based on the quantum-mechanical principles that (i) there are superpositions of several states, and that (ii) before a measurement of such a superposition, we can specify only the probability of obtaining one of these states as a result. These principles are what make quantum cryptography possible, not only in theory, but also as a practical method.

---

[6]In an extension of Bell's one-liner, those theories which, on the one hand, one cannot really (or does not want to) justify, but which, on the other hand, agree well with experimental results and are very useful for all practical purposes, are called *fapp* theories. Quantum mechanics may be such a theory, if one regards it only as a tool (or judges it primarily by its usefulness) and is not willing (or able) to reflect upon its meaning.

[7]However, the movie title 'Quantum of Solace' promises not a 'quantum jump', but rather a minimum in terms of comfort for James Bond—quantum solace, so to speak.

[8]We have already seen that this is not always true, e.g. in the algebraic approach, where the basic ideas can be formulated using simple vector algebra.

[9]This term is short and to the point, but also a bit misleading. As we shall see shortly, quantum mechanics does not help to encrypt a message, but rather ensures that the key cannot be discovered by a spy.

[10]For this reason, the topic is also very well suited for discussion at the school level.

10.3.1   Introduction

Cipher texts and encryptions were already common in pre-Christian cultures. One of the most famous old encryption methods is attributed to Caesar, and is still called the *Caesar cipher*. Here, the text is encoded by replacing each letter with for example the third letter which follows it in the alphabet. Thus, 'cold' becomes 'frog' and 'bade' becomes 'edgh'.

Of course, nowadays it is a no-brainer to crack this ciphering method—it suffices that one knows very precisely for each language the frequency of occurrence of each letter. Modern cryptography has developed much more elaborate processes. It enjoyed an enormous boom in both world wars, where it also provided a strong impetus to the development of electronic computers. One of the first, called *Colossus*, was built at the end of the Second World War and was used for decoding purposes.

A word on nomenclature: One encodes, encrypts or ciphers an unencrypted or plain text by means of a cipher or key. The result is a encrypted text or cipher text. If it is decoded, decrypted or deciphered, one again recovers the plain text.

10.3.2   One-Time Pad

This encryption method was developed in 1917. Gilbert Vernam is usually named as its author. In 1949, Claude Shannon proved the *absolute security* of the method. In this process, it is known how to encrypt and decrypt. Its security is based exclusively on the fact that the *key is secret* (and only if this is guaranteed is the process absolutely secure).

The method works as follows: First, the alphabet (and some major punctuation marks, etc.) is converted into numbers. As an example, we might have:

A	B	C	D	E	...	X	Y	Z	,	.	?
00	01	02	03	04	...	23	24	25	26	27	28	29

as our pool of 30 characters. If the message consists of N characters, then the key must also consist of N characters. They are pulled out of the pool *at random*. Compared to 'normal' text, this has the advantage that each character occurs on average with equal frequency. Thus, even if pieces of the key are known, it cannot be reconstructed.

As a concrete example, we choose the key 06/29/01/27/ .... Encrypting the message 'BADE' leads to:

B	A	D	E	Message, plain text T
01	00	03	04	Plain text T, numbers
06	29	01	27	Key S
07	29	04	01	$V = (T + S) \pmod{30}$, numbers
H	?	E	B	Cipher text V

and decrypting leads to

H	?	E	B	Cipher text V
07	29	04	01	Ciphertext V, numbers
06	29	01	27	Key S
01	00	03	04	$T = (V - S) \pmod{30}$, numbers
B	A	D	E	Message, plain text T

Some remarks on practical procedures:

- The cipher text V is transmitted *publicly*. The security depends entirely on the fact that the key is known only to the sender and the recipient.
- The procedure is absolutely safe if each key is used only *once*. Hence the name 'pad'—one can imagine the sender and receiver each having an identical (writing) pad, and there is just one key on each sheet. After encrypting and decrypting the key is obsolete; the top sheet of the pad is stripped off and thrown away. The next page on the pad contains the next key.
- In binary notation, the method is basically the same, but more adapted to computers. This could be as follows $(1 + 1 = 0)$:

Text T	0	1	1	0	1	0	0	1	0	1	1	1
Key S	1	0	1	0	0	1	0	0	0	1	1	0
T + S	1	1	0	0	1	1	0	1	0	0	0	1
S	1	0	1	0	0	1	0	0	0	1	1	0
$\Rightarrow$ T = T + S $\pm$ S	0	1	1	0	1	0	0	1	0	1	1	1

- Especially in the English literature, certain names have become firmly entrenched. The sender is called 'Alice', the recipient 'Bob'. We will consider in addition a third person, namely a spy. For the name of the spy, one could think that it should now begin with 'C', e.g. 'Charlotte' (and in French texts, one does indeed find this name); but the English word 'eavesdropping' suggests immediately the name 'Eve', and that is how the spy is usually named—not alphabetically, but gender-specifically correct.
- The one-time pad method is thus based on a *public* exchange of the encrypted message, while the key is transmitted *secretly*. The problem of safe and secret transfer of keys between Alice and Bob is called *key distribution*. The great difficulty here is: How can we be sure that Eve has not read the key in secrecy, without leaving traces on the paper or on the CD, or has photographed it? There is a kind of

mnemonic in cryptography which describes this classical dilemma ironically: 'You can communicate completely secretly, provided you can communicate completely secretly.'

Here, quantum mechanics enters the scene, and brings with it several methods. All have in common that they associate the key distribution to quantum-mechanical characteristics and thus secure it. This is called *quantum key distribution*. A particularly simple method is the so-called BB84 protocol (Bennett and Brassard, 1984).[11] It is essentially based on the idea of a Ph.D. student in the sixties. Stephen Wiesner at that time devised a method of making counterfeit-proof banknotes using polarized photons (so to speak 'quantum money'). Although the practical implementation of this idea is not readily possible even today, in retrospect it is not really understandable why Wiesner's attempts to publish this idea around 1970 were rejected rigorously by the journal reviewers. Wiesner had to wait more than ten years before he could describe his proposal in the literature.[12] In any case—at least Charles Bennett, a friend of Wiesner's, recognized the cryptographic potential of his idea and developed, together with Brassard, the *BB84 protocol*.

10.3.3 BB84 Protocol Without Eve

In the following, the information is transmitted by polarized photons, where we will consider only linear polarization. As usual, we denote the horizontally- and vertically-polarized states by $|h\rangle$ and $|v\rangle$.

As we said above, the secure and confidential transmission of the key is all-important. Alice could now send a key by forwarding to Bob a random sequence of $|h\rangle$ and $|v\rangle$. However, she must tell Bob the orientation of the polarizer (e.g. by phone), and when Eve overhears this communication, she could listen safely without Alice or Bob being aware of her. To increase security, we must use quantum mechanics; more precisely, projection and the superposition principle.

And this is how it works: Alice chooses randomly one of two polarization directions: horizontal/vertical or diagonally left/right, symbolized by ⊞ and ⊠, where the crosses in the squares mark the polarization planes.[13] We can represent the states as $|h\rangle$ and $|v\rangle$ plus $|\backslash\rangle$ and $|/\rangle$ for the 'diagonal' measurements. The superposition principle is expressed by the fact that the 'diagonal' states are linear combinations of the 'linear' ones, $[|h\rangle \pm |v\rangle]/\sqrt{2}$. So if one measures with a 'linear' polarizer a 'diagonal' state, one obtains $|h\rangle$ and $|v\rangle$, each with probability $\left(1/\sqrt{2}\right)^2 = 1/2$.

To keep the notation transparent, we assign values to the states:

---

[11] Another method, called the E91 protocol (the 'E' designates Artur Ekert), works with *entangled* photons (for this concept see Chap. 20, Vol. 2).

[12] Unfortunately, one must not be too far ahead of one's time. Depicting blue horses in the 15th century probably caused (at most) some head-shaking. That applies also in science.

[13] The ⊠ plane is of course the ⊞ plane, rotated by 45°. Moreover, the ⊞ states are the eigenvectors of σ_z, and the ⊠ states, up to a sign, are the eigenvectors of σ_x; cf. Chap. 4.

$$\frac{1 \,\hat{=}\, |h\rangle \; 1 \,\hat{=}\, |\backslash\rangle}{0 \,\hat{=}\, |v\rangle \; 0 \,\hat{=}\, |/\rangle}$$

The exact choice of this mapping plays no role,[14] but it must be agreed upon between Alice and Bob. Similarly, the orientation of the polarizers (= basis) is *publicly* known.

The BB84 protocol operates as follows:

1. Alice and Bob fix the start and the end of the key transmission and the timing with which the photons are sent, for example one photon every tenth of a second.
2. Alice dices (i.e. generates at random) a basis and a value, i.e. ⊞ or ⊠ and 1 or 0. The bit thus described[15] is sent to Bob as a polarized photon.
3. Of course, Bob does not know the basis and the value which Alice has sent. He dices a basis and measures the photon in this basis. He may or may not choose (by chance, with probability 1/2) the same basis as Alice. In the first case, he always measures *the same value* as Alice—this is crucial for the functioning of the method. If the bases do not match, there is only a probability of 1/2 that Bob measures the correct value. Up to this point the whole thing looks, for example, like this:

A basis	⊞	⊠	⊠	⊠	⊞	⊠	⊞	⊠	⊞	⊠
A value	1	0	0	1	1	0	0	1	0	1
B basis	⊠	⊠	⊞	⊠	⊞	⊠	⊞	⊞	⊞	⊞
B possible measurements	1 0	0	1 0	1	1	0	0	1 0	0	1 0
B actual measurement	1	0	1	1	1	0	0	0	0	1

For the first photon, Bob did not choose the basis used by Alice. His measurement can then be 1 or 0; we have inserted 1 as a concrete example.[16] By the way, the results obtained with different bases used by A and B do not matter for the key transmission, as we shall see in a moment.

4. In this way, the necessary number of photons is transmitted, while Alice and Bob record their bases and values. The transfer process is then completed. The next step is a *public* exchange: Bob tells Alice which basis he used for each photon, and Alice tells Bob whether it was the right one. It is important that the value (i.e. 0 or 1) is *not made public*. After that, Alice and Bob remove all values for which the polarization orientations do not match. This also applies to all

---

[14]For example, the mapping $0 \,\hat{=}\, |h\rangle$ and $1 \,\hat{=}\, |v\rangle$ would be just as good.

[15]By a *bit*, one denotes a quantity that can take on only two values, here 0 and 1.

[16]We remark that Bob, in his measurements with a 'wrong' basis, may of course also obtain other values, and these with equal probability. The last row in the table above is a concrete example of a total of 16. Other possibilities for Bob's actual measurements are e.g. | 0 | 0 | 0 | 1 | 1 | 0 | 0 | 1 | 0 | 0 | or | 1 | 0 | 0 | 1 | 1 | 0 | 0 | 0 | 0 | 0 |.

measurements or times at which Alice did not send a photon or Bob did not detect one although one was underway (dark counts). Since Alice and Bob always get the same values for the same basis, the remaining values make up the key. In this eavesdropper-free scenario, it is known to no-one other than Alice and Bob. In our example, the key is

$$\boxed{\text{Key}} - \boxed{0} - \boxed{1}\boxed{1}\boxed{0}\boxed{0} - \boxed{0} - \;\rightarrow\; 011000.$$

However, the world is not so simple, and eavesdroppers and spies are everywhere. How do we deal with this problem?

10.3.4 BB84 Protocol with Eve

The situation is as follows: Alice sends one photon per time interval, and Eve intercepts each photon or a certain portion of them (of course without Alice and Bob being able to perceive this by ordinary means of observation), using e.g. a PBS, and transmits them on to Bob. This may seem simple, but actually it is not so easy for Eve to carry out this interception. One of the possible applications is, for example, to send keys from the earth (summit stations) to satellites. If one is really dealing with single-photon processes, it is impossible for Eve to intercept individual photons in transit and remain unnoticed, without in this case the whole world being able to look at her. For other types of transmission (via fiber-optic cable, etc.), espionage techniques are possible, but certainly not easy to implement.

But we assume in the following (for the purpose of a conservative estimate) that Eve can overcome this problem. However, quantum mechanics ensures that she still cannot listen without being recognized.

The argument runs like this: Since Eve never knows which basis Alice has set, she must choose her basis, just like Bob, randomly with a hit rate of 50%. When using the wrong basis, Eve will not measure the value chosen by Alice in 50% of the cases. Bob in turn measures, if he has chosen at random the same basis as Eve, the same value as she does; or otherwise, with probability 1/2, the value 0 or the value 1. This could for example look like this:

A basis	⊞	⊠	⊠	⊠	⊞	⊠	⊞	⊠	⊞	⊠
A value	1	0	0	1	1	0	0	1	0	1
E basis	⊠	⊞	⊞	⊠	⊞	⊞	⊠	⊠	⊞	⊞
E possible measurements	1/0	1/0	1/0	1	1	1/0	1/0	1	0	1/0
E actual measurement	1	0	1	1	1	0	1	1	0	0
B basis	⊠	⊠	⊞	⊠	⊞	⊠	⊞	⊞	⊞	⊞
B possible measurements	1	1/0	1	1	1	1/0	1/0	1/0	0	0
B actual measurement	1	0	1	1	1	1	1	0	0	0

Alice and Bob again compare their bases for each photon and keep only the values for which the bases coincide. For example:

Alice	–	0	–	1	1	0	0	–	0	–
Bob	–	0	–	1	1	1	1	–	0	–

And here we see the great advantage of quantum cryptography. The difference in the keys of Alice and Bob makes it *in principle detectable* that Eve was spying! Quantum-mechanical methods of key distribution make it virtually impossible for Eve to remain unnoticed. In order to detect the spy, Alice and Bob have to compare publicly parts of their keys, and cannot use the whole key directly. But since one can transmit very large amounts of information very simply with photons, it is not a particular disadvantage for Alice and Bob to discard parts of their keys upon consultation. The essential details of the procedure can be found in Appendix P, Vol. 1.

We want to pursue the question of which level of certainty can be achieved for Eve's unmasking. To quantify the issue, we assume that Alice and Bob have chosen the same basis (the other photons are eliminated anyway). Eve can then chose randomly (and with probability 1/2) the same basis, in which case Alice's value is passed on, or the other basis, in which case there are four different possibilities. In detail, they are as follows:

Alice's basis	⊞	⊞	⊞	⊞	⊞
Alice's value	1	1	1	1	1
Eve's basis	⊞	⊠	⊠	⊠	⊠
Eve's value	1	1	1	0	0
Bob's basis	⊞	⊞	⊞	⊞	⊞
Bob's value	1	1	0	1	0
probability	1/2	1/8	1/8	1/8	1/8

After the elimination of the results of different bases used by Alice and Bob, we have the following situation: (a) Eve has 75 agreement with the values of Alice, and (b) in a quarter of the cases, a different value results at the corresponding position of the keys of Alice and Bob. Thus, there is a chance of $1 - 1/4$ per photon that Eve remains undetected. If Eve has spied on a total of N photons of the key, the chance of discovering her is given by $p_{\text{discover}} = \left(1 - [1 - 1/4]^N\right)$. For a very short key or very few measurements, Eve may be lucky and stay undetected (e.g. for the first five photons in the above example), but uncovering her is practically certain with even a moderately long key. Here are some numerical values:

N	10	10^2	10^3	10^4
$1 - p_{\text{discover}}$	$10^{-1.25} = 0.056$	$10^{-12.5}$	10^{-125}	10^{-1249}

Compared to this, the chance to win the lottery (6 out of 49) is relatively high; as is well known, it is $1/\binom{49}{6} = 1/13983816 \approx 10^{-7.1}$. Even for only a moderately large N of the order of 100 or 1000, it is virtually impossible that Eve can listen in without being recognized.

If Eve spies on each photon, this manifests itself in an average error rate of 25% when the keys of Alice and Bob are compared. If she spies on every second photon, it is 12.5%, etc. So, when Alice and Bob compare their keys, they see not only *whether* Eve has been spying, but also can estimate *how many* photons she has eavesdropped on. However, errors can also arise due to noise and other processes which, for example, unintentionally change the polarization. By comparing, Alice and Bob can determine which part of the key Eve knows *at most*. If the error rate is too high, say well over 10%, the key is discarded and a new key is transmitted.

Now one could imagine that Eve calmly replicates the photons sent by Alice, transmits the original to Bob and performs appropriate measurements on the copies. But this does not work, as is guaranteed by another peculiarity of quantum mechanics: Namely, the *no-cloning theorem* of quantum mechanics states that one cannot copy an arbitrary state, but only a state that is *already known*, as well as the states orthogonal to it. We will discuss this point in Chap. 26, Vol. 2 (quantum information). In the context of our current considerations, the theorem applies, since the two non-mutually-orthogonal basis systems ⊞ and ⊠ are used.

Up to this point we have considered the contributions of quantum mechanics. What follows are classical, not quantum-mechanical methods; they are outlined in Appendix P, Vol. 1.

A final remark: We have assumed idealized conditions—all detection devices work with one hundred percent efficiency, there is no noise (behind which Eve could try to hide), and so on. So the question is whether the method is also suitable for actual, practical use. One can investigate this issue theoretically, and finds a positive answer. But here it is perhaps more interesting to note that the method indeed works well in practice. In fact, a number of quantum cryptographic experiments have been performed to date. Among others, the world's first quantum-encrypted money transfer was carried out on April 21st, 2004 in Vienna. The photons were guided through a 1500 m long fiber-optic cable that connected the city hall with a bank. Furthermore, there was an experiment in 2002 using a telescopic connection, i.e. without expensive fiber-optic cables. Here, the photon travelled through the clear mountain air from the summit station of the Karwendelbahn a distance of 23.4 km to the Max Planck hut on the Zugspitze.[17] But even in the polluted air of an urban area (Munich), the procedure has been successfully tested[18]; the photons travelled a free distance of 500 m. The transfer rate was around 60 kbit/s; the system was operated continuously and stably

---

[17]C. Kurtsiefer et al., 'A step towards global key distribution', *Nature* 419 (2002), p. 450.

[18]See the webpage 'Experimental Quantum Physics', http://xqp.physik.uni-muenchen.de/.

for 13 h. A much longer transmission distance was attained in 2007, when a quantum key was transferred over 144 km, namely between the Canary Islands of La Palma and Tenerife.[19] In principle, it therefore appears possible to use satellites for such secure and encrypted signaling, e.g. for transatlantic connections.[20]

---

[19]R. Ursin et al., 'Entanglement-based quantum communication over 144 km', *Nature Physics* 3 (2007), p. 481.

[20]See e.g. S. Liao et al., Satellite-to-ground quantum key distribution, *Nature* 549, 43–47, https://doi.org/10.1038/nature23655 (Sep 2017), where a quantum key distribution over a distance of up to 1,200 km is reported. Quantum keys may also be distributed in optical fibers over remarkable distances of up to 100 km; see e.g. K.A. Patel et al., Coexistence of high-bit-rate quantum key distribution and data on optical fiber, *Phys. Rev.* X 2, 041010 (2012)), or Paul Jouguet et al., Experimental demonstration of long-distance continuous-variable quantum key distribution, *Nature Photonics* (2013), https://doi.org/10.1038/nphoton.2013.63. In addition, the feasibility of BB84 quantum key distribution between an aircraft moving at 290 km/h at a distance of 20 km was recently proven for the first time; see: Sebastian Nauert et al., Air-to-ground quantum communication, *Nature Photonics* (2013), https://doi.org/10.1038/nphoton.2013.46.

Chapter 11
Abstract Notation

We are now starting to bring together the analytical and the algebraic approaches to quantum mechanics. In this chapter, we first consider the vector space of solutions of the SEq in more detail. After a brief excursion into matrix mechanics, we treat the abstract representation of quantum mechanics, which is formulated in terms of the familiar bras and kets.

In Chap. 10, we saw that the analytic and the algebraic approaches lead to very similar formulations. We deepen this parallelism in the following sections by showing that the expression $\int \Phi^* \Psi dV$ is a scalar product. With some additional assumptions, it follows that the vector spaces of both the algebraic and analytic approaches are Hilbert spaces. With this background, we can then formulate a representation-independent, i.e. an *abstract* notation.

All the spectra which we consider in this chapter are discrete and non-degenerate.

11.1 Hilbert Space

11.1.1 Wavefunctions and Coordinate Vectors

In Chap. 10, we ventured the guess that $\int \Phi^* \Psi dV$ is a scalar product. We now want to provide some additional motivation for this assumption.

We start with a Hamiltonian H (with a discrete and non-degenerated energy spectrum). Its eigenfunctions $\varphi_n (\mathbf{r})$, i.e. the solutions of the stationary SEq

$$H\varphi_n (\mathbf{r}) = E_n \varphi_n (\mathbf{r}); \ n = 1, 2, \ldots \tag{11.1}$$

are known and form a CONS. Because of the completeness of $\varphi_n (\mathbf{r})$, we can write any solution $\psi (\mathbf{r}, t)$ of the time-dependent SEq as

© Springer Nature Switzerland AG 2018
J. Pade, *Quantum Mechanics for Pedestrians 1*, Undergraduate Lecture
Notes in Physics, https://doi.org/10.1007/978-3-030-00464-4_11

$$\psi\left(\mathbf{r}, t\right) = \sum_n c_n \varphi_n\left(\mathbf{r}\right) e^{-i\frac{E_n t}{\hbar}}; \ c_n \in \mathbb{C}. \tag{11.2}$$

To save writing, we restrict the following considerations to the initial state (i.e. we freeze time at $t = 0$):

$$\psi\left(\mathbf{r}, 0\right) = \sum_n c_n \varphi_n\left(\mathbf{r}\right). \tag{11.3}$$

The total time evolution can be determined easily from (11.2). Due to the orthonormality of the eigenfunctions, the coefficients c_n are specified uniquely by the initial condition $\psi\left(\mathbf{r}, 0\right)$:

$$c_n = \int \varphi_n^*\left(\mathbf{r}\right) \psi\left(\mathbf{r}, 0\right) \mathrm{d}V. \tag{11.4}$$

We can understand this situation a little differently. To this end, we take into account the fact that the eigenfunctions $\{\varphi_n\left(\mathbf{r}\right)\}$ represent an orthonormal basis of the vector space $\mathcal{V}$ of solutions of the SEq—analogous to the three unit vectors $\mathbf{e}_x, \mathbf{e}_y$ and $\mathbf{e}_z$ in the visual space or $\mathbb{R}^3$. In the latter space, we can represent a general vector $\mathbf{v}$ as $\mathbf{v} = v_x \mathbf{e}_x + v_y \mathbf{e}_y + v_z \mathbf{e}_z$, where the components or expansion coefficients v_x, v_y, v_z are usually called the *coordinates* of $\mathbf{v}$. It makes no difference whether we specify $\mathbf{v}$ or v_x, v_y, v_z—we can calculate $\mathbf{v}$ uniquely from v_x, v_y, v_z and *vice versa*, if the unit vectors are known.

The situation described in (11.3) and (11.4) is quite analogous - only we are dealing with the function $\psi\left(\mathbf{r}, 0\right)$ instead of the vector $\mathbf{v}$, the eigenfunctions φ_n instead of the unit vectors $\mathbf{e}_i$, and the constants c_n instead of the coordinates v_x. For example, the c_n can be determined uniquely (for known φ_n), if $\psi\left(\mathbf{r}, 0\right)$ is given, and *vice versa*. We can thus denote the expansion coefficients c_n as coordinates and have the same information (always assuming that φ_n is known), whether we are given $\psi\left(\mathbf{r}, 0\right)$ or the coordinate vector

$$\mathbf{c} = \begin{pmatrix} c_1 \\ c_2 \\ \vdots \end{pmatrix} \tag{11.5}$$

We now consider two wavefunctions $\psi = \sum_i c_i \varphi_i$ and $\chi = \sum_j d_j \varphi_j$. Because of the orthonormality of the eigenfunctions, we have

$$\int \psi^* \chi \mathrm{d}V = \sum_{ij} c_i^* d_j \int \varphi_i^* \varphi_j \mathrm{d}V = \sum_{ij} c_i^* d_j \delta_{ij} = \sum_i c_i^* d_i. \tag{11.6}$$

We find exactly the same result when we take the dot product of the two coordinate vectors $\mathbf{c}$ and $\mathbf{d}$; it is

$$\mathbf{c}^\dagger \mathbf{d} = \begin{pmatrix} c_1^* & c_2^* & \dots \end{pmatrix} \cdot \begin{pmatrix} d_1 \\ d_2 \\ \vdots \end{pmatrix} = \sum_i c_i^* d_i \qquad (11.7)$$

Comparing (11.6) and (11.7), we see that the expression $\int \psi^* \chi dV$ is clearly a scalar product.[1]

11.1.2 The Scalar Product

The formal confirmation that $\int \psi^* \chi dV$ is a scalar product is found in mathematics. There, the scalar product is generally defined as a rule which assigns to two elements x and y of a vector space a scalar (x, y), where the following properties must apply: (x, y) is (i) positive definite: $(x, x) \geq 0$ and $(x, x) = 0 \leftrightarrow x = 0$; (ii) linear: $(x, \alpha y + \beta z) = \alpha (x, y) + \beta (x, z)$; (iii) Hermitian or conjugate symmetric: $(x, y) = (y, x)^*$ (see also Appendix F, Vol. 1). *Any* rule that meets these requirements is a scalar product (also called *Hermitian form*).

In order to test $\int f^* g dV$ for these properties, we do not choose the notation (f, g), but refer instead to the algebraic approach $\langle f| g \rangle$, i.e.

$$\langle f| g \rangle := \int f^* g dV. \qquad (11.8)$$

At this point is not clear how a ket $|g\rangle$ and a bra $\langle f|$ are specifically defined; we address this question in Chap. 12. But notwithstanding this, we can easily show that $\int f^* g dV$ is a scalar product - even though the expression may not have looked like one when we wrote it down for the first time in Chap. 5. For it is immediately apparent that $\int f^* g dV$ assigns a number to two elements.[2] Furthermore, $\int f^* g dV$ is

1. positive definite: $\langle f| f \rangle = \int f^* f dV \geq 0, \in \mathbb{R}$ where $\langle f| f \rangle = 0 \leftrightarrow f \equiv 0$.
2. linear[3]: $\langle f| \alpha g + \beta h \rangle = \int f^* (\alpha g + \beta h) dV = \alpha \int f^* g dV + \beta \int f^* h dV = \alpha \langle f| g \rangle + \beta \langle f| h \rangle$.
3. Hermitian or conjugate symmetric: $\langle f| g \rangle = \int f^* g dV = \left(\int f g^* dV \right)^* = \langle g| f \rangle^*$.

---

[1] We see, by the way, that the scalar product is independent of the representation. The left-hand sides of (11.6) and (11.7) are two different representations of the same expression.

[2] We remark again that in general, we do not specify the integration limits for integrals as in (11.8). It is tacitly assumed that one integrates over the entire domain of definition of the integrand. Contrary to the initial impression, these integrals are *definite* integrals - in other words, scalars (which may be time dependent).

[3] More precisely, semi-linear in the first and linear in the second component (also denoted as anti-linear or conjugate linear in the first argument and linear in the second argument). Therefore, the form is not called bilinear, but sesquilinear. In mathematics, the form is usually defined the other way around, as antilinear in the second argument.

By means of the scalar product $\int f^* g dV$, we can therefore not only define as usual the length or norm of wavefunctions by $||\varphi|| = \sqrt{\langle \varphi | \varphi \rangle} = \sqrt{\int \varphi^* \varphi dV}$ and the orthogonality of two wave functions (as elements of the vector space) by $\langle \varphi | \psi \rangle = \int \varphi^* \psi dV = 0$, but also we can use general statements about scalar products (e.g. the Schwarz and the triangle inequalities) without further ado.

Thus, the solutions of the SEq span a complex vector space in which a scalar product is defined. Such spaces are called *unitary spaces*, as we know already from the algebraic approach (Chap. 4). At this point it is perhaps possible to understand somewhat better why the eigenfunctions φ_m, φ_n, obeying $\int \varphi_m^* \varphi_n dV = \delta_{nm}$, are called orthonormal. On the one hand, one can call an element of a vector space a 'vector'—an eigenfunction as an element of a vector space is an eigenvector. On the other hand, $\int \varphi_m^* \varphi_n dV = 0$ for $n \neq m$ means in the sense of a scalar product that the eigenfunctions (as eigenvectors) are pairwise orthogonal,[4] and $\int \varphi_n^* \varphi_n dV = 1$ means that they have length 1 or are normalized.

11.1.3 Hilbert Space

The way we constructed it, our unitary space is *separable*.[5] This means, essentially, that there is a CONS of at most *countably infinite dimension*. Any vector ψ can be expanded in terms of this CONS (*expansion theorem*):

$$\psi(\mathbf{r}, t) = \sum_n d_n(t) \varphi_n(\mathbf{r}) \text{ with } d_n(t) = \int \varphi_n^*(\mathbf{r}) \psi(\mathbf{r}, t) \, dV. \tag{11.9}$$

This sum and others like $|\psi(\mathbf{r}, t)|^2 = \sum_n |d_n(t)|^2$ must of course be meaningful, i.e. they must converge to an element which itself belongs to the vector space. We therefore require that the vector space be *complete*, which means that sequences[6] have limits which are themselves elements of the vector space.[7] A space with all these ingredients is called a (separable)[8] *Hilbert space* $\mathcal{H}$. A quantum-mechanical state is an element of $\mathcal{H}$ and thus may be denoted, as stated above, as a vector, even if it is in fact a function in the concrete representation.

---

[4]As said above, this does not mean that the graphs of the functions are orthogonal to each other or something similar. The statement refers only to the (abstract) angle between two vectors in the vector space.

[5]The term 'separable' which occurs here has nothing to do with the requirement of 'separability,' which means that a system (function) is separable into functions of space and of time.

[6]The technical term is Cauchy sequences, see Appendix G, Vol. 1.

[7]The requirement of completeness has no straightforward physical meaning, but it occurs in many proofs of laws concerning Hilbert spaces.

[8]There are also non-separable Hilbert spaces (for example, in the quantization of fields). But in 'our' quantum mechanics, they play no role, so here 'Hilbert space' means in general 'separable Hilbert space.'

We see in retrospect that the unitary spaces which we considered in the algebraic approach are also separable Hilbert spaces. Now, the punchline of this story is that *all* Hilbert spaces of the same dimension are *isomorphic*, i.e. there are reversible unique (one-to-one, bijective) mappings between them. That is why we also often speak of *the* Hilbert space $\mathcal{H}$ of dimension N, for which there are various realizations or representations. In particular, the space of solutions (11.2) of the SEq, as spanned by $\{\varphi_n (\mathbf{r})\}$, and the space of the coordinate vectors $\mathbf{c}$ are isomorphic; that is they are just different representations of the same systems. Of course, the question arises as to whether there is a *representation-independent*, i.e. *abstract* formulation of these systems. We take up this issue below.

If we disregard the technical issues (which we do not consider to a greater extent in the following), then $\mathcal{H}$ is basically a very intuitive structure. As we have already indicated above, we can in principle imagine everything as in $\mathbb{R}^3$, despite the possibly much higher dimensionality of the Hilbert space and its use of complex numbers. In both spaces, mutually orthogonal and normalized vectors, i.e. unit vectors, constitute a basis and span the entire space; any vector can therefore be represented as a linear combination of basis vectors. In addition, we also have an inner product in both spaces, which automatically defines a norm. We can imagine an intuitive analog to the time-dependent (normalized) state vector $\psi (\mathbf{r}, t) \in \mathcal{H}$ of (11.9): In $\mathbb{R}^3$ this would be a vector of length 1, which moves in the course of time. A state with a sharp energy $\in \mathcal{H}$ corresponds to a circular motion in $\mathbb{R}^3$, because the time dependence in $\mathcal{H}$ is given by $\exp(-i\omega t)$.[9]

11.2 Matrix Mechanics

We have seen that we obtain the same information if we specify the vector $\mathbf{c}$ instead of the wavefunction $\psi (\mathbf{r}, 0)$. We will now apply this 'algebraization' to eigenvalue problems, also.

In fact, in the early days of quantum mechanics there were two competing formulations: *Matrix mechanics* (associated with the name W. Heisenberg, corresponding essentially to our algebraic approach), and *wave mechanics* (linked to the name Schrödinger, corresponding essentially to our analytic approach). Quite soon it became clear that these formulations, for the same initial physical situation, were just two different descriptions of the same facts, which hence could be converted one-to-one into each other. This can be shown rather simply in a way similar to the above provisional representation using coordinates.

We start from the formulation of an eigenvalue problem of wave mechanics, considering a wavefunction $\Psi (x)$ and an operator A with a discrete and non-degenerate spectrum:

---

[9]We note that approaching quantum mechanics by means of the Hilbert space is not the only possible option. As a starting point, one could for example consider a C^*-algebra (see Appendix G, Vol. 1), or the aforementioned replacement process $\{,\}_{\text{Poisson}} \rightarrow \frac{1}{i\hbar} [,]_{\text{commutator}}$. This method is called canonical quantization, see Appendix W, Vol. 2.

$$A\Psi(x) = a\Psi(x).\tag{11.10}$$

This problem can be written as a *matrix equation* (that is, among other things: with no spatial dependence). To demonstrate this, in a first step we expand the wavefunction in terms of the eigenfunctions $\{\varphi_i\}$ of the Hamiltonian (or any other basis system in $\mathcal{H}$) as $\Psi = \sum_n c_n \varphi_n(x)$, and obtain

$$A\sum_n c_n \varphi_n(x) = a\sum_n c_n \varphi_n(x).\tag{11.11}$$

Then we multiply by φ_m^* and take the scalar product:

$$\sum_n c_n \int \varphi_m^* A \varphi_n dV = \sum_n c_n \int \varphi_m^* \varphi_n dV = ac_m\tag{11.12}$$

The integral $\int \varphi_m^* A \varphi_n dV$ is a *number* that depends on n and m. We call this number A_{mn}:

$$\sum_n c_n A_{mn} = ac_m.\tag{11.13}$$

The expression on the left side is simply the product of the matrix $\{A_{mn}\} \equiv \mathbb{A}$ with the column vector $\mathbf{c}$:

$$\mathbb{A}\mathbf{c} = a\mathbf{c}.\tag{11.14}$$

The column vector $\mathbf{c}$ is of course just the coordinate vector introduced above. In this way, we can formulate quantum mechanics as matrix mechanics, representing operators as matrices and states as column vectors. In practice, this is not done in general, but only in cases where this approach is particularly well-suited (e.g. in lower-dimensional systems).

11.3   Abstract Formulation

Thus far, we have met up with various formulations of states, which at first glance seem to have little in common. In the analytical approach, we started from the wavefunction $\psi(\mathbf{r})$, but instead we could have chosen the coordinate vector $\mathbf{c}$. Moreover, there are other possibilities, e.g. the Fourier transform of $\varphi(\mathbf{k}) = \int \psi(\mathbf{r}) e^{i\mathbf{kr}} d^3r$, which provides the same information as the wavefunction itself. In the algebraic approach, we worked with kets, for which the various representations in the form of a column vector are possible. An analogous consideration applies to different formulations of operators: We have just seen that we can likewise write the operators of the analytical approach as matrices. In the algebraic approach, we defined operators as dyadic products, or represented them as matrices. All together, we have quite different but equivalent ways at hand to describe the same facts.

The circumstance that Hilbert spaces of the same dimension are isomorphic is beyond the parallelism of the two approaches established in Chap. 10, and it shows that the analytical and the algebraic approaches are actually just different manifestations of the same facts. For whether we work in a Hilbert space of one or the other approach (and in which one) is irrelevant, insofar as there are one-to-one transformations between all Hilbert spaces of the same dimension. But if we have very different representations for the same facts, there *must* exist a representation-independent, i.e. abstract core.[10]

The fact that we can represent one and the same (physical) situation in many different ways is known to us in a similar form e.g. from $\mathbb{R}^3$. There, a vector also has different representations (components), depending on how we define our coordinate system in space. We can specify this vector abstractly, i.e. in a coordinate-free manner, by writing it not just as a column vector with several components, but instead by denoting it as $\mathbf{a}$ or $\vec{a}$ or something similar. That will not only suffice for many formulations (e.g. $\mathbf{l} = \mathbf{r} \times \mathbf{p}$), but it also facilitates them or renders them expressible in a compact form; think of e.g. the Maxwell equations. For concrete calculations, however, one often has to specify the vectors in some particular representation.[11]

Here, we are in a similar situation, seeking an abstract designation for the elements of the Hilbert space. The notations $\mathbf{a}$ or $\vec{a}$ are 'used up' and also too strongly suggest a column vector. Instead, the convention of writing an abstract vector of the Hilbert space as a ket, $|\Psi\rangle$ has been adopted. It is for this reason that we denoted states as kets from the start in the algebraic approach[12] (a justification in retrospect, so to speak).

In this way, we can write for example an eigenvalue equation like

$$A_{\text{spatial}} \Psi(x) = a \Psi(x) \, ; \, A_{\text{matrix}} \mathbf{c} = a \mathbf{c} \tag{11.15}$$

in the abstract formulation as

$$A_{\text{abstract}} |\Psi\rangle = a |\Psi\rangle . \tag{11.16}$$

Some remarks are in order:

---

[10]Only the dimension of the state space matters here. The physical system can take a variety of forms. The electronic spin with its two orientations, the polarization of a photon, e.g. with horizontally and vertically linearly-polarized states, the MZI with the basis states $|H\rangle$ and $|V\rangle$, in a certain sense the ammonia molecule (NH_3, where the N atom can tunnel through the H_3 plane and occupy two states with respect to it) are some examples of physically different systems which all 'live' in a two-dimensional Hilbert space.

[11]In fact, the notation $\mathbf{a}$ is very abstract—it does not reveal anything about the dimension nor the individual components. We know nothing more than simply that it is a vector. Nevertheless, this notation is often not perceived as particularly abstract. This is probably due to the fact that one was introduced to it at the beginning of physics courses and it now seems familiar.

[12]This is also why we chose the symbol $\cong$ to distinguish between an abstract ket and its representation as a column vector.

1. It is not known at this point how the ket $|\Psi\rangle$ is formulated in detail[13]; it is just an *abstract notation*, comparable to the designation of a vector by the symbol **a**. Similarly, the form of the operator A_{abstract} is not known at this point; this is comparable to the use of the abstract symbol $\mathbb{A}$ for a general matrix.[14]
2. For the sake of a better distinction, we have denoted the nature of the operators in (11.15) and (11.16) by an index. But it is quite common[15] to use the same symbol for the operators in different concrete and abstract representations, that is to write simply A in all three equations:

$$A\Psi(x) = a\Psi(x); \; A\mathbf{c} = a\mathbf{c}; A|\Psi\rangle = a|\Psi\rangle. \tag{11.17}$$

Strictly speaking, this is evidently wrong, because A refers to quite different mathematical objects e.g. in the expressions $A\Psi(x)$ and $A|\Psi\rangle$. That this 'nonchalant' notation is rather ambiguous may seem annoying, but it is widespread and can, if one is used to it, even be quite practical. Of course, it must be clear from the context what is precisely meant where necessary.

In the algebraic approach, we introduced the symbol $\cong$ to emphasize the difference between an abstract ket and its representation as a column vector. In the following, we will relax this rule and often use $=$ instead of $\cong$, thus following common practice.

3. The relationship between $\Psi(x)$ and $|\Psi\rangle$ will be addressed in Chap. 12.

The following paragraph is merely a repetition of the facts already discussed in the preceding (even-numbered) chapters. We recall that the adjoint (of a column vector) means the transposed and complex conjugated vector. The adjoint[16] of a ket is a *bra*:

$$(|\Psi\rangle)^{\dagger} \equiv \langle\Psi| \tag{11.18}$$

(accordingly, the adjoint of a column vector is the row vector with complex conjugate elements). The adjoint of an operator A is written as $A^{\dagger}$, where $AA^{\dagger} = A^{\dagger}A$ holds. Because the application of an operator A to a ket $|\Psi\rangle$ gives another ket, one also writes

$$A|\Psi\rangle \equiv |A\Psi\rangle. \tag{11.19}$$

The adjoint of a number is its complex conjugate $c^{\dagger} = c^*$. In particular, we have for the scalar product $\langle f|g\rangle$:

---

[13]It is in any case not a column vector (even if this idea sometimes proves to be helpful).

[14]To avoid misunderstandings: **a** is an abstract or general *column vector*, whereas $|\Psi\rangle$ is an *abstract state* which can be represented, where appropriate, as a column vector, but for which also other representations exist. Quite analogously, $\mathbb{A}$ denotes a general matrix and A_{abstract} an abstract operator, which can, where appropriate, be represented as a matrix.

[15]However, there are books that distinguish them quite consistently.

[16]Note that here 'adjoint' means the Hermitian adjoint $a^{\dagger}$, as always in non-relativistic quantum mechanics. In relativistic quantum mechanics, one uses instead the Dirac adjoint $a^{\dagger}\gamma_o$.

$$\langle f \mid g \rangle^{\dagger} = \langle f \mid g \rangle^{*} = \langle g \mid f \rangle. \tag{11.20}$$

In the adjoint of a compound expression, the order of the constituents is reversed. We give some examples for the adjoint:

$$
\begin{aligned}
(c \mid \Psi \rangle)^{\dagger} &= c^{*} \langle \Psi \mid = \langle \Psi \mid c^{*} \\
(A \mid \Psi \rangle)^{\dagger} &= \mid A\Psi \rangle^{\dagger} = \langle A\Psi \mid = \langle \Psi \mid A^{\dagger} \\
\langle \Phi \mid A \mid \Psi \rangle^{\dagger} &= \langle \Psi \mid A^{\dagger} \mid \Phi \rangle.
\end{aligned}
\tag{11.21}
$$

Expressions of the form $\langle \varphi \mid A \mid \psi \rangle$ are called *matrix elements* . Finally, we note again the equations defining a Hermitian operator: In the formulation with integrals, we have

$$\int \Psi_1^{*} A \Psi_2 dV = \int (A\Psi_1)^{*} \Psi_2 dV. \tag{11.22}$$

With $\int f^{*} g dV = \langle f \mid g \rangle$, this is written in the bra-ket notation as

$$
\begin{aligned}
\int \Psi_1^{*} A \Psi_2 dV &= \langle \Psi_1 \mid A\Psi_2 \rangle = \langle \Psi_1 \mid A \mid \Psi_2 \rangle \\
\int (A\Psi_1)^{*} \Psi_2 dV &= \langle A\Psi_1 \mid \Psi_2 \rangle = \langle \Psi_1 \mid A^{\dagger} \mid \Psi_2 \rangle.
\end{aligned}
\tag{11.23}
$$

Comparing the right-hand sides exhibits a familiar result: For a Hermitian operator, it holds that $A = A^{\dagger}$; the operator is self-adjoint.[17]

Experience shows that it is rather difficult to imagine something quite abstract (only kidding!). So here, we give the hint to think of a column (row) vector in the case of a ket (bra), and of a matrix in the case of an operator (and not to forget that this is an *auxiliary notion*). In this way, many 'calculation rules' and statements become quite familiar, e.g. the rule that operators do not commute in general—which applies also to matrices.

Although the relation between e.g. $\Psi(x)$ and $\mid\Psi\rangle$ still needs to be clarified, we can 'play around' a little with the abstract notation. As an example, we consider a CONS $\{\varphi_n(x)\}$. From the expansion theorem, it follows for each wavefunction $\Psi(x)$ that

$$\Psi(x) = \sum_n c_n \varphi_n(x). \tag{11.24}$$

With

$$c_m = \int \varphi_m^{*}(x) \Psi(x) dx = \langle \varphi_m \mid \Psi \rangle, \tag{11.25}$$

we find due to

$$\int \varphi_m^{*}(x) \varphi_n(x) dx = \delta_{nm} \tag{11.26}$$

---

[17] In fact, there may be a difference between self-adjoint and Hermitian (see Chap. 13 and Appendix I, Vol. 1). Among the problems considered here, this difference is not noticeable.

the equation

$$\langle \Psi | \Psi \rangle = \int \Psi^*(x)\Psi(x)\mathrm{d}x \underset{(11.24)}{=} \int \sum_{n,m} c_n^* c_m \varphi_n^*(x)\varphi_m(x)\mathrm{d}x \underset{(11.26)}{=}$$
$$\underset{(11.26)}{=} \sum_n c_n^* c_n \underset{(11.25)}{=} \sum_n \langle \Psi | \varphi_n \rangle \langle \varphi_n | \Psi \rangle . \tag{11.27}$$

Comparing the right-hand and left-hand sides, we obtain the completeness relation in the abstract notation

$$\sum_n |\varphi_n\rangle \langle \varphi_n| = 1, \tag{11.28}$$

i.e. a result which we knew already from the algebraic approach.

11.4 Concrete: Abstract

Finally, we make a remark about the relationship of the abstract formulation to concrete representations.

In the algebraic approach, the 'de-abstracting' of kets is not problematic. We can associate (and have done so occasionally) a ket to a representation as a column vector such as e.g. $|h\rangle \cong \begin{pmatrix} 1 \\ 0 \end{pmatrix}$. Since we formulate operators in this approach as sums over dyadic products, the concrete representation of operators may be easily formulated. Thus, there are no difficulties with the algebraic formulation at this level.

The situation is similar in the analytical approach when we have a discrete spectrum. Again, as we have just explored, we can represent states and operators that depend on local variables by vectors and matrices.

In order to illustrate these relationships through examples, we start with a Hamiltonian $H = -\frac{\hbar^2}{2m}\nabla^2 + V(\mathbf{r})$ with a discrete and nondegenerate spectrum. We want to derive the matrix representation and the abstract formulation for the stationary SEq. (Similar considerations for the time-dependent SEq can be found in the exercises.) The eigenvalue problem (stationary SEq) reads

$$H\psi(\mathbf{r}) = E\psi(\mathbf{r}). \tag{11.29}$$

The eigenfunctions $\varphi_n(\mathbf{r})$ and the eigenvalues E_n of the Hamiltonian are known:

$$H\varphi_n(\mathbf{r}) = E_n\varphi_n(\mathbf{r}); n = 1, 2, \ldots \tag{11.30}$$

Every state $\psi(\mathbf{r})$ can be written as a linear combination of the eigenfunctions (which form a CONS):

$$\psi(\mathbf{r}) = \sum_n c_n\varphi_n(\mathbf{r}); c_n = \int \varphi_n^*(\mathbf{r})\psi(\mathbf{r})\,\mathrm{d}V. \tag{11.31}$$

In the matrix representation, we represent the state by the column vector of the coefficients c_n:

$$\mathbf{c} = \begin{pmatrix} c_1 \\ c_2 \\ \vdots \end{pmatrix} \tag{11.32}$$

We identify as follows:

$$\varphi_1(\mathbf{r}) \rightarrow \begin{pmatrix} 1 \\ 0 \\ \vdots \end{pmatrix}; \ \varphi_2(\mathbf{r}) \rightarrow \begin{pmatrix} 0 \\ 1 \\ \vdots \end{pmatrix} \ \text{etc.} \tag{11.33}$$

Now we have to replace H by a matrix. For this, we repeat the above reasoning by inserting (11.31) into (11.29), multiplying by $\varphi_m^*(\mathbf{r})$, and integrating:

$$\sum_n c_n \int \varphi_m^*(\mathbf{r}) H\varphi_n(\mathbf{r}) \ dV = E \sum_n c_n \int \varphi_m^*(\mathbf{r}) \varphi_n(\mathbf{r}) \ dV. \tag{11.34}$$

On the left-hand side, we employ (11.30), and on both sides we make use of the orthonormality of the eigenfunctions. It follows that

$$E_m c_m = E c_m. \tag{11.35}$$

In other words, the Hamiltonian H is replaced by a diagonal matrix H_{matrix}:

$$H_{\text{matrix}} = \begin{pmatrix} E_1 & 0 & \cdots \\ 0 & E_2 & \cdots \\ \vdots & \vdots & \ddots \end{pmatrix} \tag{11.36}$$

and the stationary SEq in the matrix representation reads

$$H_{\text{matrix}}\mathbf{c} = E\mathbf{c}. \tag{11.37}$$

From this equation, we can reconstruct (11.29)—but as said above, only if we know the eigenfunctions $\varphi_n(\mathbf{r})$ which do not appear in (11.37).

In order to arrive at the abstract notation, we interpret the vector $\begin{pmatrix} 1 \\ 0 \\ \vdots \end{pmatrix}$ as a representation of the ket $|\varphi_1\rangle$, and analogously for the other components. Then it follows that

$$|\varphi_1\rangle \langle\varphi_1| \cong \begin{pmatrix} 1 \\ 0 \\ \vdots \end{pmatrix}(1 \ 0 \ \ldots) = \begin{pmatrix} 1 & 0 & \cdots \\ 0 & 0 & \cdots \\ \vdots & \vdots & \ddots \end{pmatrix} \tag{11.38}$$

and we find for the abstract representation of the Hamiltonian[18]:

$$H_{\text{abstract}} = \sum_n |\varphi_n\rangle \langle \varphi_n| E_n. \tag{11.39}$$

11.5   Exercises

1. Show that the equation

$$\sum_i c_i A_{ji} = a c_j \tag{11.40}$$

   may be written in the matrix representation as

$$\mathbb{A}\mathbf{c} = a\mathbf{c} \tag{11.41}$$

   with the matrix $\{A_{ji}\} \equiv \mathbb{A}$ and the column vector $\mathbf{c}$. Is the equation valid also for non-square matrices?
2. Do the functions of one variable which are continuous in the interval $[0, 1]$ form a Hilbert space?
3. The space $l^{(2)}$ consists of all vectors $|\varphi\rangle$ with infinitely many components (coordinates) $c_1, c_2, \ldots$, such that

$$\||\varphi\rangle\|^2 = \sum_n |c_n|^2 < \infty. \tag{11.42}$$

   Show that also the linear combination of two vectors $|\varphi\rangle$ and $|\chi\rangle$ belongs to this space, and that the scalar product $\langle \varphi | \chi \rangle$ is defined.
4. Given the operator A and the equation

$$i \frac{\mathrm{d}}{\mathrm{d}t} |\psi\rangle = A |\psi\rangle, \tag{11.43}$$

   which condition must A fulfill so that the norm of $|\psi\rangle$ is conserved?
5. Given the operator A, derive the equation

$$i\hbar \frac{\mathrm{d}}{\mathrm{d}t} \langle A \rangle = \langle [A, H] \rangle + i\hbar \langle \dot{A} \rangle \tag{11.44}$$

   in the bra-ket formalism.
6. Given a Hamiltonian H with a discrete and nondegenerate spectrum, (a) in the formulation with space variables, and (b) as an abstract operator; what is in each case the matrix representation of the time-dependent SEq?

---

[18] This form is called *spectral representation*; we discuss it in more detail in Chap. 13.

Chapter 12
Continuous Spectra

In this chapter we start by considering continuous spectra, which we have neglected thus far. Then we investigate the relationship between $\psi(x)$ and $|\psi\rangle$. With these results, the unification of the analytic and the algebraic approaches to quantum mechanics is completed.

So far, we have excluded *continuous spectra* from the discussion, e.g. by placing our quantum object between infinitely high potential walls,[1] thus discretizing the energy spectrum. The fact that we adopted this limitation had less to do with physical reasons, but rather almost exclusively with mathematical ones.[2] From a physical point of view, a continuous spectrum (e.g. the energy of a free quantum object) makes perfect sense. But we have the problem that the corresponding eigenfunctions are not square integrable, and therefore we cannot properly define a scalar product. This hurdle may be circumvented or alleviated, as we shall show in a moment, by the construction of *eigendifferentials* which leads to *improper vectors*. Finally, we examine the question of how a ket $|\Psi\rangle$ is related to the corresponding wavefunction $\Psi(\mathbf{r})$, or how we can transform abstract equations into 'concrete' equations, e.g. in the position or the momentum representation.

---

[1] Another possibility would be the introduction of periodic boundary conditions.

[2] Below the Planck scale ($\sim 10^{-35}$ m, $\sim 10^{-44}$ s), neither space nor time may exist, so that ultimately these variables would become 'grainy' or discrete. (On this scale, space is thought to be something like a foam bubbling with tiny black holes, continuously popping in and out of existence.) There are attempts to determine whether space is truly grainy, but results have so far been inconclusive. Experimentally, these orders of magnitude are still very far from being directly accessible (if they ever will be); the currently highest-energy accelerator, the LHC at CERN in Geneva, attains a spatial resolution of 'only' $\sim 10^{-19}$ m. Recently, however, indirect methods were proposed; see Jakob D. Bekenstein, 'Is a tabletop search for Planck scale signals feasible?', http://arxiv.org/abs/1211.3816 (2012); or Igor Pikovski et al., 'Probing Planck-scale physics with quantum optics', *Nature Physics* 8, 393–397 (2012). For a recent paper see e.g. V. Faraoni, 'Three new roads to the Planck scale', *American Journal of Physics* 85, 865 (2017); https://doi.org/10.1119/1.4994804

© Springer Nature Switzerland AG 2018
J. Pade, *Quantum Mechanics for Pedestrians 1*, Undergraduate Lecture
Notes in Physics, https://doi.org/10.1007/978-3-030-00464-4_12

A remark on notation: discrete spectra are often written with Latin, continuous spectra with Greek letters (exceptions to this are position x and momentum k, as well as the energy E, which can have discrete and/or continuous values). For example, if the spectrum of the Hamiltonian is discrete or continuous, we write

$$H \,|\varphi_l\rangle = E_l \,|\varphi_l\rangle \quad \text{or} \quad H \,|\varphi_\lambda\rangle = E_\lambda \,|\varphi_\lambda\rangle, \qquad (12.1)$$

or

$$H \,|E_l\rangle = E_l \,|E_l\rangle \quad \text{or} \quad H \,|E_\lambda\rangle = E_\lambda \,|E_\lambda\rangle. \qquad (12.2)$$

In addition, a 'direct' terminology is common, in which a state of quantum number n or λ (the system 'has' the quantum number n or λ) is written as

$$\begin{aligned} &|n\rangle : \ \text{discrete quantum number } n \\ &|\lambda\rangle : \ \text{continuous quantum number } \lambda. \end{aligned} \qquad (12.3)$$

12.1 Improper Vectors

Free motion (cf. Chap. 5) is a simple example of the continuous case. We have[3] $\varphi_k(x) = \frac{1}{\sqrt{2\pi}}e^{ikx}$, and with our notation

$$\int \varphi_{k'}^*(x)\,\varphi_k(x)\,\mathrm{d}x \equiv \langle \varphi_{k'} \,|\varphi_k\rangle \equiv \langle k' \,|k\rangle \qquad (12.4)$$

for the scalar product, it follows that:

$$\langle k' \,|k\rangle = \frac{1}{2\pi} \int\limits_{-\infty}^{\infty} e^{ix(k-k')}\,\mathrm{d}x = \delta(k'-k). \qquad (12.5)$$

The difficulty lies in the fact that the physical problem is not properly formulated: The Heisenberg uncertainty principle tells us that a state with a definite, sharp momentum has an infinite position uncertainty, as may indeed be read off directly from the function e^{ikx}. Mathematically, this is expressed by the fact that the integral in (12.5) is not defined 'properly'—the integral does not exist in the usual sense, but is a functional, namely the delta function $\delta(k'-k)$.[4] This means that the ket $|\varphi_k\rangle \equiv |k\rangle$ is well defined, but not the bra $\langle \varphi_k| \equiv \langle k|$. In other words, the state

---

[3]The factor $\frac{1}{\sqrt{2\pi}}$ is due to the normalization of the function, see below.

[4]It is clear that the delta function cannot be a function. That it is still denoted as one may be due to the often rather nonchalant or easygoing approach of physicists to mathematics. More is given on the delta function in Appendix H, Vol. 1.

$|\varphi_k\rangle$ is not normalizable. Such state vectors are commonly called *improper* states[5] (i.e. not square-integrable states), in contrast to the proper states, which are square integrable.[6]

We want to illustrate the problem by means of a simple example. The Hilbert space consists of all functions of x defined on the interval $-1 \leq x \leq 1$; the scalar product is given by $\int_{-1}^{1} u^*(x)v(x)dx$. The problem is that the position operator x is indeed self-adjoint, but has no eigenvalues (so we cannot measure a local value). For if we want to solve the eigenvalue equation $xu_{x_0}(x) = x_0 u_{x_0}(x)$ for the eigenvalue x_0, we find $(x - x_0)u_{x_0}(x) = 0$, so that for $x \neq x_0$, the trivial solution $u_{x_0}(x) = 0$ is always obtained. The choice $u_{x_0}(x) = \delta(x - x_0)$ does not help, because the delta function is not square integrable and therefore not part of the Hilbert space. One can express this fact as mentioned above, by noting that such 'precise' measurements are not compatible with the uncertainty principle (and thus can be understood only in an idealized formulation).

We emphasize that for us, the problems with continuous spectra are based not primarily on mathematics (e.g. on the fact that such eigenfunctions depart from the mathematical structure of Hilbert space), but rather that *unphysical states* like the delta function states appear, which physically are not permissable in the context of quantum mechanics.[7] Since these unphysical states are not square integrable, the previously-developed probability concept of quantum mechanics cannot work with them (or at least not readily)—that is the essential difficulty.

The basic idea for getting the problem under control is to discretize[8] the continuous variable and then to let the gaps go to zero. We consider the following manipulations for general improper states $|\lambda\rangle$, which fulfill the equation (the 'substitute' of the ON relation for proper vectors)[9]:

$$\langle \lambda' | \lambda \rangle = \delta(\lambda' - \lambda). \tag{12.6}$$

As indicated in Fig. 12.1, we divide the continuum into fixed intervals of width $\Delta\lambda$ (the process is demonstrated here in one dimension; it works analogously in higher dimensions).[10] $|\lambda\rangle$ can be integrated within such an interval:

---

[5]These states are also called *Dirac states*.

[6]The strict mathematical theory of continuous spectra is somewhat elaborate (keywords e.g. *rigged Hilbert space* or *Gel'fand triple*). We content ourselves here with a less rigorous and more heuristic approach.

[7]The electron is a point object, but not its wavefunction—that would be in contradiction to the uncertainty principle.

[8]Discretizations of continuous variables are used also in other areas, e.g. in lattice gauge theories or in the numerical treatment of differential equations. Moreover, a discrete space is taken as a basis for an alternative derivation/motivation of the SEq (hopping equation; see Appendix J, Vol. 1).

[9]Whoever wishes may keep in mind $\frac{1}{\sqrt{2\pi}}e^{i\lambda x}$ instead of $|\varphi_\lambda\rangle \equiv |\lambda\rangle$ — this is not quite correct, but may be helpful here and is preferable to the auxiliary notion of a column vector.

[10]Note that we cover the axis completely with non-overlapping intervals $\Delta\lambda$.

Fig. 12.1 Discretization of
the continuous variable λ

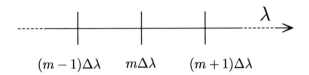

$$(m-1)\Delta\lambda \qquad m\Delta\lambda \qquad (m+1)\Delta\lambda$$

$$|\lambda_m, \Delta\lambda\rangle := \frac{1}{\sqrt{\Delta\lambda}} \int\limits_{\lambda_m}^{\lambda_m + \Delta\lambda} |\lambda'\rangle \, d\lambda', \tag{12.7}$$

where λ_m is an integral multiple of the grid size $\Delta\lambda$: $\lambda_m = m\Delta\lambda$ with $m \in \mathbb{Z}$. The expression $|\lambda_m, \Delta\lambda\rangle$ is called an *eigendifferential*. Eigendifferentials, in contrast to continuous functions, are completely 'well-behaved'. In particular, the bras belonging to them exist, and they form an ON system:

$$\begin{aligned}
\langle\lambda_n, \Delta\lambda \,|\, \lambda_m, \Delta\lambda\rangle &= \frac{1}{\Delta\lambda} \int\limits_{\lambda_n}^{\lambda_n + \Delta\lambda} d\alpha \, \langle\alpha| \int\limits_{\lambda_m}^{\lambda_m + \Delta\lambda} d\beta \, |\beta\rangle \\
&= \frac{1}{\Delta\lambda} \int\limits_{\lambda_n}^{\lambda_n + \Delta\lambda} d\alpha \int\limits_{\lambda_m}^{\lambda_m + \Delta\lambda} d\beta \, \langle\alpha|\beta\rangle \\
&= \frac{1}{\Delta\lambda} \int\limits_{\lambda_n}^{\lambda_n + \Delta\lambda} d\alpha \int\limits_{\lambda_m}^{\lambda_m + \Delta\lambda} d\beta \, \delta(\alpha - \beta) \\
&= \frac{1}{\Delta\lambda} \int\limits_{\lambda_m}^{\lambda_m + \Delta\lambda} d\beta \, \delta_{\lambda_n \lambda_m} = \delta_{\lambda_n \lambda_m}.
\end{aligned} \tag{12.8}$$

Since we have covered the entire λ axis, the states $|\lambda_m, \Delta\lambda\rangle$ are complete and therefore form a CONS. Thus, the eigendifferentials constitute a basis, with which any ket $|\Psi\rangle$ can be represented as

$$|\Psi\rangle = \sum_{\lambda_m} |\lambda_m, \Delta\lambda\rangle \, \langle\lambda_m, \Delta\lambda \,|\, \Psi\rangle. \tag{12.9}$$

This is an approximation of the continuous system which gets better and better with increasingly finer subdivision of the intervals, i.e. with decreasing $\Delta\lambda$.[11] For sufficiently small $\Delta\lambda$, we can approximate the eigendifferential (12.7) using the mean value theorem for integration (see Appendix D, Vol. 1):

---

[11]A remark on the summation index: the range of values of λ runs through all integral multiples of the grid size $\Delta\lambda$.

$$\left|\lambda_m, \Delta\lambda\right\rangle \approx \frac{1}{\sqrt{\Delta\lambda}} \left|\lambda_\mu\right\rangle \Delta\lambda = \sqrt{\Delta\lambda} \left|\lambda_\mu\right\rangle; \quad m \le \mu \le m+1, \tag{12.10}$$

and it follows that

$$\left|\Psi\right\rangle \approx \sum_{\lambda_\mu} \left|\lambda_\mu\right\rangle \left\langle\lambda_\mu \mid \Psi\right\rangle \Delta\lambda. \tag{12.11}$$

We now go to the limiting case[12]:

$$\left|\Psi\right\rangle = \lim_{\Delta\lambda \to 0} \sum_{\lambda_\mu} \left|\lambda_\mu\right\rangle \left\langle\lambda_\mu \mid \Psi\right\rangle \Delta\lambda = \int \left|\lambda\right\rangle \left\langle\lambda \mid \Psi\right\rangle d\lambda \tag{12.12}$$

or

$$\int \left|\lambda\right\rangle \left\langle\lambda\right| d\lambda = 1, \tag{12.13}$$

so that we can also expand each state in a series of improper vectors. Although this process is mathematically not clearly defined, in view of the square integrability, we can permit it in the sense that (12.12) is an abbreviation for (12.9) (simply as an imagined limiting process).

The reason for this approach is that it is possibly much easier to work with improper vectors than with proper ones. It may well be very useful to describe a quantum object using plane waves[13] e^{ikx}, although they—being infinitely extended and everywhere equal in magnitude—certainly cannot represent real physical objects. The same is true for the delta function. One can imagine a wavefunction concentrated at a point and let it tend to a delta function in the mathematical limit—but it is impossible to realize such a state physically.

In short, delta functions and plane waves are, where appropriate, very practical tools for mathematical formulations, but one must not forget that a physical state is always represented only by a square-integrable wavefunction.

With this caveat, we also accept series expansions of improper vectors $\left|\varphi_\lambda\right\rangle \equiv \left|\lambda\right\rangle$. One speaks in this context of the *extended Hilbert space* (i.e. the set of proper and improper state vectors). Generally, the extended Hilbert space is also denoted by $\mathcal{H}$, i.e. by the same symbol as the proper Hilbert space. In summary, this means that we can work with improper vectors just as with proper ones, but we have to accept the occurrence of functionals such as the delta function. The orthonormality of proper and improper vectors is expressed by the equations

$$\left\langle\varphi_n \mid \varphi_{n'}\right\rangle = \delta_{nn'} \text{ and } \left\langle\varphi_{\lambda'} \mid \varphi_\lambda\right\rangle = \delta\left(\lambda - \lambda'\right) \tag{12.14}$$

---

[12]This is nothing more than the transition from a sum to an integral, well-known from school mathematics. One lets the length of the subdivisions tend to zero, so that the upper sum and lower sum approach and converge to the integral in the limit, i.e. for infinitesimal interval length. This process is reflected in the integral sign $\int$—it is simply a stylized 'S', for 'sum'.

[13]We repeat the remark that e^{ikx} is actually an *oscillation* in space. But one always refers in this context to a *wave*, because one keeps the time-dependent factor $e^{i\omega t}$ in mind, so to speak.

and their completeness by[14]:

$$\sum |\varphi_n\rangle \langle \varphi_n| = 1 \text{ and } \int |\varphi_\lambda\rangle \langle \varphi_\lambda| \, d\lambda = 1. \tag{12.15}$$

The expansion theorem reads

$$|\Psi\rangle = \sum |\varphi_n\rangle \langle \varphi_n| \Psi\rangle \text{ and } |\Psi\rangle = \int |\varphi_\lambda\rangle \langle \varphi_\lambda |\Psi\rangle \, d\lambda. \tag{12.16}$$

Thus, we can transfer our previous statements for the discrete case to the continuous case if we perform the following substitutions:

$$n \to \lambda; \sum \to \int ; \delta_{nn'} \to \delta(\lambda - \lambda'). \tag{12.17}$$

Indeed, we may make life even easier by introducing a new symbol,[15] namely $\sum\!\!\!\!\!\!\int$. The expansion theorem is then written as

$$|\Psi\rangle = \sum\!\!\!\!\!\!\int |\alpha_j\rangle \langle \alpha_j| \Psi\rangle \tag{12.18}$$

with

$$\sum\!\!\!\!\!\!\int |\alpha_j\rangle \langle \alpha_j| \Psi\rangle = \begin{cases} \sum_j \\ \int dj \\ \sum_j + \int dj \end{cases} \text{ for } \begin{cases} \text{proper} \\ \text{improper} \\ \text{proper and improper} \end{cases} \text{states,} \tag{12.19}$$

and the completeness relation reads

$$\sum\!\!\!\!\!\!\int |\alpha_j\rangle \langle \alpha_j| \, dj = 1. \tag{12.20}$$

Similarly, orthonormality can be expressed more compactly with the following new symbol (*extended* or *generalized Kronecker symbol:*)

$$\delta(i, j) = \begin{cases} \delta_{ij} & \text{for } i, j = \text{discrete} \\ \delta(i - j) & \text{for } i, j = \text{continuous,} \end{cases} \tag{12.21}$$

namely as

$$\langle \alpha_i | \alpha_j \rangle = \delta(i, j). \tag{12.22}$$

---

[14]We recall that the increment of the sum is 1 (so we have $\Delta n = 1$). With this, one can emphasize the formal similarity between sums and integrals even more, e.g. in the form $\sum |\varphi_n\rangle \langle \varphi_n| \Delta n = 1$.
[15]There are other notations; e.g. Schwabl uses the symbol $\mathcal{S}$.

12.2   Position Representation and Momentum Representation

We will now address the question of how the ket $|\Psi\rangle$ is related to the wavefunction $\Psi(x)$.

First a preliminary remark: We have just considered why and how one can accept improper states in quantum mechanics. This allows the following formulation, using the example of position measurements considered above: Assume there is a quantum object at a point x in space,[16] i.e. with regard to its position, it is in the (abstract, improper) state $|x\rangle$. The measurement of its position can be symbolized by the position operator X, and we then have:

$$X|x\rangle = x|x\rangle. \tag{12.23}$$

In words: If we measure the state $|x\rangle$ (i.e. if we apply the position operator X to $|x\rangle$), then we find the number x as the measured value. $|x\rangle$ is an improper vector with

$$\langle x|x'\rangle = \delta(x - x') \quad \text{(ON)} \quad \text{and} \quad \int |x\rangle \langle x|\,\mathrm{d}x = 1 \text{ (C)}. \tag{12.24}$$

After this remark, we go to a proper Hilbert space, i.e. a space spanned by a CONS of proper vectors $\varphi_n(x)$. A wavefunction $\Psi(x)$ can be expanded in terms of the CONS:

$$\Psi(x) = \sum_n c_n \varphi_n(x). \tag{12.25}$$

For the coefficients, we find:

$$c_n = \int \varphi_n^*(x)\,\Psi(x)\,\mathrm{d}x. \tag{12.26}$$

Due to the orthonormality of $\{\varphi_n(x)\}$, we have

$$\int \Psi^*(x)\,\Psi(x)\,\mathrm{d}x = \sum_n c_n^* c_n. \tag{12.27}$$

We replace the coefficients by using (12.26), and obtain

$$\int \Psi^*(x)\,\Psi(x)\,\mathrm{d}x = \sum_n \int \mathrm{d}x'\,\varphi_n(x')\,\Psi(x')\int \mathrm{d}x\,\varphi_n^*(x)\,\Psi^*(x)$$
$$= \int \mathrm{d}x'\int \mathrm{d}x\,\Psi(x')\,\Psi^*(x)\sum_n \varphi_n(x')\,\varphi_n^*(x). \tag{12.28}$$

---

[16]It is clear that this is an idealized assumption which is not compatible with the uncertainty principle. But we can proceed on this assumption in terms of the above considerations concerning the eigendifferential.

Comparing the right- and left-hand sides, we see that the following equation must hold:

$$\sum_n \varphi_n \left(x'\right) \varphi_n^* \left(x\right) = \delta \left(x' - x\right). \tag{12.29}$$

We note that the expression on the left-hand side is not a scalar product. Recalling our analogy ($\varphi \rightarrow$ column vector) introduced earlier, as well as ($\varphi^\dagger \rightarrow$ row vector), we can suppose that we have an expression of the form $\sum |\varphi_n\rangle \langle\varphi_n|$. In fact, we derived such an expression in the last chapter. Starting with $\langle\Psi| \Psi\rangle$ and carrying out exactly the same procedure (but only in the abstract space),[17] we obtained there the completeness relation in the form (see (12.15))

$$\sum_n |\varphi_n\rangle \langle\varphi_n| = 1. \tag{12.30}$$

In other words, the (12.29) and (12.30) describe the same facts (completeness of the basis set), just with different notations. We see this more clearly if we transform (12.30) using (12.24):

$$\langle x| \left(\sum_n |\varphi_n\rangle \langle\varphi_n|\right) |x'\rangle = \sum_n \langle x| \varphi_n\rangle \langle\varphi_n |x'\rangle = \langle x| x'\rangle = \delta \left(x - x'\right). \tag{12.31}$$

The comparison of this equation with (12.29) suggests the following identification:

$$\varphi_n \left(x\right) = \langle x| \varphi_n\rangle \tag{12.32}$$

$\varphi_n \left(x\right)$ is called the *position representation* of the ket $|\varphi_n\rangle$. Formally, it is a scalar product of two abstract vectors, whose result is—as always—a scalar, and to which we can apply the (now) well-known rules of calculation. For example, we have

$$\langle\varphi_n| x\rangle = \langle x| \varphi_n\rangle^\dagger = \langle x| \varphi_n\rangle^* = \varphi_n^* \left(x\right). \tag{12.33}$$

So far so good. We now can 'play around' with this notation a bit. One question might be: If $\langle x| \varphi\rangle = \varphi \left(x\right)$ is the position representation of $|\varphi\rangle$, then what is the position representation of $|x'\rangle$? This state is characterized by the fact that a position measurement returns the result x'—the quantum object is at x', and only there can we

---

[17]We repeat this derivation briefly by writing the first line of (12.28) with bra-kets (remember: $\int f^* g \, dx = \langle f |g\rangle$):

$$\int \Psi^* \left(x\right) \Psi \left(x\right) dx = \langle\Psi |\Psi\rangle = \sum_n \langle\varphi_n |\Psi\rangle \langle\Psi |\varphi_n\rangle$$

$$= \sum_n \langle\Psi |\varphi_n\rangle \langle\varphi_n |\Psi\rangle = \langle\Psi| \left(\sum_n |\varphi_n\rangle \langle\varphi_n|\right) |\Psi\rangle .$$

find it.[18] In fact, we have already answered the question with (12.24): The position representation of $|x'\rangle$ is $\langle x |x'\rangle = \delta (x - x')$.

Another question: We have worked with states which are characterized not by a sharp position, but by a sharp momentum (or $k = p/\hbar$). In the abstract notation, this is the ket $|k\rangle$, whose position representation we already know—it is a plane wave[19] (the prefactor is due to the normalization):

$$\langle x| k \rangle = \frac{1}{\sqrt{2\pi}} e^{ikx}. \tag{12.34}$$

By taking the adjoint, we obtain immediately the *momentum representation* of a state with a sharply-defined position:

$$\langle k| x \rangle = \frac{1}{\sqrt{2\pi}} e^{-ikx}, \tag{12.35}$$

and therefore

$$\langle k |k'\rangle = \delta (k - k') \text{ (ON)} \quad \text{and} \quad \int dk\, |k\rangle \langle k| = 1 \text{ (C)}. \tag{12.36}$$

In short, the improper vectors $|k\rangle$ also form a CONS.

We can now write the ket $|\Psi\rangle$ in both the position and the momentum representations, namely as[20]

$$\langle x |\Psi\rangle = \Psi(x): \text{position representation}$$
$$\langle k |\Psi\rangle = \hat{\Psi}(k): \text{momentum representation.} \tag{12.37}$$

How are these two representations related? We multiply by 1 and obtain

$$\langle x |\Psi\rangle = \langle x| \int dk\, |k\rangle \langle k |\Psi\rangle = \int dk\, \langle x |k\rangle \langle k |\Psi\rangle$$
$$\langle k |\Psi\rangle = \langle k| \int dx\, |x\rangle \langle x |\Psi\rangle = \int dx\, \langle k |x\rangle \langle x |\Psi\rangle, \tag{12.38}$$

---

[18]We note again that this is an idealized formulation.

[19]Again, the above statement on oscillations and waves applies.

[20]Because these are two representations of the same ket $|\Psi\rangle$, sometimes the same symbol is used for both representations, i.e. $\Psi(x)$ and $\Psi(k)$, although these two functions are not the same (as mapping, that is, in the sense that one does not obtain $\Psi(k)$ by simply replacing x by k in $\Psi(x)$). What is precisely meant has to be inferred from the context. To avoid confusion, we use the notation $\hat{\Psi}(k)$.

or in the 'usual' notation:

$$\Psi(x) = \frac{1}{\sqrt{2\pi}} \int dk \; e^{ikx} \hat{\Psi}(k)$$

$$\hat{\Psi}(k) = \frac{1}{\sqrt{2\pi}} \int dx \; e^{-ikx} \Psi(x). \tag{12.39}$$

We see that the position and momentum representations of a ket are *Fourier transforms* of each other.[21]

Finally the question arises as to how to derive equations and operators in the two representations. We consider an abstract eigenvalue equation of the form

$$A \, |\Psi\rangle = a \, |\Psi\rangle \tag{12.40}$$

and wish to write it in the position representation. For this purpose, we multiply first with a bra $\langle x|$:

$$\langle x| \, A \, |\Psi\rangle = a \, \langle x \, |\Psi\rangle \tag{12.41}$$

and then multiply the left-hand side by the identity:

$$\langle x| \, A \int dx' \, |x'\rangle\langle x' \, |\Psi\rangle = a \, \langle x \, |\Psi\rangle \rightarrow$$
$$\int dx' \, \langle x| \, A \, |x'\rangle\langle x' \, |\Psi\rangle = a \, \langle x \, |\Psi\rangle. \tag{12.42}$$

Now it depends on the matrix element $\langle x| \, A \, |x'\rangle$ how to continue. A significant simplification of the equation is obtained only if the following applies:

$$\langle x| \, A \, |x'\rangle = \delta(x - x')A(x). \tag{12.43}$$

In this case, one says that A is *diagonal* in the position representation, or that the operator A is a *local* operator.[22] With this understanding, (12.42) apparently may be written as

$$\int dx' \delta(x - x')A(x)\langle x' \, |\Psi\rangle = A(x) \langle x \, |\Psi\rangle = a \, \langle x \, |\Psi\rangle, \tag{12.44}$$

or, in the familiar notation with a position variable,

$$A\Psi(x) = a\Psi(x), \tag{12.45}$$

---

[21] For an introduction to Fourier transforms, see Appendix H, Vol. 1.

[22] Quasi-local operators are defined via the derivative of the delta function:

$$A(x, y) = a(x)\,\delta(x - y) \quad \text{local operator}$$
$$B(x, y) = b(x)\,\delta'(x - y) \quad \text{quasi-local operator}$$

where A now stands for the position representation of the operator. We point out again that the operators A in (12.40) and (12.45) are not identical. Though one usually writes the same symbol, they are quite different mathematical objects.

Finally, an example: We want to derive the position representation of the momentum operator p_{op} (which we know already, of course). For better readability, we indicate the operator for the moment by an index op.

We start with the abstract eigenvalue equation

$$p_{op} \left| k \right\rangle = \hbar k \left| k \right\rangle . \tag{12.46}$$

For the following considerations, we know nothing about the momentum operator apart from this eigenvalue equation. $\left| k \right\rangle$ is a state of well-defined momentum (note $p = \hbar k$), and $\hbar k$ is its eigenvalue or measured value. Multiplication with $\left\langle x \right|$ and insertion of the identity leads to:

$$\int \left\langle x \right| p_{op} \left| x' \right\rangle \left\langle x' \middle| k \right\rangle \mathrm{d}x' = \hbar k \int \left\langle x \middle| x' \right\rangle \left\langle x' \middle| k \right\rangle \mathrm{d}x' = \hbar k \left\langle x \middle| k \right\rangle = \frac{\hbar k}{\sqrt{2\pi}} e^{ikx}.$$
$$\tag{12.47}$$

It follows that:

$$\int \left\langle x \right| p_{op} \left| x' \right\rangle e^{ikx'} \mathrm{d}x' = \hbar k \int \delta(x - x') e^{ikx'} \mathrm{d}x' = \frac{\hbar}{i} \int \delta(x - x') \frac{\partial}{\partial x'} e^{ikx'} \mathrm{d}x'.$$
$$\tag{12.48}$$

Comparing the left- and right-hand sides yields

$$\left\langle x \right| p_{op} \left| x' \right\rangle = \delta(x - x') \frac{\hbar}{i} \frac{\partial}{\partial x}. \tag{12.49}$$

Thus, the momentum operator is diagonal in the position representation, and takes the well-known form—as was to be expected.

An example of a non-diagonal operator in the position representation (projection operator) can be found in the exercises. We have thus filled in a gap as promised in previous chapters.

12.3   Conclusions

We started in Chaps. 1 and 2 with two (at first sight completely different) descriptions of states, namely as position-dependent wavefunctions $\psi\,(\mathbf{r})$ (analytical approach, odd chapters) on the one hand, and as kets $\left| \varphi \right\rangle$ (algebraic approach, even chapters), with their representations as column vectors, on the other hand. After travelling long route, which encompassed necessarily a lot of other material (in large part, the path was also our destination),[23] we have combined the two approaches in this chapter

---

[23]"Caminante no hay camino, se hace camino al andar…" Antonio Machado, Spanish poet.

Fig. 12.2 The same physical
situation permits various
representations

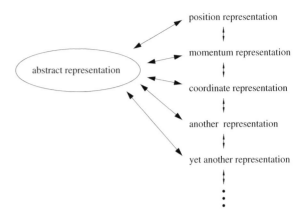

and have seen that in the end, they are simply different formulations of the same
facts; this also allows for other representations, cf. Fig. 12.2. We will take advantage
of this circumstance in the following and will switch back and forth between the two
formulations as proves be convenient for us—wave or matrix mechanics, or the
abstract notation.

However, it is clear that wave mechanics cannot describe various properties, e.g.
the spin (or, beyond the scope of this book, strangeness, charm, etc.). In other words,
for any position-dependent wavefunction, there is a ket, but the converse is not
true. But this is now no longer a problem, as we have extended wave mechanics to
a general formalism in this chapter.

12.4 Exercises

1. Given an eigenstate $|k\rangle$ of the momentum operator; how is this state described
   in the position representation?
2. Show by using $\langle x|\, k\rangle = \frac{1}{\sqrt{2\pi}}e^{ikx}$ that the improper vectors $|k\rangle$ form a CONS.
3. Given an improper vector $|\varphi_\lambda\rangle$, what is the associated eigendifferential $\left|\varphi_{\lambda,\Delta\lambda}\right\rangle$?
4. Given the state $|k\rangle$ with the sharply-defined momentum k; we have $\langle x|\, k\rangle = \frac{1}{\sqrt{2\pi}}e^{ikx}$.

   (a) What is the (abstract) eigendifferential?
   (b) How is the eigendifferential expressed in the position representation?
   (c) Show that the eigendifferentials of (b) are orthonormal.

5. Given the SEq in the abstract formulation

$$i\hbar\frac{\mathrm{d}}{\mathrm{d}t}\,|\psi\rangle = H\,|\psi\rangle, \qquad (12.50)$$

(a) Formulate this equation in the position representation and in the momentum representation.

(b) How can one calculate the matrix element $\langle k| H |k'\rangle$, if H is known in the position representation?

6. Given a CONS $\{|\varphi_n\rangle\}$; formulate the projection operator

$$P_1 = |\varphi_1\rangle \langle\varphi_1| \tag{12.51}$$

in the position representation.

7. A and B are self-adjoint operators with $[A, B] = i\hbar$, and $|a\rangle$ is an eigenvector of A with the eigenvalue a. Then we have

$$\langle a |[A, B]| a\rangle = \langle a |AB - BA| a\rangle = (a - a) \langle a |B| a\rangle = 0. \tag{12.52}$$

On the other hand, we also have:

$$\langle a |[A, B]| a\rangle = \langle a |i\hbar| a\rangle = i\hbar \neq 0. \tag{12.53}$$

Question: where is the flaw in this argument?

Chapter 13
Operators

In this chapter, we assemble some basic properties of the most important types of operators in quantum mechanics.

As we have seen, the states of quantum mechanics are defined on an (extended) Hilbert space $\mathcal{H}$. Changes of these states are caused by operators: This can be, for example, the time evolution of the system itself, or the filtering of certain states out of a general state. We have already met up with the zoo of operators of quantum mechanics (Hermitian, unitary and projection operators). But given the central role of operators in quantum mechanics, we want to discuss in this chapter some of their properties in more detail, taking the abstract formulation as a basis.[1]

With one exception, the operators considered in this book are linear. An operator A is called *linear* if for any two states and any two numbers $\alpha, \beta \in \mathbb{C}$, it holds that:

$$A \left(\alpha \left| \varphi \right\rangle + \beta \left| \psi \right\rangle \right) = \alpha A \left| \varphi \right\rangle + \beta A \left| \psi \right\rangle. \tag{13.1}$$

For the exception, namely an *antilinear* operator B, it holds that:

$$B \left(\alpha \left| \varphi \right\rangle + \beta \left| \psi \right\rangle \right) = \alpha^* B \left| \varphi \right\rangle + \beta^* B \left| \psi \right\rangle. \tag{13.2}$$

An anti-linear map is for example the complex conjugation, and thus also the scalar product with respect to the first component, since then $\left\langle \lambda a \, | \, b \right\rangle = \lambda^* \left\langle a \, | \, b \right\rangle$. Furthermore, the time-reversal operator is anti-linear (see Chap. 21, Vol. 2).

An operator is called *bounded* if there is a constant C which does not depend on the states $\left| \varphi \right\rangle \in \mathcal{H}$, so that for all states, it holds that

$$\| A \left| \varphi \right\rangle \| \leq C \, \| \left| \varphi \right\rangle \|. \tag{13.3}$$

---

[1] Further material on operators is found in Appendix I, Vol. 1.

© Springer Nature Switzerland AG 2018
J. Pade, *Quantum Mechanics for Pedestrians 1*, Undergraduate Lecture
Notes in Physics, https://doi.org/10.1007/978-3-030-00464-4_13

The *domain of definition* (or briefly the domain) of an operator A is the set of all vectors $|\varphi\rangle \in \mathcal{H}$, such that $A|\varphi\rangle$ also belongs to $\mathcal{H}$. One can show that the domain of definition of A is the whole Hilbert space, if and only if A is bounded.

If two operators A and B commute, one says that they are *simultaneously* measurable. However, this notion is not defined by any sort of time consideration, but is simply a short form for the fact that the measurement result is independent of the chronological order in which we measure A and B.

13.1 Hermitian Operators, Observables

We can distinguish three levels: First, there is the measurable physical variable A_{phys}, which is modelled in quantum mechanics by a Hermitian operator $A_{op} = A_{op}^{\dagger}$. This abstract operator can be expressed, if necessary, in a concrete representation as A_{repr}. As an example, we consider the angular momentum. The measurable physical variable is $\mathbf{l}_{\text{phys}}$, the corresponding operator $\mathbf{l}_{op} = \mathbf{r} \times \mathbf{p}$, and in the position representation, it is $\mathbf{l}_{\text{repr}} = \frac{\hbar}{i}\mathbf{r} \times \nabla$, as is well known. As mentioned above, often the same notation is used for all three objects (in the example $\mathbf{l}$), since usually the context makes clear what is meant. We will proceed essentially in this way.

We start from the eigenvalue equation (the spectrum is assumed to be discrete and not degenerate):

$$A|\varphi_n\rangle = a_n|\varphi_n\rangle; \quad n = 1, 2, \ldots ; A = A^{\dagger}. \tag{13.4}$$

The possible result of a measurement of the measurable physical variable A is one of the eigenvalues of the operator A. Because of the importance of this fact there is a special name, namely *observable*.[2] We mean by this a Hermitian operator that represents a consistently measurable physical quantity. Some remarks on the concept 'observable' are found in Appendix I, Vol. 1.

We point out that we use 'self-adjoint' and 'Hermitian' as equivalents, which applies for all of the systems we consider. In fact, under certain conditions the two terms are not identical; in infinite-dimensional vector spaces, Hermiticity does not necessarily imply self-adjoint. More on this topic may be found in Appendix I, Vol. 1.

Two remarks are appropriate here:

1. An operator A is called *anti-Hermitian* if $A^{\dagger} = -A$. Each operator C can be broken into a Hermitian and an anti-Hermitian part[3]:

---

[2]The term is not defined in the same way everywhere, and is sometimes rather avoided. The reason for this rejection stems in part from the fact that the name 'observable' suggests that without an observer (perhaps even a human), physical quantities cannot become real. We explicitly point out that for us, the term observable does not imply this problem, but is simply a technical term in the above sense.

[3]Much in the way that each function can be decomposed into a mirror-symmetric and a point-symmetric part.

$$C = C_{\text{Hermitian}} + C_{\text{anti-Hermitian}} = \frac{C + C^\dagger}{2} + \frac{C - C^\dagger}{2}. \qquad (13.5)$$

2. The product of an operator with its adjoint is a Hermitian operator: $\left(AA^\dagger\right)^\dagger = A^{\dagger\dagger} A^\dagger = AA^\dagger$. In addition, $AA^\dagger$ is a *positive operator*; that is, for all $|\varphi\rangle$, it holds that $\langle\varphi| AA^\dagger |\varphi\rangle \geq 0$. This follows from the fact that $\langle\varphi| AA^\dagger |\varphi\rangle$ is the square of a norm, since we have $\langle\varphi| AA^\dagger |\varphi\rangle = \left\| A^\dagger |\varphi\rangle \right\|^2$.[4]

Finally, a word about the symmetrization discussed in Chap. 3. For example, in classical mechanics we have $xp_x = p_x x$, but for the corresponding quantum-mechanical quantities, $xp_x \neq p_x x$. For this reason, we introduced the symmetrized form $\frac{1}{2}(xp_x + p_x x)$. We can now deliver the reasoning: Given two Hermitian operators A and B with $[A, B] \neq 0$. The product AB is not Hermitian (so it cannot correspond to a measurable variable), for $(AB)^\dagger = BA \neq AB$. But we can construct a Hermitian operator by taking the symmetrized form $C = \frac{1}{2}(AB + BA)$, because it is $C^\dagger = \frac{1}{2}(AB + BA)^\dagger = C$. According to the above considerations, it cannot be guaranteed offhand that this symmetrized operator represents an observable.

13.1.1 Three Important Properties of Hermitian Operators

In the following, we want to prove three important properties of Hermitian operators using the bra-ket formalism. We know already two of them, namely that the eigenvalues are real and that the eigenfunctions are pairwise orthogonal (we assume that the spectrum is not degenerate). In addition, we will show that commuting Hermitian operators have a common CONS.

13.1.1.1 Eigenvalues Are Real

Since measurements of physical quantities always mean measurements of real numbers (lengths, angles, arc degrees etc.), we require that the eigenvalues of the modelling operators also be real. This is indeed the case for Hermitian operators, as we now show (again).

The operator A is Hermitian, $A^\dagger = A$; its eigenvalue equation is

$$A |\varphi_n\rangle = a_n |\varphi_n\rangle; \quad n = 1, 2, \ldots \qquad (13.6)$$

with eigenvectors $|\varphi_n\rangle$. We multiply from the left by a bra:

$$\langle\varphi_n |A| \varphi_n\rangle = a_n \langle\varphi_n |\varphi_n\rangle = a_n. \qquad (13.7)$$

---

[4] We note that the term 'positive operator' is common but *not negative* or *positive-semidefinite* would be more correct. However, one can make the distinction between positive (≥ 0) and strictly positive (> 0).

We then have:

$$a_n^\dagger = a_n^* = \langle \varphi_n \,|A|\, \varphi_n \rangle^\dagger = \left(\varphi_n \left| A^\dagger \right| \varphi_n \right) = \langle \varphi_n \,|A|\, \varphi_n \rangle = a_n. \qquad (13.8)$$

Thus, the eigenvalues of a Hermitian operator are real.

13.1.1.2 Eigenvectors Are Orthogonal

Next, we want to show that we have $\langle \varphi_m \,|\varphi_n \rangle = 0$ for $n \neq m$, provided that the spectrum is not degenerate (degenerate spectra are discussed further below). We start with

$$A \,|\varphi_n\rangle = a_n \,|\varphi_n\rangle \ \text{ and } \ \langle \varphi_m| \, A = a_m \,\langle \varphi_m|, \qquad (13.9)$$

since A is Hermitian and hence has real eigenvalues. It follows that

$$\langle \varphi_m \,|A|\, \varphi_n \rangle = a_n \,\langle \varphi_m \,|\varphi_n \rangle \ \text{ and } \ \langle \varphi_m \,|A|\, \varphi_n \rangle = a_m \,\langle \varphi_m \,|\varphi_n \rangle. \qquad (13.10)$$

Subtracting the two equations leads to

$$(a_m - a_n) \,\langle \varphi_m \,|\varphi_n \rangle = 0. \qquad (13.11)$$

Therefore, it must hold (since we have assumed non-degeneracy) that $\langle \varphi_m| \, \varphi_n \rangle = 0$ for $n \neq m$. If we take into account also the normalization of the eigenfunctions, we find, as expected:

$$\langle \varphi_m \,|\varphi_n \rangle = \delta_{nm}. \qquad (13.12)$$

Thus, the eigenfunctions of a (nondegenerate) Hermitian operator always form an orthonormal system.

13.1.1.3 Commuting Hermitian Operators Have a Common CONS

Given two Hermitian operators A and B (with nondegenerate spectra). They commute if and only if they have a common CONS of eigenvectors. To prove the claim, two steps are necessary: Step 1: $[A, B] = 0 \rightarrow$ common system; Step 2: common system $\rightarrow [A, B] = 0$.

Step 1. We start with

$$A \,|\varphi_i\rangle = a_i \,|\varphi_i\rangle \qquad (13.13)$$

where $\{|\varphi_i\rangle\}$ is a CONS. It follows that

$$BA \,|\varphi_i\rangle = \begin{cases} Ba_i \,|\varphi_i\rangle = a_i B \,|\varphi_i\rangle \\ AB \,|\varphi_i\rangle, \text{ since } [A, B] = 0 \end{cases} \qquad (13.14)$$

or, in summary,

$$AB\,|\varphi_i\rangle = a_i B\,|\varphi_i\rangle. \qquad (13.15)$$

Comparing this equation with (13.13) shows (because in both cases the same eigenvalue a_i appears) that $B\,|\varphi_i\rangle$ must be a multiple of the eigenfunction $|\varphi_i\rangle$:

$$B\,|\varphi_i\rangle \sim |\varphi_i\rangle. \qquad (13.16)$$

We call the proportionality constant b_i. It follows that

$$B\,|\varphi_i\rangle = b_i\,|\varphi_i\rangle, \qquad (13.17)$$

i.e. the operator B has the CONS $\{\varphi_i\}$, also. But as the roles of A and B can be interchanged in this argument; it follows that both operators have exactly the same CONS. Thus step 1 is completed.

Step 2. On condition that a common CONS exists, it will be shown that the commutator $[A, B]$ vanishes. Thus we assume:

$$A\,|\varphi_i\rangle = a_i\,|\varphi_i\rangle \text{ and } B\,|\varphi_i\rangle = b_i\,|\varphi_i\rangle. \qquad (13.18)$$

It follows that

$$BA\,|\varphi_i\rangle = a_i B\,|\varphi_i\rangle = a_i b_i\,|\varphi_i\rangle \text{ and } AB\,|\varphi_i\rangle = b_i A\,|\varphi_i\rangle = b_i a_i\,|\varphi_i\rangle. \qquad (13.19)$$

The right-hand sides of these equations are equal, hence also the left-hand sides must be equal, so we find

$$(AB - BA)\,|\varphi_i\rangle = [A, B]\,|\varphi_i\rangle = 0. \qquad (13.20)$$

This equation does not tell us that the commutator vanishes, but only that its application to an *eigenvector* gives zero. On the other hand, we know that the system $\{\varphi_i\}$ is complete, meaning that *any* vector can be represented as

$$|\Psi\rangle = \sum_i d_i\,|\varphi_i\rangle. \qquad (13.21)$$

With this, we have for any vector $|\Psi\rangle$

$$[A, B]\,|\Psi\rangle = \sum_i d_i\,[A, B]\,|\varphi_i\rangle = 0 \qquad (13.22)$$

and hence the statement $[A, B] = 0$ is true in the *entire* Hilbert space.

Commuting observables thus have a common system of eigenvectors. A remark: time-independent observables which commute with the Hamiltonian are *conserved quantities*, see Chap. 9.

13.1.2 Uncertainty Relations

13.1.2.1 For Two Hermitian Operators

In Chap. 9, we defined the standard deviation or uncertainty ΔA by

$$(\Delta A) = \sqrt{\langle A^2 \rangle - \langle A \rangle^2}. \tag{13.23}$$

Starting from this, one can derive the uncertainty relation (or uncertainty principle) for Hermitian operators A and B. This is carried out in Appendix I, Vol. 1; we note here only the result:

$$\Delta A \cdot \Delta B \geq \frac{1}{2} |\langle [A, B] \rangle|. \tag{13.24}$$

This general uncertainty relation for two Hermitian operators is particularly popular for the pair x and p_x. Because of $[x, p_x] = i\hbar$, we have

$$\Delta x \cdot \Delta p_x \geq \frac{\hbar}{2}. \tag{13.25}$$

This is sketched in Fig. 13.1.[5]

13.1.2.2 When Does the Uncertainty Relation Hold
and What Does It Mean?

We emphasize that the derivation of the uncertainty relation assumes ideal (error-free) measuring instruments. In fact, the experimental errors of measuring instruments in real experiments are usually much larger than the quantum uncertainties. Accordingly, the uncertainty relation is *not* a statement about the accuracy of measuring instruments, but rather the description of a pure quantum effect.

We have seen in Chap. 9 that expressions such as Δx are *state-dependent* averaging processes. If one does not take this into account, one can deduce everything possible and impossible, and this is also true for the uncertainty relation. The notation $(\Delta x)_\psi$ or $\Delta_\psi x$ is less common, but its use would prevent this kind of misunderstanding.

---

[5]One can show that violating the uncertainty principle implies that it is also possible to violate the second law of thermodynamics; see Esther Hänggi & Stephanie Wehner, 'A violation of the

Fig. 13.1 Different realizations of the uncertainty relation (13.25). It is like pressing a balloon—pressed in one direction, the balloon evades the pressure and expands in the other direction

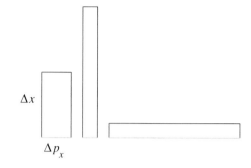

In addition, the uncertainty relation (13.24) makes sense only for those states which are in the domain of definition of A and B as well as in those of the products of the operators occurring in the derivation. For all other states, the uncertainty principle is irrelevant. For example, there are functions which are in the Hilbert space of square integrable functions, but outside the domains of definition of the (unbounded) operators x and p. For these functions, one cannot establish the inequality (13.25). Examples can be found in the exercises and in Appendix I, Vol. 1. Operator equations, and generally statements about operators, do not apply to all states, but only to those that are in the domain of definition of the operators.[6]

Regarding the meaning of the uncertainty relation (13.24), we encounter a typical situation in quantum mechanics. The theoretical formulations and derivations are quite 'straightforward' and uncontroversial. Problems occur only when one asks what all this 'really' means. We illustrate the situation with the example of the two positions that we have already briefly mentioned in Chap. 2.

1. The first position assumes that the relation (13.24) applies only to an *ensemble*. It is, therefore, about the statistical distribution in the measurement results, if one measures both A and B in a large number of identically-prepared systems (i.e. in each system, either A or B). For $[A, B] \neq 0$, the measurements are indeed incompatible, but because they are carried out on *different* systems or ensemble members, these measurements can in no way mutually interfere. In this case, the uncertainty principle has nothing to do with the possibility of performing simultaneous measurements of two quantities. It can be interpreted at best as a fundamental limitation in preparing a state (or the corresponding ensemble) as accurately as possible. In this case, the standard deviation ΔA is a straightforward concept.

2. The second position assumes that the relation (13.24) applies to *single* events. Of course, ΔA has nothing to do in this case with a statistical distribution as in the ensemble position just mentioned. As we have seen and will develop further, a central position of this view of quantum mechanics is that it does not make sense

---

uncertainty principle implies a violation of the second law of thermodynamics', Nature Communications 4, Article number 1670 (2013), https://doi.org/10.1038/ncomms2665.

[6]It is not just about the comb, so to speak, but also about the hair that is combed.

for a typical quantum state to say that A has any value at all (which does not mean that A has a value which we do not know). Under this assumption, ΔA can be interpreted, as we have seen in Chap. 9, as a numerical measure of the extent to which the property A is *not* owned by the system, since e.g. $\Delta_\psi A = 0$ means that $|\psi\rangle$ is an eigenstate of A. The same applies to $\Delta_\psi B$. The uncertainty relation is then a statement as to what extent a system can have or cannot have the properties A and B at the same time.

In addition to these two positions or interpretations of quantum mechanics, there are several more, as we shall see in later chapters.[7] Which of these is the 'correct' one is not (yet) clear at present. We can just say at this point that the formalism of quantum mechanics is unique, but its interpretation is anything but uncontroversial. In Chap. 14, and Chap. 27, Vol. 2, and especially in Chap. 28, Vol. 2, we will address these questions again.

13.1.2.3 Uncertainty Relation for Time and Energy

In the uncertainty relation (13.24), we cannot insert the time directly into A or B, since it is not an operator in quantum mechanics, but a simple *parameter* (one cannot say e.g. that "a quantum object has a well-defined time"). Nevertheless, one can formulate a statement that links time and energy. For this we consider a not explicitly time-dependent Hermitian operator A. As described in Sect. 9.3, we have

$$i\hbar\frac{\mathrm{d}}{\mathrm{d}t}\langle A\rangle = \langle[A,H]\rangle. \tag{13.26}$$

Together with the uncertainty relation, it follows that

$$\Delta A \cdot \Delta H \geq \frac{1}{2}|\langle[A,H]\rangle| = \frac{\hbar}{2}\left|\frac{\mathrm{d}}{\mathrm{d}t}\langle A\rangle\right|. \tag{13.27}$$

We define a time interval $\Delta\tau$ by

$$\Delta\tau = \frac{\Delta A}{\left|\frac{\mathrm{d}}{\mathrm{d}t}\langle A\rangle\right|}. \tag{13.28}$$

This is a measure of the time during which the value of A changes by ΔA. For example, we have for a conserved quantity $\frac{\mathrm{d}}{\mathrm{d}t}\langle A\rangle = 0$, hence $\Delta\tau = \infty$. With these concepts, we find

---

[7] Quite apart from the literary process, such as with David Foster Wallace in *Infinite jest*: "The mind says, a box-and-forest-meadows-mind can move with quantum-speed and be anytime anywhere and hear in symphonic sum of the thoughts of the living ... The mind says: It does not really matter whether Gately knows what the term quanta means. By and large, it says there are ghosts ... in a completely different Heisenberg dimension of exchange rates and time courses."

$$\Delta H \cdot \Delta \tau \geq \frac{\hbar}{2} \tag{13.29}$$

which often is written as

$$\Delta E \cdot \Delta t \gtrsim \frac{\hbar}{2} \tag{13.30}$$

This can be interpreted as a correlation between lifetime and variation in energy.

13.1.3 Degenerate Spectra

We have essentially confined ourselves to observables with nondegenerate spectra. In the discrete case, the eigenvalue equation is given by

$$A \left| \varphi_n \right\rangle = a_n \left| \varphi_n \right\rangle, \quad n = 1, 2, \ldots \tag{13.31}$$

Since A is an observable, the set $\{|\varphi_n\rangle\}$ of the eigenvectors is a basis for $\mathcal{H}$, in terms of which any state can be represented by an expansion:

$$\left| \psi \right\rangle = \sum_n c_n \left| \varphi_n \right\rangle. \tag{13.32}$$

We can express this also by saying that every eigenvector spans a *one-dimensional* subspace.

Let us now turn to the case of a *degenerate discrete spectrum*. The eigenvalue equation is

$$A \left| \varphi_{n,r} \right\rangle = a_n \left| \varphi_{n,r} \right\rangle, \quad n = 1, 2, \ldots; \ r = 1, 2, \ldots, g_n. \tag{13.33}$$

Here, g_n is the *degree of degeneracy* of the eigenvalue a_n; for $g_n = 1$, a_n is not degenerate. The $\left| \varphi_{n,r} \right\rangle$ are g_n linearly independent eigenvectors (for given n); they span the subspace (eigenspace) $\mathcal{H}_n$ of dimension g_n of the eigenvalue a_n. The eigenvectors $\left| \varphi_{n,r} \right\rangle$ with the same index n are not necessarily orthogonal to each other (we can always assume, however, that they are normalized), but with the usual methods of orthogonalization one can construct an orthogonal system in terms of them. The subspaces of *different* indices n are mutually orthogonal, $\mathcal{H}_n \perp \mathcal{H}_m$ for $n \neq m$.[8]

Finally, the expansion of a state is given by:

$$\left| \psi \right\rangle = \sum_n \sum_{r=1}^{g_n} c_{n,r} \left| \varphi_{n,r} \right\rangle, \quad n = 1, 2, \ldots; \ r = 1, 2, \ldots, g_n \tag{13.34}$$

---

[8]Two subspaces $\mathcal{H}_n$ and $\mathcal{H}_m$ are mutually orthogonal if any vector in $\mathcal{H}_n$ is orthogonal to any vector in $\mathcal{H}_m$.

In the case of a continuous spectrum, we obtain the corresponding statements and formulations by the usual replacements: discrete index → continuous variable; summation → integration; Kronecker symbol → Delta function; cf. Chap. 12. See also the chapter 'Discrete-continuous' in Appendix T, Vol. 1.

13.2 Unitary Operators

Before we repeat the definition of a unitary operator, we consider the definition of an inverse operator. Thus, we assume that an operator A with $Af = g$ exists. If the inverse of this mapping also exists, $f = A^{-1}g$, then A^{-1} is the inverse operator of A. We have $AA^{-1} = A^{-1}A$. A *unitary operator* U is defined by[9]

$$U^\dagger = U^{-1} \longleftrightarrow U^\dagger U = UU^\dagger = 1. \tag{13.35}$$

For the eigenvalues of a unitary operator, with $U |u\rangle = u |u\rangle$ and $\langle u| U^\dagger = u^* \langle u|$, the equation $|u|^2 = 1$ applies. Hence, the eigenvalues of a unitary operator are on the unit circle.

13.2.1 Unitary Transformations

With unitary operators we can define *unitary transformations* of states and operators. Common notations are

$$U |\Psi\rangle = |\Psi'\rangle \quad \text{and} \quad UAU^\dagger = A'. \tag{13.36}$$

The interesting thing about unitary transformations is that they leave important properties and quantities unchanged, namely the lengths of vectors and the 'angle' between them, and thus scalar products, as well as matrix elements and eigenvalues (see exercises). In this respect, a unitary transformation is an analogue of the rotation in elementary vector calculus. We can visualize it as a transition from one basis to another basis or coordinate system. Suppose that there two CONS $\{|\varphi_n\rangle\}$ and $\{|\psi_n\rangle\}$. Then we have

$$|\psi_n\rangle = \sum_m |\varphi_m\rangle \langle \varphi_m| \psi_n\rangle = \sum_m U_{mn} |\varphi_m\rangle, \tag{13.37}$$

---

[9]To be exact, there is the second requirement, $U\alpha |\varphi\rangle = \alpha U |\varphi\rangle$. For *antiunitary* operators T, it holds also that $TT^\dagger = T^\dagger T = 1$, but in contrast to the unitary operators, $T\alpha |\varphi\rangle = \alpha^* T |\varphi\rangle$. Anti-unitary operators appear, apart from the complex conjugation, in quantum mechanics only in connection with time reversal (see Chap. 21, Vol. 2). So the equation $UU^\dagger = U^\dagger U = 1$ almost always refers to unitary operators.

and it follows that

$$\delta_{n'n} = \langle \psi_{n'} | \psi_n \rangle = \sum_{m'm} U^*_{m'n'} \langle \varphi_{m'} | \varphi_m \rangle U_{mn} = \sum_m U^*_{mn'} U_{mn} = \left(U^\dagger U \right)_{n'n},$$

(13.38)

which is just another notation for $U^\dagger U = 1$.

13.2.2 Functions of Operators, the Time-Evolution Operator

For a given general operator A, we can define powers of A, or construct other expressions such as power series of the form

$$\sum_{n=0}^{\infty} a_n A^n$$

(13.39)

as we have already done several times in previous chapters. An example is $\sum_{n=0}^{\infty} a_n \frac{d^n}{dx^n}$.

Of course, in view of such expressions there is generally the question whether a series converges at all (i.e. whether it makes sense to write it). The answer depends on the coefficients a_n and the functions to which we apply the operator.

We want to examine an example in more detail. Consider a time-independent Hamiltonian H

$$i\hbar \frac{d}{dt} |\Psi(t)\rangle = H |\Psi(t)\rangle.$$

(13.40)

We wish to show that the *time-evolution operator*

$$U(t) = \sum_{n=0}^{\infty} \left(-i \frac{t}{\hbar} \right)^n \frac{H^n}{n!} \equiv e^{-i \frac{Ht}{\hbar}}$$

(13.41)

is unitary and transforms the initial state $|\Psi(0)\rangle$ into the state $|\Psi(t)\rangle$.[10] To that end, we assume that the state vector can be expanded in a power series about $t = 0$:

$$|\Psi(t)\rangle = |\Psi(0)\rangle + \frac{t}{1!} \left(\frac{d}{dt} |\Psi(t)\rangle \right)_{t=0} + \frac{t^2}{2!} \left(\frac{d^2}{dt^2} |\Psi(t)\rangle \right)_{t=0} + \cdots$$

$$= \sum_n \frac{t^n}{n!} \left(\frac{d^n}{dt^n} |\Psi(t)\rangle \right)_{t=0}.$$

(13.42)

---

[10]Since it, so to say, impels or propagates the state $|\Psi\rangle$ through time, it is also called *propagator*.

The time derivatives can be expressed with the help of the SEq as powers of H[11]:

$$i\hbar\tfrac{d}{dt}\,|\Psi(t)\rangle = H\,|\Psi(t)\rangle$$

$$(i\hbar\tfrac{d}{dt})^2\,|\Psi(t)\rangle = i\hbar\tfrac{d}{dt} H\,|\Psi(t)\rangle = H^2\,|\Psi(t)\rangle\cdots \qquad (13.43)$$

$$(i\hbar\tfrac{d}{dt})^n\,|\Psi(t)\rangle = i\hbar\tfrac{d}{dt} H^{n-1}\,|\Psi(t)\rangle = H^n\,|\Psi(t)\rangle.$$

We replace the time derivatives in (13.42) by these expressions and find

$$|\Psi(t)\rangle = \sum_n \frac{t^n}{n!}\left(\frac{d^n}{dt^n}\,|\Psi(t)\rangle\right)_{t=0} = \sum_{n=0}^{\infty}\left(-i\frac{t}{\hbar}\right)^n \frac{H^n}{n!}\,|\Psi(0)\rangle. \qquad (13.44)$$

On the right-hand side, we have an operator which acts on the initial state $|\Psi(0)\rangle$:

$$U(t) = \sum_{n=0}^{\infty}\left(-i\frac{t}{\hbar}\right)^n \frac{H^n}{n!} = e^{-i\frac{Ht}{\hbar}}, \qquad (13.45)$$

and can thus write the time evolution compactly as

$$|\Psi(t)\rangle = U(t)\,|\Psi(0)\rangle \qquad (13.46)$$

or more generally, $|\Psi(t_2)\rangle = U(t_2 - t_1)\,|\Psi(t_1)\rangle$.

We note that (13.45) and (13.46) are equivalent to the SEq in the form (13.40). Ultimately, it is just a matter of personal preference or habit, which one of the two formulations one uses. In any case, we can see very clearly in (13.46) the deterministic nature of the SEq: specifying an initial condition determines uniquely its solution for all times, as we have already derived in an example in Chap. 5.

Finally, we want to show that the time evolution operator U is unitary. We extend the proof and show that in general, the following relation holds:

$$\text{For } \hat{U} = e^{iA} \text{ with } A = A^\dagger \text{ follows } \hat{U}^{-1} = \hat{U}^\dagger. \qquad (13.47)$$

For the proof, we use the power series of the exponential function:

$$\hat{U}^\dagger = \left(e^{iA}\right)^\dagger = \sum \left(\frac{i^n}{n!}A^n\right)^\dagger = \sum \frac{(-i)^n}{n!}(A^n)^\dagger$$

$$= \sum \frac{(-i)^n}{n!}(A^\dagger)^n = \sum \frac{(-i)^n}{n!}A^n = e^{-iA} = U^{-1}. \qquad (13.48)$$

---

[11]Note that H does not depend on time, which is why we obtain such simple formulations. Propagators for time-dependent Hamiltonians can also be formulated, but this is somewhat more complicated.

It follows $\hat{U}^{\dagger}\hat{U} = e^{iA}e^{-iA} = 1$; hence, the operator is unitary and $U^{-1} = e^{-iA}$. The general formulation of this fact is found in the theorem of Stone, see Appendix I, Vol. 1.

We finally mention in passing that the propagator can be written as an integral operator, which is advantageous in some contexts. More on this topic in Appendix I, Vol. 1. Note that the propagator plays a dominant role in quantum field theory.

13.3 Projection Operators

An expression of the form $P = |\varphi_1\rangle \langle\varphi_1|$ is the simplest *projection operator*. Applying it to a vector, the operator projects it (as the name suggests) onto a subspace; in this example, that subspace which is spanned by $|\varphi_1\rangle$.

Generally, an *idempotent* operator is defined by

$$P^2 = P. \tag{13.49}$$

If P is Hermitian in addition, it is called a projection operator. In a Hilbert space of dimension N, we can define, for example, the projection operator

$$P = \sum_{n\leq N'} |\varphi_n\rangle \langle\varphi_n|, \tag{13.50}$$

where $\{\varphi_n\}$ is an ON system of dimension $N' \leq N$. That this is indeed a projection operator can be seen from

$$P^2 = \sum_n |\varphi_n\rangle \langle\varphi_n| \sum_m |\varphi_m\rangle \langle\varphi_m| = \sum_n \sum_m |\varphi_n\rangle \delta_{mn} \langle\varphi_m| = \sum_n |\varphi_n\rangle \langle\varphi_n| = P. \tag{13.51}$$

The eigenvalue equation for a projection operator P reads

$$P |p\rangle = p |p\rangle \tag{13.52}$$

with the eigenvectors $|p\rangle$ and the eigenvalues p.[12] Multiplication by P yields

$$P^2 |p\rangle = Pp |p\rangle = pP |p\rangle = p^2 |p\rangle. \tag{13.53}$$

On the other hand, because of $P^2 = P$, it follows that

$$P^2 |p\rangle = P |p\rangle = p |p\rangle, \tag{13.54}$$

and therefore,

---

[12]Here the notation p has, of course, nothing to do with the momentum, but with p as projection.

$$p^2 = p \text{ or } p = 0 \text{ and } 1. \tag{13.55}$$

As we well know, the completeness relation of a CONS provides us with a special projection operator, namely, a projection onto 'everything'.

$$P_{\text{whole space}} = \sum_n |\varphi_n\rangle \langle\varphi_n| = 1. \tag{13.56}$$

In words: this projection operator projects onto the whole space.[13] As we have seen, the completeness relation is not only often a very useful tool in conversions, but is also very simple to handle—because we insert just an identity (and this, by the way, gives the procedure its name). A simple example:

$$|\Psi\rangle = 1 \cdot |\Psi\rangle = \sum_n |\varphi_n\rangle \langle\varphi_n |\Psi\rangle = \sum_n |\varphi_n\rangle c_n. \tag{13.57}$$

We see that $c_n = \langle\varphi_n |\Psi\rangle$ is the projection of Ψ onto φ_n.
Note: Two projection operators are called (mutually) orthogonal if $P_1 P_2 = 0$ (this also holds for the corresponding subspaces).

13.3.1 Spectral Representation

Suppose that in a Hilbert space $\mathcal{H}$, the eigenfunctions of an operator A form a CONS $\{|a_n\rangle, n = 1, 2, \ldots\}$. Then we can express the operator using the projection operators $P_n = |a_n\rangle \langle a_n|$ built with the eigenfunctions. For with

$$A |a_n\rangle = a_n |a_n\rangle, \tag{13.58}$$

and

$$\sum_n |a_n\rangle \langle a_n| = \sum_n P_n = 1, \tag{13.59}$$

it follows that

$$A |a_n\rangle \langle a_n| = a_n |a_n\rangle \langle a_n| \leftrightarrow \sum_n A |a_n\rangle \langle a_n| = \sum_n a_n |a_n\rangle \langle a_n| \tag{13.60}$$

and therefore

$$A = \sum_n a_n |a_n\rangle \langle a_n| = \sum_n a_n P_n. \tag{13.61}$$

---

[13]To avoid misunderstandings, we repeat the remark that the last equation is an *operator equation*, i.e. simply two different representations of one operator.

This is called the *spectral representation* of an operator.[14] The spectral representation in the degenerate case is treated in the exercises.

An operator C whose eigenfunctions are not $|a_n\rangle$ can be expressed similarly. It is

$$C = \sum_n |a_n\rangle \langle a_n| \, C \sum_m |a_m\rangle \langle a_m| = \sum_{n,m} c_{nm} |a_n\rangle \langle a_m|, \tag{13.62}$$

with $c_{nm} = \langle a_n| \, C \, |a_m\rangle$.

13.3.2 Projection and Properties

Using projectors, we can make a connection to the term *property* of a system. We again start from an operator A with the CONS $\{|a_n\rangle\}, n = 1, 2, \ldots\}$, with a non-degenerate spectrum, and from the projection operators $P_n = |a_n\rangle \langle a_n|$. For these operators, the eigenvalue equation holds:

$$P_n |a_m\rangle = |a_n\rangle \langle a_n |a_m\rangle = \delta_{nm} \cdot |a_m\rangle. \tag{13.63}$$

That means that $|a_n\rangle$ is an eigenvector of P_n with eigenvalue 1; all other states $|a_m\rangle$ with $n \neq m$ are eigenvectors of P_n with the eigenvalue 0. In other words: P_n projects onto a *one-dimensional* subspace of $\mathcal{H}$.

We now consider a system in the normalized state $|\psi\rangle = \sum_n c_n |a_n\rangle$ and the projection operator $P_k = |a_k\rangle \langle a_k|$. We have

$$P_k |\psi\rangle = \sum_n c_n P_k |a_n\rangle = \sum_n c_n |a_k\rangle \langle a_k |a_n\rangle = c_k |a_k\rangle. \tag{13.64}$$

This means that the state $|\psi\rangle$ is an eigenstate of P_k with the eigenvalue 1 iff $c_n = \delta_{kn}$ ($c_k = 1$, because of the normalization of the state), and with the eigenvalue 0 iff $c_k = 0$:

$$\begin{aligned} P_k |\psi\rangle = 1 \cdot |\psi\rangle &\Leftrightarrow |\psi\rangle = c_k |a_k\rangle \, ; c_k = 1 \\ P_k |\psi\rangle = 0 \cdot |\psi\rangle &\Leftrightarrow |\psi\rangle = \sum_n c_n |a_n\rangle \, ; c_k = 0. \end{aligned} \tag{13.65}$$

So we can draw the following conclusion: $P_k = 1$ means 'if the system is in the state $|\psi\rangle$ and A is measured, then the result is a_k' or (in a slightly more casual formulation) 'A has a value of a_k' or 'the system has the property a_k'. In this sense, we can understand projection operators as representing yes-no observables, i.e. as a response to the question of whether the value of a physical quantity A is given by a_k (1: Yes, the quantum system has the property a_k) or not (0: No, the quantum system does not have the property a_k).[15]

---

[14]We have found it already as an example in an exercise of Chap. 11.

[15]We have here a connection to logic (via '1 $\hat{=}$ true' and '0 $\hat{=}$ false'). In classical physics, such a statement (the quantity A has the value a_k) is either true or false; in quantum mechanics or quantum logic, the situation may be more complex.

For if a state has a property a_k (in the sense that it had it before the measurement and the measurement makes us aware of this previously unknown value, e.g. 'horizontally linear-polarized'), then $P_k = 1$ and all other projections equal zero, $P_{n \neq k} = 0$.[16]

13.3.3 Measurements

We have formulated the measurement process already in previous chapters with the help of projection operators—a measurement corresponds to the projection of a state onto a particular subspace, which is one- or multidimensional, as the case may be. Also, the 'production' of an initial state at time $t = 0$ can be regarded as a kind of measurement, because here a superposition state is projected onto a certain subspace. However, this is generally not called 'measurement' but rather *preparation* of a state.[17] We want to set down in this section once again the essential terms in the case of degeneracy.

We start at the initial time with a state that evolves unitarily according to the Schrödinger equation until the time of measurement. We assume that this state is a superposition of basis states, as is described by the expansion theorem. Immediately before the measurement, we can for example write

$$|\Psi\rangle = \sum_n \sum_{r=1}^{g_n} c_{n,r} |\varphi_{n,r}\rangle , \qquad (13.66)$$

with

$$\langle \varphi_{m,s} | \varphi_{n,r} \rangle = \delta_{m,n} \delta_{r,s}. \qquad (13.67)$$

Through the measurement, the state vector is changed; if we measure the system in the state m (with or without degeneracy; the denominators are due to the normalization), we obtain

$$|\Psi\rangle \stackrel{\text{measurement}}{\longrightarrow} \frac{c_m |\varphi_m\rangle}{|c_m|} \text{ or } |\Psi\rangle \stackrel{\text{measurement}}{\longrightarrow} \frac{1}{\sqrt{\sum_{r=1}^{g_n} |c_{m,r}|^2}} \sum_{r=1}^{g_m} c_{m,r} |\varphi_{m,r}\rangle \qquad (13.68)$$

(reduction of the wave packet, collapse of the wave function). The vector $\sum_{r=1}^{g_n} c_{m,r} |\varphi_{m,r}\rangle$ is none other than the projection of $|\Psi\rangle$ onto the subspace belonging to m; the projection operator is

----

[16]More on this topic in Chap. 27, Vol. 2.

[17]Some remarks on terms that arise in connection with 'measurements' are given in Appendix S, Vol. 1.

$$P_m = |\varphi_m\rangle\langle\varphi_m| \text{ or } P_m = \sum_{r=1}^{g_m} |\varphi_{m,r}\rangle\langle\varphi_{m,r}|, \tag{13.69}$$

so that we can write (13.68) compactly (i.e. irrespective of whether or not there is degeneracy) as

$$|\Psi\rangle \xrightarrow{\text{measurement}} \frac{P_m|\Psi\rangle}{\sqrt{\langle\Psi|P_m|\Psi\rangle}} = \frac{P_m|\Psi\rangle}{\sqrt{\langle P_m\rangle}}. \tag{13.70}$$

Therefore, a measurement can be understood in this sense as a projection onto a corresponding subspace.[18]

13.4 Systematics of the Operators

For greater clarity, we want to discuss briefly the 'family tree' of the operators used here (see Fig. 13.2); a similar pedigree for matrices can be found in Appendix F, Vol. 1). They are all linear (with the aforementioned exception of the complex conjugation and time reversal operations) and *normal*. An operator A is called normal if it fulfills $AA^\dagger = A^\dagger A$.

We can easily convince ourselves that the operators which are important for quantum mechanics (Hermitian, positive, projection, unitary operators) are all normal:

$$\begin{aligned} A \text{ Hermitian: } A = A^\dagger &\to AA^\dagger = A^\dagger A \\ U \text{ unitary: } U^{-1} = U^\dagger &\to UU^\dagger = 1 = U^\dagger U. \end{aligned} \tag{13.71}$$

The interest in normal operators is, among other things, that they can be diagonalized. Actually, we find more generally: An operator can be diagonalized by a

Fig. 13.2 The family tree of linear operators

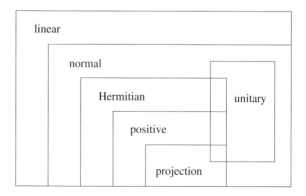

---

[18]We use here the fact that all states $e^{i\alpha}|\Psi\rangle$ are physically equivalent for arbitrary real α. See also Chap. 14.

unitary transformation iff it is normal.[19] Non-normal operators can also be diagonalized under some circumstances—but not by a unitary (i.e. length-conserving) transformation. An example is given in the exercises.

Because of the diagonalizability of the operators or matrices occurring in quantum mechanics, we can always expand in terms of eigenfunctions without having to worry about Jordan normal forms or the like. This contributes significantly to the well-behaved character of quantum mechanics.

13.5 Exercises

1. Let A be a linear and B an anti-linear operator; $|\varphi\rangle$ is a state. Compute or simplify $A\,(i\,|\varphi\rangle)$ and $B\,(i\,|\varphi\rangle)$.
2. Show that the complex conjugation $\mathcal{K}$ is an anti-linear operator.
3. Show that the commutator $C = [A, B]$ of two Hermitian operators A and B is anti-Hermitian.
4. The Hermitian operators A and B fulfill $[A, B] \neq 0$. Consider the operator $Q = c\,[A, B]$. For which c is Q a Hermitian operator?
5. Consider the operator $Q = AB$, where A and B are Hermitian matrices. Under what conditions is Q a Hermitian operator?
6. Show in the bra-ket representation that:

   (a) Hermitian operators have real eigenvalues.
   (b) The eigenfunctions of Hermitian operators are pairwise orthogonal (assuming the spectrum is not degenerate).

7. Show that the mean value of a Hermitian operator A is real, and the mean value of an anti-Hermitian operator B is imaginary.
8. What is the quantum-mechanical operator for the classical term $\mathbf{p} \times \mathbf{l}$?
9. Calculate the mean value of σ_z for the normalized state $\begin{pmatrix} a \\ b \end{pmatrix}$.
10. Given the time-independent Hamiltonian H; what is the associated time evolution operator $U(t)$?
11. Let U be the operator $U = e^{iA}$, where A is a Hermitian operator. Show that U is unitary.
12. What are the eigenvalues that a unitary operator can have?
13. Show that the time evolution operator $e^{-i\frac{Ht}{\hbar}}$ is unitary.
14. Show that scalar products, matrix elements, eigenvalues and expectation values are invariant under unitary transformations.
15. P_1 and P_2 are projection operators. Under which conditions are $P = P_1 + P_2$ and $P = P_1 P_2$ projection operators?
16. Formulate the matrix representation of the operator $P = |e_1\rangle \langle e_1|$ in $\mathbb{R}^3$.
17. What is the general definition of a projection operator?

---

[19]The proof is found in Appendix I, Vol. 1.

18. Given the CONS $\{|\varphi_n\rangle\}$; for which c_n is the operator $A = \sum c_n |\varphi_n\rangle \langle\varphi_n|$ a projection operator?

19. Which eigenvalues can a projection operator have?

20. Given the CONS $\{|\varphi_n\rangle\}$ in a Hilbert space of dimension N. Consider the operator

$$P = \sum_{n \leq N} |\varphi_n\rangle \langle\varphi_n| \qquad (13.72)$$

with $N' \leq N$. Show that P is a projection operator.

21. Given the operator A with degenerate spectrum:

$$A |\varphi_{n,r}\rangle = a_n |\varphi_{n,r}\rangle ; \quad r = 1, \ldots, g_n. \qquad (13.73)$$

(a) Formulate the projection operator onto the states with subscript n.
(b) Formulate the spectral representation of A.

22. Given the operators $A = |\varphi\rangle \langle\varphi|$ and $B = |\psi\rangle \langle\psi|$. Let $\langle\varphi| \psi\rangle = \alpha \in \mathbb{C}, \alpha \neq 0$. For which α is the operator $C = AB$ a projection operator?

23. Given the operator $Q = B^\dagger B$, where B is unitary. How can Q be more simply written?

24. Given the operator $Q = B^\dagger B$, where B is not unitary. Show that the eigenvalues of Q are real and that they are not negative.

25. Given the operator $A = \beta |\varphi\rangle \langle\psi|$. Let $\langle\psi| \varphi\rangle = \alpha \neq 0$; α and β are complex constants. The states $|\varphi\rangle$ and $|\psi\rangle$ are normalized. Which conditions must $|\varphi\rangle$, $|\psi\rangle$, α and β fulfill to ensure that A is a Hermitian, a unitary, or a projection operator?

26. Given a CONS $\{|\varphi_n\rangle\}$ and an operator

$$A = \sum_{n,m} c_{nm} |\varphi_n\rangle \langle\varphi_m|; \quad c_{nm} \in \mathbb{C}. \qquad (13.74)$$

How must the coefficients c_{nm} be chosen in order that A be a Hermitian, a unitary, or a projection operator?

27. A CONS $\{|\varphi_n\rangle, n = 1, 2, \ldots, N\}$ spans a vector space $\mathcal{V}$.

(a) Show that each operator A acting in $\mathcal{V}$ can be represented as

$$A = \sum_{n,m} c_{nm} |\varphi_n\rangle \langle\varphi_m|. \qquad (13.75)$$

(b) Consider the special case $N = 3$:

$$A |\varphi_1\rangle = - |\varphi_2\rangle; \quad A |\varphi_2\rangle = - |\varphi_3\rangle; \quad A |\varphi_3\rangle = - |\varphi_1\rangle + |\varphi_2\rangle . \qquad (13.76)$$

What is the operator A? (Determine the coefficients c_{nm}, i.e. formulate A as a linear combination of products $|\varphi_i\rangle\langle\varphi_j|$.)

28. How is the generalized Heisenberg uncertainty relation formulated for each of the pairs (x, l_x), (x, l_y), (x, l_z)?
29. For the Pauli matrices, the following uncertainty relation holds:

$$\Delta\sigma_x \Delta\sigma_y \geq |\langle\sigma_z\rangle|. \tag{13.77}$$

For which normalized states $\psi=\begin{pmatrix} a \\ b \end{pmatrix}$ is the right-hand side a minimum/maximum?

30. What is the generalized uncertainty relation for H and $\mathbf{p}$?
31. The position operator in the Heisenberg picture,[20] x_H, is given by

$$x_H = e^{i\frac{tH}{\hbar}} x e^{-i\frac{tH}{\hbar}}. \tag{13.78}$$

How does this operator depend explicitly on time? The potential is assumed to be constant, $\frac{dV}{dx} = 0$. Hint: Use the equation

$$e^{iA} B e^{-iA} = B + i\,[A, B] + \frac{i^2}{2!}\,[A, [A, B]] + \frac{i^3}{3!}\,[A, [A, [A, B]]] + \cdots \tag{13.79}$$

or

$$i\hbar\frac{d}{dt}x_H = [x_H, H] \tag{13.80}$$

(or both for practice).

32. A Hamiltonian H depends on a parameter q, $H = H(q)$. In addition, $E(q)$ is a nondegenerate eigenvalue and $|\varphi(q)\rangle$ the corresponding eigenvector:

$$H(q)|\varphi(q)\rangle = E(q)|\varphi(q)\rangle. \tag{13.81}$$

Show that

$$\frac{\partial E(q)}{\partial q} = \langle\varphi(q)|\frac{\partial H(q)}{\partial q}|\varphi(q)\rangle. \tag{13.82}$$

(This equation is also called the Feynman-Hellmann theorem.)

33. $\{|n\rangle\}$ is a CONS. Every solution of the SEq may be written as

$$|\psi\rangle = \sum_l a_l |l\rangle \tag{13.83}$$

---

[20]See also Appendix Q, Vol. 1, 'Schrödinger picture, Heisenberg picture and interaction picture'.

and every operator A as

$$A = \sum_{mn} c_{mn} |n\rangle \langle m|. \tag{13.84}$$

Can the non-Hermitian operator A (i.e. $c_{mn} \neq c_{nm}^*$ for at least one pair n, m) have a real expectation value (for arbitrary states $|\psi\rangle$) under these conditions?

34. We consider the Hamiltonian $H = 1 + a\sigma_y$, already introduced in the exercises for Chap. 8.

(a) What is the expected result of the measurement of the x-component of the spin in the state $|\psi_t\rangle$ with $|\psi_0\rangle = \begin{pmatrix} 1 \\ 0 \end{pmatrix}$?

(b) What is the uncertainty Δs_x in this state?

(c) Calculate the commutator $[s_x, s_y]$ and formulate the uncertainty relation for the observables s_x and s_y for arbitrary times t.

35. Given an eigenvalue problem $A |a_m\rangle = a_m |a_m\rangle$ ($\{|a_m\rangle\}$ is a CONS); we can define a function of the operator by

$$F(A) |a_m\rangle := F(a_m) |a_m\rangle. \tag{13.85}$$

(a) Show that:

$$F(A) = \sum_m F(a_m) P_m \tag{13.86}$$

with $P_m = |a_m\rangle \langle a_m|$.

(b) Show that if $F(a)$ is real for all eigenvalues a_m, then $F(A)$ is self-adjoint.

36. What are the conditions which the elements of a two-dimensional normal matrix have to fulfill?

37. Given the matrix

$$A = \begin{pmatrix} 0 & \gamma^2 \\ 1 & 0 \end{pmatrix}; \quad \gamma \neq 0. \tag{13.87}$$

(a) Is A normal?

(b) Show that A is diagonalizable for almost all γ, but not by a unitary transformation.

38. In the derivation of the uncertainty relation, the functions must be in the domains of definition of the operators and of the operator products involved. If they are not, we do not obtain meaningful statements. As an example, we consider the function:

$$f(x) = \frac{\sin x^2}{x}. \tag{13.88}$$

(a) Is $f(x)$ square-integrable?

(b) Is $f(x)$ within the domains of definition of the operator x?

(c) Can a meaningful uncertainty relation be derived for $f(x)$?

(d) Can similar statements be made for the function $g(x) = \frac{\sin x}{x}$?

39. Given two operators A and B which commute with their commutator, $[A, [A, B]] = [B, [A, B]] = 0$. Show that:

$$\left[B, A^n\right] = n\left[B, A\right] A^{n-1}. \tag{13.89}$$

40. Show that the momentum operator is given in the coordinate representation by $p = \frac{\hbar}{i} \frac{d}{dx}$. Make use only of the commutator $[x, p] = i\hbar$ and derive, making use of the previous exercise, the result:

$$[p, f(x)] = \frac{\hbar}{i} \frac{df(x)}{dx}. \tag{13.90}$$

41. Given two operators A and B which commute with their commutator, $[A, [A, B]] = [B, [A, B]] = 0$. Show that

$$e^{A+B} = e^A e^B e^{-\frac{1}{2}[A,B]}. \tag{13.91}$$

This is a special case of the *Baker-Campbell-Hausdorff formula* (relation, theorem). The general case considers e^{A+B} for two operators, which do not have to commute with their commutator (this is used e.g. in (13.79)). By the way, these authors published their work in 1900, well before the birth of quantum mechanics.

(a) First, prove the equation

$$\left[B, e^{xA}\right] = e^{xA}\left[B, A\right] x. \tag{13.92}$$

(b) Define

$$G(x) = e^{xA} e^{xB} \tag{13.93}$$

and show that the following equation holds:

$$\frac{dG}{dx} = (A + B + [A, B] x) G. \tag{13.94}$$

Integrate this equation.

Chapter 14
Postulates of Quantum Mechanics

In this chapter, we compile our findings about quantum mechanics from the preceding chapters, insofar as they concern its structure, and we formulate some basic rules for its application. They make up the general framework for our further considerations.

In previous chapters, we have frequently mentioned structural elements of quantum mechanics, which we now summarize and present systematically, namely in the form of *postulates* or *rules*. These rules map the behavior of physical systems, or more precisely, our methods for describing that behavior.[1] In fact, there are basically only three questions to which the physical description of a system must provide answers:

1. How can we describe the state of the system at a given time?
2. Which variables of the system are measurable and how can we predict the results of measurements?
3. How is the state of the system obtained at time t from its known initial state at time t_0?

The answers to these questions differ, of course, depending on the field (classical mechanics, quantum mechanics, hydrodynamics, quantum electrodynamics, ...). In the following, we will clothe them in the form of *postulates for quantum mechanics*, whereby the word 'postulate' in this context is equivalent to a thesis, a principle or a rule which is not proven, but is quite plausible and evident. We do not aim at an absolutely rigorous system of axioms (in the sense of a minimal set of statements). It is rather a question of creating a viable and practical set of rules (also called 'quasi-axiomatic'), and for the sake of practicality we also take into account the fact that

---

[1] "The human understanding is of its own nature prone to suppose the existence of more order and regularity in the world than it finds." Francis Bacon (1561–1626), English philosopher and statesman, in *New Organon*. "Rain, snow, winds follow each other so that we do not animadvert a clear law in their order, but laws again are only conceived by us to facilitate comprehending a thing, as we create species." Georg Christoph Lichtenberg, *Scrap Books*, Vol. A (192).

© Springer Nature Switzerland AG 2018
J. Pade, *Quantum Mechanics for Pedestrians 1*, Undergraduate Lecture Notes in Physics, https://doi.org/10.1007/978-3-030-00464-4_14

one postulate might be derivable from others.[2] Moreover, the set of rules established here is only one of many possible such sets. Other formulations of the postulates can be found in Appendix R, Vol. 1.[3]

In introductions to quantum mechanics, the postulates are often presented right at the beginning, so to speak as the basis for the further development of quantum mechanics. This proceeding has the immediate advantage of conceptual clarity, since e.g. borrowing from classical mechanics (the correspondence principle, etc.) is not necessary. On the other hand, for the 'uninitiated', the postulates somehow seem to fall from the heavens—without background information, it is probably quite difficult to understand how formulations like these were arrived at in the first place.[4]

Our access to quantum mechanics was a two-pronged one in the first chapters. In the analytical approach, we began with the dynamics (SEq, question 3) and subsequently took up questions 1 and 2. In the algebraic approach, we tried to make the postulates plausible (or to anticipate their statements) in the order given here and on the basis of simple physical systems.

Finally, we remark that the postulates do not raise any new difficulties of comprehension, but nevertheless they surely sharpen our view of open problems. In previous chapters, we have already mentioned questions of this kind several times, and we will do so again at the end of this chapter.

14.1   Postulates

The numbering of the postulates refers to the numbers of the questions given above.

14.1.1   States, State Space (Question 1)

We have seen that both the solutions of the SEq as well as vectors, e.g. polarization vectors, can be linearly superposed and satisfy the axioms of a vector space. The first postulate summarizes this situation.

---

[2]Moreover, one cannot of course exclude with certainty today that the rules established in the following will be (or will have to be) modified sometime later.

[3]Indeed, there is no overall agreement about the basic facts. For instance, some of the authors listed in Appendix R, Vol. 1 treat the indistinguishability of identical quantum objects as a postulate of quantum mechanics, but others do not.

[4]The (quasi-) axiomatic approach has the great advantage that it does not need false analogies and does not implant false images into the minds of students; thus, it has been proposed as a possible way to teach quantum mechanics in schools. This is feasible with a (suitably adapted) form of the algebraic approach. With the exception of Postulate 3, the postulates can be deduced, or at least motivated, if one confines oneself essentially to the two-dimensional case. It is not least for this reason that the algebraic approach is of great didactic interest.

Postulate 1: The state of a quantum system at a given moment is completely defined by giving its *state vector* (ket), $|\varphi\rangle$. The state vector is an element of the *Hilbert space* $\mathcal{H}$, which also is called the *state space*.

Remarks:

1. In contrast to e.g. classical mechanics, quantum mechanics describes the states by elements of a *vector space*, i.e. by vectors. The abstract state vector or ket, $|\varphi\rangle$, is the mathematical representation of the information we have about the physical state of the system.

2. Because $\mathcal{H}$ is a vector space, the *superposition principle* holds; it is characteristic of the linearity of the theory. As a dominant principle of quantum mechanics, it is responsible for many of those phenomena of quantum mechanics which seem so strange to our everyday understanding.

3. Because of the linearity of the theory, we can always assume that the state vectors are *normalized*. If this is not the case, we must, where necessary, normalize them *post hoc* by dividing by the norm.

4. In anticipation, we note that only eigenvalues and absolute values of scalar products such as $|\langle \varphi | \psi \rangle|$ are relevant to a measurement. This means that the states $|\varphi\rangle$ and $|\varphi'\rangle = e^{i\alpha} |\varphi\rangle$ with $\alpha \in \mathbb{R}$, differing only in their phase, are physically equivalent (which fact we have used in considering the infinite potential well). Strictly speaking, a (normalized) physical state is therefore not represented by a vector, but rather by a *ray* in $\mathcal{H}$, i.e. the set $\{e^{i\alpha} |\varphi\rangle, \alpha \in \mathbb{R}\}$. This fact is called the 'independence of the physics from the *global phase*'. Changing the *relative phases* naturally leads to different states; $c_1 |\varphi\rangle + c_2 |\psi\rangle$ and $c_1 e^{i\alpha} |\varphi\rangle + c_2 |\psi\rangle$ are physically different for $\alpha \neq 2n\pi$.

   As we shall see, the difference between a ray and a vector is hereafter (luckily) only once of real importance, namely in the consideration of symmetry under time reversal (see Chap. 21, Vol. 2). Apart from this exception, we can work with state vectors (and ignore rays).

5. In the examples considered so far, any vector in $\mathcal{H}$ is a *physically realizable* state. This is not necessarily the case in all situations, as we shall see later in the treatment of identical particles, where there is no superposition of the states of fermions and bosons. The nonexistence of such (superposed) states is reflected in *superselection rules*.

6. It is still controversial just what the state vector 'really' means. The opinion that the state vector describes the physical reality of an *individual* quantum system is shared by many (and is also the position taken in this book), but it is by far not the only one. For more on this issue see Chap. 28, Vol. 2 on the interpretations of quantum mechanics. We stress once again that the state vector does not have a direct and concrete (everyday) meaning.

14.1.2 Probability Amplitudes, Probability (Question 2)

We have shown, e.g. by considering the polarization, that the absolute square of an amplitude gives the probability of finding the system in the respective state. This fact is generalized in Postulate 2.1.

Postulate 2.1: If a system is described by the vector $|\varphi\rangle$, and $|\psi\rangle$ is another state, then a probability amplitude exists for finding the system in state $|\psi\rangle$, and it is given by the scalar product $\langle \psi \,|\varphi\rangle$ in $\mathcal{H}$. The probability that the system is in the state $|\psi\rangle$ is the absolute square $|\langle \psi \,|\varphi\rangle|^2$ of the probability amplitude.

Remarks:

1. The vectors must be normalized to ensure that the probability concept is inherently consistent.
2. Using the projection operator $P_\psi = |\psi\rangle \langle \psi|$, the term $|\langle \psi \,|\varphi\rangle|^2$ can be written as $\langle \varphi| \, P_\psi \, |\varphi\rangle$.
3. The probability statements of this postulate provide a direct link to the term 'expectation value' or 'mean value.'
4. Probabilities usually indicate that the necessary information is not completely available. Hence, the idea arose quite early that quantum mechanics is not a complete theory, and hidden variables (hidden to us) must be added. But this is not the case according to present knowledge, at least not in the sense that the hidden variables have the simple and familiar properties of classical physics. We will address this issue in later chapters in Vol. 2.
5. This postulate is also called *Born's rule*.

14.1.3 Physical Quantities and Hermitian Operators (Question 2)

We have seen that a measurable physical quantity such as the momentum is represented by a Hermitian operator. The next postulate generalizes this relationship.

Postulate 2.2: Every measurable *physical quantity* is described by a *Hermitian operator* A acting in $\mathcal{H}$; this operator is an observable.[5] If a physical quantity is measured, the result can be only one of the eigenvalues of the corresponding observable A.

Remarks:

1. Quantum mechanics describes physical quantities by *operators* (in contrast to classical mechanics).
2. These operators are *observables*, i.e. Hermitian operators that represent a consistently measurable physical quantity.[6] In this way, we take into account that

---

[5]Recall that we defined 'observables' as those Hermitian operators which represent a consistent measurable physical quantity, cf. Sect. 13.1.

[6]We note again that the word *observable* does not imply the existence of a (human) observer.

14.1 Postulates   191

not every self-adjoint operator (with reasonable eigenfunctions) must necessarily represent a physical observable.[7] More on this topic in Vol. 1 Appendix I.

3. Because the operators are Hermitian, measurements always yield *real* values.
4. Not all physically measurable quantities are associated with non-trivial operators. Mass and charge, for example, are and remain simple numbers.

14.1.4 Measurement and State Reduction (Question 2)

If a right circular-polarized photon $|r\rangle = (|h\rangle + i\,|v\rangle)/\sqrt{2}$ passes through a PBS, we obtain with a probability $1/2$ either a horizontally or a vertically linear-polarized photon, i.e. $|h\rangle$ or $|v\rangle$. The next postulate formalizes and generalizes this fact.

Postulate 2.3: Suppose that the measurement of A on a system which was originally in the state $|\varphi\rangle$ yielded the value a_n. Then, immediately after the measurement, the state of the system is the normalized projection of $|\varphi\rangle$ onto the eigenspace belonging to a_n (see Chap. 13)[8]

$$|\varphi\rangle \rightarrow |\psi\rangle = \frac{P_n\,|\varphi\rangle}{\sqrt{\langle\varphi|\,P_n\,|\varphi\rangle}} \tag{14.1}$$

The state $|\psi\rangle$ is normalized:

$$\|P_n\,|\varphi\rangle\|^2 = \langle\varphi|\,P_n^\dagger P_n\,|\varphi\rangle = \langle\varphi|\,P_n\,|\varphi\rangle \tag{14.2}$$

Remarks:

1. This postulate assumes an *ideal measurement*; meaning, among other things, that further measurements on the quantum object must be possible.[9] Immediately after measuring, the state of the system is always an eigenvector of A with eigenvalue a_n. Any immediately following further measurement must, of course, yield the same result.[10] The transition from a superposition state to a single state is called *state reduction* or *collapse of the wavefunction*. It is an *irreversible* evolution which marks a direction in time.[11]

---

[7]By the way, the practical implementation of arbitrary operators often raises some difficulties. However, for us this is not a strong constraint, because we need essentially only the well-known operators such as position, momentum, etc., or combinations of them.

[8]Any remaining phase plays no physical role; see the remark following Postulate 1 about states and rays.

[9]The only change in the measured system is the collapse of the wavefunction. In particular, the spectrum remains unchanged. One speaks in this context also of 'recoilless'. See also Appendix S, Vol. 1, where several remarks pertinent to the topic of 'measurement' can be found.

[10]Thereby one can prevent a change of state by measuring it repeatedly. The associated keyword is 'quantum Zeno effect'. The concept is summarized by the handy phrase 'a watched pot never boils'. More about this topic in Appendix L, Vol. 1.

[11]This applies only if the initial state is not already an eigenstate of the operator.

2. The position underlying this postulate is not sensitive to the details of the measurement process, but rather assumes the measuring apparatus to be a kind of black box. A more detailed analysis of the measurement process, including interactions of the quantum system with the measuring apparatus and the environment, shows that one can interpret Postulate 2.3 as a consequence of Postulates 2.1 and 2.2. However, this postulate is *fapp*, i.e. a useful 'working tool' for all the usual applications of quantum mechanics. We will address this question again in Chap. 24 (decoherence) and in Chap. 28 (interpretations), both in Vol. 2.

3. Measurement in quantum mechanics is obviously something very different than in classical physics. Classically, a (single) value of a physical quantity is measured, which already existed before the measurement (pre-existence). In quantum mechanics, this is the case only when the system is initially in an eigenstate of the measured observable; otherwise, there is no well-defined measurement value before the measurement.[12] This fact is also called the *eigenvector-eigenvalue rule*: A state has the value a of a property represented by the operator A if and only if the state is an eigenvector of A with eigenvalue a.[13] In this case, we can say that the system has the property a (For more cautious formulations of this relation, see Chap. 13.).

4. The spreading of the measured values is sometimes attributed to the fact that the measurement disturbs the measured quantity (e.g. the spin) uncontrollably. But this is wrong from the perspective of Postulate 2.3. For if the system is in an eigenstate of the measured quantity before the measurement, it will not be disturbed by the measurement. If it is not in an eigenstate, the measured value does not exist as such before the measurement—and what does not exist, cannot be disturbed.[14]

5. For degenerate and continuous cases, (14.1) must be appropriately modified.

6. This postulate is also called the projection postulate, the Neumann projection postulate, the postulate of Neumann–Lüders, etc.

14.1.5 Time Evolution (Question 3)

So far, our discussion was limited to a fixed time—now we start the clock. We recall that we have restricted ourselves to time-independent interactions.

Postulate 3: The temporal evolution of the state vector $|\psi(t)\rangle$ of an *isolated* quantum system is described by the equation (evolution equation, *Schrödinger equation*):

$$i\hbar\frac{d}{dt}|\psi(t)\rangle = H|\psi(t)\rangle. \tag{14.3}$$

---

[12] See also the corresponding remarks in Chap. 13 (projection operators).

[13] Thus we have here a translation rule which connects physical quantities to mathematical objects.

[14] This remark is of course a bit shortened and flippant. The point is that the value of a variable is determined by the measurement, in general. Before the measurement, the value does not exist and thus can not be disturbed, for perturbation means changing an *existing* value into another one. More on this issue e.g. in Chap. 20, Vol. 2.

The Hermitian operator H which is associated with the total energy of the system is called the *Hamiltonian*.[15]

Remarks:

1. We consider isolated systems which do not interact with their environments. Their realization is anything but trivial, which is one of the obstacles to the rapid development of quantum computers. If, on the other hand, there is a coupling (observed or unobserved) to the degrees of freedom of the environment, one speaks of an *open* quantum system. More on this issue is to be found in Chap. 24, Vol. 2 (decoherence) and in Appendix S, Vol. 1.

2. The postulate tells us nothing about the specific form of the Hamiltonian. This is determined by the physical problem and the accuracy with which one wants to describe it. Further considerations are found below in the 'Concluding Remarks' section.

3. We had already stated the main characteristics of the SEq: it is among other things (a) complex, (b) linear, (c) of first order with respect to time. Stochastic components do not occur, hence the SEq is deterministic. Stationary states (eigenstates of energy E) have the time behavior of $|\varphi(t)\rangle = e^{-i\frac{tE}{\hbar}} |\varphi(0)\rangle$.

4. Since H is Hermitian, the time evolution is unitary and thus *reversible* and norm-preserving:

$$\frac{d}{dt} \langle \psi(t) | \psi(t) \rangle = \langle \dot{\psi}(t) | \psi(t) \rangle + \langle \psi(t) | \dot{\psi}(t) \rangle = 0. \tag{14.4}$$

In contrast, the measurement process according to Postulate 2.3 is in general not unitary and is therefore irreversible; the norm is not conserved, but instead, one has to normalize the new measurement result. In order that the SEq $i\hbar |\dot{\psi}(t)\rangle = H |\psi(t)\rangle$ holds, e.g. between two measurements, the system must be isolated. During the measurement, however, the system is not isolated.

We can also formulate the time evolution with the help of the propagator rather than the differential form (14.3), and thus express Postulate 3 in another form:

Postulate 3′: The state vector at the initial time $|\varphi(t_0)\rangle$ is transferred into the state $|\varphi(t)\rangle$ at time t by a *unitary operator* $U(t, t_0)$, called the *time evolution operator* or *propagator*:

$$|\varphi(t)\rangle = U(t, t_0) |\varphi(t_0)\rangle. \tag{14.5}$$

Remarks:

1. The unitarity of the propagator ensures the conservation of the norm.
2. For time-independent H, the propagator can be represented as $U = e^{-i\frac{Ht}{\hbar}}$.[16]

---

[15] We note that (14.3) applies also to a time-dependent $H(t)$. But since we restrict ourselves in this whole book to the consideration of time-independent H, we formulate this postulate only for that case.

[16] For the formulation of the propagator as an integral operator, see Appendix I, Vol. 1; an example for the case of free motion is found in Chap. 5, Exercise 11.

3. With the propagator, the reversibility of the time evolution can be seen in a particularly simple manner, since we have $|\varphi(t_0)\rangle = U^{-1}(t, t_0) |\varphi(t)\rangle$.
4. Postulates 3 and $3'$ are equivalent for our purposes. Strictly speaking, however, there is a difference, since U is bounded even if H is not bounded. In this respect, the propagator U appears to be more fundamental than the Hamiltonian H.

We want to stress here again a fundamental difference between classical mechanics and quantum mechanics. While classical mechanics describes the time evolution of the *factual*, quantum mechanics (or the SEq) describes the time evolution of the *possible*. In other words, the possibility structure of our universe is not fixed, but is a dynamically evolving structure.

14.2 Some Open Problems

As stated in the introduction to this chapter, we summarize here once again problems of comprehension which are essentially centered around the concept of measurement, a completely innocuous notion in classical physics.[17] In contrast, measurement seems to play a very special role in quantum mechanics. This was already clear in the early days of quantum mechanics, and even today the problem is not solved in depth, but remains the subject of current discussions.

One can of course avoid all these problems by adopting the *instrumentalist* or *pragmatic* point of view; namely, that we live in a classical world, and that the postulates are simply computational tools or instructions that work well without asserting the claim of representing reality. Niels Bohr put it this way: "There is no quantum world. There is only an abstract physical description. It is wrong to think that the task of physics is to find out how nature is. Physics concerns what we can say about nature." With this position (also called the *minimal interpretation*), one need not worry about the problems listed below, let alone about trying to solve them.

However, many people are dissatisfied with the idea that the fundamental description of the world should be a handful of rules which are closed to debate. The *realistic* position assumes that the quantum systems of the theory have real counterparts in one way or the other. The postulates, together with this point of view, are called the *standard interpretation* (or standard representation) by many authors.

---

[17]In classical mechanics, the properties of a system are always well defined (where we always assume a non-pathological phase space). They can be described as functions of the phase space variables (i.e. points in phase space) and thus always have a direct three-dimensional spatial significance. In quantum mechanics, properties are not always well defined and we cannot represent them mathematically as functions of point sets. Small table:

	Classical mechanics	Quantum mechanics
State space	Phase space (point set)	Hilbert space
States	Points	Vectors
Properties	Functions of points	Eigenvalues of operators

Regardless of the question of pragmatic *vs.* realistic, the problematic concepts play a fundamental role in the formulation of quantum mechanics (or its postulates), and therefore certainly deserve a deeper understanding, or at least a deeper awareness. That is why we want to address briefly the key issues in the following.[18]

To avoid misunderstandings, we emphasize one remark: On the formal level—technically, so to speak—quantum mechanics works perfectly, with often impressively accurate results. In fact, quantum mechanics is one of the most carefully examined and well-tested physical theories in existence; it has yet to be falsified experimentally.

So the problem has to do with the level of *understanding* of quantum mechanics. What does all this imply, what does it mean? If we take the above postulates as a basis, some open and interrelated questions arise concerning the measurement process, which we summarize here briefly. The discussion is continued in Chap. 24 (decoherence) and Chap. 28 (interpretations), both in Vol. 2.[19]

1. **Status of the Measurement Process**: It is perhaps time to define more precisely the concept of *measurement*. By measurement, we understand the performance of an irreversible operation on a system which determines the status of one (or more) physical quantities, namely as a storable number. More remarks on the term 'measurement' (or on different but related concepts such as preparation, testing, maximum test etc.) can be found in Appendix S, Vol. 1.

   The special status of the measurement has nothing to do with whether one assumes pragmatically that values of a physical quantity have a meaning only as the result of a measurement, or if one asserts that the postulates are valid also for individual systems (realistic position). With regard to this relationship between the measurement and the values of physical quantities, the pragmatically-oriented must explain why the concept of 'measurement' plays such a fundamental role in quantum mechanics.[20] But also the realistically-oriented assign a prominent role to the measurement, since it causes the transition from the possible to the actual. It remains to investigate when an *interaction* between two systems A and B is the *measurement* of a physical quantity of A by B, and, in that context, whether and how one can describe the measurement in a quantum-mechanical way, i.e. including the measuring apparatus.

2. **Probability**: This term enters into the theory through Postulate 2.1. Before the measurement, one can in general specify only a probability that a particular

---

[18]"But our present (quantum-mechanical) formalism is not purely epistemological; it is a peculiar mixture describing in part realities of Nature, in part incomplete human information about Nature - all scrambled up by Heisenberg and Bohr into an omelette that nobody has seen how to unscramble. Yet we think that the unscrambling is a prerequisite for any further advance in basic physical theory. For, if we cannot separate the subjective and objective aspects of the formalism, we cannot know what we are talking about; it is just that simple." E.T. Jaynes in: *Complexity, Entropy and the Physics of Information* (ed. Zurek, W.H.) 381 (Addison-Wesley, 1990).

[19]In anticipation of our further discussion, we want to point out here that there are not answers to all the questions, at least not unique answers.

[20]This is especially true in the traditional view, according to which the measurement apparatus is to be regarded as a classical system.

measurement result will appear. This is true even if the maximum information about the system is known. Through the measurement, one and only one of the available options is realized. This means, in other words, that an observable generally does not have a definite value *before* a measurement. Correspondingly, measurement does not determine the value of an observable which it already has, but rather this value is created by the measurement itself—the measurement determines the reality, and not *vice versa*.

As explained in Chap. 2 and later on several times, the occurrence of probabilities in classical physics means that we do not have sufficient information at our disposal in order to calculate certain properties explicitly. In quantum mechanics, the situation is different. Here, the term 'probability' or 'objective chance' is literally a structural element of the theory and stands for the fact that quantum mechanics is concerned with possibilities, one of which is then realized by the measurement process. From the classical point of view, one would assume that beneath our level of formulation there are further, thus far hidden variables, knowledge of which would allow us to avoid the use of probabilities. But this assumption has been disproved experimentally (at least in its local or non-contextual, i.e. intuitively plausible form, see Chaps. 20 and 27, Vol. 2); apparently, we cannot avoid the concept of objective chance.

3. **Collapse**: How can we explain the change from a superposition to a single state as described in Postulate 2.3? What is the mechanism, what is its time frame? Is it a fundamental effect or only a pragmatic approximation to the description of a quantum system and a measuring apparatus which can be derived in principle from the existing formalism? That would obviously be particularly important if one does not want to attribute an essential meaning to the 'measurement', but would rather like to see it as simply one of many possible interactions.

This is not an exotic, constructed effect. We think e.g. of the right circular-polarized photon, already invoked several times, which we sent through a linear analyzer, only to find it afterwards in the state of e.g. horizontal linear polarization. In this process of measurement, the superposition $\frac{|h\rangle+i|v\rangle}{\sqrt{2}}$ is collapsed into the state $|h\rangle$.

Of course the answer to the question also depends on what we mean by 'state'.[21] Does it entail a direct description of the system, or only of our knowledge of the system? In the latter case, the collapse would represent a change in our knowledge, by adding more information. Otherwise, there would have to be a way to formulate the state reduction in direct physical terms.

As we will see in the discussion on entangled systems later on, the collapse of states is a non-local (i.e. superluminal) effect.

---

[21] For example, 'state' can have the meanings:—an *individual* quantum system A,—our *knowledge* of the properties of the system A,—the result of a measurement that has been or could be performed on system A,—an *ensemble* E (real or hypothetical) of identically prepared copies of a system,—our *knowledge* of the properties of the ensemble E,—the results of *repeated* measurements that have been or could be performed on the ensemble E.

4. **Two Time Evolutions**: Through the measurement (observation), the wave function collapses. This change of state is discontinuous, irreversible, not deterministic in principle and therefore is in contrast to the continuous and reversible time evolution of the SEq. Accordingly, we have two very different processes or dynamics. This raises the following questions: Are the rules of quantum mechanics really different for observed and for unobserved systems? If so, why? What is an observer, must it be a human observer? Is the observer also subject to the laws of quantum mechanics? If so, how there can be irreversible evolutions that contradict the SEq? We will later find answers to some of these questions, especially in Chap. 24, Vol. 2.

5. **The Boundary Between Classical Mechanics and Quantum Mechanics**: If the measurement is not included in the SEq, does this mean that measurement is a non-quantum-mechanical process, so that the measuring apparatus is not subject to the rules of quantum mechanics? If that is the case—which rules apply instead? It is a common notion that the measuring apparatus obeys classical rules. Thus there would be two areas, the classical domain and quantum mechanics. But where is the cut (also called the *Heisenberg cut*) between quantum mechanics and classical mechanics, what begins where, what ends where? From which size, from which particle number on is a system no longer described by quantum mechanics, but instead by classical mechanics?

Of course, the idea of starting with small quantum mechanical systems and studying larger and larger ones to see whether and if so, how, they become 'more classical' is obvious. This fails, however, due to the fact that the description of larger quantum-mechanical systems with an increasing number of degrees of freedom becomes very quickly enormously complicated, so that a clear relationship between classical mechanics and quantum mechanics is difficult to elucidate.

One can in principle think of three possibilities: (1) classical mechanics includes quantum mechanics; (2) classical mechanics and quantum mechanics are on equal footing; and (3) quantum mechanics includes classical mechanics; see Fig. 14.1. The majority of the physics community favors the third option, i.e. that classical mechanics is based on quantum mechanics. But here, also, it has to be clarified

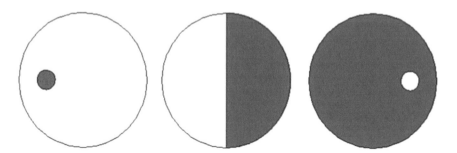

Fig. 14.1 Schematic figure of possible boundaries quantum mechanics–classical mechanics. *White*: classical mechanics, *black*: quantum mechanics

where the cut is (if it exists at all). In addition, there is the question: If quantum mechanics is the basic theory—why do we never see certain quantum effects in the macroscopic world, such as the superposition of states?

Let us consider the problem again from a different angle. The measuring apparatus consists of atoms and is thus itself a quantum system. In fact, it can interact only in this way with the measured quantum object. On the other hand, the measuring apparatus has to react as a classical system when it finally returns a result. Quantum system *and* classical system—these are two requirements for the measuring apparatus, which seem difficult to reconcile.

These problems of demarcation are particularly clear-cut in quantum cosmology, which attempts to describe the entire universe as a *single* quantum system. If one assumes that the universe is isolated and its dynamics (as a giant quantum system) are described by a *single* SEq (which is deterministic), then this poses the question of how measurement can be a process that is performed on a quantum system from the *outside*.

In any case, in practice, the distinction between a quantum-mechanical system and a classical measurement apparatus has been very successful (*fapp*). Whoever works only in a results-oriented manner may be satisfied with the minimal interpretation, i.e. with the argument that the wavefunction is not a description of real objects, but only a tool by which one can obtain the relevant results.[22]

We repeat that we will take up these issues again in later chapters. Especially the theory of decoherence (Chap. 24, Vol. 2) will alleviate most of these problems.

---

[22]Discussing the nature of the wavefunction is not an ivory-tower topic, but rather the subject of ongoing research. For instance, as said above there is the view that the wavefunction reflects the partial knowledge which an experimenter has about the system. But such a view is wrong if one follows a recently published theorem; it is a no-go theorem which states that if the quantum state represents merely information about the real physical state of a system, then experimental predictions will be obtained which contradict those of quantum theory. However, this theorem depends on the crucial assumption that quantum systems have an objective underlying physical state—an assumption which is controversial. See e.g. Matthew F. Pusey et al., 'On the reality of the quantum state', *Nature Physics 8*, 475–478 (2012) or S. Mansfield, 'Reality of the quantum state: Towards a stronger ψ-ontology theorem', *Phys. Rev. A 94*, 042124 (Oct 2016).

Apart from that, we note that even though there is no explanation of the wavefunction in everyday terms, it is measurable. To date, the experimental determination of wavefunctions (i.e. modulus and phase or real and imaginary parts) has been accomplished by means of certain indirect methods (called tomographic methods). However, recently a method was presented for measuring wavefunctions *directly*. In it, a special technique is used, called *weak measurement*. See Jeff S. Lundeen et al., 'Direct measurement of the wave function', Nature 474, 188–191 (2011). We note that these measurements are performed on an ensemble; it is impossible to determine the completely unknown wavefunction of a *single* system. For other experimental methods see e.g. G.C. Knee, 'Towards optimal experimental tests on the reality of the quantum state', *New Journal of Physics* 19, 023004, https://doi.org/10.1088/1367-2630/aa54ab (Feb 2017).

14.3 Concluding Remarks

14.3.1 Postulates of Quantum Mechanics as a Framework

We have distilled out the postulates from considerations of simple example systems. This is possible precisely because the postulates do not depend on the specific system, but constitute something like the general framework or the general rules of quantum mechanics. In other words, the postulates are valid for all possible systems (referred to our frame of consideration), simple as well as complicated ones.

Hence it is clear that the postulates cannot act as instructions for the practical calculation of a physical problem. In order to do this, one has to determine the state space $\mathcal{H}$ and the Hamiltonian H for the system under consideration. Of course, it then becomes relevant which concrete physical system is selected, how it is modelled physically, which precision is required, and so on.

We have described e.g. neutrino oscillations without further ado in a two-dimensional space. This is obviously a very crude model, but it is entirely adequate for the intended purpose. In the analytical approach, we obtained the Hamiltonian by using the correspondence principle,[23] i.e. by replacement of the classical quantities $(\mathbf{r}, \mathbf{p})$ which occur in the energy by $\left(\mathbf{r}, \frac{\hbar}{i}\nabla\right)$. This allows us to represent a non-relativistic quantum object in a scalar potential, while vector potentials, interactions between a number of quantum objects, relativistic effects such as spin, etc. are not considered. We can see our simple model as the beginning of a 'hierarchy of models'; we address this point briefly again in Chap. 17, Vol. 2.

In short, the choice of $\mathcal{H}$ and H always means that one operates with certain models and approximations.[24] The postulates, however, are strict. This is schematically indicated in Fig. 14.2.

14.3.2 Outlook

The postulates in the form in which we have presented them give the foundation of quantum mechanics, but there are some complements needed. In subsequent chapters in Vol. 2, we will become acquainted with the following three:

1. We will extend the concept of *state* and also look at states that are no longer represented as vectors of an (extended) Hilbert space (so-called mixed states, keyword *density operator*, Chap. 22, Vol. 2).

---

[23] We repeat the remark that this principle has a mainly heuristic value. A more convincing method is e.g. the introduction of position and momentum operators by means of symmetry transformations (see Chap. 21, Vol. 2, and Appendix L, Vol. 2).

[24] "Although this may be seen as a paradox, all exact science is dominated by the idea of approximation." (Bertrand Russell).

Fig. 14.2 The postulates as
framework for the quantum
mechanical description of
physical systems

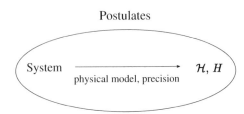

2. The systems considered so far are isolated, i.e. they are not coupled to any environment whatsoever. We will consider in the following also systems that are composed of several interacting subsystems (keyword *open systems*, Chap. 24, Vol. 2).
3. By an extension to open systems, we want to trace the separate role of the measurement (i.e. the projection postulate 2.3) back to the other postulates (keyword *decoherence*, Chap. 24, Vol. 2).

But first, in the initial chapters of Vol. 2, we will fill in the conceptual framework given above with some applications, some specialized subjects and extensions.

14.4 Exercises

1. Given are an observable A and a state $|\varphi\rangle$. Show by means of Postulates (2.1) and (2.2) that the expected result of a measurement of A is given by $\langle A\rangle = \langle\varphi| A |\varphi\rangle$. To simplify the discussion, we consider an observable A whose eigenvalues are discrete and nondegenerate and whose eigenvectors form a CONS, $A |n\rangle = a_n |n\rangle$.
2. Show that the operator $s_x + s_z$ is Hermitian, but does not represent a measurable physical quantity if understood literally, i.e., as the instruction to measure the x-component plus (and) the z-component of the spin. The spin matrices s_i are related to the Pauli matrices σ_i by $s_i = \frac{\hbar}{2}\sigma_i$.
3. (An example concerning projections, probabilities and expectation values.) The angular momentum operator **L** for angular momentum 1 can be represented in the vector space $\mathbb{C}^3$ by the following matrices (see Chap. 16, Vol. 2):

$$L_x = \frac{\hbar}{\sqrt{2}}\begin{pmatrix} 0 & 1 & 0 \\ 1 & 0 & 1 \\ 0 & 1 & 0 \end{pmatrix}; \quad L_y = \frac{\hbar}{\sqrt{2}}\begin{pmatrix} 0 & -i & 0 \\ i & 0 & -i \\ 0 & i & 0 \end{pmatrix}; \quad L_z = \hbar\begin{pmatrix} 1 & 0 & 0 \\ 0 & 0 & 0 \\ 0 & 0 & -1 \end{pmatrix}$$
(14.6)

(a) Which measured results are possible in a measurement of L_i ($i = x, y, z$)?
(b) What are the corresponding eigenvectors for L_z?
(c) What are the probabilities of measuring the results $+\hbar, 0, -\hbar$ on the state

$$|\psi\rangle = \begin{pmatrix} 1 \\ i \\ -2 \end{pmatrix}?$$
(14.7)

4. Given the state

$$|\psi\rangle_v = \frac{|x_1\rangle\, e^{-i\omega t} + |x_2\rangle\, e^{-2i\omega t}}{\sqrt{2}} \tag{14.8}$$

with normalized and mutually orthogonal states $|x_i\rangle$. We measure the x_1 component of $|\psi\rangle_v$. After the measurement, we have

$$|\psi\rangle_n = |x_1\rangle\, e^{-i\omega t} \tag{14.9}$$

Illustrate this state reduction by considering the change in the real or imaginary part of $|\psi\rangle$.

Appendix A
Abbreviations and Notations

For a better overview, we collect here some abbreviations and specific notations.

Abbreviations

ala	Algebraic approach
ana	Analytical approach
ClM	Classical mechanics
CONS	Complete orthonormal system
CSCO	Complete system of commuting observables
DEq	Differential equation
EPR	Einstein–Podolsky–Rosen paradox
fapp	'Fine for all practical purposes'
MZI	Mach–Zehnder interferometer
ONS	Orthonormal system
PBS	Polarizing beam splitter
QC	Quantum computer
QM	Quantum mechanics
QZE	Quantum Zeno effect
SEq	Schrödinger equation

Operators

There are several different notations for an operator which is associated with a physical quantity A; among others: (1) A, that is the symbol itself, (2) $\hat{A}$, notation with hat (3) $\mathcal{A}$, calligraphic typefont, (4) A_{op}, notation with index. It must be clear from the context what is meant in each case.

For special quantities such as the position x, one also finds the uppercase notation X for the corresponding operator.

© Springer Nature Switzerland AG 2018
J. Pade, *Quantum Mechanics for Pedestrians 1*, Undergraduate Lecture
Notes in Physics, https://doi.org/10.1007/978-3-030-00464-4

The Hamiltonian and the Hadamard Transformation

We denote the Hamiltonian by H. With reference to questions of quantum information, H stands for the Hadamard transformation.

Vector Spaces

We denote a vector space by $\mathcal{V}$, a Hilbert space by $\mathcal{H}$.

Appendix B
Units and Constants

B.1 Systems of Units

Units are not genuinely natural (although some are called that), but rather are man-made and therefore in some sense arbitrary. Depending on the application or scale, there are various options that, of course, are fixed precisely in each case by convention.

Those unit systems in which some fundamental constants are set equal to 1 and are dimensionless are generally referred to as *natural unit systems*. As we mentioned, the word 'natural' here is understood as part of the name and not as a descriptive adjective. We consider the following natural units: Planck units, the unit system of high-energy physics (theoretical units, also called natural units), and the unit system of atomic physics (atomic units).

B.1.1 Planck Units

Here, the speed of light c, the Planck constant $\hbar$, the gravitational constant G as well as the Boltzmann constant k_B and the electric field constant (multiplied by 4π), $4\pi\varepsilon_0$, are all set equal to 1. Their relations to the SI values are shown in the following table:

Quantity	Expression	Value (SI)
Charge	$q_P = \sqrt{4\pi\varepsilon_0 c\hbar}$	$1.876 \times 10^{-18}\,\mathrm{C}$
Length	$l_P = \sqrt{\frac{G\hbar}{c^3}}$	$1.616 \times 10^{-35}\,\mathrm{m}$
Mass	$m_P = \sqrt{\frac{c\hbar}{G}}$	$2.177 \times 10^{-8}\,\mathrm{kg}$
Temperature	$T_P = \frac{m_P c^2}{k_B}$	$1.417 \times 10^{32}\,\mathrm{K}$
Time	$t_P = \frac{l_P}{c}$	$5.391 \times 10^{-44}\,\mathrm{s}$

© Springer Nature Switzerland AG 2018
J. Pade, *Quantum Mechanics for Pedestrians 1*, Undergraduate Lecture Notes in Physics, https://doi.org/10.1007/978-3-030-00464-4

The Planck scale probably marks a limit to the applicability of the known laws of physics. Distances substantially smaller than the Planck length cannot be considered as meaningful. The same holds true for processes that are shorter than the Planck time. Because of $l_P = ct_P$, such a process would take place on a distance scale that would be smaller than the Planck length. By comparison, the LHC accelerator has a spatial resolution of about 10^{-19} m; its accessible energy is of the order of $E_{\text{LHC}} = 10\,\text{TeV}$.

B.1.2 Theoretical Units (Units of High-Energy Physics)

Here, c and $\hbar$ are set equal to 1; the other constants remain unchanged. The unit of energy is not determined by the choice of c and $\hbar$; it is usually expressed in eV (or MeV, GeV etc.). Energy and mass have the same unit; this applies also to space and time.

Quantity	Unit	Expression	Value (SI)
Energy	eV		1.602×10^{-19} J
Length	1/eV	$c\hbar$/eV	1.973×10^{-7} m
Mass	eV	eV/c^2	1.783×10^{-36} kg
Temperature	eV	eV/k_B	1.160×10^4 K
Time	1/eV	$\hbar$/eV	6.582×10^{-16} s

In SI units, $c\hbar = 3.1616 \cdot 10^{-26}$ Jm $= 0.1973\,\text{GeVfm}$. Since $c\hbar = 1$ in theoretical units, we have the rule of thumb

$$1 \text{ fm (SI)} \,\hat{=}\, 5/\text{GeV (TE)} \tag{B.1}$$

B.1.3 Atomic Units

In atomic units, $e = m_e = \hbar = 1$. These units, which are related to properties of the electron and the hydrogen atom, are mainly used in atomic and molecular physics. All quantities are formally dimensionless which are multiples of the basic units. If they are not dimensionless in SI units, they are generally marked by the formal 'unit character' a.u. (the dots are part of the unit symbol).

The Hartree unit of energy is twice the ionization potential of the hydrogen atom.

Quantity	Atomic unit	Value (SI)
Angular momentum	Planck constant $\hbar$	1.055×10^{-34} Js
Charge	Elementary charge e	1.602×10^{-19} C
Energy	Hartree energy E_h, H	4.360×10^{-18} J
Length	Bohr radius $a_0 = \frac{\hbar}{mc\alpha}$	5.292×10^{-11} m
Mass	Electron mass m_e	9.109×10^{-31} kg
Time unit	Atomic time unit, $1\,a.t.u. = \frac{\hbar}{E_h}$	2.419×10^{-17} s

B.1.4 Units of Energy

Energy is a central concept of physics, and this manifests itself among other things in the multitude of units used. The most common are summarized in the following table. Thereby, one uses $E = \hbar\omega = h\nu$ and $c = \lambda\nu$.

Unit	Conversion factor	Comment
eV	1	
Joule	1.602×10^{-19} J	
Kilowatt hour	4.451×10^{-26} kWh	
Calorie	3.827×10^{-20} cal	
Wavelength in nanometer	1239.85 nm	From $E = hc/\lambda$
Frequency in Hertz	2.41797×10^{14} Hz	From $E = h/T$ (T = time)
Wave number	8065.48 cm^{-1}	From $E = hc/\lambda$
Temperature	11,604.5 K	From $E = k_B T$ (T = temperature)
Rydberg	0.07350 Ry	Ionization potential of the H atom
Hartree	0.03675 H	
Energy equivalent mass, E/c	1.783×10^{-36} kg	

B.2 Some Constants

Derived units in the SI:

$$1 \text{ N (Newton)} = 1 \text{ kgms}^{-2}; \quad 1 \text{ W (Watt)} = 1 \text{ Js}^{-1}; \quad 1 \text{ C (Coulomb)} = 1 \text{ As}$$
$$1 \text{ F (Farad) 0 } 1 \text{ AsV}^{-1}; \quad 1 \text{ T (Tesla)} = 1 \text{ Vsm}^{-2}; \quad 1 \text{ WB (Weber)} = 1 \text{ Vs}$$

Important constants in eV:

$$h = 4,1357 \cdot 10^{-16} \text{ eVs}; \quad \hbar = 6,5821 \cdot 10^{-16} \text{ eVs}$$
$$m_e c^2 = 0,511 \text{ MeV}; \quad m_e c^2 \alpha^2 = 2 \cdot 13,6 \text{ eV}; \quad m_e c^2 \alpha^4 = 1,45 \cdot 10^{-3} \text{ eV}$$

Quantity	Symbol	Value	Unit
Speed of light in vacuum	c	299, 792, 458 (exact)	ms^{-1}
Magnetic field constant	μ_o	$4\pi \times 10^{-7}$ (exact)	TmA^{-1}
Electric field constant	ε_o	8.85419×10^{-12}	Fm^{-1}
Planck constant	h	6.62618×10^{-34}	Js
(Reduced) Planck constant	$\hbar$	1.05459×10^{-34}	Js
Elementary charge	e	1.60219×10^{-19}	C
Newtonian gravitational constant	G	6.672×10^{-11}	$\text{m}^3\,\text{kg}^{-1}\text{s}^{-2}$
Boltzmann constant	k_B	1.381×10^{-23}	JK^{-1}
Rest mass of the electron	m_e	9.10953×10^{-31}	kg
Rest mass of the proton	m_p	1.67265×10^{-27}	kg
Fine structure constant	α	$1/137.036$	
Rydberg constant	R	2.17991×10^{-18}	J
Bohr radius	a_o	5.29177×10^{-11}	m
Magnetic flux quantum	Φ_o	2.068×10^{-15}	Wb
Stefan–Boltzmann constant	σ	5.671×10^{-8}	$\text{Wm}^{-2}\text{K}^{-4}$
Magnetic moment of the electron	μ_e	9.28483×10^{-24}	JT^{-1}
Magnetic moment of the proton	μ_p	1.41062×10^{-26}	JT^{-1}

Large and small: Size comparisons

Size of the universe 10^{28} m; diameter of an atomic nucleus 10^{-15} m; Planck length 10^{-35} m

Ratio universe/Planck length 10^{63}

B.3 Dimensional Analysis

One advantage of physics over mathematics lies in the existence of physical units. Thus, one can test results, conjectures etc. by checking their units. Can e.g. the expression $T = 2\pi\sqrt{l \cdot g}$ be correct? No, the unit of the left-hand side is s, while at the right hand side it is m/s. This principle can be applied constructively (so-called dimensional analysis, Buckingham π-theorem). As an example, consider once more the pendulum. The system data are the mass of the pendulum bob, the length of the string and the acceleration of gravity. A time (oscillation time or period) can be represented only by the combination $\sqrt{l/g}$. So we *must* have $T \sim \sqrt{l/g}$.

In addition, the physical units show that an expression such as e^{ir} (so long as r has the unit m) can not be right; alone for dimensional reasons, it must read e^{ikr}, where k has the unit m^{-1}.

B.4   Powers of 10 and Abbreviations

deci, d	−1	deka, da	1
centi, c	−2	hecto, h	2
milli, m	−3	kilo, k	3
micro, μ	−6	mega, M	6
nano, n	−9	giga, G	9
pico, p	−12	tera, T	12
femto, f	−15	peta, P	15
atto, a	−18	exa, E	18
zepto, z	−21	zetta, Z	21
yocto, y	−24	yotta, Y	24

B.5   The Greek Alphabet

Name	Lower case	Upper case	Name	Lower case	Upper case
alpha	α	A	nu	ν	N
beta	β	B	xi	ξ	Ξ
gamma	γ	Γ	omicron	o	O
delta	δ	Δ	pi	π	Π
epsilon	ε, ϵ	E	rho	ρ	P
zeta	ζ	Z	sigma	σ, ς (coda)	Σ
eta	η	H	tau	τ	T
theta	ϑ, θ	Θ	upsilon	υ	Y
iota	ι	I	phi	φ, ϕ	Φ
kappa	κ	K	chi	χ	X
lambda	λ	Λ	psi	ψ	Ψ
mu	μ	M	omega	ω	Ω

Appendix C
Complex Numbers

C.1 Calculating with Complex Numbers

In the algebraic representation, a complex number z is given by

$$z = a + ib \qquad (C.1)$$

where $a \in \mathbb{R}$ is the real part and $b \in \mathbb{R}$ the imaginary part of z (and *not* ib); $a = \mathrm{Re}\,(z)$ and $b = \mathrm{Im}\,(z)$. The number i is the imaginary unit defined by $i^2 = -1$, for which the 'normal' rules of calculation hold, e.g. $ib = bi$. The *conjugate complex* number z^* is defined by[1]

$$z^* = a - ib. \qquad (C.2)$$

Accordingly, a real number u can be characterized by $u = u^*$, an imaginary number v by $v = -v^*$.

Addition, subtraction and multiplication of two complex numbers $z_k = a_k + ib_k$ follow familiar rules:

$$\begin{aligned} z_1 \pm z_2 &= a_1 \pm a_2 + i\,(b_1 \pm b_2) \\ z_1 \cdot z_2 &= a_1 a_2 - b_1 b_2 + i\,(a_1 b_2 + a_2 b_1) \end{aligned} \qquad (C.3)$$

and in particular for $c \in \mathbb{R}$, we have

$$c \cdot z_2 = ca_2 + icb_2. \qquad (C.4)$$

For division, we use the complex conjugate:

$$\frac{z_1}{z_2} = \frac{z_1\,z_2^*}{z_2\,z_2^*} = \frac{a_1 a_2 + b_1 b_2 + i\,(-a_1 b_2 + a_2 b_1)}{a_2^2 + b_2^2}. \qquad (C.5)$$

---

[1] The notation $\bar{z}$ is also common.

© Springer Nature Switzerland AG 2018
J. Pade, *Quantum Mechanics for Pedestrians 1*, Undergraduate Lecture
Notes in Physics, https://doi.org/10.1007/978-3-030-00464-4

Fig. C.1 Algebraic
representation in the
complex plane

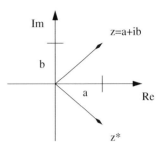

Fig. C.2 Polar
representation in the
complex plane

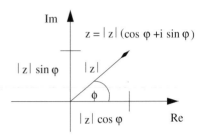

Complex numbers can be represented in an intuitive manner in the *Gaussian plane* (complex plane), see Fig. C.1. For example, we see that the conjugate complex number z^* is the reflection of z through the real axis.

In many cases (and almost always in quantum mechanics), the algebraic form (C.1) is not particularly useful. A different representation is more suitable, namely polar coordinates in the complex plane. A complex number z is then determined by the length of the radius vector and the angle φ between the radius vector and the positive real axis; cf. Fig. C.2. The length of the radius vector of a complex number $z = a + ib$ is called its modulus $|z|$ or complex norm[2]

$$|z| = \sqrt{a^2 + b^2} \geq 0. \tag{C.6}$$

We have the relations:

$$z \cdot z^* = |z|^2; \quad |z_1 \cdot z_2| = |z_1| \cdot |z_2|. \tag{C.7}$$

With $|z|$ and φ, we can write[3]

$$z = |z|\,(\cos \varphi + i \sin \varphi). \tag{C.8}$$

Evidently (and quite intuitively), the complex number does not change if we add a multiple of 2π to φ, $\varphi \rightarrow \varphi + 2m\pi$ with $m \in \mathbb{Z}$. All angles are allowed, of

---

[2] Also called its absolute value. $|z|^2$ is often called the absolute square.

[3] Also called trigonometric form of the complex numbers.

course, but the *principal value* of the angle is confined to the interval $-\pi < \varphi \leq \pi$. The ambiguity of the angle is typical of complex numbers (there is nothing similar in $\mathbb{R}$), and is used constructively, for example in the theory of functions (of complex variables) and other areas.

With the help of the fundamental equation[4]

$$e^{ix} = \cos x + i \sin x \tag{C.9}$$

we obtain finally the *exponential representation* of a complex number z

$$z = |z| \, e^{i\varphi}. \tag{C.10}$$

We know how to determine the modulus of a complex number $z = a + ib$. The determination of the angle φ, called the *phase* or *argument*, is somewhat more complicated. Equation (C.8) suggests the relation $\varphi = \arctan \frac{b}{a}$ (and it is often given in collections of formulas, etc.). But this cannot always hold true, since otherwise $z_1 = 3 + 4i$ and $z_2 = -3 - 4i$ would have the same phase, which is obviously wrong. The correct relation can be formulated differently; a possibility is[5]

$$\varphi = \arctan \frac{b}{a} + \frac{|b|}{b} \frac{1 - \frac{|a|}{a}}{2} \pi; \ a, b \neq 0$$
$$\varphi = \frac{1 - \frac{|a|}{a}}{2} \pi \ \text{for} \ a \neq 0, b = 0 \ \text{and} \ \varphi = \frac{|b|}{b} \frac{\pi}{2} \ \text{for} \ a = 0, b \neq 0. \tag{C.11}$$

Of course, one can add $2m\pi$ to the phase if necessary. The only number without a well-defined phase is the complex number 0. It has the value 0, while its phase is indeterminate.

We want to point out some relations which are sometimes quite handy. Namely, as is seen from (C.9), we have

$$i = e^{i\pi/2}; \ -1 = e^{i\pi}; \ 1 = e^{2i\pi} \tag{C.12}$$

where, of course, $2im\pi$ can be added the exponent. Due to (C.12), a factor i can be interpreted as a phase (or phase shift) of $\pi/2$ or $90°$; for -1, we have accordingly π or $180°$.

We can utilize the ambiguity of the phase in a constructive manner, e.g. to find the roots of a number. We demonstrate this by means of a concrete example: Find all the numbers z for which

$$z^3 = 7 \tag{C.13}$$

holds. Taking the modulus on both sides leads to $|z|^3 = 7$, with the solution $|z| = 7^{1/3}$. We can therefore write $z = |z| \, e^{i\varphi} = 7^{1/3} e^{i\varphi}$ and thus obtain

---

[4]The Feynman Lectures on Physics, 5th Edition, 1970, Vol I, p.22–10, "We summarize with this, the most remarkable formula in mathematics: $e^{i\theta} = \cos \theta + i \sin \theta$. This is our jewel."

[5]For a positive real part, we have $\varphi = \arctan \frac{b}{a}$; for a negative real part, we have to add or subtract π, depending on the sign of the imaginary part, to obtain the principal value.

Fig. C.3 Third roots of 1

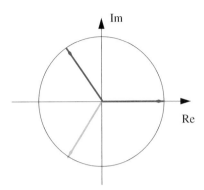

$$e^{3i\varphi} = 1. \tag{C.14}$$

For the right-hand side, we write down all the complex possibilities that exist for 1, namely

$$1 = e^{2im\pi}; \; m \in \mathbb{Z}. \tag{C.15}$$

It follows that

$$e^{3i\varphi} = e^{2im\pi}; \; m \in \mathbb{Z} \tag{C.16}$$

and this leads to

$$\varphi = 0, \pm\frac{2}{3}\pi, \pm\frac{4}{3}\pi, \pm\frac{6}{3}\pi, \dots \tag{C.17}$$

or, more compactly,

$$\varphi = m\pi \bmod(2\pi). \tag{C.18}$$

If we restrict ourselves to the principal values, we obtain the three solutions

$$\varphi = 0, \pm\frac{2}{3}\pi \text{ or } z_1 = 7^{1/3}, \; z_{2,3} = 7^{1/3}e^{\pm i2\pi/3}. \tag{C.19}$$

In the complex plane, this can be understood quite readily: Taking the third or nth root means—with regard to the phase—dividing the full circle by 3 or n. In the example of the 'third root', we obtain the angle $0°$ and $\pm120°$ (or $0°$, $120°$, $240°$); see Fig. C.3. In this context, we recall the *fundamental theorem* of algebra which states that *each polynomial of order n has n zeros.*

Finally, we note that the number system is complete with the complex numbers— no arithmetic operation leads out of it.[6] Expressions such as i^i or $(a+ib)^{(c+id)}$ may be unfamiliar, but they are reasonable and calculable (see exercises).

---

[6]In contrast, e.g. subtraction leads out of the natural numbers and division out of whole numbers.

C.2   Are Complex Numbers Less Intuitive than Real Numbers?

The resistance to complex numbers[7] is often justified by the claim that complex numbers are not intuitive, are unreal and too abstract. This is hard to comprehend, if one thinks of the perfectly clear description of taking the roots in the complex plane. Presumably, this claim has to do less with the facts themselves and more with psychology—perhaps, since the adjectives 'complex' and 'imaginary' suggest that these numbers are difficult (complex) and not really existent (imaginary).

Several hundred years ago, when the child 'complex number' was baptized, it was perhaps even wise to choose those names to avoid discussions with the more conservative. Another example of this prudence is the question of whether the earth orbits the sun. Today we know that the sun does not move around the earth, but we still say, 'the sun rises', knowing that it is an outdated convention of speech (which also has its own beauty), far from taking it literally. And just as the sun does not rise in reality, complex numbers are not difficult or counterintuitive. If a point on the real line is intuitively clear, then so is a point in the complex plane.

The real problem probably lies somewhere else, but we do not perceive it anymore, perhaps because we have become accustomed to it. It is that we can not think of a *point in the mathematical sense*, i.e. an entity with dimension zero. A number (on a numerical axis) corresponds to a point. And this point may be hiding an incredible amount of information, for instance in a rational number (i.e. a fraction), let alone in an irrational number.[8] The following little digression is intended to show this.

Suppose we want to save the contents of all the books in all the libraries of the world with minimum space requirements. To this end, we encode all existing characters, numbers, punctuation marks, Chinese characters, hieroglyphics, cuneiform characters—everything used for writing. If we assume that the number of all those diverse characters and symbols is less than a million, we can characterize each of them by a six-digit decimal number (digits 0–9).[9] And now we translate one book after another into our new code, simply by inserting the appropriate numbers for the characters; the codes for different books are simply written down one after another. When we are finished, we have a very long number $\mathcal{N}$ before us.[10] If we want to

---

[7]By the way, complex numbers are not new-fangled stuff; they have been in use for more than 400 years. Apparently, the northern Italian mathematician Rafael Bombelli (1526–1572) in his work *L'Algebra* was the first to introduce imaginary numbers.

[8]We can grasp small natural numbers directly *as a set*, but to distinguish immediately between 39, 40 and 41 will overcharge most of us (in mother ducks, this limit seems to be reached at 6 or 7 ducklings). Large numbers are completely beyond our comprehension (how much time would it take to count to a million or a billion?). Fractions are also difficult, even very simple ones ('just give me the smaller half').

[9]Of course, a binary or hexadecimal notation would work as well. And if there are more than a million symbols, we need seven decimal digits or more. But this changes nothing in our main argument.

[10]For a rough estimate of the order of magnitude, we assume that a line encompasses 70 characters and a page 50 lines. Hence, a book of 300 pages contains about a million characters. In a library with

write it down the 'usual' way, we need more or less six times as much space as for the originals.

A more space-saving procedure would be the following: ahead of $\mathcal{N}$, we write a zero and a decimal point, thus obtaining a number between 0 and 1, namely $\mathcal{M} = 0.\mathcal{N}$ which we mark exactly on a ruler—if we could in fact do this. It is essential at this point that we can save within a rational number (i.e. a fraction!) between 0 and 1 the content of *all* the libraries of the world. And of course there is still more: In a neighborhood of $\mathcal{M}$, there exists a number where every 'i' is exchanged with an 'o', and another where every book is encoded backwards except the first 1,000 positions; another which contains all the books that ever appeared or will appear in the future. This holds analogously for encoding music. Is there a number that contains all the works that Mozart would have written if he had lived for 10, 20, 30 years longer? Is there a number that includes all the books or music that have not been written to date, and another that includes everything that will never be written?

Who says that there is no poetry in numbers? And in an irrational number, which consists of an infinite number of decimal places without a period, arbitrarily more information (in fact, unimaginably more) can of course be stored.

The real problem lies in the fact that we can imagine small spots, but not a mathematical point. Given this, it is not really understandable why it should be so much harder to imagine a point not on a straight line, but on a plane. Complex numbers are not more 'difficult' or 'counterintuitive' than real numbers—rather the opposite, because operations like taking roots have a clear and intuitive meaning in the complex plane.

C.3 Exercises

1. Given $z_1 = 3 - i, z_2 = 3 + i, z_3 = 1 - 3i, z_4 = 1 + 3i$; sketch the points in the complex plane and calculate their complex norms.
2. Given $z_1 = 3 - 4i$ and $z_2 = -1 + 2i$; calculate $|z_1|, |z_2|, z_1 \pm z_2, z_1 \cdot z_2, \frac{z_1}{z_2}, \frac{1}{z_1}$.
3. Given

$$z = \frac{3 - 4i}{6 + i\sqrt{2}} \cdot (8 - 7i) + 6i. \tag{C.20}$$

   Determine z^*.
4. Write the following complex numbers in the form $\rho e^{i\varphi}$:

---

10 million books, we must therefore encode around 10^{13} characters and will obtain a number of $6 \cdot 10^{13}$ digits. Suppose that on the world average, there is one such library per 40,000 inhabitants. Then, assuming a world population of eight billion, we have globally about 200,000 libraries. (Of course, this is certainly an overoptimistic estimate, not only in view of the situation in developing countries. But here, we are considering only rough orders of magnitude.) This would result in the 'literary number' $\mathcal{N} = 1.2 \cdot 10^{19}$—hence something around 10^{19} (although of course many books would be repeated numerous times, and their copies could be eliminated to reduce the number). For comparison, the Loschmidt constant N_L, which specifies the number of molecules per unit volume of an ideal gas under normal conditions, is $N_L = 2.7 \cdot 10^{19}/\text{cm}^3$.

$$3 + 4i; \ 3 - 4i; \ -3 + 4i; \ -3 - 4i. \tag{C.21}$$

5. Given $z = \frac{-1 \pm i\sqrt{3}}{2}$. Represent the numbers in the complex plane. Calculate z^3.
6. What is the polar representation of $z = \frac{1 \pm i\sqrt{3}}{2}$ and $z = \frac{-1 \pm i\sqrt{3}}{2}$?
7. Show that all complex numbers of the form $e^{i\varphi}$ lie on the unit circle around the origin.
8. Show that the multiplication of a complex number by i means rotating this number by $\frac{\pi}{2}$ or $90°$.
9. Show that $e^{i\varphi} = \cos\varphi + i\sin\varphi$ by means of the power series of the trigonometric functions.
   Solution:

$$
\begin{aligned}
e^{i\varphi} &= \sum_{n=0}^{\infty} \frac{i^n \varphi^n}{n!} = \sum_{n=0}^{\infty} \frac{i^{2n} \varphi^{2n}}{(2n)!} + \sum_{n=0}^{\infty} \frac{i^{2n+1} \varphi^{2n+1}}{(2n+1)!} \\
&= \sum_{n=0}^{\infty} \frac{(-1)^n \varphi^{2n}}{(2n)!} + i \sum_{n=0}^{\infty} \frac{(-1)^n \varphi^{2n+1}}{(2n+1)!} = \cos\varphi + i\sin\varphi.
\end{aligned}
\tag{C.22}
$$

10. Show that $e^{i\varphi} = \cos\varphi + i\sin\varphi$ by taking derivatives.
    Solution:

$$\left(e^{i\varphi}\right)' = ie^{i\varphi} = i \ \cos\varphi - \sin\varphi = (\cos\varphi + i\sin\varphi)'. \tag{C.23}$$

11. Show that

$$\cos x = \frac{e^{ix} + e^{-ix}}{2}; \quad \sin x = \frac{e^{ix} - e^{-ix}}{2i}. \tag{C.24}$$

    Solution:

$$e^{ix} = \cos x + i\sin x; \ e^{-ix} = \cos x - i\sin x; \rightarrow e^{ix} + e^{-ix} = 2\cos x. \tag{C.25}$$

12. Given a function

$$f = (a + ib)e^{ikx} - (a - ib)e^{-ikx}. \tag{C.26}$$

    This function may be brought into the form

$$f = A\sin B. \tag{C.27}$$

    Determine A and B.
    Solution: With $a + ib = \sqrt{a^2 + b^2}e^{id}$ and $d = \arctan\frac{b}{a}$, it follows that[11]:

---

[11] This value for d holds for $a > 0$; otherwise, d is given by the more complex expression C.11.

$$f = \sqrt{a^2 + b^2}\,e^{i(kx+d)} - \sqrt{a^2 + b^2}\,e^{-i(kx+d)} = 2i\sqrt{a^2 + b^2}\,\sin(kx + d).$$
$$(C.28)$$

13. Using only $e^{ix} = \cos x + i \sin x$, show that

$$\sin 2x = 2 \sin x \cdot \cos x; \cos 2x = \cos^2 x - \sin^2 x \qquad (C.29)$$

is valid.
Solution:

$$2 \sin x \cdot \cos x = 2\frac{e^{ix} - e^{-ix}}{2i}\frac{e^{ix} + e^{-ix}}{2} = \frac{e^{2ix} - e^{-2ix}}{2i} = \sin 2x. \qquad (C.30)$$

14. Using $e^{ix} = \cos x + i \sin x$, determine the coefficients a and b in the equation

$$\cos^3 \varphi = a \cos \varphi + b \cos 3\varphi. \qquad (C.31)$$

15. Show that:

$$\sin^2 x = \frac{1}{2}\left(1 - \cos 2x\right); \cos^2 x = \frac{1}{2}\left(1 + \cos 2x\right). \qquad (C.32)$$

16. Show that:

$$\sin 3x = 3 \sin x - 4 \sin^3 x; \cos 3x = 4 \cos^3 x - 3 \cos x. \qquad (C.33)$$

17. Is the equation
$$(\cos x + i \sin x)^n = \cos nx + i \sin nx \qquad (C.34)$$

correct?
Solution: Yes, since

$$(\cos x + i \sin x)^n = \left(e^{ix}\right)^n = e^{inx} = \cos nx + i \sin nx. \qquad (C.35)$$

18. We start from
$$e^{ix} = A \cos x + B \sin x \qquad (C.36)$$

where A and B are to be determined (rule of the game: we have only this equation and do not know at this point that in fact, $e^{ix} = \cos x + i \sin x$ holds). First show that $A = 1$. Then show that $B = \pm i$ must hold, using complex conjugation. How can one fix the sign of B?
Solution: For $x = 0$, we have

$$e^{i \cdot 0} = e^0 = 1 = A \cdot 1 + B \cdot 0 = A. \qquad (C.37)$$

In addition, we have with $A = 1$

$$e^{-ix} = e^{i(-x)} = \cos x - B \sin x. \qquad (C.38)$$

It follows that

$$1 = e^{ix}e^{-ix} = (\cos x + B \sin x)(\cos x - B \sin x) = \cos^2 x - B^2 \sin^2 x. \qquad (C.39)$$

With $\cos^2 x + \sin^2 x = 1$ (Pythagoras), it follows that $B^2 = -1$, and therefore $B = \pm i$. For the decision as to which sign holds, additional information is needed (power series, derivatives).

19. Calculate $e^{i\frac{\pi}{2}m}$ for $m \in \mathbb{Z}$.
20. Given

$$z^8 = 16; \quad z^3 = -8; \qquad (C.40)$$

calculate all solutions.

21. Calculate i^i and $(a + ib)^{(c+id)}$.
    Solution: With $i = e^{i\left(\frac{\pi}{2}+2\pi m\right)}$; $m = 0, \pm 1, \pm 2, \ldots$, it follows that

$$i^i = \left(e^{i\left(\frac{\pi}{2}+2\pi m\right)}\right)^i = e^{-\left(\frac{\pi}{2}+2\pi m\right)}; m = 0, \pm 1, \pm 2, \ldots \qquad (C.41)$$

Appendix D
Calculus I

In the following, some general basic relations from calculus are compiled.

D.1 One Real Independent Variable

D.1.1 The Taylor Expansion

If a function is differentiable sufficiently often, we can write it as a *Taylor series*,[12] i.e. the function at a point $a + x$ can be expressed as the sum of the functions and derivatives at the neighboring point a:

$$f(a+x)$$
$$= f(a) + \tfrac{x}{1!}f^{(1)}(a) + \tfrac{x^2}{2!}f^{(2)}(a) + \cdots + \tfrac{x^n}{n!}f^{(N)}(a) + \tfrac{x^N}{(N+1)!}f^{(N+1)}(a+\lambda x)$$
$$= \sum_{n=0}^{N} \tfrac{x^n}{n!}f^{(n)}(a) + R_N.$$

(D.1)

The term R_N is called remainder term or Lagrange remainder; it is $0 < \lambda < 1$.

Under suitable conditions, the remainder term vanishes for $N \to \infty$ and the sum converges, so that we may write

$$f(a+x) = \sum_{n=0}^{\infty} \frac{x^n}{n!} f^{(n)}(a)$$

(D.2)

We thus have written the function as *power series* $\sum_{k=0}^{\infty} c_k x^k$ (with $c_n = f^{(n)}(a)/n!$).

---

[12] Brook Taylor, British mathematician, 1685–1731.

© Springer Nature Switzerland AG 2018
J. Pade, *Quantum Mechanics for Pedestrians 1*, Undergraduate Lecture Notes in Physics, https://doi.org/10.1007/978-3-030-00464-4

In general, a power series does not converge for all x, but only for $|x| < \rho$. The *convergence radius* ρ can be determined by

$$\rho = \lim_{n\to\infty}\left|\frac{c_n}{c_{n+1}}\right|; \; \rho = \lim_{n\to\infty}\frac{1}{\sqrt[n]{|c_n|}} \tag{D.3}$$

if the limits exist. For $x = \rho$ and $x = -\rho$, the power series may converge or diverge.

The three 'most important' functions e^x, $\cos x$ and $\sin x$ have power series with infinite convergence radii and are therefore particularly well behaved:

$$e^x = \sum_{n=0}^{\infty}\frac{x^n}{n!}; \; \cos x = \sum_{n=0}^{\infty}(-1)^n\frac{x^{2n}}{(2n)!}; \; \sin x = \sum_{n=0}^{\infty}(-1)^n\frac{x^{2n+1}}{(2n+1)!}. \tag{D.4}$$

In other words, in the exponent of the exponential function, we can insert 'anything' for x, and this expression is always defined by the power series as long as x^n exists. For example, $e^{\mathbf{M}}$ for a square matrix $\mathbf{M}$ is defined as $e^{\mathbf{M}} = \sum_{n=0}^{\infty}\frac{\mathbf{M}^n}{n!}$, while the exponential function of a non-square matrix is not defined.

Examples of power series with finite radius of convergence ($\rho = 1$) are $(1 + x)^{\alpha}$ as well as $\ln(1 + x)$ and $\arctan x$:

$$(1 + x)^{\alpha} = 1 + \frac{\alpha}{1!}x + \frac{\alpha(\alpha-1)}{2!}x^2 + \cdots = \sum_{n=0}^{\infty}\binom{\alpha}{n}x^n; \; \begin{array}{l}|x| \leq 1 \text{ for } \alpha > 0 \\ |x| < 1 \text{ for } \alpha < 0\end{array} \tag{D.5}$$

and

$$\ln(1 + x) = x - \frac{x^2}{2} + \frac{x^3}{3} - \cdots = \sum_{n=1}^{\infty}(-1)^{n+1}\frac{x^n}{n}; \; -1 < x \leq 1 \tag{D.6}$$

and

$$\arctan x = x - \frac{x^3}{3} + \frac{x^5}{5} - \cdots = \sum_{n=0}^{\infty}(-1)^n\frac{x^{2n+1}}{2n+1}; \; -1 < x < 1. \tag{D.7}$$

By means of the power series, one can also find very practical approximations for functions, if the x values are sufficiently small; for example

$$e^x \approx 1 + x; \; \cos x \approx 1 - \tfrac{x^2}{2}; \; \sin x \approx x - \tfrac{x^3}{6}$$
$$(1 + x)^{\alpha} \approx 1 + \alpha x; \; \ln(1 + x) \approx x - \tfrac{x^2}{2}; \; \arctan x \approx x - \tfrac{x^3}{3} \tag{D.8}$$

For sufficiently small x, often the first term is sufficient. For instance, we have $\sin x \approx x$ in the interval $|x| < 0.077$ (corresponding to an angle of $4.4°$) with an accuracy of less than or equal to one part per thousand.

D.1.2   L'Hôpital's Rule

This rule concerns indefinite expressions such as $\frac{0}{0}$ or $\frac{\infty}{\infty}$. If we have e.g. $\lim\limits_{x \to x_0} f(x) = 0$ and $\lim\limits_{x \to x_0} g(x) = 0$, then the expression $\lim\limits_{x \to x_0} \frac{f(x)}{g(x)}$ has this form. The rule of L'Hôpital[13] states that under this assumption, it holds that:

$$\lim_{x \to x_0} \frac{f(x)}{g(x)} = \lim_{x \to x_0} \frac{f'(x)}{g'(x)}. \tag{D.9}$$

One can easily prove this by substituting the corresponding Taylor expansions around x_0 for the functions. If the right-hand side of the equation is again an indefinite term, we apply the rule again. Example:

$$\lim_{x \to 0} \frac{\sin x}{x} = \lim_{x \to 0} \frac{\cos x}{1} = 1; \quad \lim_{x \to 0} \frac{e^x - 1 - x}{x^2} = \lim_{x \to 0} \frac{e^x - 1}{2x} = \lim_{x \to 0} \frac{e^x}{2} = \frac{1}{2}. \tag{D.10}$$

In indefinite terms of other types one has to rearrange accordingly. We sketch this only symbolically:

$$0 \cdot \infty = 0 \cdot \frac{1}{0} \text{ or } \frac{1}{\infty} \cdot \infty; \quad \infty - \infty = \infty \left(1 - \frac{\infty}{\infty} \right). \tag{D.11}$$

Example:

$$\lim_{x \to 0} x \ln x = \lim_{x \to 0} \frac{\ln x}{1/x} = \lim_{x \to 0} \frac{1/x}{-1/x^2} = - \lim_{x \to 0} x = 0. \tag{D.12}$$

In terms of the form 0^0 or similar expressions, one takes the logarithm. Example:

$$\lim_{x \to 0} x^x = \lim_{x \to 0} e^{x \ln x} = \lim_{x \to 0} e^{-x} = 1. \tag{D.13}$$

D.1.3   Mean Value Theorem for Integration

We consider the definite integral

$$I = \int_a^b f(x) \mathrm{d}x \tag{D.14}$$

---

[13]Guillaume François Antoine, Marquis de L'Hôpital (also written L'Hospital), French mathematician, 1661–1704.

Fig. D.1 On the mean value
theorem of integration

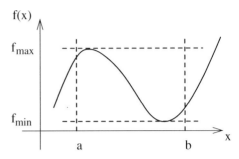

where the function $f(x)$ is sufficiently well behaved. Then it holds that

$$I = \int_a^b f(x)\mathrm{d}x = f(\xi)\,(b-a)\ \ with\ \xi \in [a,b]. \tag{D.15}$$

We know about ξ only that it lies within the interval $[a, b]$; where, exactly, is not determined by the theorem.

The reason for the theorem is that if $f_{\min}$ and $f_{\max}$ are the minimal and maximal values of $f(x)$ in the interval, then we have (see Fig. D.1):

$$f_{\min} \cdot (b - a) \leq I \leq f_{\max} \cdot (b - a). \tag{D.16}$$

Accordingly, the exact value of I must exist for some intermediate value $f_{\min} \leq f(\xi) \leq f_{\max}$, i.e. for $a \leq \xi \leq b$.

D.2 Several Independent Variables

D.2.1 Differentiation

The *partial derivative* of a function of several independent variables $f(x_1, x_2, \ldots)$ with respect to e.g. x_1 is defined as

$$\frac{\partial f(x_1, x_2, \ldots)}{\partial x_1} = \lim_{\varepsilon \to 0} \frac{f(x_1 + \varepsilon, x_2, \ldots) - f(x_1, x_2, \ldots)}{\varepsilon} \tag{D.17}$$

In this definition, the variables $x_2, x_3, \ldots$ play the role of constants.

The use of the symbol ∂ has been adopted to make clear from the outset that it is a partial derivative. Besides $\frac{\partial}{\partial x}$, there are also notations such as ∂_x or the like; instead of $\frac{\partial f}{\partial x}$, one can also write f_x or $f_{|x}$.

The term $\frac{\partial f(x_1, x_2, \dots)}{\partial x_1}$ denotes the change of the function if x_1 is varied and all other independent variables stay fixed. By varying all variables simultaneously, we obtain the total change of the function, which can be expressed as the *total derivative*:

$$df(x_1, x_2, \dots) = \frac{\partial f(x_1, x_2, \dots)}{\partial x_1} dx_1 + \frac{\partial f(x_1, x_2, \dots)}{\partial x_2} dx_2 + \cdots . \qquad (D.18)$$

Higher derivatives are defined accordingly; for example, the term

$$\frac{\partial^2 f(x_1, x_2, \dots)}{\partial x_i \partial x_j} \equiv \partial_{x_i} \partial_{x_j} f(x_1, x_2, \dots) \qquad (D.19)$$

means that we first have to take the derivative of the function with respect to x_j, and then with respect to x_i (execution of the steps from right to left). The order of the derivatives does not play a role iff the first and second partial derivatives of f are continuous; in this case, we have $\partial_{x_i} \partial_{x_j} f(x_1, x_2, \dots) = \partial_{x_j} \partial_{x_i} f(x_1, x_2, \dots)$. We will always assume that all functions are sufficiently smooth and thus satisfy this condition, so that we never have to pay attention to the order of the derivatives. A counterexample is found in the exercises.

By the way, we generally assume in quantum mechanics that we can interchange limit processes (differentiation, integration, summation). As a concrete example, we consider the equation

$$\frac{d}{dt} \int_{-\infty}^{\infty} \rho(x, t)\, dx = \int_{-\infty}^{\infty} \frac{\partial}{\partial t} \rho(x, t)\, dx \qquad (D.20)$$

which we used in Chap. 7 (time invariance of the total probability). We can take the differentiation into the integral iff ρ is continuous with respect to x and is differentiable with respect to t, and $\partial \rho / \partial t$ is continuous with respect to x.

Similar considerations would have to be made in other cases; but we will not do so, presuming on quantum mechanics' 'good nature' (and, of course, knowing that others have already provided the necessary proofs).

D.2.2 Taylor Series

For a function of several variables, the Taylor series reads

$$f(x_1 + a_1, \dots, x_n + a_n) = \sum_{j=0}^{\infty} \frac{1}{j!} \left[\sum_{k=1}^{n} a_k \frac{\partial}{\partial x_k} \right]^j f(x_1, \dots, x_n) \qquad (D.21)$$

or, in compact form (for the definition of ∇, see below):

$$f(\mathbf{r} + \mathbf{a}) = \sum_{j=0}^{\infty} \frac{1}{j!} (\mathbf{a} \cdot \nabla)^j f(\mathbf{r}).$$ (D.22)

The first terms of the expansion are

$$f(\mathbf{r} + \mathbf{a}) = f(\mathbf{r}) + (\mathbf{a} \cdot \nabla) f(\mathbf{r}) + \frac{1}{2} (\mathbf{a} \cdot \nabla) (\mathbf{a} \cdot \nabla) f(\mathbf{r}) + \cdots$$ (D.23)

D.2.3 Vector Algebra

A closer look at the total derivative (D.18) shows that the right-hand side can be expressed as a scalar product[14]:

$$df = \frac{\partial f}{\partial x_1} dx_1 + \frac{\partial f}{\partial x_2} dx_2 + \cdots = \left(\frac{\partial f}{\partial x_1}, \frac{\partial f}{\partial x_2}, \ldots \right) \cdot (dx_1, dx_2, \ldots)$$ (D.24)

The second vector can be written as $d\mathbf{r} = (dx_1, dx_2, \ldots)$. For the first vector, the notation

$$\left(\frac{\partial f}{\partial x_1}, \frac{\partial f}{\partial x_2}, \ldots \right) = \nabla f$$ (D.25)

has become established; here the nabla operator (or briefly just nabla) is formally a vector with the components[15]

$$\nabla = \left(\frac{\partial}{\partial x_1}, \frac{\partial}{\partial x_2}, \ldots \right)$$ (D.26)

Being a vector operator, nabla can be applied to scalar functions $f(x_1, x_2, \ldots)$ and vector functions $\mathbf{F}(x_1, x_2, \ldots)$. The application to f is referred to as *gradient* of f and is also written as grad f:

$$\nabla f = \left(\frac{\partial f}{\partial x_1}, \frac{\partial f}{\partial x_2}, \ldots \right) = \text{grad } f.$$ (D.27)

---

[14]Here, we use a row vector $(dx_1, dx_2, \ldots)$ and not the corresponding column vector; this is only for typographical convenience.

[15]The symbol ∇ is not a Hebrew letter, but an upside down Delta. This sign was named 'nabla' in the nineteenth century, because it resembles an ancient harp (névél in Hebrew, nábla in Greek). The 'D' (as in Delta) in the Hebrew alphabet is the 'Daleth' ד, the 'N' (as in nabla) the 'Nun' נ.

The application to **F** is called the *divergence* and is also written as div **F**:

$$\mathbf{\nabla} \cdot \mathbf{F} = \frac{\partial F_1}{\partial x_1} + \frac{\partial F_2}{\partial x_2} + \cdots = \text{div } \mathbf{F}. \tag{D.28}$$

In three dimensions, one can also take the vector product of $\mathbf{\nabla}$ with a vector function; this is called the curl:

$$\mathbf{\nabla} \times \mathbf{F} = \left(\frac{\partial F_3}{\partial x_2} - \frac{\partial F_2}{\partial x_3}, \frac{\partial F_1}{\partial x_3} - \frac{\partial F_3}{\partial x_1}, \frac{\partial F_2}{\partial x_1} - \frac{\partial F_1}{\partial x_2} \right) = \text{curl } \mathbf{F}. \tag{D.29}$$

We note the different character of the operations:

$$\begin{aligned} \text{gradient:} \quad & \mathbf{\nabla} \, \text{scalar} \rightarrow \textbf{vector} \\ \text{divergence:} \quad & \mathbf{\nabla} \cdot \textbf{vector} \rightarrow \text{scalar} \\ \text{curl:} \quad & \mathbf{\nabla} \times \textbf{vector} \rightarrow \textbf{vector}. \end{aligned} \tag{D.30}$$

The two notations with $\mathbf{\nabla}$ and with grad $-$ div $-$ curl are equivalent; each one has advantages and disadvantages.

Multiple Applications

For appropriate combinations, multiple applications of the nabla operator are defined:

$\mathbf{\nabla} f$	$\mathbf{\nabla} (\mathbf{\nabla} \times f) = \text{div grad } f = \nabla^2 f$	$\mathbf{\nabla} \times (\mathbf{\nabla} f) = \text{curl grad } f = 0$
$\mathbf{\nabla} \times \mathbf{F}$	$\mathbf{\nabla} (\mathbf{\nabla} \times \mathbf{F}) = \text{grad div } \mathbf{F}$	$\mathbf{\nabla} \times (\mathbf{\nabla} \times \mathbf{F})$ not defined
$\mathbf{\nabla} \times \mathbf{F}$	$\mathbf{\nabla} \times (\mathbf{\nabla} \times \mathbf{F}) = \text{div curl } \mathbf{F} = 0$	$\mathbf{\nabla} \times (\mathbf{\nabla} \times \mathbf{F}) = \text{curl curl } \mathbf{F} =$
		$= \mathbf{\nabla} (\mathbf{\nabla} \times \mathbf{F}) - \nabla^2 \mathbf{F} = \text{grad div } \mathbf{F} - \nabla^2 \mathbf{F}$

Here, ∇^2 is the Laplacian (Laplace's differential operator), $\nabla^2 = \frac{\partial^2}{\partial x_1^2} + \frac{\partial^2}{\partial x_2^2} + \frac{\partial^2}{\partial x_3^2}$. As we see from the table, it is defined by $\mathbf{\nabla} \cdot (\mathbf{\nabla} f) = \nabla^2 f$.

Integral Theorems

For completeness, we note the three main integral theorems in short form.

The line integral of a gradient field on a curve C depends only on its endpoints:

$$\int_{\mathbf{r}_1, C}^{\mathbf{r}_2} \mathbf{\nabla} f(\mathbf{r}) \, d\mathbf{r} = f(\mathbf{r}_2) - f(\mathbf{r}_1). \tag{D.31}$$

Given a volume V which is enclosed by a surface S; the orientation of the surface is such that its normal points outwards. Then the *Gaussian integral theorem* (or Gauss-Ostrogradski, also called the divergence theorem) reads:

$$\int_V (\mathbf{\nabla} \cdot \mathbf{F}(\mathbf{r})) \, dV = \int_S \mathbf{F}(\mathbf{r}) \cdot d\mathbf{S} \tag{D.32}$$

Fig. D.2 Polar coordinates

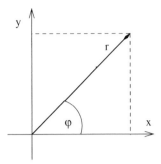

where the left-hand side is a volume integral and the right-hand side is a surface integral.

Given an oriented surface S which is enclosed by a curve C (the sense of rotation is chosen so that it forms a right-hand helix with the surface normal). Then the *Stokes integral theorem* (also called the curl theorem) reads

$$\int_S (\nabla \times \mathbf{F}(\mathbf{r})) \cdot d\mathbf{S} = \int_C \mathbf{F}(\mathbf{r}) \cdot d\mathbf{r} \qquad (D.33)$$

where the left-hand side is a surface integral and the right-hand side is a line integral.

D.3 Coordinate Systems

D.3.1 Polar Coordinates

The polar coordinates (r, φ) are related to the Cartesian coordinates (x, y) by

$$\begin{aligned} x &= r \cos \varphi \\ y &= r \sin \varphi \end{aligned} \quad 0 \le r;\ 0 \le \varphi \le 2\pi \qquad (D.34)$$

Here, r is the distance from the origin, see Fig. D.2.

As a simple application, we derive the transformation equations for an active rotation. If we rotate a point described by (r, φ) through an angle ψ, its new coordinates are

$$\begin{aligned} x &= r \cos \varphi \\ y &= r \sin \varphi \end{aligned} \rightarrow \begin{aligned} x' &= r \cos (\varphi + \psi) \\ y' &= r \sin (\varphi + \psi). \end{aligned} \qquad (D.35)$$

It follows from the addition theorems of trigonometric functions[16] that

---

[16] $\sin(\alpha + \beta) = \sin \alpha \cos \beta + \cos \alpha \sin \beta,\ \cos(\alpha + \beta) = \cos \alpha \cos \beta - \sin \alpha \sin \beta.$

$$x' = r \cos \varphi \cos \psi - r \sin \varphi \sin \psi = x \cos \psi - y \sin \psi$$
$$y' = r \sin \varphi \cos \psi + r \cos \varphi \sin \psi = y \cos \psi + x \sin \psi \qquad (D.36)$$

or, in compact form,

$$\begin{pmatrix} x' \\ y' \end{pmatrix} = \begin{pmatrix} \cos \psi & -\sin \psi \\ \sin \psi & \cos \psi \end{pmatrix} \begin{pmatrix} x \\ y \end{pmatrix} \qquad (D.37)$$

as the representation of an active rotation through the angle ψ.

D.3.2 Cylindrical Coordinates

Cylindrical coordinates (ρ, φ, z) are not used in this text, but we show them here for the sake of completeness. They are related to the Cartesian coordinates (x, y, z) by

$$\begin{aligned} x &= \rho \cos \varphi \\ y &= \rho \sin \varphi \quad 0 \leq \rho; \ 0 \leq \varphi \leq 2\pi \\ z &= z. \end{aligned} \qquad (D.38)$$

Here, ρ is the distance from the z axis (cylinder axis), see Fig. D.3.

The transformation between the two coordinate systems is carried out with the help of e.g.

$$\frac{\partial}{\partial \rho} = \frac{\partial x}{\partial \rho} \frac{\partial}{\partial x} + \frac{\partial y}{\partial \rho} \frac{\partial}{\partial y} = \cos \varphi \frac{\partial}{\partial x} + \sin \varphi \frac{\partial}{\partial y} \qquad (D.39)$$

and analogously for the other variables.

For the unit vectors, it follows that:

$$\mathbf{e}_\rho = \begin{pmatrix} \cos \varphi \\ \sin \varphi \\ 0 \end{pmatrix}; \quad \mathbf{e}_\varphi = \begin{pmatrix} -\sin \varphi \\ \cos \varphi \\ 0 \end{pmatrix}; \quad \mathbf{e}_z = \begin{pmatrix} 0 \\ 0 \\ 1 \end{pmatrix}. \qquad (D.40)$$

Fig. D.3 Cylinder coordinates

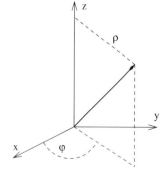

Fig. D.4 Spherical
coordinates

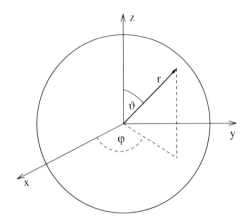

D.3.3 Spherical Coordinates

The spherical coordinates[17] (r, ϑ, φ) are related to the Cartesian coordinates (x, y, z)
by

$$
\begin{aligned}
x &= r \cos \varphi \sin \vartheta \\
y &= r \sin \varphi \sin \vartheta \quad 0 \le r; \ 0 \le \vartheta \le \pi; \ 0 \le \varphi \le 2\pi \\
z &= r \cos \vartheta .
\end{aligned}
\tag{D.41}
$$

Here, r is the distance from the origin, see Fig. D.4.
    The reversed relations are given by

$$
\begin{aligned}
r &= \sqrt{x^2 + y^2 + z^2} \\
\vartheta &= \arccos \frac{z}{\sqrt{x^2+y^2+z^2}} \\
\varphi &= \arctan \frac{x}{y}.
\end{aligned}
\tag{D.42}
$$

The transformation between the two coordinate systems is carried out with the
help of e.g.

$$
\frac{\partial}{\partial r} = \frac{\partial x}{\partial r}\frac{\partial}{\partial x} + \frac{\partial y}{\partial r}\frac{\partial}{\partial y} + \frac{\partial z}{\partial r}\frac{\partial}{\partial z} = \cos \varphi \sin \vartheta \frac{\partial}{\partial x} + \sin \varphi \sin \vartheta \frac{\partial}{\partial y} + \cos \vartheta \frac{\partial}{\partial z}
\tag{D.43}
$$

and analogously for the other variables. For convenience, we give the transformation
matrix:

---

[17] Also called spherical polar coordinates.

$$\frac{\partial r}{\partial x} = \sin \vartheta \cos \varphi \quad \frac{\partial r}{\partial y} = \sin \vartheta \sin \varphi \quad \frac{\partial r}{\partial z} = \cos \vartheta$$

$$\frac{\partial \vartheta}{\partial x} = \frac{\cos \vartheta \cos \varphi}{r} \quad \frac{\partial \vartheta}{\partial y} = \frac{\cos \vartheta \sin \varphi}{r} \quad \frac{\partial \vartheta}{\partial z} = -\frac{\sin \vartheta}{r} \tag{D.44}$$

$$\frac{\partial \varphi}{\partial x} = -\frac{\sin \varphi}{r \sin \vartheta} \quad \frac{\partial \varphi}{\partial y} = \frac{\cos \varphi}{r \sin \vartheta} \quad \frac{\partial \varphi}{\partial z} = 0.$$

For the unit vectors, it follows that:

$$\mathbf{e}_r = \begin{pmatrix} \cos \varphi \sin \vartheta \\ \sin \varphi \sin \vartheta \\ \cos \vartheta \end{pmatrix}; \quad \mathbf{e}_\vartheta = \begin{pmatrix} \cos \varphi \cos \vartheta \\ \sin \varphi \cos \vartheta \\ -\sin \vartheta \end{pmatrix}; \quad \mathbf{e}_\varphi = \begin{pmatrix} -\sin \varphi \\ \cos \varphi \\ 0 \end{pmatrix}. \tag{D.45}$$

The components of a vector $\mathbf{A}$ which is written as $\mathbf{A} = A_x \mathbf{e}_x + A_y \mathbf{e}_y + A_z \mathbf{e}_z$ in Cartesian coordinates are:

$$\mathbf{A} = A_r \mathbf{e}_r + A_\vartheta \mathbf{e}_\vartheta + A_\varphi \mathbf{e}_\varphi \tag{D.46}$$

with

$$A_r = \mathbf{A} \cdot \mathbf{e}_r = A_x \cos \varphi \sin \vartheta + A_y \sin \varphi \sin \vartheta + A_z \cos \vartheta \tag{D.47}$$

and analogously for the other components.

Volume Elements, Surface Elements

In Cartesian coordinates, the infinitesimal volume element is

$$dV = dx \, dy \, dz. \tag{D.48}$$

In cylindrical coordinates, we have

$$dV = d\rho \, \rho d\varphi \, dz = \rho \, d\rho \, d\varphi \, dz; \tag{D.49}$$

and in spherical coordinates (see Fig. D.5),

$$dV = dr \, rd\vartheta \, r \sin \vartheta d\varphi = r^2 \sin \vartheta \, dr \, d\vartheta \, d\varphi. \tag{D.50}$$

In particular, for radially symmetrical functions, we have

$$\int f(r) dV = \int_0^\infty dr \int_0^\pi d\vartheta \int_0^{2\pi} d\varphi \, r^2 \sin \vartheta \, f(r) = 4\pi \int_0^\infty dr \, r^2 f(r). \tag{D.51}$$

For a surface element in spherical coordinates, it follows with (D.5) that

$$df = r^2 \sin \vartheta \, d\vartheta \, d\varphi \tag{D.52}$$

Fig. D.5 Volume element in spherical coordinates

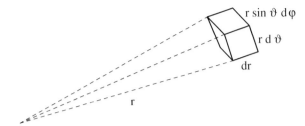

and, making use of $df = r^2 d\Omega$, the solid angle $d\Omega$ is given by

$$d\Omega = \sin\vartheta \, d\vartheta \, d\varphi. \tag{D.53}$$

The Gradient and Laplace Operators in Spherical Coordinates

Gradient:

The gradient of a function $f(\mathbf{r})$ can be written in Cartesian coordinates as

$$\nabla f(\mathbf{r}) = \frac{\partial f}{\partial x}\mathbf{e}_x + \frac{\partial f}{\partial y}\mathbf{e}_y + \frac{\partial f}{\partial z}\mathbf{e}_z. \tag{D.54}$$

With the above transformations, we obtain in spherical coordinates

$$\nabla f(\mathbf{r}) = \frac{\partial f}{\partial r}\mathbf{e}_r + \frac{1}{r}\frac{\partial f}{\partial \vartheta}\mathbf{e}_\vartheta + \frac{1}{r\sin\vartheta}\frac{\partial f}{\partial \varphi}\mathbf{e}_\varphi. \tag{D.55}$$

In particular, for a function $g(r)$ depending on r only, we have

$$\nabla g(r) = \frac{dg(r)}{dr}\mathbf{e}_r = \frac{dg(r)}{dr}\frac{\mathbf{r}}{r}. \tag{D.56}$$

The Laplacian:

The Laplacian in Cartesian coordinates reads[18]:

$$\nabla^2 = \frac{\partial^2}{\partial x^2} + \frac{\partial^2}{\partial y^2} + \frac{\partial^2}{\partial z^2}. \tag{D.57}$$

With the transformations (D.43), it can be converted to spherical coordinates:

---

[18] When there are different sets of coordinates, it is standard to index the nabla operators accordingly:

$$\nabla_{\mathbf{r}}^2 = \frac{\partial^2}{\partial x^2} + \frac{\partial^2}{\partial y^2} + \frac{\partial^2}{\partial z^2}; \ \nabla_{\mathbf{r}'}^2 = \frac{\partial^2}{\partial x'^2} + \frac{\partial^2}{\partial y'^2} + \frac{\partial^2}{\partial z'^2}.$$

$$\nabla^2 = \frac{\partial^2}{\partial r^2} + \frac{2}{r}\frac{\partial}{\partial r} + \frac{1}{r^2}\left[\frac{1}{\sin\vartheta}\frac{\partial}{\partial\vartheta}\left(\sin\vartheta\frac{\partial}{\partial\vartheta}\right) + \frac{1}{\sin^2\vartheta}\frac{\partial^2}{\partial\varphi^2}\right]. \tag{D.58}$$

Using the angular momentum operator $\mathbf{l}$, this expression can be written more compactly. Because of $l_x = \frac{\hbar}{i}\left(y\frac{\partial}{\partial z} - z\frac{\partial}{\partial y}\right)$ etc., we have in spherical coordinates

$$l_x = \frac{\hbar}{i}\left(-\sin\varphi\frac{\partial}{\partial\vartheta} - \cot\vartheta\cos\varphi\frac{\partial}{\partial\varphi}\right)$$

$$l_y = \frac{\hbar}{i}\left(\cos\varphi\frac{\partial}{\partial\vartheta} - \cot\vartheta\sin\varphi\frac{\partial}{\partial\varphi}\right) \tag{D.59}$$

$$l_z = \frac{\hbar}{i}\frac{\partial}{\partial\varphi}$$

and therefore

$$\mathbf{l}^2 = -\hbar^2\left[\frac{1}{\sin\vartheta}\frac{\partial}{\partial\vartheta}\left(\sin\vartheta\frac{\partial}{\partial\vartheta}\right) + \frac{1}{\sin^2\vartheta}\frac{\partial^2}{\partial\varphi^2}\right]. \tag{D.60}$$

Thus, the Laplacian can be written as

$$\nabla^2 = \frac{\partial^2}{\partial r^2} + \frac{2}{r}\frac{\partial}{\partial r} - \frac{\mathbf{l}^2}{\hbar^2 r^2}. \tag{D.61}$$

For the sum of the first two terms, there are also other common expressions:

$$\frac{\partial^2}{\partial r^2} + \frac{2}{r}\frac{\partial}{\partial r} = \frac{1}{r^2}\frac{\partial}{\partial r}r^2\frac{\partial}{\partial r} = \frac{1}{r}\frac{\partial^2}{\partial r^2}r. \tag{D.62}$$

One can also introduce the radial momentum p_r. It is defined as

$$p_r = \frac{\hbar}{i}\frac{1}{r}\frac{\partial}{\partial r}r = \frac{\hbar}{i}\left(\frac{\partial}{\partial r} + \frac{1}{r}\right) \tag{D.63}$$

and it gives:

$$\nabla^2 = -\frac{p_r^2}{\hbar^2} - \frac{\mathbf{l}^2}{\hbar^2 r^2} \tag{D.64}$$

and

$$[r, p_r] = i\hbar. \tag{D.65}$$

D.4 Exercises

1. Calculate the following limiting values

$$\lim_{x \to \infty} x^n e^{-x}; \quad \lim_{x \to \infty} x^{-n} \ln x; \quad \lim_{x \to 0} \frac{\sin x}{x}; \quad \lim_{x \to 0} \frac{\sin kx - kx}{kx \, (1 - \cos kx)}. \qquad (D.66)$$

2. Given a function $h(x)$ and a function $g(x^2)$; write the derivatives of the following functions:

$$f(x) = \frac{1}{h(x)}; \quad f(x) = h^2(x); \quad f(x) = e^{h(x)};$$

$$f(x) = x \cdot g(x^2); \quad f(x) = e^{g(x^2)}. \qquad (D.67)$$

3. Determine the Taylor series around $x = 0$ for the functions $(a \in \mathbb{R})$

$$(1 + x)^a; \quad \ln(1 + x); \quad \arctan x. \qquad (D.68)$$

4. Given the operator $e^{\frac{d}{dx}}$, determine $e^{\frac{d}{dx}} e^x$.
5. Find the first partial derivatives with respect to x, y, z of

$$r; \quad \frac{1}{r}; \quad r^a; \quad \mathbf{r}; \quad \hat{\mathbf{r}}. \qquad (D.69)$$

Define $\mathbf{r} = (x, y, z)$, $r = |\mathbf{r}|$; $\hat{\mathbf{r}}$ is the unit vector in the direction of $\mathbf{r}$.
6. Show that:

$$\frac{\partial^2}{\partial r^2} + \frac{2}{r} \frac{\partial}{\partial r} = \frac{1}{r^2} \frac{\partial}{\partial r} r^2 \frac{\partial}{\partial r} = \frac{1}{r} \frac{\partial^2}{\partial r^2} r. \qquad (D.70)$$

7. Given a function $g(r)$, depending only on the norm (magnitude) of r, for which $\nabla^2 g(r) = 0$ holds. Determine $g(r)$ using the result of the last exercise.
8. Given a scalar function $f(\mathbf{r})$ and a vector function $\mathbf{F}(\mathbf{r})$; which of the following expressions are meaningful?

$$\text{grad } f; \quad \text{div } f; \quad \text{curl } f; \quad \text{grad } \mathbf{F}; \quad \text{div } \mathbf{F}; \quad \text{curl } \mathbf{F}; \quad \nabla f; \quad \nabla \cdot \mathbf{F}; \quad \nabla \times \mathbf{F}. \quad (D.71)$$

Write these expressions using the nabla ∇.
9. Calculate ∇r^a, $\nabla^2 x r^a$ and $\nabla \hat{\mathbf{r}}$.
10. Given the plane wave $\mathbf{F}(\mathbf{r}, t) = \mathbf{A} e^{i(\mathbf{kr} - \omega t)}$, with $\mathbf{A}$ and $\mathbf{k}$ constant vectors.

   (a) Determine the first time derivative as well as the divergence and the curl of $\mathbf{F}(\mathbf{r}, t)$.
   (b) Assume div $\mathbf{F}(\mathbf{r}, t) = 0$. What does this mean physically?
   (c) Determine $(\mathbf{k} \cdot \nabla) \mathbf{F}$ and $\mathbf{k}(\nabla \cdot \mathbf{F})$.

11. Show that:

$$\text{div grad } f = \nabla^2 f; \quad \nabla \cdot (\nabla f) = \nabla^2 f$$
$$\text{curl grad } f = 0; \quad \nabla \times \nabla f = 0$$
$$\text{div curl } \mathbf{F} = 0; \quad \nabla \cdot (\nabla \times \mathbf{F}) = 0$$
$$\text{curl curl } \mathbf{F} = \text{grad div } \mathbf{F} - \nabla^2 \mathbf{F}; \quad \nabla \times (\nabla \times \mathbf{F}) = \nabla (\nabla \cdot \mathbf{F}) - \nabla^2 \mathbf{F}$$

(D.72)

Assume that the partial derivatives commute, $\frac{\partial^2 f}{\partial x_i \partial x_j} = \frac{\partial^2 f}{\partial x_j \partial x_i}$.

12. Given a homogeneously-charged nonconducting sphere of radius R, with a total charge of Q. Using the divergence theorem, determine its electric field $\mathbf{E}$. Derive the potential Φ.

13. Given two point masses with the spherical coordinates $(r, \vartheta_1, \varphi_1)$ and $(r, \vartheta_2, \varphi_2)$. Calculate their distance d

   (a) for $\vartheta_1 = \vartheta_2$ and $\varphi_1 \neq \varphi_2$;
   (b) for $\vartheta_1 \neq \vartheta_2$ and $\varphi_1 = \varphi_2$
   One of the results contains ϑ and φ; the other, only ϑ. Give an intuitive explanation of this. Check (a) the special cases $(\varphi_1, \varphi_2) = (0, \pi)$ and $(0, \pi/2)$; and (b) $(\vartheta_1, \vartheta_2) = (0, \pi)$ and $(0, \pi/2)$.
   Hint: $\cos (a - b) = \cos a \cos b + \sin a \sin b$, and $1 - \cos a = 2 \sin^2 \frac{a}{2}$. And there is, of course, Pythagoras.

14. Show for the functions

$$f(x, y) = \frac{x^3 y - x y^3}{x^2 + y^2} \tag{D.73}$$

that the derivatives $\frac{\partial}{\partial x}$ and $\frac{\partial}{\partial y}$ do not commute at the origin, $\frac{\partial}{\partial x}\frac{\partial f}{\partial y} \neq \frac{\partial}{\partial y}\frac{\partial f}{\partial x}$.
Solution: Away from the origin, $f(x, y)$ is arbitrarily often continuously differentiable, so there the derivatives always commute. The only problem is at the origin. First, we find the first derivatives:

$$\frac{\partial f}{\partial x} = y \frac{x^4 + 4x^2 y^2 - y^4}{\left(x^2 + y^2\right)^2}; \quad \frac{\partial f}{\partial y} = -x \frac{y^4 + 4y^2 x^2 - x^4}{\left(y^2 + x^2\right)^2}. \tag{D.74}$$

Both derivatives have removable discontinuities at the origin with the value 0. The mixed derivative $\frac{\partial}{\partial y}\frac{\partial f}{\partial x}$ is not continuous at the origin. This appears in the inequality of the mixed derivatives. In fact, we have

$$\frac{\partial}{\partial y}\frac{\partial f(x, y)}{\partial x}\bigg/_{x=0, y=0} = \frac{\partial}{\partial y} y \frac{0 - y^4}{\left(0 + y^2\right)^2}\bigg/_{y=0} = -\frac{\partial}{\partial y} y \bigg/_{y=0} = -1 \quad \text{(D.75)}$$

and

$$\frac{\partial}{\partial x}\frac{\partial f(x, y)}{\partial y}\bigg/_{x=0, y=0} = -\frac{\partial}{\partial x} x \frac{0 - x^4}{\left(x^2 + 0\right)^2}\bigg/_{x=0} = \frac{\partial}{\partial x} x \bigg/_{y=0} = 1. \quad \text{(D.76)}$$

We can also show this using polar coordinates in order to get an intuitive picture.
The second derivative

$$\frac{\partial}{\partial y} \frac{\partial f\,(x,\,y)}{\partial x} = \frac{x^6 + 9x^4 y^2 - 9x^2 y^4 - y^6}{\left(x^2 + y^2\right)^3} \tag{D.77}$$

in polar coordinates reads (after some calculation):

$$\partial_y \partial_x f = 2 \sin 2\varphi \sin 4\varphi + \cos 2\varphi \cos 4\varphi. \tag{D.78}$$

We see that the result is independent of r and depends only on the angle φ. Thus, this derivative is not defined at $r = 0$. It follows, for example, that

$$\partial_y \partial_x f = 1 \text{ for } \varphi = 0; \ \partial_y \partial_x f = -1 \text{ for } \varphi = \pi/2 \tag{D.79}$$

i.e. the same result as above.

Appendix E
Calculus II

E.1 Differential Equations: Some General Remarks

A large part of physics is formulated using differential equations—classical and quantum mechanics, hydro- and electrodynamics, string theory and general relativity, and so on. Differential equations occur also in other areas: Climatology, oceanography, biology (population dynamics), chemistry (reaction kinetics), economics (growth processes) and many more. In short, differential equations are a very important means for the mathematical description of our environment.

Unfortunately, there is no general method of solving differential equations. In fact, even questions about the mere existence of solutions of certain differential equations cannot be answered to the present day, e.g. of the Navier–Stokes equations of hydrodynamics.

We briefly discuss below some of the basics.

Differential equations (DEq) are equations that link a function f with its derivatives ∂f. If the function depends on one variable only, they are called *ordinary differential equations*; if it depends on several independent variables (and if there are partial derivatives with respect to more than one variable), they are called *partial differential equations*.

The highest derivative of f which occurs in the equation determines the *order* of the DEq; the highest power of f and its derivatives which occurs determines the *degree*. The *integration of the differential equation* is another term for finding its solution. The *general solution* of a differential equation of order n has n free parameters (*integration constants*); if these free parameters are fixed by n conditions, we have a *particular* or *special* solution. These conditions may be initial and/or boundary conditions. One refers to an *initial condition* when one of the variables is the time (then at $t = 0$). If the differential equation is of order m with respect to time, then the specification of m (suitably chosen) initial conditions determines uniquely the time evolution of the solution (deterministic evolution). The *boundary conditions* refer to the boundaries ∂G of a spatial domain G within which the solution of the differential equation is considered.

© Springer Nature Switzerland AG 2018
J. Pade, *Quantum Mechanics for Pedestrians 1*, Undergraduate Lecture
Notes in Physics, https://doi.org/10.1007/978-3-030-00464-4

Fig. E.1 Family tree of
solutions of differential
equations

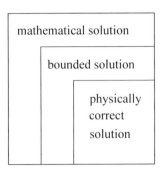

If the unknown function f occurs in each term of the differential equation, is
called a *homogeneous* differential equation, otherwise the differential equation is
inhomogeneous.

In this text we deal (almost) exclusively with *linear differential equations*, where
the function f and its derivatives occur only linearly (i.e. with the power 1). The
basic property of linear differential equations is that linear combinations of solutions
are again solutions. In essence, this means that the solutions span a vector space.
This fact (and the underlying linearity of the SEq) is central to quantum mechanics.

Before we look at those differential equations which are important in the framework of quantum mechanics, we make a general comment: Compared to mathematics, we have in physics the advantage that we can sort out, on the basis of general
considerations, mathematically absolutely correct but physically irrelevant solutions.
For example, we require physically relevant solutions to be bounded in the domain
of definition; so we can omit unbounded solutions. But even bounded solutions do
not always fulfill the requirements; suppose, for example, that a differential equation
has as its solutions two plane waves, one running from right to left, the other running
oppositely. If it is clear from physical reasons that there can be, for example, only
the wave running from right to left, we have to eliminate the other wave, although it
is mathematically a completely valid solution. Figure E.1 illustrates this situation.

E.2   Ordinary Differential Equations

To oversimplify somewhat, we need only two representatives of ordinary differential
equations. We formulate them for the independent variable t; of course, the results
hold analogously for the independent variable x.

The first differential equation is the general differential equation of first order
(which occurs in radioactive decay, absorption of radiation, etc.):

$$\dot{f}(t) = g(t)f(t) \tag{E.1}$$

with a given function $g(t)$. The solution reads

$$f(t) = Ce^{\int g(t)dt} \tag{E.2}$$

where C is the free integration constant ($n = 1$).

The second DEq, which is especially important (not only for quantum mechanics), is the differential equation of second order:

$$\ddot{f}(t) = z^2 f(t); \; z \in \mathbb{C} \tag{E.3}$$

with the general solution

$$f(t) = c_1 e^{zt} + c_2 e^{-zt}. \tag{E.4}$$

The integration constants c_1 and c_2 may be fixed by initial conditions, e.g.

$$f(0) = f_0; \; \dot{f}(0) = \dot{f}_0. \tag{E.5}$$

It follows that

$$f(t) = \frac{zf_0 + \dot{f}_0}{2z} e^{zt} + \frac{zf_0 - \dot{f}_0}{2z} e^{-zt}. \tag{E.6}$$

The cases $z \in \mathbb{R}$ and $z \in \mathbb{I}$ are of fundamental importance; they are customarily written as:

$$\ddot{f}(t) = \omega^2 f(t) \text{ and } \ddot{f}(t) = -\omega^2 f(t); \; \omega \in \mathbb{R} \tag{E.7}$$

with the solutions

$$f(t) = c_1 e^{\omega t} + c_2 e^{-\omega t} \text{ and } f(t) = c_1 e^{i\omega t} + c_2 e^{-i\omega t}. \tag{E.8}$$

The first equation describes exponential behavior, the second a harmonic oscillation. With x instead of t, the differential equations read[19]

$$g''(x) = k^2 g(x) \text{ and } g''(x) = -k^2 g(x); \; k \in \mathbb{R} \tag{E.9}$$

with the solutions

$$g(x) = c_1 e^{kx} + c_2 e^{-kx} \text{ and } g(x) = c_1 e^{ikx} + c_2 e^{-ikx}. \tag{E.10}$$

---

[19]For a clearer distinction, one often writes κ for exponential behavior and k for oscillatory behavior:

$$g''(x) = \kappa^2 g(x) \text{ and } g''(x) = -k^2 g(x); \; \kappa, k \in \mathbb{R}$$

with the solutions
$$g(x) = c_1 e^{\kappa x} + c_2 e^{-\kappa x} \text{ and } g(x) = c_1 e^{ikx} + c_2 e^{-ikx}.$$

E.3 Partial Differential Equations

Apart from the continuity equation used in Chap. 7,

$$\frac{\partial \rho}{\partial t} + \nabla \mathbf{j} = 0 \tag{E.11}$$

(derived in Appendix N, Vol. 1), the partial differential equations of interest to us in the framework of quantum mechanics are of second order with respect to the spatial variables. An external characteristic is the appearance of the Laplacian $\nabla^2 = \partial_x^2 + \partial_y^2 + \partial_z^2$. Another feature of this differential equation is its linearity. For completeness, we also cite in the following some differential equations which are not used elsewhere in the text. All the functions which occur are functions of $\mathbf{r}$.

Laplace's (homogeneous) equation

$$\nabla^2 \varphi = 0 \tag{E.12}$$

is a special case of the (inhomogeneous) Poisson equation

$$\nabla^2 \varphi = f. \tag{E.13}$$

With $f = -\frac{1}{\varepsilon_0}\rho$, this is the conditional equation of a potential φ for given charge density ρ in electrostatics.

In electrodynamics, this equation is replaced by the potential equation

$$\left(\nabla^2 - \frac{1}{c^2}\frac{\partial^2}{\partial t^2} \right) \varphi = \Box\varphi = -\frac{1}{\varepsilon_0}\rho \tag{E.14}$$

where we have used the d'Alembertian (also called quabla) $\Box = \nabla^2 - \frac{1}{c^2}\frac{\partial^2}{\partial t^2}$.[20] We find analogous equations for the vector potential $\mathbf{A}$ and the current density $\mathbf{j}$ by inserting the replacement $\varphi, \rho \rightarrow A_i, j_i/c^2$ with $i = 1, 2, 3$ or, more compactly, $\varphi, \rho \rightarrow \mathbf{A}, \mathbf{j}/c^2$.

For $\rho = 0$, the following homogeneous wave equation results from E.13:

$$\frac{1}{c^2}\frac{\partial^2}{\partial t^2}\varphi = \nabla^2\varphi. \tag{E.15}$$

The differential equations of second order in the spatial coordinates considered so far are of zeroth or second order in time, and thus require no or two initial conditions. In contrast, one initial condition suffices for the heat flow equation:

$$\frac{\partial}{\partial t}T = \lambda\nabla^2 T \tag{E.16}$$

---

[20]The d'Alembertian is defined by some authors with the opposite sign as $\Box = \frac{1}{c^2}\frac{\partial^2}{\partial t^2} - \nabla^2$.

and for the time-dependent SEq

$$i\hbar\frac{\partial}{\partial t}\psi = H\psi = \left(-\frac{\hbar^2}{2m}\nabla^2 + V\right)\psi. \tag{E.17}$$

We note the great similarity of these two equations—the main difference is 'just' the occurrence of the factor i in the SEq.[21] Both equations are deterministic in the sense that the specification of the initial conditions $T(\mathbf{r}, 0)$ or $\psi(\mathbf{r}, 0)$ uniquely determines the solutions $T(\mathbf{r}, t)$ and $\psi(\mathbf{r}, t)$ for all times.

With the separation *ansatz*

$$\psi(\mathbf{r}, t) = \varphi(\mathbf{r}) e^{-i\frac{Et}{\hbar}}, \tag{E.18}$$

we obtain the stationary SEq from (E.17):

$$E\varphi = H\varphi = \left(-\frac{\hbar^2}{2m}\nabla^2 + V\right)\varphi \tag{E.19}$$

This equation is an eigenvalue problem. In general, solutions exist only for certain values of E. These values E are called the *eigenvalues*, and the associated solutions *eigenfunctions* or *eigenvectors*. The set of all eigenvalues is called the *spectrum*. The spectrum can contain a finite or an infinite number of elements. The eigenvalues can be countable (*discrete spectrum*) or uncountable (*continuous spectrum*). Spectra may also contain both discrete and continuous elements; these two components can furthermore overlap.

An eigenvalue is called *degenerate* if there are a number of different eigenfunctions belonging to this eigenvalue. The simplest case of the eigenvalue problem (E.19) is a non-degenerate, discrete spectrum; in this case,

$$H\varphi_n = E_n\varphi_n; \quad n = 1, 2, \ldots \tag{E.20}$$

applies. In the case of degeneracy, we have

$$H\varphi_{n,r} = E_n\varphi_{n,r}; \quad n = 1, 2, \ldots; \quad r = 1, 2, \ldots, g_n \tag{E.21}$$

where g_n is the degree of degeneracy.

Closed analytical solutions of the stationary SEq exist only for a handful of potentials. In particular, the free three-dimensional problem

$$E\varphi(\mathbf{r}) = -\frac{\hbar^2}{2m}\nabla^2\varphi(\mathbf{r}) \tag{E.22}$$

---

[21] As noted in the text, due to this 'small difference' i, there are worlds between the solutions of the heat-conduction equation and the Schrödinger equation.

has the solutions

$$\varphi\left(\mathbf{r}\right) = \sum_{l,m} [a_l\, j_l\left(kr\right) + b_l n_l\left(kr\right)]\, Y_l^m\left(\vartheta, \varphi\right);\ \ k^2 = \frac{2m}{\hbar^2} E. \tag{E.23}$$

The $j_l\left(kr\right)$ and $n_l\left(kr\right)$ are spherical Bessel functions, the $Y_l^m\left(\vartheta, \varphi\right)$ spherical harmonics. For these functions and other analytical solutions of the SEq, see Appendix B, Vol. 2.

E.4 Exercises

1. Given the eigenvalue problem

$$\frac{d^2}{dx^2} f\left(x\right) = -k^2 f\left(x\right);\ \ k > 0;\ \ 0 \leq x \leq a \tag{E.24}$$

with the boundary condition

$$f\left(0\right) = f\left(a\right) = 0, \tag{E.25}$$

determine the allowed values of k and the associated eigenfunctions.

2. Given the differential equations

$$f''\left(x\right) + k^2 f\left(x\right) = 0 \text{ and } f''\left(x\right) - k^2 f\left(x\right) = 0 \tag{E.26}$$

with $k \in \mathbb{R}$; what are the general solutions of these equations?

3. Given the differential equation

$$y^{(n)} = \frac{d^n}{dx^n} y\left(x\right) = y\left(x\right); \tag{E.27}$$

what is its general solution?

4. Show that the linear combination of solutions of the SEq (time-dependent and stationary) are again themselves solutions.

5. Given the wave equation

$$\partial_t^2 f\left(\mathbf{r}, t\right) = c^2 \nabla^2 f\left(\mathbf{r}, t\right). \tag{E.28}$$

The initial conditions $f\left(\mathbf{r}, 0\right)$ and $\dot{f}\left(\mathbf{r}, 0\right)$ are known. Formulate the general solution.

6. The heat conduction equation

$$\partial_t T\left(\mathbf{r}, t\right) = D\nabla^2 T\left(\mathbf{r}, t\right) \tag{E.29}$$

is solved by

$$T(\mathbf{r}, t) = e^{tD\nabla^2} T(\mathbf{r}, 0).$$ (E.30)

Determine the solution $T(\mathbf{r}, t)$ for the initial condition $T(\mathbf{r}, 0) = T_0 + T_1 \cos$ (**kr**). Discuss the result; is it physically plausible?
Solution: First we show that (E.30) satisfies the heat conduction equation. We have

$$\partial_t T(\mathbf{r}, t) = \partial_t e^{tD\nabla^2} T(\mathbf{r}, 0) = D\nabla^2 e^{tD\nabla^2} T(\mathbf{r}, 0) = D\nabla^2 T(\mathbf{r}, t).$$ (E.31)

Next, we determine $e^{tD\nabla^2} (T_0 + T_1 \cos(\mathbf{kr}))$. It is

$$e^{tD\nabla^2}(T_0 + T_1 \cos(\mathbf{kr})) = \sum_{n=0}^{\infty} \frac{t^n D^n \nabla^{2n}}{n!}(T_0 + T_1 \cos(\mathbf{kr}))$$

$$= T_0 + T_1 \sum_{n=0}^{\infty} \frac{t^n D^n \nabla^{2n}}{n!} \cos(\mathbf{kr}).$$ (E.32)

Because of $\nabla^2 \cos(\mathbf{kr}) = -\mathbf{k}^2 \cos(\mathbf{kr})$, it follows that

$$T(\mathbf{r}, t) = e^{tD\nabla^2}(T_0 + T_1 \cos(\mathbf{kr}))$$

$$= T_0 + T_1 \sum_{n=0}^{\infty} \frac{t^n D^n (-\mathbf{k}^2)^n}{n!} \cos(\mathbf{kr}) = T_0 + T_1 \cos(\mathbf{kr}) e^{-D\mathbf{k}^2 t}.$$ (E.33)

The initial condition is a starting temperature T_0 with a superposed variation $\sim T_1$ which levels off more and more in the course of time, according to $T_1 e^{-D\mathbf{k}^2 t}$.
7. Show that

$$F(x, t) = \frac{1}{(at)^{1/2}} e^{-b\frac{x^2}{t}}$$ (E.34)

is a solution of the one-dimensional heat conduction equation. Determine the constants a and b. What would be a similar solution of the SEq?
8. Show that

$$\Phi(\mathbf{r}) = \frac{1}{4\pi\varepsilon_0} \int \frac{\rho(\mathbf{r}')}{|\mathbf{r} - \mathbf{r}'|} d^3 r'$$ (E.35)

is a solution of

$$\nabla^2 \Phi = -\frac{1}{\varepsilon_0} \rho(\mathbf{r}).$$ (E.36)

Hint: Use Fourier transformation (see Appendix H, Vol. 1):

$$\nabla_\mathbf{r}^2 \frac{1}{|\mathbf{r} - \mathbf{r}'|} = -4\pi\delta(\mathbf{r} - \mathbf{r}').$$ (E.37)

Remark: If several sets of coordinates occur, it is not clear in the notation ∇^2 with respect to which coordinates the differentiation should be carried out. In this case, one frequently writes the corresponding coordinates as an index: $\nabla_\mathbf{r}^2$ means the derivative with respect to the corresponding components of $\mathbf{r}$.

Solution: In the following, the difference between $\mathbf{r}$ and $\mathbf{r}'$ is essential. We have

$$\nabla_\mathbf{r}^2 \Phi (\mathbf{r}) = \frac{1}{4\pi\varepsilon_0} \nabla_\mathbf{r}^2 \int \frac{\rho(\mathbf{r}')}{|\mathbf{r}-\mathbf{r}'|} d^3 r' = \frac{1}{4\pi\varepsilon_0} \int \rho(\mathbf{r}')\nabla_\mathbf{r}^2 \frac{1}{|\mathbf{r}-\mathbf{r}'|} d^3 r'$$

$$= -\frac{1}{4\pi\varepsilon_0} \int \rho(\mathbf{r}')4\pi\delta(\mathbf{r}-\mathbf{r}')d^3 r' = -\frac{1}{\varepsilon_0}\rho(\mathbf{r}). \tag{E.38}$$

9. Given a function $g(r)$ with $\nabla^2 g(r) = 0$. Determine $g(r)$.
10. Solve the equation

$$\left(\frac{d^2}{dr^2} + \frac{2}{r}\frac{d}{dr} + 1 - \frac{l(l+1)}{r^2} \right) f_l(r) = 0 \tag{E.39}$$

by means of a power series expansion. Write down explicitly the regular and the irregular solution for $l = 0$.

Appendix F
Linear Algebra I

F.1 Vectors (Real, Three Dimensional)

In this section, we consider 'physical' vectors, i.e. triples of (real) measurable quantities, which are referred to a coordinate system so that a change of the coordinate system leads to an analogous change of the components of the vector. These vectors are initially written as row vectors; the distinction between the column and row vector is introduced later on in the section on matrix calculus. In print, vectors are frequently denoted by boldface type, $\mathbf{r}$; handwritten, by an arrow, $\vec{r}$. The prototype of a 'physical' vector is the position vector

$$\mathbf{r} = (x, y, z). \tag{F.1}$$

A general vector is given by

$$\mathbf{v} = \left(v_x, v_y, v_z\right) \text{ or } \mathbf{v} = (v_1, v_2, v_3) \tag{F.2}$$

or a similar notation. The norm (magnitude, value) of this vector reads

$$|\mathbf{v}| = v = \sqrt{v_x^2 + v_y^2 + v_z^2}. \tag{F.3}$$

If $|\mathbf{v}| = 1$, the vector is said to be *normalized*. The space spanned by the set of all these vectors is denoted by $\mathbb{R}^3$ ($\mathbb{R}$ for real, 3 means the dimension).

F.1.1 Basis, Linear Independence

With the help of the Cartesian unit vectors

$$\mathbf{e}_x = (1, 0, 0); \ \ \mathbf{e}_y = (0, 1, 0); \ \ \mathbf{e}_z = (0, 0, 1) \tag{F.4}$$

© Springer Nature Switzerland AG 2018
J. Pade, *Quantum Mechanics for Pedestrians 1*, Undergraduate Lecture
Notes in Physics, https://doi.org/10.1007/978-3-030-00464-4

each vector can be written as

$$\mathbf{v} = a\mathbf{e}_x + b\mathbf{e}_y + c\mathbf{e}_z. \tag{F.5}$$

The terms a, b, c are called the components or coordinates of the vector. Unit vectors are frequently written with a hat: $\mathbf{e}_x \equiv \hat{\mathbf{x}}$, etc.

These unit vectors have important properties: namely, they are *linearly independent* and they form a *complete* set. A set of vectors $\{\mathbf{v}_1, \mathbf{v}_2, \ldots\}$ is linearly independent if the equation

$$\lambda_1 \mathbf{v}_1 + \lambda_2 \mathbf{v}_2 + \cdots = 0 \tag{F.6}$$

can be satisfied only for $\lambda_1 = \lambda_2 = \cdots = 0$. The completeness of the system $\{\mathbf{e}_x, \mathbf{e}_y, \mathbf{e}_z\}$ implies that every vector (F.2) can be represented in the form (F.5). In other words: the Cartesian unit vectors (F.4) form a *basis* of $\mathbb{R}^3$.

F.1.2   Scalar Product, Vector Product

The scalar product (inner product, dot product) of two vectors $\mathbf{v} = (v_1, v_2, v_3)$ and $\mathbf{w} = (w_1, w_2, w_3)$ is a number and is defined by[22]

$$\mathbf{vw} = \mathbf{v} \cdot \mathbf{w} = v_1 w_1 + v_2 w_2 + v_3 w_3. \tag{F.7}$$

Another representation is

$$\mathbf{vw} = vw \cos \varphi \tag{F.8}$$

where φ is the angle between the two vectors (intermediate angle), see Fig. F.1. This relation shows at once that two vectors are orthogonal iff $\mathbf{vw} = 0$. Indeed, the scalar product is closely linked to the term *projection*. The perpendicular projection of $\mathbf{w}$ onto $\mathbf{v}$ (i.e. the component of $\mathbf{w}$ parallel to $\mathbf{v}$) clearly has the length $w \cos \varphi$; the component of $\mathbf{w}$ which is parallel to $\mathbf{v}$ is thus given by $\mathbf{w}' = w \cos \varphi \cdot \hat{\mathbf{v}} = \frac{\mathbf{vw}}{\mathbf{vv}} \cdot \mathbf{v}$, and the scalar product is then $\mathbf{vw} = \mathbf{vw}'$. Of course, this reasoning could be repeated with the roles reversed.

The vector product (cross product or skew product) of two three-dimensional vectors is defined by

$$\mathbf{v} \times \mathbf{w} = (v_2 w_3 - v_3 w_2, \, v_3 w_1 - v_1 w_3, \, v_1 w_2 - v_2 w_1). \tag{F.9}$$

Another formulation uses the Levi–Civita symbol (permutation symbol, epsilon tensor) ε_{ijk}:

---

[22]The definition of the scalar product for complex vectors is given in the Sect. F.2.

Fig. F.1 Projection of
w onto **v**

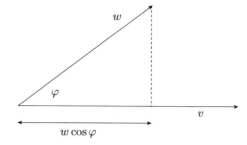

$$\varepsilon_{ijk} = \begin{cases} 1 & ijk \text{ is an even permutation of 123} \\ -1 & \text{if } ijk \text{ is an odd permutation of 123} \\ 0 & \text{otherwise.} \end{cases} \qquad (\text{F.10})$$

Then the vector product is written as[23]

$$(\mathbf{v} \times \mathbf{w})_j = \sum_{k,m=1}^{3} \varepsilon_{jkm} v_k w_m. \qquad (\text{F.11})$$

The norm of the vector product is given by

$$|\mathbf{v} \times \mathbf{w}| = vw |\sin \varphi| . \qquad (\text{F.12})$$

As can be seen, the vector product vanishes for $\varphi = 0$, i.e. for collinear vectors.

F.1.3 Polar and Axial Vectors

Vectors can be distinguished according to how they react to transformation of the spatial coordinates $(x, y, z) \rightarrow (-x, -y, -z)$ or $\mathbf{r} \rightarrow -\mathbf{r}$ (parity transformation, coordinate inversion).

A *polar vector* transforms like a position vector, for example the momentum (because of $\mathbf{p} = m\dot{\mathbf{r}}$):

$$\mathbf{p} \rightarrow -\mathbf{p} \qquad (\text{F.13})$$

An *axial vector* (= *pseudovector*) transforms like the angular momentum, according to

$$\mathbf{l} \rightarrow \mathbf{l}, \qquad (\text{F.14})$$

---

[23]With Einstein's summation convention (repeated indices are implicitly summed over, i.e. without explicitly noting the summation), this is written $(\mathbf{v} \times \mathbf{w})_j = \varepsilon_{jkm} v_k w_m$. We will not use this convention, however.

Since $\mathbf{l} = \mathbf{r} \times \mathbf{p} \rightarrow \mathbf{l} = (-\mathbf{r}) \times (-\mathbf{p}) = \mathbf{r} \times \mathbf{p}$. It generally applies that the vector product of two polar vectors is a pseudovector.

These definitions allow us, moreover, to make the distinction between *scalars* such as $\mathbf{r} \cdot \mathbf{p}$, which do not change under the transformation $\mathbf{r} \rightarrow -\mathbf{r}$, and *pseudoscalars*, which change their signs. All scalar products of an axial and a polar vector are pseudoscalars; an example is $\mathbf{l} \cdot \mathbf{p}$.

The distinction between polar and axial vectors plays a role in e.g. the study of parity violation, occurring e.g. in beta decay and generally in the weak interactions.

F.2 Matrix Calculus

One can modify a vector by multiplying it by a number (a scalar), resulting in a change in length of the vector, but not in a change of its direction. For other transformations such as the rotation of a vector, one requires matrices.

Matrices have rows and columns, so we have to distinguish between column and row vectors. Therefore, in this section, vectors are no longer denoted by boldface type, but rather are written as column and row vectors. Furthermore, we no longer limit ourselves to real numbers, but use more generally complex numbers. In addition, we consider arbitrary dimensions; in this sense, the vectors occurring in the following are no longer the 'physical' vectors of the last section, but vectors in the general sense of linear algebra.

Matrices can be represented as rectangular arrays of numbers with m rows and n columns:

$$A = \begin{pmatrix} a_{11} & a_{12} & \ldots & a_{1n} \\ a_{21} & a_{22} & \ldots & a_{2n} \\ \vdots & \vdots & \ddots & \vdots \\ a_{m1} & a_{m2} & \ldots & a_{mn} \end{pmatrix} = (a_{mn}). \tag{F.15}$$

This is called an $m \times n$ matrix. The set of all $m \times n$ matrices is denoted by $K^{m \times n}$ or $M(m \times n, K)$, where K is the underlying field of numbers. We restrict ourselves in the following to complex matrices $K = \mathbb{C}$ (or possibly to the subset of real matrices $K = \mathbb{R}$).

Multiplying a matrix by a scalar c (scalar multiplication) means that all its elements are multiplied by c; the addition of two matrices (which have to be of the same dimensions, of course) is term by term:

$$cA = \begin{pmatrix} ca_{11} & ca_{12} & \ldots & ca_{1n} \\ ca_{21} & ca_{22} & \ldots & ca_{2n} \\ \vdots & \vdots & \ddots & \vdots \\ ca_{m1} & ca_{m2} & \ldots & ca_{mn} \end{pmatrix} = (ca_{mn}) \tag{F.16}$$

and

$$(a_{mn}) + (b_{mn}) = (a_{mn} + b_{mn}). \tag{F.17}$$

The product of two matrices A and B (matrix multiplication) can be carried out if the number of columns of A is equal to the number of rows of B. If A is a $k \times m$ matrix and B is an $m \times n$ matrix, then $A \cdot B$ is a $k \times n$ matrix. The calculation is performed according to the rule 'row times column':

$$A \cdot B = (c_{kn}); \quad c_{ij} = \sum_{l=1}^{m} a_{il} b_{lj}. \tag{F.18}$$

Some examples: First the product of a 2×3 and a 3×2 matrix:

$$\begin{pmatrix} 1 & 2 & 3 \\ a & b & c \end{pmatrix} \begin{pmatrix} 4 & 7 \\ 5 & 8 \\ 6 & 9 \end{pmatrix} = \begin{pmatrix} 1 \cdot 4 + 2 \cdot 5 + 3 \cdot 6 & 1 \cdot 7 + 2 \cdot 8 + 3 \cdot 9 \\ a \cdot 4 + b \cdot 5 + c \cdot 6 & a \cdot 7 + b \cdot 8 + c \cdot 9. \end{pmatrix} \tag{F.19}$$

Column vectors can be understood as $n \times 1$ matrices, row vectors as $1 \times n$ matrices:

$$\begin{pmatrix} 1 & 2 \\ 3 & 4 \end{pmatrix} \begin{pmatrix} a \\ b \end{pmatrix} = \begin{pmatrix} a + 2b \\ 3a + 4b \end{pmatrix} \tag{F.20}$$

$$(a \ b) \begin{pmatrix} 1 & 2 \\ 3 & 4 \end{pmatrix} = (a + 3b \ 2a + 4b). \tag{F.21}$$

The product of a row vector and a column vector is a number (scalar product), the product of a column and a row vector a matrix (dyadic product)

$$(c \ d) \begin{pmatrix} a \\ b \end{pmatrix} = ca + db \tag{F.22}$$

$$\begin{pmatrix} a \\ b \end{pmatrix} (c \ d) = \begin{pmatrix} ac & ad \\ bc & bd \end{pmatrix}. \tag{F.23}$$

Multiple products are also defined, where applicable:

$$(c \ d) \begin{pmatrix} 1 & 2 \\ 3 & 4 \end{pmatrix} \begin{pmatrix} a \\ b \end{pmatrix} = (c \ d) \begin{pmatrix} a + 2b \\ 3a + 4b \end{pmatrix} = c(a + 2b) + d(3a + 4b). \tag{F.24}$$

Even if the product AB exists, the product BA is not automatically defined (see exercises). But this always holds true for square matrices, i.e. $n \times n$ matrices. However, in this case, the product is not commutative as a rule, and we have $AB \neq BA$ in general. Example:

$$\begin{pmatrix} 1 & 2 \\ 3 & 4 \end{pmatrix} \begin{pmatrix} 0 & 1 \\ 1 & 0 \end{pmatrix} = \begin{pmatrix} 2 & 1 \\ 4 & 3 \end{pmatrix}; \quad \begin{pmatrix} 0 & 1 \\ 1 & 0 \end{pmatrix} \begin{pmatrix} 1 & 2 \\ 3 & 4 \end{pmatrix} = \begin{pmatrix} 3 & 4 \\ 1 & 2 \end{pmatrix}. \tag{F.25}$$

For the remainder of this section, we restrict ourselves to square matrices.

The *identity matrix* (unit matrix) is the matrix which has 1 for each element on the principal diagonal and 0 for all other elements. It is denoted by E, E_n, $\mathbb{I}$, Id, $\mathbf{1}$ or the like; often just by 1. The zero matrix has, according to its name, only zeroes as entries; it is usually denoted by 0.

For a square matrix A, any power A^n with $n \in \mathbb{N}$ is defined (A^0 is the identity matrix). For this reason, we can also insert matrices into polynomials or power series, e.g. in exponential functions such as

$$e^A = \sum_{n=0}^{\infty} \frac{1}{n!} A^n. \tag{F.26}$$

The power A^m of a square matrix A can be the zero matrix (in contrast to complex numbers z; z^n is always unequal to zero for $z \neq 0$). In this case the matrix is called nilpotent with index m. The simplest example is the matrix A with index 2:

$$A = \begin{pmatrix} 0 & 1 \\ 0 & 0 \end{pmatrix}; \quad A^2 = \begin{pmatrix} 0 & 0 \\ 0 & 0 \end{pmatrix}. \tag{F.27}$$

Every square matrix A is associated with two scalar parameters, its trace $tr(A) = tr A$ and its determinant $\det A$. The trace of a square matrix is defined as the sum of all its diagonal elements:

$$tr(A) = tr(a_{nn}) = \sum_{j=1}^{n} a_{jj}. \tag{F.28}$$

We have $tr(AB) = tr(BA)$, even if the matrices A and B do not commute. It follows that the trace is invariant under cyclic permutations, e.g. $tr(ABC) = tr(BCA) = tr(CAB)$.

The determinant of a square matrix (an alternating multilinear form) is also a number ($\in K$). For a 2×2 matrix, it is given by

$$\det \begin{pmatrix} a & b \\ c & d \end{pmatrix} = \begin{vmatrix} a & b \\ c & d \end{vmatrix} = ad - bc. \tag{F.29}$$

Determinants of higher-dimensional matrices may be calculated by means of the Laplace expansion of determinants; two equivalent formulations are

$$\det A = \sum_{j=1}^{n} (-1)^{i+j} a_{ij} \cdot \det A_{ij} \text{ expansion with respect to row } i$$

$$\det A = \sum_{i=1}^{n} (-1)^{i+j} a_{ij} \cdot \det A_{ij} \text{ expansion with respect to column } j \tag{F.30}$$

where A_{ij} denotes that $(n-1) \times (n-1)$-matrix which arises from A by deleting the ith row and jth column. An example can be found in the exercises.

Determinants are zero iff rows (or columns) are linearly dependent; this holds true in particular if two rows (or two columns) are identical.

The determinant of matrix products can be reduced to the individual determinants:

$$\det (A \cdot B) = \det A \cdot \det B. \tag{F.31}$$

Finally, we note a relation between the trace and the determinant:

$$\det e^A = e^{tr(A)}. \tag{F.32}$$

F.2.1 Special Matrices

From now on we use the bra-ket notation: $|a\rangle$ denotes a column vector, $\langle b|$ a row vector.

The importance of matrices in mathematics and mathematical physics is reflected among other things by the fact that there are a number of special matrices. Before we go into details, we will define the important terms transposed and adjoint.

The *transposed* matrix A^T of a given matrix A is obtained by interchanging the roles of its rows and columns:

$$A = \begin{pmatrix} a_{11} & \cdots & a_{1n} \\ \vdots & \ddots & \vdots \\ a_{m1} & \cdots & a_{mn} \end{pmatrix} ; \quad A^T = \begin{pmatrix} a_{11} & \cdots & a_{m1} \\ \vdots & \ddots & \vdots \\ a_{1n} & \cdots & a_{mn}. \end{pmatrix} \tag{F.33}$$

Taking in addition the complex conjugate of all the matrix elements, we obtain the *adjoint* matrix $A^\dagger$:

$$A = \begin{pmatrix} a_{11} & \cdots & a_{1n} \\ \vdots & \ddots & \vdots \\ a_{m1} & \cdots & a_{mn} \end{pmatrix} ; \quad A^\dagger = \begin{pmatrix} a_{11}^* & \cdots & a_{m1}^* \\ \vdots & \ddots & \vdots \\ a_{1n}^* & \cdots & a_{mn}^*. \end{pmatrix} \tag{F.34}$$

For the determinants, we then find

$$\det A^T = \det A; \quad \det A^\dagger = \det A^*. \tag{F.35}$$

The adjoint of a column vector is thus a row vector with complex conjugated entries and *vice versa*:

$$
\begin{pmatrix} a_1 \\ a_2 \\ \vdots \end{pmatrix}^\dagger = \begin{pmatrix} a_1^* & a_2^* & \dots \end{pmatrix} ; \quad \begin{pmatrix} a_1 & a_2 & \dots \end{pmatrix}^\dagger = \begin{pmatrix} a_1^* \\ a_2^* \\ \vdots \end{pmatrix} \tag{F.36}
$$

or, compactly:

$$
|a\rangle^\dagger = \langle a| ; \quad \langle a|^\dagger = |a\rangle . \tag{F.37}
$$

The product of a row vector and a column vector (i.e. the scalar product) is written[24]

$$
\langle a| \, b \rangle = \begin{pmatrix} a_1^* & a_2^* & \dots \end{pmatrix} \begin{pmatrix} b_1 \\ b_2 \\ \vdots \end{pmatrix} = a_1^* b_1 + a_2^* b_2 + \cdots \tag{F.38}
$$

This expression generalizes the formulations of the last section for real vectors.

In contrast, the dyadic product $|a\rangle \, \langle b|$ is itself a matrix:

$$
|a\rangle \, \langle b| = \begin{pmatrix} a_1 \\ a_2 \\ \vdots \end{pmatrix} \begin{pmatrix} b_1^* & b_2^* & \dots \end{pmatrix} = \begin{pmatrix} a_1 b_1^* & a_1 b_2^* & \dots \\ a_2 b_1^* & a_2 b_2^* & \dots \\ \vdots & \vdots & \vdots \end{pmatrix} \tag{F.39}
$$

In the rest of this section, we confine ourselves to square matrices. We list some important types of matrices; a graphical summary is given in Fig. F.2.

If the determinant of a matrix A is not equal to zero, A is said to be *regular*. In this case there exists another matrix A^{-1}, such that[25] $AA^{-1} = A^{-1}A = E$. The matrix A^{-1} is called the inverse matrix to A. Matrices with vanishing determinants are called *singular*.

A matrix A is termed *diagonalizable* if there is a regular matrix B such that $D = BAB^{-1}$ is a diagonal matrix. A subset of diagonalizable matrices are *normal* matrices that commute with their adjoints: $AA^\dagger = A^\dagger A$.

Very important for physical description are two types of matrices with special symmetry. A matrix is called *symmetric* if $A = A^T$, and it is called *Hermitian* if $A = A^\dagger$.

A real matrix A is called *orthogonal*, if $A^{-1} = A^T$ or $AA^T = E$. These matrices represent e.g. rotations in an n dimensional space. A complex matrix is called *unitary* if $A^{-1} = A^\dagger$ or $AA^\dagger = E$. These can be thought of as rotations in the n dimensional complex space.

For a *projection*(matrix), we have $A^2 = A$. We call such matrices *idempotent*. If the projection is in addition Hermitian, it is called a *Hermitian projection* (or normal projection, orthogonal projection or projector).

---

[24]One does not write $\langle a| \, |b\rangle$, but instead $\langle a| \, b\rangle$, saving a vertical line in this and similar expressions.

[25]In finite-dimensional spaces, the left inverse is equal to the right inverse. For dim $= \infty$ this is not necessarily the case.

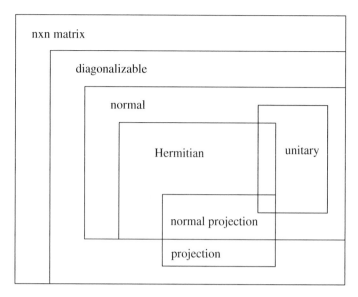

Fig. F.2 Family tree of square matrices

Unitary and Hermitian (normal) matrices are related in various ways. For example, a normal matrix A can be *diagonalized unitarily*, i.e. there is a unitary matrix U such that UAU^{-1} is diagonal (for the proof see the exercises). In fact, the following applies: A can be diagonalized unitarily iff A is normal. This 'iff' is valid only for unitary diagonalization—non-unitary diagonalizability may also occur for non-normal operators (see exercises).

Hermitian and unitary matrices and Hermitian projections (which are all normal matrices, i.e. diagonalizable) and their generalizations for corresponding operators play an important role in quantum mechanics.

Furthermore, in quantum mechanics, matrices with a countably infinite number of columns or rows are found. To multiply them, one has to impose additional conditions on the components, because the sums which occur are infinite series and do not necessarily converge. These issues are treated in more detail e.g. in functional analysis.

F.2.2 The Eigenvalue Problem

If A is an $n \times n$ matrix and v an n-dimensional vector, then the eigenvalue problem has the form[26]:

---

[26]In this section, we drop the bra-ket notation, since only column vectors occur (denoted by v), and no row vectors.

$$Av = \lambda v \qquad (F.40)$$

where λ is an (in general complex) number. Thus, we have to find vectors which are mapped by A onto the λ-fold of themselves. These vectors are called *eigenvectors*, and the corresponding numbers λ are their *eigenvalues*. The eigenvectors indicate the directions in which A acts as a multiplication by λ (i.e. a number), while in other directions, Av is no longer proportional to v.

We first calculate the eigenvalues. We rewrite (F.40)[27] to give

$$(\lambda E - A)\, v = 0. \qquad (F.41)$$

In order that this system have other solutions besides just the trivial solution $v = 0$, the following condition must be fulfilled:

$$\det(\lambda E - A) = 0. \qquad (F.42)$$

If we write this determinant out in full according to the rules for the $n \times n$ matrix A, we see that it is a polynomial of order n in λ. This polynomial $p_n(\lambda)$ is called the *characteristic polynomial* of A

$$p_n(\lambda) = \det(\lambda E - A). \qquad (F.43)$$

The determination of the eigenvalues is thus equivalent to finding of the zeros of $p_n(\lambda)$. The fundamental theorem of algebra (see also Appendix C, Vol. 1) says that each polynomial of order n has n zeros (which are complex, in general). Thus, we can write the polynomial as product of linear factors (polynomial factorization)

$$p_n(\lambda) = (\lambda - \lambda_1)(\lambda - \lambda_2)\ldots(\lambda - \lambda_n). \qquad (F.44)$$

Multiple zeros occur in this factorization a number of times according to their multiplicity. The set of all eigenvalues is called the *spectrum*. For an example see the exercises.

We note in addition that trace and determinant of a matrix A are directly related to its eigenvalues, namely by

$$tr\,(A) = \sum_j \lambda_j; \quad \det(A) = \prod_j \lambda_j. \qquad (F.45)$$

Now that we know the eigenvalues, we have to determine the eigenvectors. For this, the eigenvalues are inserted into (F.40), where one usually indexes from the outset:

$$Av_i = \lambda_i v_i; \; i = 1, \ldots, n. \qquad (F.46)$$

---

[27] Often this is simply written as $(\lambda - A)\, v = 0$.

This system of linear equations is now solved by the usual techniques. Example:

$$A = \begin{pmatrix} 0 & 1 \\ 1 & 0 \end{pmatrix}; \quad \lambda_1 = 1; \quad \lambda_2 = -1. \tag{F.47}$$

Written out in full, we have e.g. for the eigenvalue λ_1

$$\begin{pmatrix} 0 & 1 \\ 1 & 0 \end{pmatrix} \begin{pmatrix} v_{1,1} \\ v_{1,2} \end{pmatrix} = \begin{pmatrix} v_{1,1} \\ v_{1,2} \end{pmatrix} \tag{F.48}$$

or

$$\begin{aligned} v_{1,2} &= v_{1,1} \\ v_{1,1} &= v_{1,2} \end{aligned} \tag{F.49}$$

It follows that:

$$v_1 = \begin{pmatrix} 1 \\ 1 \end{pmatrix} v_{1,1} \text{ and analogously } v_2 = \begin{pmatrix} 1 \\ -1 \end{pmatrix} v_{2,1} \tag{F.50}$$

where $v_{1,1}$ and $v_{2,1}$ are arbitrary complex numbers. *All* vectors v_1 of this form are solutions of the eigenvalue equation (F.48). In other words, these vectors span a one-dimensional subspace which is called the *eigenspace* of the eigenvalue λ_1 (the zero vector is not considered to be an eigenvector, but is element of the eigenspace). That one nevertheless speaks of *the* eigenvector (and not of *an* eigenvector) is due to the fact that one refers in general to *normalized* vectors; in the example these are

$$v_1 = \frac{1}{\sqrt{2}} \begin{pmatrix} 1 \\ 1 \end{pmatrix} \text{ and } v_2 = \frac{1}{\sqrt{2}} \begin{pmatrix} 1 \\ -1 \end{pmatrix}. \tag{F.51}$$

If there are multiple eigenvalues, the situation may be more complicated; in this case one speaks of *degeneracy*. For simplicity, we restrict ourselves to the case of normal matrices, which is relevant to quantum mechanics. If an eigenvalue λ occurs d-fold, it is d-fold degenerate (degree of degeneracy d). In this case, the eigenspace of λ has the dimension d.

F.2.3 A Remark on Hermitian Matrices

Measured variables are represented in quantum mechanics by Hermitian matrices (or, more generally, by Hermitian operators). In particular in Chap. 13, the properties of these operators are discussed. Among other things, it is found that their eigenvalues are real, that eigenvectors belonging to different eigenvalues are orthogonal to each other, and that two commuting Hermitian operators possess a common CONS.

For an Hermitian (or more generally normal) matrix A, one can always find a unitary matrix so that the operator $U^{-1}AU$ is diagonal, i.e. $U^{-1}AU = D$ applies. As the columns of U, one can choose the eigenvectors; the diagonal elements of D are the eigenvalues that occur as frequently as their degree of degeneracy indicates.[28] The explicit calculation is given in the exercises. By the way, the spectral representation in Chap. 13 is just another formulation of this fact.

From this, the rule follows that commuting Hermitian operators can be diagonalized simultaneously (since they have a common CONS). Because of the diagonalizability of the operators or matrices occurring in quantum mechanics, one can always expand in terms of the eigenfunctions, without having to worry about Jordan normal forms or the like.

F.3   Exercises

1. Given $x = (4, -2, 5)$; determine a, b, c, d so that the three vectors x, $y = (-1, a, b)$ and $z = (-1, c, d)$ are pairwise orthogonal.
2. x and y are three-dimensional vectors. Show that x and $x \times y$ as well as y and $x \times y$ are mutually orthogonal.
3. Given the vectors

$$\mathbf{a} = \begin{pmatrix} 1 \\ 2 \\ 3 \end{pmatrix}; \ \mathbf{b} = \begin{pmatrix} 3 \\ 2 \\ 1 \end{pmatrix}; \ \mathbf{c} = \begin{pmatrix} 0 \\ A \\ B \end{pmatrix}, \tag{F.52}$$

calculate the scalar product $\mathbf{a} \cdot \mathbf{b}$, the vector product $\mathbf{a} \times \mathbf{b}$, and the dyadic product $\mathbf{ab}$. For which A, B are the three vectors $\mathbf{a}$, $\mathbf{b}$, $\mathbf{c}$ linearly independent?
4. Given the Coriolis force $\mathbf{F}_C = 2m\,(\mathbf{v} \times \boldsymbol{\omega})$; in what direction does it act on a freely falling body?
Solution: The Earth's rotation is counterclockwise (seen from above the north pole), and is thus mathematically positive, i.e. $\boldsymbol{\omega} = (0, 0, \omega)$ with $\omega > 0$. The speed of a body falling to Earth's surface is $\mathbf{v} = (v, 0, 0)$ with $v < 0$; for the sake of simplicity, we have assumed that the mass falls along the x axis (i.e. at the equator); for $v > 0$, the mass would be moving away from the Earth's surface. It follows that $\mathbf{F}_C = 2m\,(\mathbf{v} \times \boldsymbol{\omega}) = 2m\,(0, -v\omega, 0)$. Because of $v < 0$, the term $-v\omega$ is positive and a deflection towards positive values of y results, i.e. to the east.
5. Given the matrix M

$$M = \begin{pmatrix} 0 & 1 \\ 1 & 0 \end{pmatrix}, \tag{F.53}$$

calculate e^M, e^{iM}, $\cos M$ and $\sin M$.

---

[28]The geometric multiplicity of these eigenvalues equals their algebraic multiplicity.

6. Given $A = \begin{pmatrix} 1 & 0 & 2 \\ 0 & -2i & 1 \end{pmatrix}$ and $B = \begin{pmatrix} 1 & 0 \\ 5i & 2 \\ -i & 1 \end{pmatrix}$; determine, if defined, A^2, AB, BA, B^2, and $(AB)^2$.

7. Given

$$A = \begin{pmatrix} 1 & 0 & 2i \\ 3 & -2i & -2 \end{pmatrix};$$
(F.54)

find A^T and $A^\dagger$. Solution:

$$A^T = \begin{pmatrix} 1 & 3 \\ 0 & -2i \\ 2i & -2 \end{pmatrix}; \quad A^\dagger = \begin{pmatrix} 1 & 3 \\ 0 & 2i \\ -2i & -2 \end{pmatrix}.$$
(F.55)

8. Given $A = \begin{pmatrix} 1 & 2 \\ -2 & -1 \end{pmatrix}$. Is A normal? Answer the same question for $B = \begin{pmatrix} 1 & 2 \\ 3 & 4 \end{pmatrix}$.

Solution: With $A^\dagger = \begin{pmatrix} 1 & -2 \\ 2 & -1 \end{pmatrix}$, it follows that

$$\begin{pmatrix} 1 & 2 \\ -2 & -1 \end{pmatrix}\begin{pmatrix} 1 & -2 \\ 2 & -1 \end{pmatrix} = \begin{pmatrix} 5 & -4 \\ -4 & 5 \end{pmatrix}; \quad \begin{pmatrix} 1 & -2 \\ 2 & -1 \end{pmatrix}\begin{pmatrix} 1 & 2 \\ -2 & -1 \end{pmatrix} = \begin{pmatrix} 5 & 4 \\ 4 & 5 \end{pmatrix}.$$
(F.56)

Hence, A is not normal.

9. Show that the matrix

$$A = \begin{pmatrix} 0 & 1 \\ a & 0 \end{pmatrix}; \quad a \in \mathbb{R}, a \neq 0$$
(F.57)

is diagonalizable, but for $a \neq 1$ not normal.

Solution: To see if A is normal, we calculate

$$AA^\dagger = \begin{pmatrix} 0 & 1 \\ a & 0 \end{pmatrix}\begin{pmatrix} 0 & a \\ 1 & 0 \end{pmatrix} = \begin{pmatrix} 1 & 0 \\ 0 & a^2 \end{pmatrix}; \quad A^\dagger A = \begin{pmatrix} 0 & a \\ 1 & 0 \end{pmatrix}\begin{pmatrix} 0 & 1 \\ a & 0 \end{pmatrix} = \begin{pmatrix} a^2 & 0 \\ 0 & 1 \end{pmatrix}.$$
(F.58)

Hence, the matrix is not normal for $a^2 \neq 1$.

Concerning the diagonalization, we note that A has the eigenvalues $\lambda_1 = \sqrt{a}$ and $\lambda_2 = -\sqrt{a}$ and the eigenvectors $v_1 = c_1 \begin{pmatrix} 1 \\ \sqrt{a} \end{pmatrix}$ and $v_2 = c_2 \begin{pmatrix} 1 \\ -\sqrt{a} \end{pmatrix}$.

Hence, for the transformation matrix T we can use the *ansatz*

$$T = \begin{pmatrix} 1 & 1 \\ \sqrt{a} & -\sqrt{a} \end{pmatrix}; \quad T^{-1} = \frac{1}{2\sqrt{a}}\begin{pmatrix} \sqrt{a} & 1 \\ \sqrt{a} & -1 \end{pmatrix}$$
(F.59)

The check gives

$$T^{-1}AT = \frac{1}{2\sqrt{a}} \begin{pmatrix} \sqrt{a} & 1 \\ \sqrt{a} & -1 \end{pmatrix} \begin{pmatrix} 0 & 1 \\ a & 0 \end{pmatrix} \begin{pmatrix} 1 & 1 \\ \sqrt{a} & -\sqrt{a} \end{pmatrix} = \begin{pmatrix} \sqrt{a} & 0 \\ 0 & -\sqrt{a} \end{pmatrix}. \quad \text{(F.60)}$$

Due to

$$T^{\dagger} = \begin{pmatrix} 1 & \sqrt{a^*} \\ 1 & -\sqrt{a^*} \end{pmatrix}, \quad \text{(F.61)}$$

we have $T^{-1} \neq T^{\dagger}$. Thus, the transformation is not unitary (as is the case with normal matrices).

10. Show that a Hermitian matrix A is diagonalizable by a unitary transformation—in other words that there is a unitary matrix U with $U^{-1}AU = D$.
    Solution: The (nondegenerate) eigenvalue problem is $Av_n = c_n v_n$. If we denote the components of the vector v_n by $v_{n|m}$, we can write this using $A = (a_{ij})$ as:

$$Av_n = c_n v_n \text{ or } \sum_j a_{lj} v_{n|j} = c_n v_{n|l}. \quad \text{(F.62)}$$

For the transforming unitary matrix, we choose U with the components

$$u_{kj} = v_{j|k}. \quad \text{(F.63)}$$

Thus, the columns of this matrix are the eigenvectors.
We have to check whether

$$U^{-1}AU = D \text{ or } AU = UD \quad \text{(F.64)}$$

holds, where D is a diagonal matrix with the entries $d_{ij} = d_{jj}\delta_{ij}$. We have

$$(AU)_{ij} = \sum_k a_{ik} u_{kj} = \sum_k a_{ik} v_{j|k} = c_j v_{j|i} \quad \text{(F.65)}$$

and

$$(UD)_{ij} = \sum_k u_{ik} d_{kj} = \sum_k u_{ik} d_{jj}\delta_{kj} = u_{ij} d_{jj} = d_{jj} v_{j|i} \quad \text{(F.66)}$$

and clearly the last two results are identical for the choice $d_{jj} = c_j$.
To show that U is unitary, we use the fact that the columns of U are the eigenvectors, which are mutually orthogonal and can be assumed to be normalized. It follows then that

$$(U^{\dagger}U)_{ij} = \sum_k u_{ki}^* u_{kj} = \sum_k v_{i|k}^* v_{j|k} = \delta_{ij} \quad \text{(F.67)}$$

and

$$\left(UU^\dagger\right)_{ij} = \sum_k u_{ik}u_{jk}^* = \sum_k v_{k|i}^* v_{k|j} = \delta_{ij}. \tag{F.68}$$

In case of degeneracy, the proof is somewhat more extensive, but analogous.

11. Hermitian matrices are unitarily diagonalizable. Using this fact, prove that also normal matrices are unitarily diagonalizable.

Solution: A matrix A is called normal, if $\left[A, A^\dagger\right] = 0$. We see that the two matrices $B = A + A^\dagger$ and $C = i\left(A - A^\dagger\right)$ are Hermitian; they commute and therefore can be diagonalized simultaneously by a unitary transformation. This we can write as

$$\begin{aligned} UBU^{-1} &= UAU^{-1} + UA^\dagger U^{-1} = D \\ UCU^{-1} &= iUAU^{-1} - iUA^\dagger U^{-1} = D' \end{aligned} \tag{F.69}$$

where D and D' are diagonal matrices. Because of

$$2UAU^{-1} = D - iD', \tag{F.70}$$

we have demonstrated the proposition.

12. Calculate the determinant of $A = \begin{pmatrix} 1 & 2 & 3 \\ 4 & 5 & 6 \\ 7 & 8 & 9 \end{pmatrix}$.

Solution: For practice, we calculate the determinant twice. The upper calculation is an expansion in terms of the first row, the lower of the second column. The results are of course identical.

$$\det\begin{pmatrix} 1 & 2 & 3 \\ 4 & 5 & 6 \\ 7 & 8 & 9 \end{pmatrix} = \begin{cases} 1\cdot\det\begin{pmatrix} 5 & 6 \\ 8 & 9 \end{pmatrix} - 2\cdot\det\begin{pmatrix} 4 & 6 \\ 7 & 9 \end{pmatrix} + 3\cdot\det\begin{pmatrix} 4 & 5 \\ 7 & 8 \end{pmatrix} \\ -2\cdot\det\begin{pmatrix} 4 & 6 \\ 7 & 9 \end{pmatrix} + 5\cdot\det\begin{pmatrix} 1 & 3 \\ 7 & 9 \end{pmatrix} - 8\cdot\det\begin{pmatrix} 1 & 3 \\ 4 & 6 \end{pmatrix} \end{cases}$$

$$= \begin{cases} 45 - 48 - 2\,(36 - 42) + 3\,(32 - 35) \\ -2\,(36 - 42) + 5\,(9 - 21) - 8\,(6 - 12) \end{cases} = 0. \tag{F.71}$$

13. Determine the eigenvalues and the polynomial factorization for the matrix

$$A = \begin{pmatrix} i\sqrt{3} & 0 & 0 \\ 0 & 1 & 2 \\ 0 & -2 & -1 \end{pmatrix} \tag{F.72}$$

as well as its trace and determinant.

Solution: We have

$$p_3\left(\lambda\right) = \det\left(\lambda E - A\right) = \begin{vmatrix} \lambda - i\sqrt{3} & 0 & 0 \\ 0 & \lambda - 1 & -2 \\ 0 & 2 & \lambda + 1 \end{vmatrix} = \left(\lambda - i\sqrt{3}\right)\left(\lambda^2 + 3\right). \tag{F.73}$$

The zeros of $p\left(\lambda\right)$ follow from this as

$$\lambda_1 = i\sqrt{3}; \quad \lambda_2 = i\sqrt{3}; \quad \lambda_3 = -i\sqrt{3}, \tag{F.74}$$

and the polynomial factorization of the characteristic polynomial reads

$$p_3\left(\lambda\right) = \left(\lambda - i\sqrt{3}\right)^2\left(\lambda + i\sqrt{3}\right). \tag{F.75}$$

We find the double zero $i\sqrt{3}$ and the simple zero $-i\sqrt{3}$.
Trace and determinant are given by

$$tr\left(A\right) = i\sqrt{3} + 1 - 1 = i\sqrt{3}; \quad \sum_j \lambda_j = \lambda_1 = i\sqrt{3} + i\sqrt{3} - i\sqrt{3} = i\sqrt{3} \tag{F.76}$$

and

$$\det(A) = i\sqrt{3}\left(-1 + 4\right) = 3i\sqrt{3}; \quad \prod_j \lambda_j = \lambda_1 = i\sqrt{3}\cdot i\sqrt{3}\cdot\left(-i\sqrt{3}\right) = 3i\sqrt{3}. \tag{F.77}$$

14. Given the eigenvalue problem

$$Mv = \lambda v \tag{F.78}$$

with the matrix

$$M = \begin{pmatrix} 0 & -2i \\ 2i & 3 \end{pmatrix}, \tag{F.79}$$

determine the eigenvalues and the associated normalized eigenvectors. Are the two eigenvectors orthogonal?
Solution: The eigenvalues are given as solutions of the secular equation:

$$\det\begin{pmatrix} -\lambda & -2i \\ 2i & 3 - \lambda \end{pmatrix} = 0 \rightarrow \lambda\left(\lambda - 3\right) - 4 = 0 \rightarrow \lambda_1 = 4; \quad \lambda_2 = -1. \tag{F.80}$$

The eigenvectors $v_j = \begin{pmatrix} a_j \\ b_j \end{pmatrix}$ are solutions of

$$\begin{pmatrix} 0 & -2i \\ 2i & 3 \end{pmatrix}\begin{pmatrix} a_j \\ b_j \end{pmatrix} = \lambda_j\begin{pmatrix} a_j \\ b_j \end{pmatrix} \tag{F.81}$$

and are given by (without normalization):

$$v_1 = a_1 \begin{pmatrix} 1 \\ 2i \end{pmatrix}; \quad v_2 = a_2 \begin{pmatrix} 1 \\ -\frac{i}{2} \end{pmatrix}; \tag{F.82}$$

the normalized eigenvectors are:

$$v_1 = \frac{1}{\sqrt{5}} \begin{pmatrix} 1 \\ 2i \end{pmatrix}; \quad v_2 = \frac{2}{\sqrt{5}} \begin{pmatrix} 1 \\ -\frac{i}{2} \end{pmatrix}. \tag{F.83}$$

For the scalar product, we find:

$$v_1^\dagger v_2 = \frac{2}{5} \begin{pmatrix} 1 & -2i \end{pmatrix} \begin{pmatrix} 1 \\ -\frac{i}{2} \end{pmatrix} = 0. \tag{F.84}$$

Thus, the eigenvalues are orthogonal—as they must be, because M is Hermitian.

Appendix G
Linear Algebra II

Quantum mechanics operates in complex vector spaces with scalar products. In this appendix, we compile some basic concepts.

G.1 Groups

Groups are important structures not only in linear algebra; they also occur in many areas of mathematical physics. They consist of a set of elements (finite or infinite), which can be combined by a calculation rule or operation, frequently written as $+$, $*$, $\circ$ or $\times$. Here, the notation does not necessarily imply 'the usual' arithmetic addition or multiplication.

Given a non-empty set of elements G and a binary operation $*$, whereby the combination of two elements of the set is again an element of G (closure); then the pair $(G, *)$ is called a group if it satisfies

- $a * (b * c) = (a * b) * c$: the operation is associative.
- There is a neutral element e with $a * e = e * a = a$.
- For each $a \in G$, there is an inverse element $a^{-1} \in G$ with $a * a^{-1} = a^{-1} * a = e$.
- If in addition, it holds that $a * b = b * a$, the group is called Abelian or commutative.

If the operation is addition/multiplication, the group is called additive/multiplicative; the neutral element is then the zero/the one, and the inverse element is $-a/\frac{1}{a}$.

Examples of Abelian groups are the real numbers with addition as the group operation and zero as a neutral element, or with multiplication as operation and one as neutral element (in the latter case, zero must be eliminated, because it has no inverse). An example of a non-Abelian group are the invertible $n \times n$ matrices with matrix multiplication as operation.

Because of the very general definition of groups, 'all sorts of things' can form a group; well-known examples from physics and symmetry transformations are

© Springer Nature Switzerland AG 2018
J. Pade, *Quantum Mechanics for Pedestrians 1*, Undergraduate Lecture Notes in Physics, https://doi.org/10.1007/978-3-030-00464-4

rotations and reflections, or the Lorentz transformations. We consider in the following some cases explicitly.

First, an example of a discrete group (countably many elements): The parity transformation $\mathcal{P}$ has the eigenvalues ± 1 (because of $\mathcal{P}^2 = 1$). The group corresponding to $\mathcal{P}$ is the multiplicative group with the two elements 1 and -1, the group Z_2.

Continuous groups have uncountably many elements. An example is the general linear group $GL(n, K)$. It is the group of all invertible $n \times n$ matrices with elements of the field K (for us, either $\mathbb{R}$ or $\mathbb{C}$; if it is clear which set is meant, one usually omits K). Restricting this set to the matrices with determinant 1, one obtains the special linear group $SL(n)$.

Special cases of $GL(n)$ are the unitary group $U(n)$ and the orthogonal group $O(n)$, i.e. the groups of unitary and orthogonal $n \times n$ matrices. If we restrict ourselves to matrices with determinant 1, we obtain the special unitary group $SU(n)$ and the special orthogonal group $SO(n)$. To give a concrete example: $SO(3)$ is the group of all rotations in three dimensions around an axis passing through $(0, 0, 0)$.

The group $GL(n, K)$ and its subsets are groups that form a continuum, which is obvious from the (older) names continuous or continuous group. Today, however, they are usually called Lie groups.

G.2 Vector Spaces

One can imagine that the concept of vector space actually originated from the 'arrows' or the vectors of physics. But it turns out that there are many sets of very different objects that follow the same rules of calculation, i.e. have the same structure. For this reason, one abstracted from the 'arrows' and defined the structure itself.

A non-empty set $\mathcal{V}$ is called a vector space over a field K (for us almost exclusively $\mathbb{C}$), if in $\mathcal{V}$ an addition and a multiplication operation with numbers from K are defined,[29] where the usual rules of vector calculus apply. These are:

With u and v in the space, $u + v$ also belongs to the vector space. In addition, $\mathcal{V}$ contains a specific element 0, and the following rules are valid[30]:

- $u + v = v + u$: commutativity of the addition;
- $u + (v + w) = (u + v) + w$: associativity of the addition;
- $u + 0 = u$: existence of the zero element;
- $u + x = v$ has always exactly one solution x.

With $u \in \mathcal{V}, \alpha \in \mathbb{C}$, then $\alpha \cdot u$ also belongs to the vector space, and the following rules hold:

- $(\alpha + \beta) \cdot u = \alpha \cdot u + \beta \cdot u$: distributive property;

---

[29]Note that it is thus postulated that one can add two states and multiply a state by a number. This is a strong requirement which many state spaces do not meet. An example: the state space of all possible positions on a chessboard.

[30]It is an additive Abelian group.

- $\alpha \cdot (u + v) = \alpha \cdot u + \alpha \cdot v$: distributive property;
- $(\alpha \cdot \beta) \cdot u = (\alpha\beta) \cdot u$: associative property of the multiplication;
- $1 \cdot u = u$: existence of the unit element.

Elements of V are called *vectors*, elements of the field K *scalars*.

There are many concrete examples of vector spaces; we mention the space of $n \times n$ matrices, the space of polynomials of degree n, the space of the functions continuous within the interval $0 \leq x \leq 1$, the space of the solutions of a linear differential equation such as the wave equation or the Schrödinger equation, the space of sequences $x = (x_0, x_1, x_2, \ldots)$. In these spaces, addition and multiplication are the usual operations.

We point out again that in this context (so to speak, in the algebraic sense), all possible objects may be called vectors, insofar as they are elements of a vector space—functions, polynomials, matrices, etc., and also the solutions of the SEq. This is not because they were set up like a column vector, but simply because they are elements of a vector space. It is certainly advisable to distinguish between the meanings of 'physical' and 'algebraic' vectors.

G.3   Scalar Product

Considering the concepts of scalar product (angle), norm (length) and metric (distance), one can also imagine that they arose in the context of 'arrows'; in connection with the abstraction process leading to the vector space, they were correspondingly abstracted from the actual objects and the structure was set up and expanded, e.g. to include complex numbers or general vectors, among other things.

A scalar product, written here as (x, y), assigns a scalar to two elements $x, y \in V$. It must meet the following requirements: The scalar product is

1. positive definite
$$(x, x) \geq 0; \ (x, x) = 0 \leftrightarrow x = 0;$$

2. linear (more exactly: semilinear in the first component, linear in the second (sesquilinearity))[31]

$$(x, \alpha y + \beta z) = \alpha (x, y) + \beta (x, z) ; \tag{G.1}$$

3. Hermitian or conjugate symmetric

$$(x, y) = (y, x)^* . \tag{G.2}$$

---

[31] A sesquilinear form is a function that assigns a number to two vectors, and which is linear in one and antilinear in the other argument. A sesqulinear form with Hermitian symmetry is called a Hermitian form.

Due to the last equation, we always have $(x, x) \in \mathbb{R}$. Clearly, the expression $\int f^* g \mathrm{d}V$ is a scalar product, $(f, g) = \int f^* g \mathrm{d}V$, or in our preferred notation, $\langle f \, | g \rangle$.[32]

There is also the notation *antilinear* in the first argument, linear in the second argument. In mathematics, this is usually defined the other way around—there, generally, the *second* element is complex conjugated, not as here the first one.

G.4 Norm

The norm intuitively means simply the length of a vector (as an element of a vector space). The properties of a (general) norm, written here as $\| \ \|$ (or with the alternative notation $| \ |$), are

1. $\|x\| \geq 0$; $\|x\| = 0 \leftrightarrow x = 0$;
2. $\|\alpha x\| = |\alpha| \cdot \|x\|$;
3. $\|x + y\| \leq \|x\| + \|y\|$ (triangle inequality).

Clearly, the expression $\sqrt{\int f^* f \mathrm{d}V}$ is a norm.

G.5 Metric

We do not need this term for our discussion of quantum mechanics, but we include it for completeness: A distance term (= metric) can be defined by $d(x, y) = \|x - y\|$. A general metric must meet the requirements:

1. $d(x, y) \in \mathbb{R}, 0 \leq d(x, y) < \infty$;
2. $d(x, y) = 0 \leftrightarrow x = y$;
3. $d(x, y) \leq d(x, z) + d(z, y)$ (triangle inequality);
4. $d(x, y) = d(y, x)$.

---

[32]The insight that $\int f^* g \mathrm{d}V$ is a scalar product is apparently only about 100 years old. So all those who did not see this at first glance may take comfort from the consideration that, evidently, this fact does not spring to everyone's eye immediately.

G.6   Schwarz's Inequality

The Schwarz inequality[33] establishes an important connection between the scalar product and the norms of two vectors. For the familiar 'arrow vectors', the scalar product is intuitively the product of the length of the first vector times the length of the vertical projection of the second vector onto the first vector (see Appendix F, Vol. 1). Hence it is clear that this quantity is smaller (or equal for parallel vectors) than the product of the lengths of the two vectors. The Schwarz inequality generalizes this relation. It reads:

$$|(x, y)| \leq ||x|| \cdot ||y|| . \tag{G.3}$$

   Proof: The inequality is fulfilled, if $x = 0$ or $y = 0$. Otherwise, it follows with $\alpha = ||y||^2 \in \mathbb{R}$ and $\beta = -(y, x)$ that:

$$\begin{aligned} 0 &\leq (\alpha x + \beta y, \alpha x + \beta y) = \alpha^2 ||x||^2 + \alpha\beta(x, y) + \alpha\beta^*(y, x) + \beta\beta^* ||y||^2 \\ &= \alpha^2 ||x||^2 - ||y||^2 |(x, y)|^2 - ||y||^2 |(x, y)|^2 + |(x, y)|^2 ||y||^2 \\ &= ||y||^4 ||x||^2 - ||y||^2 |(x, y)|^2 . \end{aligned} \tag{G.4}$$

Because of $||y|| \neq 0$, the inequality $|(x, y)|^2 \leq ||x||^2 ||y||^2$ follows.

G.7   Orthogonality

We write the Schwarz inequality for $||x||, ||y|| \neq 0$ in the form

$$\frac{|(x, y)|}{||x|| \cdot ||y||} \leq 1. \tag{G.5}$$

We assume for the moment a real vector space and carry over the only interesting result to the complex case. Thus, x and y are elements of a real vector space. Then we can write the last equation as

$$-1 \leq \frac{(x, y)}{||x|| \cdot ||y||} \leq 1. \tag{G.6}$$

---

[33]This inequality may be considered as one of the most important inequalities in mathematics. It is also called Cauchy–Schwarz, Bunyakovsky, Cauchy–Bunyakovsky–Schwarz or Cauchy–Bunyakovsky inequality. Quote (http://en.wikipedia.org/wiki/Cauchy-Schwarz_inequality, August 2012): "The inequality for sums was published by Augustin-Louis Cauchy (1821), while the corresponding inequality for integrals was first stated by Viktor Bunyakovsky (1859) and rediscovered by Hermann Amandus Schwarz (1888)". We see again how careless and unfair history may be.

This allows us to define an abstract angle $\alpha_{x,y}$ (up to multiples of 2π) between x and y (as elements of the vector space, not as functions of the position!), namely

$$\frac{(x, y)}{||x|| \cdot ||y||} = \cos \alpha_{x,y}. \tag{G.7}$$

In particular, x and y are perpendicular to each other (= are orthogonal) iff $(x, y) = 0$. We transfer this result to complex vector spaces (in fact, only this result is of interest). In other words, two vectors (or states or wavefunctions) $\Phi \neq 0$ and $\Psi \neq 0$ are orthogonal iff it holds that $(\Phi, \Psi) = 0$, i.e.

$$(\Phi, \Psi) = 0 \leftrightarrow \Phi \perp \Psi. \tag{G.8}$$

As may be seen, the zero vector is orthogonal to itself and all other vectors.

G.8 Hilbert Space

Hilbert spaces are special vector spaces, which have a high degree of structure and therefore have very useful properties for all possible calculations. We introduced them in Chap. 11 and summarize here some of their basic properties.

In general, a vector space with a scalar product is called *scalar product space* or *pre-Hilbert space*, where one distinguishes between the *Euclidean* (real case) and the *unitary* (complex case) vector space. A complete pre-Hilbert space is called a *Hilbert space* $\mathcal{H}$.[34]

A space is called *complete* if every Cauchy sequence[35] of elements of this space converges. For example, the space of rational numbers (i.e. fractions) is not complete because sequences of rational numbers can converge to real numbers, as the example of $\left(1 + \frac{1}{n}\right)^n$ shows. Here we have a rational number for any finite n, while the sequence converges towards the real (transcendental) number e which does not belong to the rational numbers.

In Hilbert spaces, the parallelogram law holds:

$$||x + y||^2 + ||x - y||^2 = 2 \left(||x||^2 + ||y||^2 \right). \tag{G.9}$$

Furthermore, we can form an *orthonormal basis* in Hilbert spaces. This is a set of normalized vectors $\{v_n \in \mathcal{H}\}$ which are pairwise orthogonal and whose linear span (the set of all their linear combinations) is the entire Hilbert space (completeness). Because of these properties, such a basis is called a complete orthonormal system, or *CONS*.

---

[34] Spaces with $\langle \varphi | \varphi \rangle < 0$ are called 'pseudo-Hilbert spaces'.

[35] A sequence $\{a_n\}$ is called a Cauchy sequence if for every $\epsilon > 0$ there is a $N(\epsilon) \in \mathbb{N}$ such that for all $n, m > N(\epsilon)$, the inequality $||a_n - a_m|| < \epsilon$ holds. Note that there is no limit in this definition.

The Hilbert spaces we consider are *separable*,[36] i.e. they have a CONS of at most countably infinite dimension.

Furthermore, according to a theorem, there are in each Hilbert space finite sets of self-adjoint operators which commute pairwise and whose common eigenvectors form a basis of the Hilbert space without degeneracy. This set is called CSCO (complete system of commuting observables); an example is given in Chap. 17, Vol. 2 (hydrogen atom).

For Hilbert spaces, the saying is apt: "If you know one, you know them all". More precisely, all Hilbert spaces which are separable and have the same dimension (finite or infinite), are *isomorphic* (i.e. geometrically identical; there are one-to-one length-preserving mappings between these spaces). Therefore, one often speaks of *the* Hilbert space of dimension n or ∞.

There are very different realizations of $\mathcal{H}$; we want to present two of them.

The prototype of all Hilbert spaces of dimension ∞ is the sequence space $l^{(2)}$. It consists of all infinite sequences of complex numbers $x = (x_1, x_2, x_3, \ldots)$ with the scalar product $(x, y) = \sum_n x_n^* y_n$ and the property that the sum of the absolute squares is finite: $(x, x) = \sum_n |x_n|^2 < \infty$. This space was introduced in 1912 by David Hilbert, after whom these spaces are named.[37] We note that the spaces $l^{(p)}$ for $p \neq 2$ are not Hilbert spaces; see the exercises.

Another space, important for quantum mechanics, is $L^{(2)}(a, b)$, the space of the functions which are square integrable in the interval (a, b) (here we restrict the discussion to one dimension). The scalar product is defined as $\int_a^b f^*(x) g(x)\,\mathrm{d}x$; and for the norm,

$$\|f\| = \left(\int_a^b |f(x)|^2 \,\mathrm{d}x \right)^{1/2} < \infty \tag{G.10}$$

must hold. If the limits of the integral are infinite, one writes $L^{(2)}(-\infty, \infty)$ or $L^{(2)}(\mathbb{R})$. An extension encompasses the $L^{(p)}$ spaces with the norm

$$\|f\| = \left(\int_a^b |f(x)|^p \,\mathrm{d}x \right)^{1/p}. \tag{G.11}$$

However, these spaces are not Hilbert spaces for $p \neq 2$; see the exercises.

Hilbert spaces occur in many different areas of mathematics and physics, such as in the spectral theory of ordinary differential equations, the theory of partial differential equations, the ergodic theory, Fourier analysis, quantum mechanics, and

[36]Non-separable Hilbert spaces occur e.g. in the quantization of fields.
[37]The axiomatic definition of the Hilbert space was given in 1927 by J. von Neumann, in the context of the mathematical treatment of quantum mechanics.

others. For example, in Fourier analysis one can expand 2π periodic functions $f(x)$ (see Appendix H, Vol. 1)

$$f(x) = \sum_{n=-\infty}^{\infty} f_n e^{inx}; \quad f_n = \frac{1}{2\pi} \int_{-\pi}^{\pi} f(x) e^{-inx} dx. \tag{G.12}$$

One can show that the functions e^{inx} form an orthogonal basis in the space L^2, i.e. that the Fourier expansion holds for all functions in this space. Other known expansions use wavelets or spherical harmonics (e.g. the multipole expansion) as CONS.

G.9 C^* Algebra

In footnotes, we have pointed out that one can formulate a (rather abstract) entry into quantum mechanics by making use of a C^* algebra. More precisely, we refer to the fact that the observables of classical mechanics (e.g. polynomials of phase-space variables) and quantum mechanics have this structure, although, of course, in different forms. For completeness and as an example of advanced formulations, we briefly give the basic definitions:

We start with a complex vector space $\mathcal{V}$, in which as above the operations $+$ and $\cdot$ are defined for $x, y \in \mathcal{V}$ and $\lambda \in \mathbb{C}$. If we define in addition for $x, y \in \mathcal{V}$ a multiplication $x \circ y$, which is associative and distributive and has a unit element, we have a *complex algebra* $\mathcal{A}$. It is:

$$(x \circ y) \circ z = x \circ (y \circ z); \quad (x + y) \circ z = x \circ z + y \circ z$$
$$\lambda (x \circ y) = (\lambda x) \circ y; \quad 1 \circ x = x \circ 1 = x. \tag{G.13}$$

A *normed algebra* is an algebra where for $x \in \mathcal{V}$, a norm $\|x\|$ is defined. A * *algebra* (pronounced star algebra) is an algebra with a mapping *, called involution, with

$$\left(x^*\right)^* = x; \quad (\lambda x)^* = \bar{\lambda} x^*; \quad (x \circ y)^* = y^* \circ x^*. \tag{G.14}$$

Here, we have denoted the complex conjugation with $\bar{\lambda}$ in order to distinguish it from the * mapping.

A *Banach algebra* is a normed * algebra with the condition $\|x\| = \|x^*\|$, and a C^* *algebra* (pronounced C-star algebra) is a Banach algebra in which the so-called C^* condition $\|x^* \circ x\| = \|x\|^2$ applies.

An example of a C^* algebra are the complex quadratic $n \times n$ matrices with a correspondingly defined norm. The * mapping means in this case taking the adjoint. In other words, the C^* algebra can be thought of as an abstraction of bounded linear operators on Hilbert spaces.

G.10 Exercises

1. Vector space; the $+$ operation is the usual addition, the field is $\mathbb{R}$. Which of the following sets is a vector space? The set of

   (a) the natural numbers;
   (b) the rational numbers;
   (c) the functions continuous on the interval $(-1, 1)$;
   (d) all 4×4-matrices?

2. Consider the 2π-periodic functions. Do they form a vector space?
3. Derive from the Schwarz inequality $|\mathbf{ab}| \leq |\mathbf{a}|\,|\mathbf{b}|$ that $\mathbf{ab} = |\mathbf{a}|\,|\mathbf{b}|\cos\varphi$ holds.
4. Show that the scalar product of two states $|x\rangle$ and $|y\rangle$ does not depend on the representation.

   Solution: Intuitively, the assertion is clear, since the scalar product is a projection of one state onto another one. In order to show this also formally, we start with two basis systems $\{|\varphi_l\rangle\,, l = 1, 2, \ldots\}$ and $\{|\psi_l\rangle\,, l = 1, 2, \ldots\}$, which we can assume to be CONS without loss of generality. The transformation between the two bases is

   $$|\psi_l\rangle = \sum_j \gamma_{lj}\,|\varphi_j\rangle. \tag{G.15}$$

   Since they are both CONS, we have

   $$\delta_{nn'} = \langle \psi_{n'} | \, \psi_n \rangle = \sum_{j'j} \gamma_{n'j'}^* \langle \varphi_{j'} | \gamma_{nj} | \varphi_j \rangle = \sum_j \gamma_{nj}^* \gamma_{nj}. \tag{G.16}$$

   The expansion of the states in terms of the two CONS reads

   $$\begin{aligned}
   |x\rangle &= \sum_l c_{xl}\,|\varphi_l\rangle = \sum_l d_{xl}\,|\psi_l\rangle \\
   |y\rangle &= \sum_l c_{yl}\,|\varphi_l\rangle = \sum_l d_{yl}\,|\psi_l\rangle .
   \end{aligned} \tag{G.17}$$

   Because of (G.15), the expansions in terms of $\{|\varphi_l\rangle\}$ and $\{|\psi_l\rangle\}$ are related by

   $$|x\rangle = \sum_m c_{xm}\,|\varphi_m\rangle = \sum_n d_{xn}\,|\psi_n\rangle = \sum_{nm} d_{xn}\gamma_{nm}\,|\varphi_m\rangle \tag{G.18}$$

   and analogously for $|y\rangle$, which leads to

   $$c_{xm} = \sum_n d_{xn}\gamma_{nm}; \; c_{ym} = \sum_n d_{yn}\gamma_{nm}. \tag{G.19}$$

   For the scalar product, it follows in the basis $\{|\varphi_l\rangle\}$ that

$$\langle x|\, y\rangle = \sum_{mm'} c^*_{xm'}\,\langle\varphi_m|\,c_{ym}\,|\varphi_m\rangle = \sum_{mm'} c^*_{xm'}c_{ym}\delta_{m'm} = \sum_m c^*_{xm}c_{ym} \quad (G.20)$$

and analogously in the basis $\{|\psi_l\rangle\}$,

$$\langle x|\, y\rangle = \sum_m d^*_{xm}d_{ym}. \quad (G.21)$$

So we have to prove the equality $\sum_m c^*_{xm}c_{ym} = \sum_m d^*_{xm}d_{ym}$. With (G.16) and (G.19), we obtain

$$\sum_m c^*_{xm}c_{ym} = \sum_m\sum_{n'} d^*_{xn'}\gamma^*_{n'm}\sum_n d_{yn}\gamma_{nm}$$
$$= \sum_{n'n} d^*_{xn'}d_{yn}\sum_n \gamma^*_{n'm}\gamma_{nm} = \sum_{n'n} d^*_{xn'}d_{yn}\delta_{n'n} = \sum_n d^*_{xn}d_{yn}. \quad (G.22)$$

5. Prove the parallelogram law which applies in a Hilbert space:

$$\|x+y\|^2 + \|x-y\|^2 = 2\left(\|x\|^2 + \|y\|^2\right). \quad (G.23)$$

Solution: We have

$$\|x+y\|^2 + \|x-y\|^2 = (x+y, x+y)^2 + (x-y, x-y)^2$$
$$= (x,x)^2 + (x,y) + (y,x) + (y,y)^2 + (x,x)^2 - (x,y) - (y,x) + (y,y)^2$$
$$= 2\left(\|x\|^2 + \|y\|^2\right). \quad (G.24)$$

6. Show that l^p with $p \neq 2$ is not a Hilbert space.
Solution: In a l^p space, the norm of $x = (x_1, x_2, x_3, \ldots)$ is defined by

$$\|x\| = \left(\sum_n |x_n|^p\right)^{1/p}. \quad (G.25)$$

We show that this norm does not satisfy the parallelogram law which applies in Hilbert spaces. For a proof we need just one counterexample; we choose $x = (1, 1, 0, 0, \ldots)$ and $y = (1, -1, 0, 0, \ldots)$. It follows then that

$$\|x+y\| = \|x-y\| = 2 \text{ and } \|x\| = \|y\| = 2^{1/p}. \quad (G.26)$$

Obviously, the parallelogram law is satisfied only for $p = 2$. Thus, the claim is proved.

Appendix H
Fourier Transforms and the Delta Function

H.1 Fourier Transforms

With the Fourier series, one can represent and analyze periodic functions; for aperiodic functions we use the *Fourier transformation*.

Definition: Given a function $f(x)$ with

$$f(x) = \frac{1}{\sqrt{2\pi}} \int_{-\infty}^{+\infty} \hat{f}(k) e^{ikx} dk. \tag{H.1}$$

Then the function $\hat{f}(k)$ can be calculated by

$$\hat{f}(k) = \frac{1}{\sqrt{2\pi}} \int_{-\infty}^{+\infty} f(x) e^{-ikx} dx. \tag{H.2}$$

One calls $f(x)$ also a function in *position space*, $\hat{f}(k)$ a function in *momentum space* (because of $p = \hbar k$).[38]

In three dimensions, the transformations are given by[39]

---

[38] Another common definition of the Fourier transform uses an asymmetric normalization:

$$f(x) = \int_{-\infty}^{+\infty} \hat{f}(k) e^{ikx} dk; \quad \hat{f}(k) = \frac{1}{2\pi} \int_{-\infty}^{+\infty} f(x) e^{-ikx} dx.$$

[39] Besides $dk_x dk_y dk_z$ and d^3k, there are other notations for volume elements, cf. Appendix D, Vol. 1.

© Springer Nature Switzerland AG 2018
J. Pade, *Quantum Mechanics for Pedestrians 1*, Undergraduate Lecture
Notes in Physics, https://doi.org/10.1007/978-3-030-00464-4

$$f(\mathbf{r}) = \frac{1}{(2\pi)^{3/2}} \int_{-\infty}^{+\infty} \hat{f}(\mathbf{k}) \, e^{i\mathbf{k}\mathbf{r}} \mathrm{d}k_x \mathrm{d}k_y \mathrm{d}k_z = \frac{1}{(2\pi)^{3/2}} \int_{-\infty}^{+\infty} \hat{f}(\mathbf{k}) \, e^{i\mathbf{k}\mathbf{r}} \mathrm{d}^3 k \qquad \text{(H.3)}$$

and

$$\hat{f}(\mathbf{k}) = \frac{1}{(2\pi)^{3/2}} \int_{-\infty}^{+\infty} f(\mathbf{r}) \, e^{-i\mathbf{k}\mathbf{r}} \mathrm{d}x \mathrm{d}y \mathrm{d}z \qquad \text{(H.4)}$$

with $\mathbf{r} = (x, y, z)$ and $\mathbf{k} = (k_x, k_y, k_z)$.

Closer inspection shows that a broad spatial distribution $f(x)$ is related to a narrow momentum distribution $\hat{f}(k)$, and *vice versa*; cf. exercises. This can be seen very clearly in the extreme case of the function $f(x) = f_0 e^{ikx}$. Here, k (and hence p) is a well-defined real number: Thus we have $\Delta p = 0$; hence the position uncertainty must be, *cum grano salis*, infinite, $\Delta x = \infty$. Indeed, $f(x) = f_0 e^{ikx}$ is 'smeared' all over a large region (in fact, it is infinite), $|f(x)| = |f_0|$ for all x. The next question is that for an object in position space with position uncertainty zero and infinite momentum uncertainty; as we shall see below, this is the so-called delta function.

Fourier transformation is a *linear* operation. The functions involved, $f(x)$ and $\hat{f}(k)$, are in general complex. For real $f(x)$, we have $\hat{f}(k) = \hat{f}^*(-k)$. The Fourier transform of $f'(x)$ is $ik\hat{f}(k)$, and a shifted function $g(x) = f(x-a)$ has the Fourier transform $\hat{g}(k) = e^{-ika} \hat{f}(k)$ (proofs of these properties are found in the exercises).

Finally, we mention the convolution theorem. A convolution of the functions f and g is an operation providing a third function h of the form

$$\int_{-\infty}^{+\infty} f(x-y) g(y) \mathrm{d}y = h(x). \qquad \text{(H.5)}$$

Such a convolution corresponds to the product of the Fourier transforms (for the proof see the exercises):

$$\hat{h}(k) = \frac{1}{\sqrt{2\pi}} \hat{f}(k) \hat{g}(k). \qquad \text{(H.6)}$$

H.2   The Delta Function

The *delta function* (also called Dirac delta function or Dirac function after its 'inventor', P.A.M. Dirac; for short, δ-function) is an important tool not only for the mathematical formulation of quantum mechanics.

H.2.1 Formal Derivation

Starting point are the equations for the Fourier transform,

$$f(x) = \frac{1}{\sqrt{2\pi}} \int_{-\infty}^{\infty} dk\, \hat{f}(k)\, e^{ikx}$$
$$\hat{f}(k) = \frac{1}{\sqrt{2\pi}} \int_{-\infty}^{\infty} dx'\, f(x')\, e^{-ikx'}. \tag{H.7}$$

This notation (quite popular in physics, but apparently less welcome in mathematics) of the integral (position of dk, dx) signifies that the integration over k acts rightward on all terms, until an addition sign, a bracket or an equals sign occurs. This is a conventional notation which, for example, allows multiple integrals to be written more concisely and is therefore often used in this context. Inserting $\hat{f}(k)$ on the right side of the first equation leads to:

$$f(x) = \frac{1}{2\pi} \int_{-\infty}^{\infty} dk \int_{-\infty}^{\infty} dx'\, f(x')\, e^{-ikx'} e^{ikx}. \tag{H.8}$$

We assume (as always) that we can interchange the integrations, which gives:

$$f(x) = \frac{1}{2\pi} \int_{-\infty}^{\infty} dx' \int_{-\infty}^{\infty} dk\, f(x')\, e^{-ikx'} e^{ikx}. \tag{H.9}$$

$f(x')$ does not depend on k and thus can be taken out of the second integral:

$$f(x) = \int_{-\infty}^{\infty} dx'\, f(x')\, \frac{1}{2\pi} \int_{-\infty}^{\infty} dk\, e^{ik(x-x')}. \tag{H.10}$$

The further considerations are based on the fact that the function f on the left side appears as $f(x)$ and on the right side under the integral as $f(x')$. The term

$$\frac{1}{2\pi} \int_{-\infty}^{\infty} dk\, e^{ik(x-x')} \tag{H.11}$$

depends only on x and x' (k is the integration variable). It is called the delta function (δ-function)[40]:

---

[40]Actually, this name is incorrect, because it is not a function in the usual sense. Below, a few comments are made on this issue.

$$\delta\left(x - x'\right) = \frac{1}{2\pi} \int_{-\infty}^{\infty} dk \, e^{ik\left(x - x'\right)}. \tag{H.12}$$

Thus, (H.10) can be written

$$f(x) = \int_{-\infty}^{\infty} dx' f\left(x'\right) \delta\left(x - x'\right). \tag{H.13}$$

Evidently, the δ-function projects the function f out of the integral, and does this at the value for which the argument of the δ-function vanishes.

H.2.2   Heuristic Derivation of the Delta Function

The δ-function can be thought of as an infinitely high and infinitely thin needle at the position $x - x' = 0$, as the following derivation shows.

We assume an interval $(x' - \varepsilon, x' + \varepsilon)$ on the x-axis, a rectangular function

$$H(x) = \begin{cases} \frac{1}{2\varepsilon} & \text{for} \quad x' - \varepsilon < x < x' + \varepsilon \\ 0 & \text{otherwise} \end{cases} \tag{H.14}$$

of area 1, and an arbitrary function $f(x)$. For the integral over the product of the two functions, we find:

$$\int_{-\infty}^{\infty} f(x) H(x) \, dx = \int_{x'-\varepsilon}^{x'+\varepsilon} f(x) H(x) \, dx = \frac{1}{2\varepsilon} \int_{x'-\varepsilon}^{x'+\varepsilon} f(x) \, dx. \tag{H.15}$$

We rearrange the last integral using the first mean value theorem for integration (see Appendix D, Vol. 1), which states that there is a value ξ (which value this is, is not revealed by the theorem), such that

$$\int_{a}^{b} g(x) \, dx = (b - a) \, g(\xi) \quad \text{with } a \leq \xi \leq b. \tag{H.16}$$

It then follows that

$$\int_{-\infty}^{\infty} f(x) H(x) \, dx = \frac{1}{2\varepsilon} \int_{x'-\varepsilon}^{x'+\varepsilon} f(x) \, dx = f(\xi); \quad x' - \varepsilon \leq \xi \leq x' + \varepsilon. \tag{H.17}$$

Fig. H.1 On the derivation
of the delta function

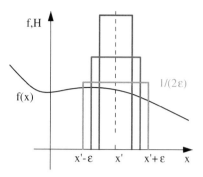

Now we let the interval boundaries approach x' (i.e. $\varepsilon \to 0$). This means that the width of the interval shrinks to 0, while the height of the rectangular function approaches infinity, since its area has the fixed value 1; see Fig. H.1. This function becomes more and more like an infinitely high and infinitely thin needle, as said above. Formally, we obtain in the limit $\varepsilon \to 0$:

$$\int_{-\infty}^{\infty} f(x')\, \delta(x - x')\, dx' = f(x). \tag{H.18}$$

This derivation starting from a rectangular function is not the only possible one; there are many others. They have in common the assumption of a 'proper' function, which in the limit (zero width, height tends to infinity, area remains constant) becomes the delta function. These functions are called *representations* of the delta function, whereby one always keeps in mind that the limit is to be taken at some point in the calculation. We give a small selection of representing functions:

$$\delta_\varepsilon(x) = \tfrac{1}{2\varepsilon} \text{ for } -\varepsilon < x < \varepsilon \text{ (rectangular function)};$$
$$\delta_\varepsilon(x) = \tfrac{1}{\varepsilon\sqrt{\pi}} e^{-x^2/\varepsilon^2} \text{ (Gaussian function)};$$
$$\delta_\varepsilon(x) = \tfrac{\varepsilon}{\pi x^2} \sin^2\left(\tfrac{x}{\varepsilon}\right);$$
$$\delta_\varepsilon(x) = \tfrac{1}{\pi} \tfrac{\varepsilon}{x^2 + \varepsilon^2};$$
$$\delta_\varepsilon(x) = \tfrac{1}{2\pi} \int_{-\infty}^{\infty} dk\, e^{ikx} e^{-\varepsilon|k|}. \tag{H.19}$$

As we said above, at some point the limit $\varepsilon \to 0$ must be taken. Knowing this, one frequently omits the convergence generating factor $e^{-\varepsilon|k|}$ in the last equation. Alternatively, one can write

$$\delta(x) = \lim_{\varepsilon \to 0} \frac{1}{\varepsilon\sqrt{\pi}} e^{-x^2/\varepsilon^2} \tag{H.20}$$

and accordingly for the other representations.

If one picks out a suitable representation, certain properties of the delta function can quite easily be proven.

H.2.3 Examples, Properties, Applications

Examples: The argument of the δ-function in the integral

$$\int_{-\infty}^{\infty} dy \left(y^2 + 4\right) \delta \left(y - 1\right) = 1 + 4 = 5 \tag{H.21}$$

vanishes for $y = 1$; accordingly, the integral has the value 5.

In the next example, the argument of the δ-function vanishes for $z = -1$:

$$\int_{-\infty}^{\infty} dz\, e^{-\gamma z^3} \delta \left(z + 1\right) = e^{-\gamma(-1)^3} = e^{\gamma} \tag{H.22}$$

Properties: Regarding the name of the delta function, inspection of equation (H.12) reveals that the δ-function cannot be a function—otherwise the integral would converge, which is clearly not the case. The δ-function is instead a map that assigns a number to a function, i.e. a *functional* (*distribution*). The fact that it is not called correctly a 'delta functional', but rather 'delta function', is due to the laxity of the physicists. Precisely because it is not in fact a function, but is only called one, the δ-function has some unusual features besides the familiar ones.

The delta function can be understood as the derivative of the Heaviside (step) function $\Theta(x)$ (or unit step function):

$$\int_{-\infty}^{x} \delta\left(x'\right) dx' = \begin{cases} 0 \\ 1 \end{cases} \text{ for } \begin{matrix} x < 0 \\ x > 0 \end{matrix} = \Theta(x); \quad \Theta'(x) = \delta(x). \tag{H.23}$$

The Heaviside function is point symmetrical, the δ-function symmetrical: $\delta(x) = \delta(-x)$. Derivatives of the δ-function may be defined by means of partial integration, that is

$$\int_{-\infty}^{\infty} f(x')\, \delta^{(n)}(x - x')\, dx' = (-1)^n\, f^{(n)}(x). \tag{H.24}$$

An important property is (see exercises)

$$\delta(ax) = \frac{1}{|a|}\delta(x) \tag{H.25}$$

from which the symmetry $\delta(-x) = \delta(x)$ follows for $a = -1$. Generalizing to a function $g(x)$ which has only simple zeros x_n, we have (see exercises):

$$\delta(g(x)) = \sum_n \frac{1}{|g'(x_n)|} \delta(x - x_n). \qquad (H.26)$$

Applications: As (H.12) shows, the delta function is an object with an infinitely large momentum uncertainty—one integrates over all momenta k. Thus, the delta function describes an object with a precisely defined position and a completely undefined momentum. This makes it the 'counterpart' of a plane wave, which indeed describes an object with a precisely defined momentum and a completely undefined position.

The use of the δ-function is not limited to quantum mechanics. For example, it can be used to describe the mass density of a classical point mass. The point mass of mass m_0 is at the position $x = x_0$. Then we find:

$$\rho(x) = m_0 \delta(x - x_0), \qquad (H.27)$$

and for the total mass, it follows that:

$$M = \int_{-\infty}^{\infty} \rho(x) \, dx = m_0 \int_{-\infty}^{\infty} \delta(x - x_0) \, dx = m_0. \qquad (H.28)$$

H.2.4 The Delta Function and the Laplace Operator

With $\mathbf{E} = -\nabla\Phi$, the first Maxwell equation $\nabla\mathbf{E} = \rho/\varepsilon_0$ can be written as $\nabla^2\Phi = -\rho/\varepsilon_0$. The potential of a point charge, whose charge density we can describe by a delta function, is known to be proportional to $1/r$. In other words, the term $\nabla^2\frac{1}{r}$ should give (so to speak, for physical reasons) essentially a delta function. This we now wish to demonstrate.

If one approaches the task directly (and a little too naively), one would simply calculate $\nabla^2\frac{1}{r}$. Since this is a radially symmetric problem, we need only consider the radial part of ∇^2, i.e. $\frac{1}{r}\frac{\partial^2}{\partial r^2}r$. We insert and obtain

$$\nabla^2\frac{1}{r} = \frac{1}{r}\frac{\partial^2}{\partial r^2}r\frac{1}{r} \stackrel{?}{=} 0 \qquad (H.29)$$

instead of a delta function. The reason is that $\frac{1}{r}$ is not defined for $r = 0$—but our charge is sitting just there.

To resolve this shortcoming, we assume a function $g_\varepsilon(r)$

$$g_\varepsilon (r) = \frac{1}{\sqrt{r^2 + \varepsilon^2}}; \quad \varepsilon \text{ arbitrary small} \tag{H.30}$$

defined everywhere. Other functions[41] would also be suitable for the following considerations: It is important only that $g_\varepsilon(r)$ is defined everywhere (and twice differentiable), and goes to $\frac{1}{r}$ in the limit $\varepsilon \to 0$.

Inserting and computing gives

$$\nabla^2 g_\varepsilon (r) = -\frac{3\varepsilon^2}{\left(r^2 + \varepsilon^2\right)^{5/2}}. \tag{H.31}$$

On the right-hand side, we have (except for the sign and possibly a multiplicative constant) another representation $\delta_\varepsilon (r)$ of the delta function (see above); the limit $\varepsilon \to 0$ leads for $r > 0$ to zero and for $r = 0$ to infinity.

Thus we have $\nabla^2 g_\varepsilon (r) = -\alpha \delta_\varepsilon (r)$. To determine the multiplicative constant α, we use $\int \delta(r) \mathrm{d}V = 1$ or $\int \delta_\varepsilon(r) \mathrm{d}V \underset{\varepsilon \to 0}{\to} 1$ and obtain

$$\begin{aligned} \alpha &= \alpha \int \delta_\varepsilon (r) \, \mathrm{d}V = \int \frac{3\varepsilon^2}{(r^2+\varepsilon^2)^{5/2}} \mathrm{d}V \\ &= 4\pi \int\limits_0^\infty \frac{3\varepsilon^2 r^2}{(r^2+\varepsilon^2)^{5/2}} \mathrm{d}r = 4\pi \left\{ \frac{r^3}{(r^2+\varepsilon^2)^{3/2}} \right\}_0^\infty = 4\pi. \end{aligned} \tag{H.32}$$

In summary, the result is

$$\nabla^2 \frac{1}{r} = -4\pi \delta (r). \tag{H.33}$$

Since r vanishes iff $\mathbf{r}$ vanishes, we can also write

$$\nabla^2 \frac{1}{r} = -4\pi \delta (\mathbf{r}). \tag{H.34}$$

Here $\delta (\mathbf{r})$ is defined in Cartesian coordinates by

$$\delta (\mathbf{r}) = \delta (x) \, \delta (y) \, \delta (z) \tag{H.35}$$

Finally, we note as an extension of (H.34) the equation

$$\nabla^2 \frac{1}{|\mathbf{r} - \mathbf{r}'|} = -4\pi \delta (\mathbf{r} - \mathbf{r}') \tag{H.36}$$

and (without proof)

$$\left(\nabla^2 + k^2\right) \frac{e^{\pm ik|\mathbf{r}-\mathbf{r}'|}}{|\mathbf{r} - \mathbf{r}'|} = -4\pi \delta (\mathbf{r} - \mathbf{r}') \text{ or } \left(\nabla^2 + k^2\right) \frac{e^{\pm ikr}}{r} = -4\pi \delta (\mathbf{r}). \tag{H.37}$$

---

[41] For example, $\frac{1-e^{-r/\varepsilon}}{r}$.

The solutions that we have found for the differential equations (H.34–H.37) are examples of *Green's functions* . Their general definition is: Given an inhomogeneous differential equation, a Green's function is defined as a specific (particular) solution of that differential equation, if the inhomogeneity is a delta function. For an example, see the exercises.

H.3   Fourier Series

Fourier series are restricted to the realm of periodic functions. Without loss of generality, we assume a 2π-periodic function (for other periods, the unit of length must be rescaled accordingly), which we can write as a superposition of plane waves:

$$f(x) = \sum_{n=-\infty}^{\infty} f_n e^{inx}; f_n \in \mathbb{C}, \text{ if } f(x) \text{ is } 2\pi \text{ periodic.} \tag{H.38}$$

We multiply by $\exp(-imx)$, integrate over x from $-\pi$ to π, and obtain (assuming that integration and summation commute):

$$\int_{-\pi}^{\pi} f(x)e^{-imx}\,\mathrm{d}x = \sum_{n=-\infty}^{\infty} f_n \int_{-\pi}^{\pi} e^{-imx}e^{inx}\,\mathrm{d}x$$
$$= \sum_{n=-\infty}^{\infty} f_n\, 2\pi\, \delta_{nm} = 2\pi\, f_m \tag{H.39}$$

making use of

$$\delta_{nm} = \frac{1}{2\pi} \int_{-\pi}^{\pi} e^{i(n-m)x}\,\mathrm{d}x. \tag{H.40}$$

In summary, we have for 2π-periodic functions (Fourier series):

$$f(x) = \sum_{n=-\infty}^{\infty} f_n e^{inx} \text{ and } f_n = \frac{1}{2\pi} \int_{-\pi}^{\pi} f(x)e^{-inx}\,\mathrm{d}x. \tag{H.41}$$

What is the distribution in momentum space? We use the general relations of Fourier transformation and obtain:

$$\hat{f}(k) = \frac{1}{2\pi} \int_{-\infty}^{\infty} f(x)e^{-ikx}\,\mathrm{d}x = \frac{1}{2\pi} \int_{-\infty}^{\infty} \sum_{n=-\infty}^{\infty} f_n e^{inx}e^{-ikx}\,\mathrm{d}x$$
$$= \sum_{n=-\infty}^{\infty} f_n \frac{1}{2\pi} \int_{-\infty}^{\infty} e^{i(n-k)x}\,\mathrm{d}x = \sum_{n=-\infty}^{\infty} f_n \delta(n-k) \tag{H.42}$$

Thus, in momentum space, we have a series of infinitely high, infinitely thin needles at the points $k = n$ with the respective weights f_n.

H.4 Discrete and Quantum Fourier Transforms

Until now we have assumed for the Fourier transform that the data sets being transformed are continuous. For discrete data sets, as are typically produced by experiments, the *discrete Fourier transform* (DFT) is applied.

We suppose that we have a set of N data[42] x_j, with $j = 0 \ldots N - 1$.[43] Then the DFT is

$$
y_k = \frac{1}{\sqrt{N}} \sum_{j=0}^{N-1} e^{\frac{2\pi i j k}{N}} x_j; \quad x_k = \frac{1}{\sqrt{N}} \sum_{j=0}^{N-1} e^{-\frac{2\pi i j k}{N}} y_j. \tag{H.43}
$$

We choose a simple example with $N = 2$, i.e. $x = (x_0, x_1)$. With

$$
y_0 = \frac{1}{\sqrt{2}} \sum_{j=0}^{1} x_j; \quad y_1 = \frac{1}{\sqrt{2}} \sum_{j=0}^{1} e^{i\pi j} x_j, \tag{H.44}
$$

it follows that

$$
\begin{aligned}
x = (1,0): \ y_0 = \tfrac{1}{\sqrt{2}}; \ y_1 = \tfrac{1}{\sqrt{2}} \ &\to \ y = \tfrac{1}{\sqrt{2}}(1,1) \\
x = (0,1): \ y_0 = \tfrac{1}{\sqrt{2}}; \ y_1 = -\tfrac{1}{\sqrt{2}} \ &\to \ y = \tfrac{1}{\sqrt{2}}(1,-1)
\end{aligned} \tag{H.45}
$$

or, written compactly,

$$
x = Hy; \quad H = \frac{1}{\sqrt{2}} \begin{pmatrix} 1 & 1 \\ 1 & -1. \end{pmatrix} \tag{H.46}
$$

The Hadamard matrix H plays an important role in quantum information (see Chap. 26, Vol. 2, and Appendices P to S in Vol. 2).

The discrete Fourier transformation of quantum states is called (discrete) *quantum Fourier transformation* (*QFT*, occasionally *DQFT*). We formulate it as operator in the bra-ket notation:

$$
U_{QFT} = \frac{1}{\sqrt{N}} \sum_{j,k=0}^{N-1} e^{2\pi i \frac{jk}{N}} |j\rangle \langle k| \tag{H.47}
$$

where $\{|j\rangle\}$ is a CONS. In the matrix representation, this reads

---

[42] The data can be complex.

[43] A practical example: When scanning digital music from a CD, the sampling frequency is 44.1 kHz, so 44 100 values must be processed per second.

$$U_{QFT} \cong \frac{1}{\sqrt{N}} \begin{pmatrix} 1 & 1 & 1 & \cdots & 1 \\ 1 & \omega & \omega^2 & \cdots & \omega^{N-1} \\ 1 & \omega^2 & \omega^4 & \cdots & \omega^{2(N-1)} \\ \vdots & \vdots & \vdots & & \vdots \\ 1 & \omega^{N-1} & \omega^{2(N-1)} & \cdots & \omega^{(N-1)(N-1)} \end{pmatrix} ; \ \omega = e^{\frac{2\pi i}{N}}. \tag{H.48}$$

The QFT is a unitary transformation (see exercises):

$$UU^{\dagger} = U^{\dagger}U = 1. \tag{H.49}$$

A linear combination of basis states

$$|\varphi\rangle = \sum_{n=0}^{N-1} \alpha_n |n\rangle \tag{H.50}$$

is mapped by the QFT onto

$$|\psi\rangle = U_{QFT} |\varphi\rangle = \frac{1}{\sqrt{N}} \sum_{n,j,k=0}^{N-1} \alpha_n e^{2\pi i \frac{jk}{N}} |j\rangle \langle k |n\rangle = \frac{1}{\sqrt{N}} \sum_{n,j=0}^{N-1} \alpha_n e^{2\pi i \frac{jn}{N}} |j\rangle \tag{H.51}$$

or, compactly,

$$|\psi\rangle = \frac{1}{\sqrt{N}} \sum_{n=0}^{N-1} \beta_n |n\rangle ; \ \beta_n = \sum_{j=0}^{N-1} \alpha_j e^{2\pi i \frac{nj}{N}}. \tag{H.52}$$

H.5 Exercises

1. Calculate

$$\int_{-\infty}^{\infty} e^{4x^2 \sin x} \delta (x - 3) \, dx; \quad \int_{-\infty}^{\infty} \cos 4x \ \delta (x) \, dx$$
$$\int_{-\infty}^{\infty} \delta (x) \, e^x dx; \quad \int_{-\infty}^{\infty} \delta (x - 2) \, f(x) \, dx. \tag{H.53}$$

2. Given an operator X with $\delta (x - a)$ as eigenfunction (or eigen'function'), so that $X\delta (x - a) = a\delta (x - a)$ holds; show that X is the position operator.
   Solution: For an arbitrary function $f(x)$, it holds that

$$Xf(x) = X \int da f(a) \delta (x - a)$$
$$= \int da f(a) X\delta (x - a) = \int da f(a) a\delta (x - a) = xf(x). \tag{H.54}$$

From this, it follows that the position operator multiplies an arbitrary space-dependent function by x.

3. Show that for the derivative of the delta function,

$$\int_{-\infty}^{\infty} \delta' (x - x_0) f (x) \, dx = -f' (x_0)$$ (H.55)

holds. Hint: Use partial integration.

4. Representations of the delta function are e.g.

$$\delta (x) = \lim_{\varepsilon \to 0} \frac{1}{2\pi} \frac{2\varepsilon}{x^2 + \varepsilon^2}; \quad \delta (x) = \lim_{\varepsilon \to 0} \frac{e^{-x^2/\varepsilon^2}}{\sqrt{\pi}\varepsilon}.$$ (H.56)

(a) Discuss the functions and sketch their graphs for different values of ε.

(b) Show that

$$\int_{-\infty}^{\infty} \frac{1}{2\pi} \frac{2\varepsilon}{x^2 + \varepsilon^2} dx = 1$$ (H.57)

holds. (Hint: differentiate arctan x). Is

$$\int_{-\infty}^{\infty} \frac{e^{-x^2/\varepsilon^2}}{\sqrt{\pi}\varepsilon} dx = 1$$ (H.58)

also valid? (Check with a formula tabulation.)

(c) Show with the help of a representation that

$$\delta (ax) = \frac{1}{|a|} \delta (x)$$ (H.59)

is valid.

Solution: We start from $\delta (x) = \lim_{\varepsilon \to 0} \frac{1}{\pi} \frac{\varepsilon}{x^2 + \varepsilon^2}$. We have:

$$\delta (ax) = \frac{1}{\pi} \lim_{\varepsilon \to 0} \frac{\varepsilon}{a^2 x^2 + \varepsilon^2} = \frac{1}{a^2} \frac{1}{\pi} \lim_{\varepsilon \to 0} \frac{\varepsilon}{x^2 + \varepsilon^2/a^2}.$$ (H.60)

On the right-hand side, we no longer have any information about the sign of a. With the new variable $\eta = \varepsilon/|a|$, it follows that

$$\delta (ax) = \frac{1}{|a|} \frac{1}{\pi} \lim_{\varepsilon \to 0} \frac{\varepsilon/|a|}{x^2 + \varepsilon^2/a^2} = \frac{1}{|a|} \frac{1}{\pi} \lim_{\eta \to 0} \frac{\eta}{x^2 + \eta^2} = \frac{1}{|a|} \delta (x).$$ (H.61)

5. Prove the equation

$$\delta(ax) = \frac{1}{|a|}\delta(x) \tag{H.62}$$

by means of a suitable substitution under the integral.
Solution: In the integral

$$A = \int_{-\infty}^{\infty} \delta(ax)\,f(x)\,dx \tag{H.63}$$

we substitute $y = ax$, obtaining

$$A = \int_{-\frac{|a|}{a}\infty}^{\frac{|a|}{a}\infty} \delta(y)\,f\left(\frac{y}{a}\right)\frac{1}{a}dy = \frac{1}{|a|}\int_{-\infty}^{\infty}\delta(y)\,f\left(\frac{y}{a}\right)dy = \frac{1}{|a|}f(0). \tag{H.64}$$

Because of

$$\int_{-\infty}^{\infty}\delta(ax)\,f(x)\,dx = \frac{1}{|a|}f(0) = \frac{1}{|a|}\int_{-\infty}^{\infty}\delta(x)\,f(x)\,dx, \tag{H.65}$$

the proposition follows directly.
6. Show that for a function $g(x)$ which has only simple zeros, the following relation
holds:

$$\delta(g(x)) = \sum_n \frac{1}{|g'(x_n)|}\delta(x - x_n). \tag{H.66}$$

Solution: The delta function makes a contribution only at those points where the
function $g(x)$ vanishes, i.e. at the zeros of $g(x)$. In the vicinity of the zeros, the
Taylor expansion can be applied:

$$g(x) = g(x_n) + g'(x_n)(x - x_n) + O\left((x - x_n)^2\right). \tag{H.67}$$

If we are 'very close' to the zeros (and only then does the delta function contribute), we can replace $g(x)$ by $g'(x_n)(x - x_n)$. It follows initially that

$$\delta(g(x)) = \sum_n \delta(g'(x_n)(x - x_n)). \tag{H.68}$$

Using the result of the last exercise, we can write

$$\delta(g(x)) = \sum_n \frac{1}{|g'(x_n)|}\delta(x - x_n). \tag{H.69}$$

7. Assume

$$g(\omega) = \begin{cases} G > 0 \\ 0 \end{cases} \quad \text{for} \quad \begin{matrix} 0 < \omega_1 < \omega < \omega_2 \\ \text{otherwise.} \end{matrix} \quad \text{(H.70)}$$

Determine the Fourier transform $f(t)$. What is the value of $f(t)$ at time $t = 0$? Calculate the intensity $|f(t)|^2$ and show that it depends only on the difference of the frequencies ω_1 and ω_2. Sketch $|f(t)|^2$.

8. Determine the Fourier transform $f(t)$ of the function

$$g(\omega) = \begin{cases} \alpha\omega \\ 0 \end{cases} \quad \text{for} \quad \begin{matrix} 0 \leq \omega \leq \Omega \\ \text{otherwise.} \end{matrix} \quad \text{(H.71)}$$

Determine and sketch the intensity $|f(t)|^2$.

9. Formulate the potential equations in the Fourier representation.

10. Show that for real $f(x)$, $\hat{f}(k) = \hat{f}^*(-k)$ applies.
   Solution: For real $f(x)$, we have $f(x) = f^*(x)$, and it follows that

$$\int_{-\infty}^{+\infty} \hat{f}(k) e^{ikx} dk = f(x) = \int_{-\infty}^{+\infty} \hat{f}^*(k) e^{-ikx} dk$$

$$= -\int_{\infty}^{-\infty} \hat{f}^*(-k) e^{ikx} dk = \int_{-\infty}^{+\infty} \hat{f}^*(-k) e^{ikx} dk; \quad \text{(H.72)}$$

   thus the proposition follows directly.

11. Show that the Fourier transform of $f'(x)$ is $ik\hat{f}(k)$.
   Solution: We have

$$f'(x) = \frac{1}{\sqrt{2\pi}} \frac{d}{dx} \int_{-\infty}^{+\infty} \hat{f}(k) e^{ikx} dk = \frac{1}{\sqrt{2\pi}} \int_{-\infty}^{+\infty} ik\hat{f}(k) e^{ikx} dk. \quad \text{(H.73)}$$

12. Show that a shifted function $g(x) = f(x - a)$ has the Fourier transform $\hat{g}(k) = e^{-ika}\hat{f}(k)$.
   Solution: We have

$$\hat{g}(k) = \frac{1}{\sqrt{2\pi}} \int_{-\infty}^{+\infty} f(x-a) e^{-ikx} dx = \frac{1}{\sqrt{2\pi}} \int_{-\infty}^{+\infty} e^{-ika} f(x-a) e^{-ik(x-a)} dx$$

$$= e^{-ika} \frac{1}{\sqrt{2\pi}} \int_{-\infty}^{+\infty} f(z) e^{-ikz} dz = e^{-ika} \hat{f}(k). \quad \text{(H.74)}$$

13. Show for the convolution

$$h(x) = \int_{-\infty}^{+\infty} f(x-y) g(y) dy \quad \text{(H.75)}$$

that the following relation applies (convolution theorem):

$$\hat{h}(k) = \sqrt{2\pi}\,\hat{f}(k)\,\hat{g}(k). \tag{H.76}$$

Solution: We have

$$\hat{h}(k) = \frac{1}{\sqrt{2\pi}} \int\limits_{-\infty}^{+\infty} dx \int\limits_{-\infty}^{+\infty} dy\, f(x-y)\,g(y)\,e^{-ikx}. \tag{H.77}$$

We assume that we can interchange the integrations, and first perform the integration over x:

$$\hat{h}(k) = \frac{1}{\sqrt{2\pi}} \int\limits_{-\infty}^{+\infty} dy\, g(y) \int\limits_{-\infty}^{+\infty} dx\, f(x-y)\,e^{-ikx}. \tag{H.78}$$

According to the previous exercise, the integration of the shifted function f yields

$$\frac{1}{\sqrt{2\pi}} \int\limits_{-\infty}^{+\infty} dx\, f(x-y)\,e^{-ikx} = e^{-iky}\,\hat{f}(k), \tag{H.79}$$

and thus it follows that

$$\hat{h}(k) = \int\limits_{-\infty}^{+\infty} dy\, g(y)\, e^{-iky}\,\hat{f}(k) = \sqrt{2\pi}\,\hat{f}(k)\,\hat{g}(k). \tag{H.80}$$

14. Determine the Fourier transformation of the rectangular function

$$f(x) = \begin{cases} A & \text{for} - b < x < b \\ 0 & \text{otherwise} \end{cases} \tag{H.81}$$

Solution: Start with

$$\hat{f}(k) = \frac{A}{\sqrt{2\pi}} \int\limits_{-b}^{+b} e^{-ikx} dx = \frac{A}{\sqrt{2\pi}} \left. \frac{e^{-ikx}}{-ik} \right|_{-b}^{b} \tag{H.82}$$

$$= \frac{A}{\sqrt{2\pi}} \frac{e^{-ikb} - e^{ikb}}{-ik} = \frac{A}{\sqrt{2\pi}} \frac{2i\sin kb}{-ik}$$

from which

$$\hat{f}(k) = A\sqrt{\frac{2}{\pi}} \frac{\sin kb}{k}. \tag{H.83}$$

What is happening at $k = 0$? Either we compute the integral $\int_{-b}^{+b} e^{-ikx} dx$ once more for $k = 0$, or we start with $\frac{\sin kb}{k}$ and apply l'Hôpital's rule, or we remember $\sin x \approx x$ for small x (first term of the power series expansion). Howsoever, in any case it follows that

$$\hat{f}(k = 0) = A\sqrt{\frac{2}{\pi}}b. \tag{H.84}$$

The first zero of $\hat{f}(k)$ is found at

$$kb = \pi \text{ or } k = \frac{\pi}{b}. \tag{H.85}$$

The last equation shows that the narrower we make the distribution in position space (i.e. the smaller is b), the broader is the distribution in momentum space—and *vice versa*[44] To quantify this, we choose the position of the first zero as a rough measure Δk of the width of $\hat{f}(k)$:

$$\Delta k \approx \frac{\pi}{b} \sim \text{'breadth'of } \hat{f}(k). \tag{H.86}$$

As a measure of the breadth of $f(x)$, we choose b:

$$\Delta x \approx b \sim \text{'breadth'of } f(x). \tag{H.87}$$

It follows that

$$\Delta k \Delta x \approx \pi \tag{H.88}$$

or, with $p = \hbar k$,

$$\Delta x \Delta p \approx \hbar \pi. \tag{H.89}$$

This is simply a 'raw form' of Heisenberg's uncertainty principle, which is derived exactly in Chap. 13. According to this relation, there is no quantum object to which we can attribute a precise position (i.e. $\Delta x = 0$) and at the same time a precise momentum ($\Delta p = 0$).

Note: this is *not* a statement about the quality of our measurement apparatus or something similar, but rather the statement that the concepts 'position' and 'momentum' lose their meaning in quantum mechanics, or at least do not maintain it in terms of our everyday understanding.

15. Show that the QFT

$$U = \frac{1}{\sqrt{N}} \sum_{j,k=0}^{N-1} e^{\frac{2\pi i j k}{N}} |j\rangle \langle k| \tag{H.90}$$

---

[44]This is similar to pressing a balloon—pressed in one direction, it evades and expands out in another direction.

is unitary. $\{|j\rangle\}$ is a CONS.
Solution: We have

$$UU^\dagger = \frac{1}{N} \sum_{j,k,j',k'=0}^{N-1} e^{\frac{2\pi i jk}{N}} |j\rangle \langle k| e^{-\frac{2\pi i j'k'}{N}} |k'\rangle \langle j'|$$

$$= \frac{1}{N} \sum_{j,k,j'=0}^{N-1} e^{\frac{2\pi i (j-j')k}{N}} |j\rangle \langle j'| . \tag{H.91}$$

We distinguish the cases $j = j'$ and $j \neq j'$. For $j = j'$, we find using completeness,

$$\frac{1}{N} \sum_{j,k=0}^{N-1} |j\rangle \langle j| = \frac{1}{N} \cdot N \sum_{j=0}^{N-1} |j\rangle \langle j| = 1. \tag{H.92}$$

For $j \neq j'$, it holds (geometrical series) that:

$$\sum_{k=0}^{N-1} e^{\frac{2\pi i (j-j')k}{N}} = \frac{1 - e^{2\pi i (j-j')}}{1 - e^{\frac{2\pi i (j-j')}{N}}} = 0. \tag{H.93}$$

16. Determine explicitly the QFT matrix (H.48) for the cases $N = 2, 3, 4$.
Solution: We have

$$N = 2: \ U_{QFT} \cong \frac{1}{\sqrt{2}} \begin{pmatrix} 1 & 1 \\ 1 & e^{\frac{2\pi i}{2}} \end{pmatrix} = \frac{1}{\sqrt{2}} \begin{pmatrix} 1 & 1 \\ 1 & -1 \end{pmatrix}$$

$$N = 3: \ U_{QFT} \cong \frac{1}{\sqrt{3}} \begin{pmatrix} 1 & 1 & 1 \\ 1 & e^{\frac{2\pi i}{3}} & e^{\frac{4\pi i}{3}} \\ 1 & e^{\frac{4\pi i}{3}} & e^{\frac{8\pi i}{3}} \end{pmatrix} = \frac{1}{\sqrt{3}} \begin{pmatrix} 1 & 1 & 1 \\ 1 & \frac{-1+i\sqrt{3}}{2} & \frac{-1-i\sqrt{3}}{2} \\ 1 & \frac{-1-i\sqrt{3}}{2} & \frac{-1+i\sqrt{3}}{2} \end{pmatrix}$$

$$N = 4: \ U_{QFT} \cong \frac{1}{\sqrt{4}} \begin{pmatrix} 1 & 1 & 1 & 1 \\ 1 & e^{\frac{2\pi i}{4}} & e^{\frac{4\pi i}{4}} & e^{\frac{6\pi i}{4}} \\ 1 & e^{\frac{4\pi i}{4}} & e^{\frac{8\pi i}{4}} & e^{\frac{12\pi i}{4}} \\ 1 & e^{\frac{6\pi i}{4}} & e^{\frac{12\pi i}{4}} & e^{\frac{18\pi i}{4}} \end{pmatrix} = \frac{1}{\sqrt{4}} \begin{pmatrix} 1 & 1 & 1 & 1 \\ 1 & i & -1 & -i \\ 1 & -1 & 1 & -1 \\ 1 & -i & -1 & i \end{pmatrix}. \tag{H.94}$$

17. Determine by making use of the Green's function the solution of the first Maxwell equation $\nabla \mathbf{E} = \rho/\varepsilon_0$ for the time-independent charge density $\rho(\mathbf{r})$. Use $\mathbf{E} =$

$-\nabla \Phi$, i.e. $\nabla^2 \Phi = -\rho/\varepsilon_0$.

Solution: We start with (H.36), i.e.

$$\nabla_{\mathbf{r}}^2 \frac{1}{|\mathbf{r} - \mathbf{r}'|} = -4\pi\delta\left(\mathbf{r} - \mathbf{r}'\right). \tag{H.95}$$

The index $\mathbf{r}$ denotes the variables with respect to which the differentiation is to be performed. Multiplication by $\rho\left(\mathbf{r}'\right)$ leads to

$$\Delta_{\mathbf{r}} \frac{\rho\left(\mathbf{r}'\right)}{|\mathbf{r} - \mathbf{r}'|} = -4\pi\delta\left(\mathbf{r} - \mathbf{r}'\right)\rho\left(\mathbf{r}'\right). \tag{H.96}$$

Integration with respect to $\mathbf{r}'$ yields

$$\Delta_{\mathbf{r}} \int \frac{\rho\left(\mathbf{r}'\right)}{|\mathbf{r} - \mathbf{r}'|} \mathrm{d}^3 r' = -4\pi \int \delta\left(\mathbf{r} - \mathbf{r}'\right)\rho\left(\mathbf{r}'\right)\mathrm{d}^3 r' = -4\pi\rho\left(\mathbf{r}\right). \tag{H.97}$$

Comparison with $\Delta\Phi = -\rho/\varepsilon_0$ leads immediately to

$$\Phi\left(\mathbf{r}\right) = \frac{1}{4\pi\varepsilon_0} \int \frac{\rho\left(\mathbf{r}'\right)}{|\mathbf{r} - \mathbf{r}'|} \mathrm{d}^3 r'. \tag{H.98}$$

In principle, additive terms f with $\nabla_{\mathbf{r}}^2 f = 0$ could occur; if necessary they can be excluded by considering the asymptotic bahavior.

Appendix I
Operators

We take a closer look at some issues from Chap. 13 and provide some additional material, insofar as it may be useful for understanding the text.

I.1 Norm, Domain of Definition

I.1.1 The Norm

The norm of an operator is defined by $\|A\| = \sup \frac{\|A|\varphi\rangle\|}{\||\varphi\rangle\|}$ or $\|A\| = \sup_{\||\varphi\rangle\|=1} \|A|\varphi\rangle\|$.

Here, an example in a real vector space:

$$A = \begin{pmatrix} 1 & 2 \\ 0 & -1 \end{pmatrix}; \quad |\varphi\rangle = \begin{pmatrix} a \\ b \end{pmatrix} \rightarrow$$

$$\|A\| = \sup \frac{\sqrt{(a+2b)^2 + b^2}}{\sqrt{a^2 + b^2}} = \sup \sqrt{\frac{(x+2)^2 + 1}{x^2 + 1}} \text{ with } x = \frac{a}{b}. \tag{I.1}$$

The function on the right-hand side is maximal for $x = \sqrt{2} - 1$; it follows that $\|A\| = \sqrt{3 + 2\sqrt{2}}$.

The operator norm for bounded operators is a 'proper' norm and complies with the three rules, among them the triangle inequality (see Appendix G, Vol. 1).[45]

---

[45]Interestingly, the norm of A is related to the spectral radius $\rho(A)$, which is defined as the largest absolute value of the eigenvalues of A, i.e. as $\rho(A) = \max_i |\lambda_i|$. In general, it holds that $\rho(A) \leq \|A\|$.
For normal operators $[A, A^\dagger] = 0$, the inequality is sharpened to give $\rho(A) = \|A\|$.

© Springer Nature Switzerland AG 2018
J. Pade, *Quantum Mechanics for Pedestrians 1*, Undergraduate Lecture Notes in Physics, https://doi.org/10.1007/978-3-030-00464-4

I.1.2 Bounded Operators

An operator is called *bounded* if there is a constant $C < \infty$, independent of the states $|\varphi\rangle$, such that for all states $|\varphi\rangle \in \mathcal{H}$, we have:

$$\| A \,|\varphi\rangle \| \le C \, \| |\varphi\rangle \| \quad \text{or} \quad \|A\| \le C. \tag{I.2}$$

With $A = \begin{pmatrix} 1 & 2 \\ 0 & -1 \end{pmatrix}$, we have just seen an example of a bounded operator. For an unbounded operator, we consider the Hilbert space $L^{(2)}\,[0, \infty]$ and the operator x. Its norm is given by

$$\|x\| = \sup \frac{\displaystyle\int_0^\infty f^*\,(x)\,x^2 f\,(x)\,\mathrm{d}x}{\displaystyle\int_0^\infty f^*\,(x)\,f\,(x)\,\mathrm{d}x}. \tag{I.3}$$

If we now find even one single function for which $\|x\| = \infty$, we have shown that x is an unbounded operator (in this Hilbert space). Such a function is, for example, $f(x) = \frac{\sin x^2}{x}$. Like x, p is also an unbounded operator; see the exercises.
    In a finite-dimensional Hilbert space, all operators are bounded (cf. the exercises); unbounded operators can therefore occur only in infinite-dimensional Hilbert spaces.

I.1.3 Domain of Definition

The domain of definition (or briefly, domain) $\mathcal{D}_A$ of an operator A is the set of all vectors $|\varphi\rangle \in \mathcal{H}$ such that $A\,|\varphi\rangle$ is also in $\mathcal{H}$. One can show that the domain of definition is the whole Hilbert space iff A is bounded. Hence, the problem with an unbounded operator A is that its domain of definition is *not* the whole Hilbert space.
    An example: In the case of the function $f(x) = \frac{\sin x^2}{x}$ just considered, we have seen that f is square integrable, but not $xf(x)$; in addition, the mean value $\langle x\rangle_f$ does not exist. Thus, the domain of definition of the unbounded operator x is not the whole Hilbert space $L^{(2)}\,[0, \infty]$.
    Also, problems may occur in other respects with unbounded operators. For example, in the equation $[x,\,p] = i\hbar$, the right-hand side is defined for the whole of $\mathcal{H}$, but the left-hand side only for a subset (see also the remarks on the uncertainty principle below).
    Bounded operators on a Hilbert space are very well behaved. This can be seen by—among others—the fact that a special name was given to them: The set of all bounded operators on a Hilbert space forms a C^* algebra (see Appendix G, Vol. 1; the operator norm and the adjoint must of course be defined).

I.2 Hermitian, Self-adjoint

The difference between the terms Hermitian and self-adjoint has to do with the fact that the domain of definition of unbounded operators is not the entire Hilbert space. Thus, the difference can occur only in infinite-dimensional Hilbert spaces; for finite-dimensional vector spaces, the two concepts are identical.

Basically, it is therefore necessary to identify not only the comb, but also the hair that is combed—the properties of an operator depend on its domain of definition. A simple example: In $L^{(2)}[0, \infty]$, the operator x is unbounded, but it is bounded in $L^{(2)}[0, 1]$.

The technical resources needed for the following are simple; just integration by parts as known from school.

I.2.1 Definitions and Differences

We begin with three definitions:

(1) An operator A, for which $\langle Au \,|v\rangle = \langle u \,|Av\rangle$ holds, is called *symmetric* or *Hermitian*.[46]

(2) Given an operator A. The adjoint[47] $A^\dagger$ of the operator A is defined as $\langle A^\dagger u \,|v\rangle = \langle u \,|Av\rangle$. We note that $A^\dagger$ is a distinct operator which can have its separate domain of definition. The equality $\langle A^\dagger u \,|v\rangle = \langle u \,|Av\rangle$ must apply for all vectors within the domain of definition.

(3) In general, for an unbounded Hermitian operator A, it is not true that $\mathcal{D}_A = \mathcal{D}_{A^\dagger}$, but rather $\mathcal{D}_A \subset \mathcal{D}_{A^\dagger}$ (or $\mathcal{D}_{A^\dagger} \subset \mathcal{D}_A$). In order that A be *self-adjoint*, we must have $A = A^\dagger$ and the two domains of definition must coincide.

A symmetric linear operator defined everywhere is self-adjoint. According to a theorem of functional analysis (Hellinger–Toeplitz theorem), such an operator is bounded. Conversely, it follows that an unbounded operator cannot be defined on the entire Hilbert space. The theorem combines two completely different properties, namely to be defined everywhere and to be bounded.

We now illustrate these concepts by means of two examples.

I.2.2 Two Examples

The standard operator used in the following is the (one-dimensional) momentum. For both examples, the Hilbert space is $L^2[0, 1]$.

---

[46] Actually, there is a minor difference between the two terms, which has to do with the question of whether the domain of definition of A is dense. But since this question has nothing to do with the following considerations, we will omit it here.

[47] Occasionally also called the Hermitian conjugate operator.

Example 1:

The operator is $p_0 = \frac{\hbar}{i}\frac{d}{dx}$. Its domain of definition $\mathcal{D}_{p_0}$ consists of all functions $g(x) \in L^2[0, 1]$ which are differentiable, have square-integrable derivatives and fulfill the boundary conditions $g(0) = g(1) = 0$ (it is this 0 to which the index in p_0 refers).

We consider the adjoint operator $p_0^\dagger$. It is defined by $\left\langle p_0^\dagger f \,\middle|\, g \right\rangle = \langle f | \, p_0 g \rangle$, and it follows that

$$\left\langle p_0^\dagger f \,\middle|\, g \right\rangle = \langle f | \, p_0 g \rangle = \frac{\hbar}{i} \int_0^1 f^*(x) g'(x)\,dx = \frac{\hbar}{i} \left\{ \left[f^* g \right]_0^1 - \int_0^1 f^{*\prime}(x) g(x)\,dx \right\}.$$

$$(I.4)$$

The integrated term on the right-hand side gives zero, and we have

$$\left\langle p_0^\dagger f \,\middle|\, g \right\rangle = \int_0^1 \left(\frac{\hbar}{i} \frac{d}{dx} f(x) \right)^* g(x)\,dx = \langle p_0 f | \, g \rangle. \qquad (I.5)$$

One might now think that p_0 is self-adjoint—but that is wrong, because the integrated term $f^*(1)g(1) - f^*(0)g(0)$ in (I.4) vanishes independently of the values of f at the boundary. Therefore, the domain of definition of $p_0^\dagger$ is larger than that of p_0, $\mathcal{D}_{p_0} \subset \mathcal{D}_{p_0^\dagger}$.

Example 2:

The same example—but with different boundary conditions: We allow for arbitrary boundary conditions of $g(x)$ and thus write simply p instead of p_0. For the adjoint operator $p^\dagger$,

$$\langle p^\dagger f | \, g \rangle = \langle f | \, pg \rangle = \frac{\hbar}{i} \int_0^1 f^*(x) g'(x)\,dx = \frac{\hbar}{i} \left\{ \left[f^* g \right]_0^1 - \int_0^1 f^{*\prime}(x) g(x)\,dx \right\}$$

$$= \frac{\hbar}{i} \left[f^*(1)g(1) - f^*(0)g(0) \right] + \int_0^1 \left(\frac{\hbar}{i} \frac{d}{dx} f(x) \right)^* g(x)\,dx.$$

$$(I.6)$$

In order for this equality to be valid, it must hold that $f^*(1) = f^*(0) = 0$. In other words, the domain of definition of $p^\dagger$ is smaller than that of p: $\mathcal{D}_{p_0^\dagger} \subset \mathcal{D}_{p_0}$. Thus, this operator is also not self-adjoint.

Symmetry of the Examples:

We want to check if the operators p_0, $p_0^\dagger$, p, $p^\dagger$ are symmetric. For p_0, we have

$$\langle p_0 f \mid g \rangle - \langle f \mid p_0 g \rangle = \int_0^1 \left(\frac{\hbar}{i}\frac{d}{dx} f(x) \right)^* g(x) dx - \int_0^1 f^*(x) \frac{\hbar}{i}\frac{d}{dx} g(x) dx$$
$$= -\frac{\hbar}{i} \int_0^1 \tfrac{d}{dx} f^*(x) g(x) dx - \frac{\hbar}{i} \int_0^1 f^*(x) \tfrac{d}{dx} g(x) dx = -\frac{\hbar}{i} [f^*(x)g(x)]_0^1 = 0.$$
(I.7)

The last equals sign is valid, since the domain of definition of p_0 is restricted to functions which vanish at the boundaries of the interval. Hence p_0 is symmetric.

We apply the same considerations to $p_0^\dagger$:

$$\left\langle p_0^\dagger f \,\middle|\, g \right\rangle - \langle f \mid p_0^\dagger g \rangle = \int_0^1 \left(\frac{\hbar}{i}\frac{d}{dx} f(x) \right)^* g(x) dx - \int_0^1 f^*(x) \frac{\hbar}{i}\frac{d}{dx} g(x) dx$$
$$= -\frac{\hbar}{i} \int_0^1 \tfrac{d}{dx} f^*(x) g(x) dx - \frac{\hbar}{i} \int_0^1 f^*(x) \tfrac{d}{dx} g(x) dx = -\frac{\hbar}{i} [f^*(x)g(x)]_0^1 .$$
(I.8)

The domain of definition of $p_0^\dagger$ also comprises functions which do not vanish at the boundary; hence, $p_0^\dagger$ is not a symmetric operator.

Analogous considerations show that the operator $p^\dagger$ in the second example is symmetric, but not the the operator p.

Extension of the Domain of Definition:

The example of p_0 and $p_0^\dagger$ has shown that the domains of definition of operator and adjoint operator may differ. However, one can often 'repair' this. Let us define $p_\alpha = \frac{\hbar}{i}\frac{d}{dx}$, i.e. once more the the derivative acting on the functions $g(x) \in L^2[0, 1]$ (of course, the derivatives must exist and also be square integrable). The difference w.r.t. p_0 consists in the different boundary conditions, namely $g(1) = e^{i\alpha} g(0)$ with $0 \leq \alpha < 1$ and $g(0) \neq 0$. Thus, the domain of definition of p_α differs from that of p_0 (we emphasize again that we are dealing indeed with different operators—all of them are written as $\frac{\hbar}{i}\frac{d}{dx}$, but they have different domains of definition). The operator p_α is again symmetric, but it is also self-adjoint, in contrast to p_0. This holds owing to

$$\langle f \mid p_\alpha g \rangle - \langle p_\alpha f \mid g \rangle = \frac{\hbar}{i} \left[f^*(x)g(x) \right]_0^1 = \frac{\hbar}{i} \left[f^*(1)g(1) - f^*(0)g(0) \right]$$
$$= \tfrac{\hbar}{i} \left[f^*(1)e^{i\alpha} - f^*(0) \right] g(0) = \tfrac{\hbar}{i} \left[f(1) - e^{i\alpha} f(0) \right]^* e^{i\alpha} g(0).$$
(I.9)

The right side vanishes iff $f(1) = e^{i\alpha} f(0)$. In other words, the domains of definition of p_α and $p_\alpha^\dagger$ are identical. We have achieved this by expanding the domain of definition of p_0.

In fact, with p_α we have constructed an entire class of operators, because if we choose a different constant α, we obtain a different domain of definition and thus a different operator, although they of course always refer to $\frac{\hbar}{i}\frac{d}{dx}$. For a closer look, we consider again the eigenvalue equation

$$\frac{\hbar}{i}\frac{d}{dx}g(x) = \lambda g(x) ; \text{ boundary condition } g(1) = e^{i\alpha}g(0). \tag{I.10}$$

The solution of this equation is $g(x) = ce^{\frac{i\lambda}{\hbar}x}$, with the boundary condition $g(1) = e^{i\alpha}g(0)$. It follows that $ce^{\frac{i\lambda}{\hbar}} = e^{i\alpha}c$, or

$$\lambda = \hbar(m + \alpha) ; \quad m \in \mathbb{Z} \tag{I.11}$$

Hence, the eigenvalues (i.e. the measurable quantities) of the operator p_α are different for each α, and we have correspondingly each time another operator p_α (again: although it always contains the 'same derivative' $\frac{\hbar}{i}\frac{d}{dx}$).

Furthermore, one cannot modify the domain of definition area for every symmetric operator in such a way that it becomes self-adjoint. These facts can also be demonstrated in terms of the momentum. We choose here $p_\infty = \frac{\hbar}{i}\frac{d}{dx}$; the domain of definition consists of the differentiable and square integrable functions $g(x) \in L^2[0, \infty]$ with $g(0) = g(\infty) = 0$ (e.g. $g(x) = xe^{-x}$). The operator p_∞ is symmetric because of

$$\langle f| p_\infty g\rangle - \langle p_\infty f| g\rangle = \frac{\hbar}{i}\int_0^\infty f^*(x)\frac{dg(x)}{dx}dr + \frac{\hbar}{i}\int_0^\infty \left(\frac{df(x)}{dx}\right)^* g(x)dr$$

$$= \frac{\hbar}{i}\left[f^*(x)g(x)\right]_0^\infty = 0. \tag{I.12}$$

For the adjoint operator, we see that:

$$\left\langle p_\infty^\dagger f\middle| g\right\rangle = \langle f| p_\infty g\rangle = \frac{\hbar}{i}\int_0^\infty f^*(x)\frac{dg(x)}{dx}dx = \frac{\hbar}{i}\left[f^*(x)g(x)\right]_0^\infty - \frac{\hbar}{i}\int_0^\infty \frac{df^*(x)}{dx}g(x)dx$$

$$= \frac{\hbar}{i}\left[f^*(x)g(x)\right]_0^\infty + \int_0^\infty \left(\frac{\hbar}{i}\frac{df(x)}{dx}\right)^* g(x)dx = \frac{\hbar}{i}\left[f^*(x)g(x)\right]_0^\infty + \langle p_\infty f| g\rangle .$$

$$\tag{I.13}$$

The integrated term on the right-hand side always vanishes because of $g(0) = g(\infty) = 0$, regardless of the values of f; the domain of definition of $p_\infty^\dagger$ is therefore larger than that of p_∞. It can be shown in this case that there is no adjustment which will make the domains of definition of p_∞ and $p_\infty^\dagger$ coincide.

I.2.3 A Note on Terminology

The use of the terms 'symmetric' and 'self-adjoint' in the mathematical literature is
very consistent, while 'Hermitian' appears there occasionally with different meanings. In physics, however, the terms Hermitian and self-adjoint are often used without
distinction; but one finds also Hermitian conjugate, adjoint, symmetric. The fact that
physics can allow itself to be a bit negligent with regard to these differences is mainly
due to the circumstance that we can imagine 'difficult' spaces as limiting cases of
simpler spaces—e.g. by discretization, as we have seen in Chap. 12.

In addition, we must not forget that the goal of physics is the description and the
widest possible understanding of the 'physical' world, which means, among other
things, that for us, mathematics is not an end in itself, but rather an essential and
powerful tool.

I.3 Unitary Operators; Stone's Theorem

We consider unitary operators briefly once more, along with the theorem of Stone.

As a definition, we can use the fact that an operator is *unitary* on a Hilbert space $\mathcal{H}$
if it has an inverse and if it conserves all scalar products, i.e. the equality $\langle U\varphi \,|U\psi\rangle =
\langle \varphi \,|\psi\rangle$ is valid for all vectors $\in \mathcal{H}$.

This definition is equivalent to the formulation that $UU^\dagger = U^\dagger U = 1$. We note
that in finite-dimensional spaces, the left inverse is automatically equal to the right
inverse. In infinite-dimensional spaces, this is not necessarily true, and that is why
one needs both formulations there, $UU^\dagger = 1$ and $U^\dagger U = 1$. As an example, we
consider vectors $(c_1, c_2, c_3, \ldots) \in C^\infty$, on which two operators A and B act according to $A\,(c_1, c_2, c_3, \ldots) = (c_2, c_3, c_4, \ldots)$ and $B\,(c_1, c_2, c_3, \ldots) = (0, c_1, c_2, \ldots)$.
Evidently, we have $AB = 1$ and $BA \neq 1$. In other words, B is the right inverse of
A, but not the left inverse (see also the exercises).

We note in passing that an operator which conserves the norm is called *isometric*.
In a finite-dimensional vector space, an isometry is automatically a unitary operator.

Because of the independence of the physical predictions of unitary transformations, we can conclude that the relation of physical variables with their mathematical
representations is defined only up to unitary transformations. More generally, one
could consider transformations $|\psi\rangle \rightarrow |\psi'\rangle$, for which $|\langle\psi' \,|\varphi'\rangle| = |\langle\psi \,|\varphi\rangle|$ holds for
all vectors. Such transformations apparently do not change probability statements.
However, there is no obvious reason that such transformations should be linear, let
alone that they must be unitary transformations. In this situation, Wigner's theorem
(see also Chap. 21, Vol. 2) comes to our aid; it states that there is an operator U which
is either unitary or anti-unitary, and which satisfies the equation $U\,|\varphi\rangle = |\varphi'\rangle$ for all
vectors in $\mathcal{H}$.

I.3.1 Stone's Theorem

Unitary operators occur naturally (so to speak automatically) if the system has a symmetry (see Chap. 21, Vol. 2). In this context, the theorem of Stone is of importance.[48]

It reads: A set of unitary operators U depending on a continuous parameter α satisfies the rule of an Abelian group:

$$U\left(\alpha_1 + \alpha_2\right) = U\left(\alpha_2\right) U\left(\alpha_1\right). \tag{I.14}$$

Then there exists an Hermitian operator T such that

$$U\left(\alpha\right) = e^{i\alpha T}. \tag{I.15}$$

We see that e^{iA} is unitary if A is self-adjoint.[49]

An equivalent formulation of this theorem is e.g.: If $U(\alpha)$, $\alpha \in \mathbb{R}$ satisfies the following three conditions: (1) the matrix element $\langle \varphi | U(\alpha) | \psi \rangle$ is for all vectors a continuous function of α; (2) $U(0) = 1$; (3) for all $\alpha_1, \alpha_2 \in \mathbb{R}$, it holds that $U(\alpha_1)U(\alpha_2) = U(\alpha_1 + \alpha_2)$—then there is a unique self-adjoint operator such that $U(\alpha) = e^{i\alpha A}$ and

$$iA \left| \psi \right\rangle = \lim_{\alpha \to 0} \frac{U(\alpha) - 1}{\alpha} \left| \psi \right\rangle \ \text{for all } \left| \psi \right\rangle \in \mathcal{H}. \tag{I.16}$$

In Chap. 13, we established the relation between a Hamiltonian H and a propagator $U = e^{-iHt/\hbar}$; we now see that it was practically a derivation by example of Stone's theorem.[50]

I.3.2 Unitary or Hermitian?

Finally, a word about the relation between unitary and Hermitian operators, related to the question of the boundedness of operators:

We know that Hermitian operators can cause problems if they are not bounded. We also know that the unitary operator $U\left(\alpha\right) = e^{i\alpha T}$ is bounded, even for an unbounded Hermitian operator T. Thus, one might regard the unitary operator U as more fundamental than the Hermitian operator T in this case.

---

[48] In practice, the theorem of Stone is one of the most important ways by which self-adjoint operators enter quantum mechanics (symmetry $\rightarrow$ unitary operator $\rightarrow$ self-adjoint operator).

[49] $\alpha \rightarrow U(\alpha)$ is called a unitary representation of the additive group of real numbers if for a one-parameter family of unitary operators $U(\alpha) = e^{i\alpha A}$, $\alpha \in \mathbb{R}$, the following applies: (1) $U(0) = 1$; (2) $U(\alpha_1)U(\alpha_2) = U(\alpha_1 + \alpha_2)$; (3) $U(-\alpha) = U^{-1}(\alpha)$.

[50] A similar consideration can be entertained for time-dependent Hamiltonians, but the result is somewhat more complicated, as in this case different times occur, which must be placed in the correct order (keyword: time-ordering operator).

As an example, we consider free one-dimensional motion with $-\infty < x < \infty$. The momentum operator p (and hence the Hamiltonian $\frac{p^2}{2m}$) is not bounded; its domain of definition comprises all functions whose derivatives are square integrable. We now choose the function $\psi(x, 0) = e^{-ix^2} \frac{\sin x}{x}$ which is continuous and differentiable, but does *not* belong to the domain of definition of the momentum operator, because its derivative is not square integrable. This means, strictly speaking, that the free time-dependent SEq is not meaningful for this initial condition—we cannot 'really' allow such an initial condition. But on the other hand, the time-evolution operator $U(t) = e^{-iHt/\hbar}$ is bounded (its norm is 1); its domain of definition is thus the entire Hilbert space. One can rewrite U in this case so that differential operators no longer appear in the exponent; the result can be written as an integral operator and reads (see also the exercises for Chap. 5):

$$\psi(x, t) = \sqrt{\frac{m}{2\pi i \hbar t}} \int\limits_{-\infty}^{\infty} e^{i \frac{m(x-y)^2}{2\hbar t}} \psi(y, 0) \, \mathrm{d}y. \tag{I.17}$$

In this formulation of the free SEq, the above problems do not occur. In other words, the unitary time-evolution operator is more fundamental than the Hamiltonian H.

As a further example, we consider the position-momentum commutation relation

$$[x, p] = i\hbar. \tag{I.18}$$

x and p are unbounded Hermitian operators; thus, the right side of this relation is always defined, but not necessarily also the left side. But one can rewrite this relation; its *Weyl form* reads

$$e^{i \frac{pa}{\hbar}} e^{ibx} e^{-i \frac{pa}{\hbar}} = e^{ibx} e^{iba} \tag{I.19}$$

(see Chap. 21, Vol. 2). On both sides of this equation, there are only bounded (unitary) operators; hence, this form is more universal than $[x, p] = i\hbar$.

We will not delve further into this topic. Perhaps we should make only the remark that these and similar considerations contribute to the somewhat nonchalant attitude towards mathematics among physicists: We can often treat problems with the usual instruments, although they are 'strictly speaking' not well defined. Of course this is not always true—one can fail miserably if one does not consider essential conditions. But by and large, quantum mechanics is quite well behaved.

I.4   The Uncertainty Principle

Those relations which concern two Hermitian operators A and B have to do with variances or standard deviations (see Chap. 9). With the deviation from the mean value

$$A_- = A - \langle A \rangle, \tag{I.20}$$

we obtain

$$\langle A_-^2 \rangle = \langle (A - \langle A \rangle)^2 \rangle = \langle A^2 \rangle - \langle A \rangle^2 = (\Delta A)^2. \tag{I.21}$$

We derive the uncertainty principle in two different ways.

I.4.1 Derivation 1

First, we note the general relation that the commutator of two Hermitian operators is an anti-Hermitian operator:

$$[A, B]^\dagger = (AB - BA)^\dagger = BA - AB = -[A, B]. \tag{I.22}$$

Thus, we can always write $[A, B] = iC$ with $C = C^\dagger$. Next, we consider the following norm:

$$\|(A_- + i\alpha B_-) |\psi\rangle\|^2 \geq 0, \ \alpha \in \mathbb{R}. \tag{I.23}$$

It holds that

$$\begin{aligned}\|(A_- + i\alpha B_-) |\psi\rangle\|^2 &= \langle \psi | (A_- - i\alpha B_-)(A_- + i\alpha B_-) |\psi\rangle \\ &= \langle \psi | A_-^2 + i\alpha [A_-, B_-] + \alpha^2 B_-^2 |\psi\rangle. \end{aligned} \tag{I.24}$$

Evaluation of the commutator gives

$$[A_-, B_-] = [A, B] = iC; \ C = C^\dagger. \tag{I.25}$$

With this, (I.24) can be written as

$$\|(A_- + i\alpha B_-) |\psi\rangle\|^2 = \langle A_-^2 \rangle - \alpha \langle C \rangle + \alpha^2 \langle B_-^2 \rangle = (\Delta A)^2 - \alpha \langle C \rangle + \alpha^2 (\Delta B)^2 \geq 0. \tag{I.26}$$

Since C is Hermitian, $\langle C \rangle$ is real (see Chap. 9). The last inequality must be satisfied *for all* α; hence there is at most *one* zero of the quadratic polynomial in α. This means[51] that

$$\langle C \rangle^2 - 4 (\Delta A)^2 (\Delta B)^2 \leq 0. \tag{I.27}$$

It follows that $(\Delta A)^2 (\Delta B)^2 \geq \langle C \rangle^2 / 4$, or

$$\Delta A \cdot \Delta B \geq \frac{1}{2} |\langle [A, B] \rangle|. \tag{I.28}$$

---

[51] The function $f(x) = x^2 + bx + c$ has the zeros $x_0 = \frac{-b \pm \sqrt{b^2 - 4ac}}{2a}$. If we require that $f(x)$ be non-negative (restricting ourselves to real numbers) then the radicand must satisfy $b^2 - 4ac \leq 0$.

This is the general uncertainty principle for two Hermitian operators. It is especially popular in terms of the pair x and p_x. Because of $[x, p_x] = i\hbar$, it follows that

$$\Delta x \cdot \Delta p_x \geq \frac{\hbar}{2}. \tag{I.29}$$

I.4.2 Derivation 2

For this derivation, we use $|\langle \varphi | \chi \rangle|^2 \leq \langle \varphi | \varphi \rangle \langle \chi | \chi \rangle$, i.e. the Schwarz inequality. Here, we insert $|\varphi\rangle = A_- |\psi\rangle = |A_-\psi\rangle$ and $|\chi\rangle = B_- |\psi\rangle = |B_-\psi\rangle$, where $|\psi\rangle$ is an arbitrary state. It follows that

$$|\langle A_-\psi | B_-\psi \rangle|^2 \leq \langle A_-\psi | A_-\psi \rangle \langle B_-\psi | B_-\psi \rangle. \tag{I.30}$$

Next, we use the fact that A and B are Hermitian, i.e. their mean values are real and consequently A_- and B_- are Hermitian, too.[52] This leads to

$$|\langle \psi | A_- B_-\psi \rangle|^2 \leq \langle \psi | A_-^2 \psi \rangle \langle \psi | B_-^2 \psi \rangle = \langle A_-^2 \rangle \langle B_-^2 \rangle. \tag{I.31}$$

On the right side, we already have acceptable terms, with

$$\langle A_-^2 \rangle = \langle (A - \langle A \rangle)^2 \rangle = (\Delta A)^2. \tag{I.32}$$

Now we consider the transformation of the left side. Here we use the fact that we can write any product of operators as a sum of a Hermitian and an anti-Hermitian part. We realize that the anticommutator of two Hermitian operators

$$\{A, B\} = AB + BA \tag{I.33}$$

is Hermitian, while the commutator is anti-Hermitian (if it does not vanish), as shown above:

$$[A, B]^\dagger = -[A, B]. \tag{I.34}$$

We know that the mean value of a Hermitian operator is real: it remains to show that the mean value of an anti-Hermitian operator is imaginary (see the exercises). Now we write

$$AB = \frac{1}{2} \{A, B\} + \frac{1}{2} [A, B]. \tag{I.35}$$

It follows first of all that

---

[52] In general, this statement is not satisfied for non-Hermitian operators.

$$|\langle\psi|A_-B_-\psi\rangle|^2 = \frac{1}{4}|\langle\psi|(\{A_-, B_-\} + [A_-, B_-])\psi\rangle|^2$$

$$= \frac{1}{4}|\langle\{A_-, B_-\}\rangle + \langle[A_-, B_-]\rangle|^2. \tag{I.36}$$

Due to $\{A_-, B_-\} \in \mathbb{R}$ and $\left[A_-, B_-\right] \in \mathbb{I}$, it follows that[53]

$$|\langle\psi\,|A_-B_-\psi\rangle|^2 = \frac{1}{4}|\langle\{A_-, B_-\}\rangle|^2 + \frac{1}{4}\left|\langle\left[A_-, B_-\right]\rangle\right|^2 \tag{I.37}$$

so that we can write (I.31), considering (I.32), as

$$(\Delta A)^2 (\Delta B)^2 \geq \frac{1}{4}|\langle\{A_-, B_-\}\rangle|^2 + \frac{1}{4}\left|\langle\left[A_-, B_-\right]\rangle\right|^2. \tag{I.38}$$

The second term on the right-hand side can be written as:

$$\left[A_-, B_-\right] = [A - \langle A\rangle, B - \langle B\rangle] = AB - BA = [A, B]. \tag{I.39}$$

Since there is no corresponding simplification for the anticommutator, it is simply omitted; this gives the inequality

$$(\Delta A)^2 (\Delta B)^2 \geq \frac{1}{4}|\langle[A, B]\rangle|^2. \tag{I.40}$$

Taking the root, we obtain the uncertainty principle:

$$(\Delta A)(\Delta B) \geq \frac{1}{2}|\langle[A, B]\rangle|. \tag{I.41}$$

I.4.3 Remarks on the Uncertainty Principle

The first remark concerns a common misinterpretation of the uncertainty principle, according to which the product of the uncertainties for non-commuting operators is *always greater* than zero. But this is not true, because the right side of the uncertainty principle contains not the bare commutator, but rather its expectation value—and that can vanish, even if the commutator itself is not zero. This can be seen perhaps most clearly if one notes explicitly the dependence on the state. As an example, we consider a general angular momentum **J**, whose components satisfy $\left[J_x, J_y\right] = i\hbar J_z$ (and cyclically interchanged equations); see Chap. 16, Vol. 2. Then we have

---

[53] As is well known, for $a + ib$ with $a, b \in \mathbb{R}$, $|a + ib|^2 = |a|^2 + |b|^2$.

$$\left(\Delta_\psi J_x\right)\left(\Delta_\psi J_y\right) \geq \frac{\hbar}{2}\left|\langle J_z\rangle_\psi\right|, \tag{I.42}$$

and for states $|\psi\rangle$ with $\langle J_z\rangle_\psi = 0$, there is no lower positive limit for $\left(\Delta_\psi J_x\right)\left(\Delta_\psi J_y\right)$. An explicit example is found in the exercises.

The second remark concerns the domain of validity of the uncertainty principle: For the position x, with $0 \leq x \leq 1$, we define the corresponding momentum by $p = \frac{\hbar}{i}\frac{d}{dx}$. Both operators are self-adjoint if the scalar product is defined as usual and the domain of definition of p is restricted to differentiable functions g which satisfy $g(1) = g(0)$ (see exercises). Does the uncertainty principle $\Delta x \Delta p \geq \frac{\hbar}{2}$ apply under these premises?

The answer is no. We consider first the eigenfunctions of p. They are determined by

$$\frac{\hbar}{i}\frac{d}{dx}g(x) = \lambda g(x); \ g(1) = g(0) \tag{I.43}$$

to give

$$g(x) = g_0 e^{\frac{i\lambda}{\hbar}x} \text{ and } g_0 e^{\frac{i\lambda}{\hbar}} = g_0 e^{2im\pi} \text{ or } \lambda = 2\hbar m\pi; \ m \in \mathbb{Z} \tag{I.44}$$

This means that

$$g_m(x) = g_0 e^{2im\pi x}; \ m \in \mathbb{Z} \tag{I.45}$$

We fix the constant g_0 by means of the normalization, i.e. $\int_0^1 g_m^*(x)g_m(x)dx = 1$, and obtain for the eigenfunctions[54]

$$g_m(x) = e^{i\alpha}e^{2im\pi x}; \ m \in \mathbb{Z} \tag{I.46}$$

For these states, we now calculate the quantities occurring in the uncertainty principle. We have

$$(\Delta p)^2 = \langle p^2\rangle - \langle p\rangle^2 = \langle g_m|p^2|g_m\rangle - \langle g_m|p|g_m\rangle^2$$

$$= \int_0^1 g_m^* p^2 g_m dx - \left[\int_0^1 g_m^* p g_m dx\right]^2 = \int_0^1 (2im\pi)^2 dx - \left[\int_0^1 (2im\pi)dx\right]^2 = 0 \tag{I.47}$$

and

$$(\Delta x)^2 = \langle x^2\rangle - \langle x\rangle^2 = \langle g_m|x^2|g_m\rangle - \langle g_m|x|g_m\rangle^2$$

$$= \int_0^1 g_m^* x^2 g_m dx - \left[\int_0^1 g_m^* x g_m dx\right]^2 = \int_0^1 x^2 dx - \left[\int_0^1 x dx\right]^2 = \frac{1}{3} - \left[\frac{1}{2}\right]^2 = \frac{1}{12}. \tag{I.48}$$

---

[54]By the way, this is essentially the basis of the Fourier series for periodic functions.

Following this argumentation, we should obtain $(\Delta p)\,(\Delta x) = 0$ and not $(\Delta p)\,(\Delta x) \geq$ $\frac{\hbar}{2}$. Where have we made a mistake?

Answer: The eigenfunctions g_m are not in the domain of definition of the operator *product px*, since $x g_m = x e^{i\alpha} e^{2im\pi x}$ does not satisfy the periodicity condition $g(1) = g(0)$ and therefore does not belong to the domain of definition of p.

This is an example of the fact that the uncertainty principle applies only when *all* terms are defined, including the terms appearing in intermediate calculations. This is once more an indication that in computations involving unbounded operators, one always has to be careful.

I.5   Hermitian Operators, Observables

We want to give a brief note on a terminology problem. It addresses the term *observable*, i.e. an observable and measurable quantity. Examples where it is intuitively clear that we are dealing with an observable are obvious (position, momentum, energy, angular momentum, etc.). But a unique and precise meaning of the term in the context of quantum mechanics does not exist.

For some (e.g. Schwabl), the term observable stands for 'physical quantity', and is different from the operators that are associated with them in quantum mechanics. For reasons of clarity, though, usually the same symbol is chosen for both; but in principle, in this terminology, the term 'observable A' is simply a shorthand notation for 'the physical quantity A_{meas} represented by the operator A_{op}'.

For others, observable denotes a Hermitian operator whose eigenvectors form a complete orthonormal system (CONS). One can understand this as a technical term that has nothing to do with the question of whether one can assign a corresponding physical quantity to an observable. However, if one wants to establish this relationship, it has to be realized that this definition is not a sharp criterion, because there are indeed Hermitian operators of this type which, in a certain sense, do not correspond to measurable quantities; an example can be found below.

Finally, there are still others who in the face of these difficulties and diffuseness declare the term 'observable' to be dispensable, because it is of no real interest. In fact, it appears that the use of the term is not a compelling necessity, but rather is due to convenience and has become simply a habit.

The fact that we still use the term 'observable' in this text is due to the circumstance that, despite its ambiguity, it makes us aware of two issues: First, it indicates that we are dealing not only with an abstract operator in an abstract space, but also with a physical quantity which we can concretely measure in the laboratory. On the other hand, the concept tells us that it is an Hermitian operator, with a real spectrum, orthogonal eigenvectors, etc.

Now for the example of the Hermitian operators just mentioned which do not correspond to any measured variable - at least in a certain sense. It is perhaps a rather subtle point, but the consideration may help to clarify the concepts.

We note that an operator can be seen as a measuring instruction. For instance, s_x means to measure the x-component of the spin, i.e. to measure the spin along the unit vector $(1, 0, 0)$. A difficulty arises when we combine operators in a way which does not result in a reasonable measuring instruction at first glance. Take, for example, sums of Hermitian operators that do not commute:

$$C = A + B \ ; \ A = A^\dagger \ ; \ B = B^\dagger \ ; \ [A, B] \neq 0 \tag{I.49}$$

C is Hermitian, of course. The problem is that the order in which measurements of A and B are carried out plays a role (because of $[A, B] \neq 0$); this, however, is not reflected anywhere in $C = A + B$. Specifically, we consider the spin-$\frac{1}{2}$ matrices $A = s_x$ and $B = s_z$:

$$C = \frac{\hbar}{2} \begin{pmatrix} 1 & 1 \\ 1 & -1 \end{pmatrix} \ ; \ [s_x, s_z] = -i\hbar s_y. \tag{I.50}$$

Clearly, C is a Hermitian operator; its eigenvalues are $\lambda = \pm\frac{\hbar}{\sqrt{2}}$. However, $C = s_x + s_z$ is not a measurable quantity in the sense of the measuring instruction 'measure the x-component plus (and) the z-component of the spin'. Even if we could measure s_x and s_z simultaneously, the result would not be equal to the measurement of C. Indeed, a measurement of C gives the value $\frac{\hbar}{\sqrt{2}}$ or $-\frac{\hbar}{\sqrt{2}}$,[55] while that of s_x and s_z respectively gives $\frac{\hbar}{2}$ or $-\frac{\hbar}{2}$, i.e. in sum (regardless of the order of the measurement of s_x and s_z), one of three values $\hbar$ or 0 or $-\hbar$, but never $\pm\frac{\hbar}{\sqrt{2}}$.

Thus, C is not an observable in the literal sense suggested by the notation that namely s_x and s_z are to be measured.[56] However, we can C represent as the spin operator along the vector $\widehat{xz} = (1, 0, 1)$:

$$C = \widehat{xz} \cdot \mathbf{s} = s_x + s_z \tag{I.51}$$

Usually, the spin is measured with respect to the unit vector, and we can write

$$C' = \frac{1}{\sqrt{2}} C = \frac{\widehat{xz}}{\sqrt{2}} \cdot \mathbf{s} = \frac{s_x + s_z}{\sqrt{2}} \tag{I.52}$$

A short calculation shows that C' has the eigenvalues $\pm\frac{\hbar}{2}$. Thus, C is an observable in the sense of an instruction to measure the spin along $(1, 0, 1)$ (which is *one* measurement), but not in the sense to measure the spin along the x-.axis and the z-axis (which would be *two* measurements).

A similar consideration applies, e.g., for the harmonic oscillator with $H = \frac{1}{2m} p^2 + \frac{m\omega^2}{2} x^2$. The energy eigenvalues are not related in a simple way to those of the operators

---

[55] See exercises to Chap. 14.

[56] Note that the eigenvectors of C form a basis in the state space, as the above definition of an observable requires.

$\frac{p^2}{2m}$ and $\frac{m\omega^2}{2}x^2$. Also here holds that the result of an energy measurement would not equal the sum of the separate measurements of $\frac{p^2}{2m}$ and $\frac{m\omega^2}{2}x^2$ even if we could measure position and momentum simultaneously.

In quantum field theory, there is another type (case) of Hermitian operators which are not observable. There, the Hamiltonians are expressed in terms of creation and annihilation operators which is possible in several distinct ways. But only one of these representations corresponds (for certain applications) to an observable, namely the so-called normal-ordered form; see Vol. 2 Appendix W.

I.6   Exercises

1. The action of two operators A and B on a vector $(c_1, c_2, c_3, \ldots)$ is $A(c_1, c_2, c_3, \ldots) = (c_2, c_3, c_4, \ldots)$ and $B\,(c_1, c_2, c_3, \ldots) = (0, c_1, c_2, \ldots)$. What is the matrix representation of the two operators? Determine AB and BA.
   Solution: We see that

$$
A = \begin{pmatrix} 0\ 1\ 0\ 0 \ldots \\ 0\ 0\ 1\ 0 \ldots \\ 0\ 0\ 0\ 1 \ldots \\ \vdots\ \vdots\ \vdots\ \vdots\ \ \vdots \end{pmatrix} ; \ B = \begin{pmatrix} 0\ 0\ 0\ 0 \ldots \\ 1\ 0\ 0\ 0 \ldots \\ 0\ 1\ 0\ 0 \ldots \\ \vdots\ \vdots\ \vdots\ \vdots\ \ \vdots \end{pmatrix} . \tag{I.53}
$$

   Furthermore, we have

$$
AB = \begin{pmatrix} 1\ 0\ 0\ 0 \ldots \\ 0\ 1\ 0\ 0 \ldots \\ 0\ 0\ 1\ 0 \ldots \\ \vdots\ \vdots\ \vdots\ \vdots\ \ \vdots \end{pmatrix} ; \ BA = \begin{pmatrix} 0\ 0\ 0\ 0 \ldots \\ 0\ 1\ 0\ 0 \ldots \\ 0\ 0\ 1\ 0 \ldots \\ \vdots\ \vdots\ \vdots\ \vdots\ \ \vdots \end{pmatrix} . \tag{I.54}
$$

   Hence, B is the right inverse of A, but not the left inverse. In a finite vector space, these terms always coincide.

2. Show that p is not bounded in the space $L^{(2)}$ of the functions which are defined on the interval $[0, b]$ and are continuous there. Solution: Consider e.g. the function $f(x) = x^{-a}$ with $a > 0$. In order that $\int_0^b x^{-2a} dx$ be defined, it must hold that $-2a + 1 > 0$ or $a < \frac{1}{2}$. Hence, all $f(x) = x^{-a}$ with $0 < a < \frac{1}{2}$ are in $L^{(2)}$, but not in the domain of definition of p, because we have

$$
\int_0^b f'^2 dx = a^2 \int_0^b x^{-2a-2} dx = a^2 \frac{x^{-2a-1}}{-2a-1} \Big/_0^b . \tag{I.55}
$$

Evidently, the last term exists only for $-2a - 1 > 0$, i.e. for $a < -\frac{1}{2}$.

3. Show that in a finite-dimensional Hilbert space, all operators are bounded.
   Solution: All Hilbert spaces of the same dimension are isomorphic. Thus, we can choose the space C^n (complex-valued n-tuple) as our finite-dimensional space; correspondingly, the operators are matrices.
   Define

$$
A = \begin{pmatrix} a_{11} & a_{12} & \cdots & a_{1n} \\ a_{21} & a_{22} & \cdots & a \\ \vdots & \vdots & \ddots & \vdots \\ a_{n1} & a_{n2} & \cdots & a_{nn} \end{pmatrix} ; \quad |\varphi\rangle = \begin{pmatrix} \varphi_1 \\ \varphi_2 \\ \vdots \\ \varphi_n \end{pmatrix} . \tag{I.56}
$$

Then we have

$$
\|A\,|\varphi\rangle\|^2 = \langle \varphi|\, A^\dagger A\, |\varphi\rangle = \sum_{j,k,l=1}^n \varphi_k^* a_{jk}^* a_{jl} \varphi_l = \sum_{j=1}^n \sum_{k=1}^n \varphi_k^* a_{jk}^* \sum_{l=1}^n a_{jl} \varphi_l . \tag{I.57}
$$

We define the ith column of A as $\langle a_i|$, i.e.

$$
|a_i\rangle = \begin{pmatrix} a_{i1}^* \\ a_{i2}^* \\ \vdots \\ a_{in}^* \end{pmatrix} ; \quad \langle \varphi\,|a_j\rangle = \sum_{k=1}^n \varphi_k^* a_{jk}^* . \tag{I.58}
$$

It follows that

$$
\|A\,|\varphi\rangle\|^2 = \sum_{j=1}^n \sum_{k=1}^n \varphi_k^* a_{jk}^* \sum_{l=1}^n a_{jl} \varphi_l = \sum_{j=1}^n \langle \varphi\,|a_j\rangle \langle \varphi\,|a_j\rangle^* = \sum_{j=1}^n |\langle \varphi\,|a_j\rangle|^2 . \tag{I.59}
$$

with the Schwarz inequality, we obtain

$$
\|A\,|\varphi\rangle\|^2 = \sum_{j=1}^n |\langle \varphi\,|a_j\rangle|^2 \le \sum_{j=1}^n \||\varphi\rangle\|^2 \,\||a_j\rangle\|^2 = \||\varphi\rangle\|^2 \sum_{j=1}^n \||a_j\rangle\|^2 . \tag{I.60}
$$

For the norm, it results that

$$
\|A\| = \sup \frac{\|A\,|\varphi\rangle\|}{\||\varphi\rangle\|} \le \sup \sqrt{\frac{\||\varphi\rangle\|^2 \sum_{j=1}^n \||a_j\rangle\|^2}{\||\varphi\rangle\|^2}} \tag{I.61}
$$

$$
= \sqrt{\sum_{j=1}^n \||a_j\rangle\|^2} \le \sqrt{n} \cdot \max \||a_j\rangle\| .
$$

The right-hand side is apparently finite and may be estimated by a constant C.

4. Given two Hermitian operators A and B with the commutator $[A, B] = i$, show that at least one of the two operators is not bounded.

   Solution: We first assume that one of the operators is bounded, say $\|B\| \leq 1$. Then we prove the relation

$$\left[A^n, B\right] = in A^{n-1}. \tag{I.62}$$

This is done by induction. Clearly, the base for $n = 1$ is correct. Then it holds that[57]

$$\left[A^{n+1}, B\right] = A\left[A^n, B\right] + [A, B] A^n = Ain A^{n-1} + i A^n = i(n+1) A^n, \tag{I.63}$$

whereby the proposition is proved.

Thus we have $A^n B - B A^n = in A^{n-1}$ or

$$\left\|A^n B - B A^n\right\| = n \left\|A^{n-1}\right\| \text{ or } n \left\|A^{n-1}\right\| \leq \left\|A^n B\right\| + \left\|B A^n\right\|, \tag{I.64}$$

where we have used the triangle inequality. For bounded operators, $\|AB\| \leq \|A\|\|B\|$ applies, and it follows that

$$n \left\|A^{n-1}\right\| \leq 2 \left\|A^n\right\| \|B\| \leq 2 \left\|A^n\right\|. \tag{I.65}$$

In a Hilbert space, $\left\|A^\dagger A\right\| = \|A\|^2$ for bounded operators; from this, for Hermitian operators, it follows that $\left\|A^2\right\| = \|A\|^2$. This leads to

$$n \|A\|^{n-1} \leq 2 \|A\|^n \text{ or } \frac{n}{2} \leq \|A\|, \tag{I.66}$$

i.e. a contradiction to the proposition that A is bounded.

5. Positive matrices:

   (a) Show that a positive matrix is self-adjoint.
   (b) Show that a matrix is positive iff all its eigenvalues are ≥ 0.

6. Show that the mean value of an anti-Hermitian operator is imaginary.

   Solution: For an anti-Hermitian operator A, we know that (see Chap. 9) $A^\dagger = -A$. It follows that

$$\langle A\rangle_\psi^* = \langle \psi| A |\psi\rangle^* = \langle \psi| A^\dagger |\psi\rangle = -\langle \psi| A |\psi\rangle = -\langle A\rangle_\psi, \tag{I.67}$$

and thus $\langle A\rangle_\psi$ is imaginary.

7. The domain of definition of $p = \frac{\hbar}{i}\frac{d}{dx}$ comprises all functions $g(x) \in L^{(2)}[0, 1]$ which are differentiable, whose derivatives are square integrable, and which satisfy the boundary condition $g(1) = g(0)$. Show that p is self-adjoint.

---

[57] We use the equation $[AB, C] = A[B, C] + [A, C] B$.

Solution: We see that

$$\langle p^\dagger f \,|g\rangle = \langle f \,|pg\rangle$$

$$= \int_0^1 f^* \frac{\hbar}{i} \frac{dg}{dx} dx = f(1)g(1) - f(0)g(0) + \int_0^1 \left(\frac{\hbar}{i}\frac{df}{dx}\right)^* g dx = \langle pf \,|g\rangle.$$

(I.68)

Thus p is symmetric. The integrated term on the right side vanishes for $f(1) = f(0)$; hence $p^\dagger$ has the same domain of definition as p. In other words: p is self-adjoint.

8. We consider a spin-$\frac{1}{2}$ system with the uncertainty principle

$$\left(\Delta_\psi s_x\right)\left(\Delta_\psi s_y\right) \geq \frac{\hbar}{2}\left|\langle s_z\rangle_\psi\right|.$$

(I.69)

Show that the right-hand side may vanish.

Solution: For a state $|\psi\rangle = \begin{pmatrix} a \\ b \end{pmatrix}$, it holds that

$$\langle s_z\rangle_\psi = \frac{\langle\psi| s_z |\psi\rangle}{\langle\psi |\psi\rangle} = \frac{|a|^2 - |b|^2}{|a|^2 + |b|^2}.$$

(I.70)

If we assume that this expression vanishes and that $|\psi\rangle$ is normalized, it follows that

$$|\psi\rangle = \frac{1}{\sqrt{2}} \begin{pmatrix} e^{i\alpha} \\ e^{i\beta} \end{pmatrix},$$

(I.71)

or (extracting a global phase),

$$|\psi\rangle = \frac{e^{i\alpha}}{\sqrt{2}} \begin{pmatrix} 1 \\ e^{i(\beta-\alpha)} \end{pmatrix} = \frac{e^{i\alpha}}{\sqrt{2}} \begin{pmatrix} 1 \\ e^{i\gamma} \end{pmatrix}.$$

(I.72)

For theses states, it always holds that

$$\left(\Delta_\psi s_x\right)\left(\Delta_\psi s_y\right) \geq 0.$$

(I.73)

We now determine the uncertainties on the left-hand side. First we have

$$\langle s_x \rangle_\psi = \frac{\hbar}{4} \left(1 \ e^{-i\gamma} \right) \begin{pmatrix} 0 & 1 \\ 1 & 0 \end{pmatrix} \begin{pmatrix} 1 \\ e^{i\gamma} \end{pmatrix}$$

$$= \frac{\hbar}{4} \left(1 \ e^{-i\gamma} \right) \begin{pmatrix} e^{i\gamma} \\ 1 \end{pmatrix} = \frac{\hbar}{2} \cos \gamma$$

$$\langle s_y \rangle_\psi = \frac{\hbar}{4} \left(1 \ e^{-i\gamma} \right) \begin{pmatrix} 0 & -i \\ i & 0 \end{pmatrix} \begin{pmatrix} 1 \\ e^{i\gamma} \end{pmatrix}$$

$$= \frac{\hbar}{4} \left(1 \ e^{-i\gamma} \right) \begin{pmatrix} -ie^{i\gamma} \\ i \end{pmatrix} = \frac{\hbar}{2} \sin \gamma. \tag{I.74}$$

Then we obtain with $s_x^2 = s_y^2 = \frac{\hbar^2}{4}$ and $(\Delta A)^2 = \langle A^2 \rangle - \langle A \rangle^2$

$$\left(\Delta_\psi s_x \right)^2 = \frac{\hbar^2}{4} - \frac{\hbar^2}{4} \cos^2 \gamma = \frac{\hbar^2}{4} \sin^2 \gamma$$

$$\left(\Delta_\psi s_y \right)^2 = \frac{\hbar^2}{4} - \frac{\hbar^2}{4} \sin^2 \gamma = \frac{\hbar^2}{4} \cos^2 \gamma, \tag{I.75}$$

and the uncertainty principle is reduced in this case to the inequality

$$\frac{\hbar^2}{4} |\sin \gamma \cos \gamma| = \frac{\hbar^2}{2} |\sin 2\gamma| \geq 0. \tag{I.76}$$

Depending on the choice of γ, the uncertainty vanishes and with it also the product of the two uncertainties.

Appendix J
From Quantum Hopping to the Schrödinger Equation

This alternative derivation of the SEq[58] is based on general principles, namely symmetry and superposition in combination with the idea of a discretized space.[59] The approach emphasizes the structurally simple side of quantum mechanics, and not the paradoxical side. It uses the idea that a quantum object which 'lives' only on the discrete sites of a lattice can still move quantum mechanically via the partial overlap of the position states on neighboring sites. This corresponds to the 'hopping equation' which we derive below.

Hopping Equation

We define a grid on the one-dimensional space in contiguous intervals of lengths l which we number consecutively by $n = \cdots - 3, -2, -1, 0, 1, 2, 3 \ldots$. In this one-dimensional lattice, we place detectors which determine where the quantum object is located at a time t with the resolution l; e.g. detector n reacts at time t. The expression

$$|nl, t\rangle \tag{J.1}$$

thus means that the quantum object is detected at time t at the (interval) position nl, in other words, that it is at the position nl at time t. We can think of $|nl, t\rangle$ as a column vector, where at time t, a 1 is at position n and everywhere else 0. For example, the state $|2l, t\rangle$ (see Fig. J.1) is given by:

---

[58] See also J. Pade and L. Polley, 'Quanten-Hüpfen auf Gittern', Physik in der Schule 36/11 (1998) 363, ('Quantum hopping on lattices', Physics in School 36/11 (1998) 363).

[59] In this discretization of space (which is also used in the lattice gauge theory), it is not necessarily assumed that such a lattice exists in nature, but rather that there is a limit to the degree of precision with which the position can be measured. Consequently, one always has to divide the space into a grid in practice, which means that also the laws of physics initially appear in discretized form. In general, it is expected that in the continuum limit, i.e. in the limit of infinitely fine resolution, the usual laws would be obtained.

© Springer Nature Switzerland AG 2018
J. Pade, *Quantum Mechanics for Pedestrians 1*, Undergraduate Lecture
Notes in Physics, https://doi.org/10.1007/978-3-030-00464-4

Fig. J.1 The state $|2l, t\rangle$

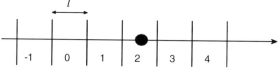

Fig. J.2 Possible
movements during the
period τ

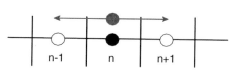

$$|2l, t\rangle = \begin{pmatrix} \vdots \\ 0 \\ 1 \\ 0 \\ \vdots \end{pmatrix} \begin{matrix} \text{Position} \\ \vdots \\ n = 3 \\ n = 2 \\ n = 1 \\ \vdots \end{matrix} \qquad \text{(J.2)}$$

The normalization condition[60]

$$\langle nl, t| \, nl, t\rangle \overset{!}{=} 1 \qquad \text{(J.3)}$$

must be satisfied for all times. Since the quantum object cannot be detected simultaneously at two places, it must hold (orthogonality) that:

$$\langle nl, t| \, n'l, t\rangle = 0; \; n \neq n'. \qquad \text{(J.4)}$$

Thus, the states form an orthonormal system (ONS), $\langle nl, t| \, n'l, t\rangle = \delta_{nn'}$.

In (J.1), we assume that the state $|nl, t\rangle$ of a (spineless) quantum object is com-
pletely determined by specifying its position at a given time. This is a clear contrast to
classical mechanics, where we need in general *two* quantities in order to characterize
the state of a particle, such as its position and its velocity. Thus, we cannot assign a
direction of motion to a quantum object in the state $|nl, t\rangle$.[61] To avoid this dilemma,
we employ the superposition principle, which allows us to superpose different states
so that the object is moving so to speak in all directions at once. We formulate this
as follows: After a short time step τ, the quantum object is either still in the same
place or it has moved on to the next adjacent interval (for sufficiently small τ, we
can rule out a move to next-nearest neighboring positions; cf. Fig. J.2).

---

[60] As usual, we denote the adjoint row vector by $\langle |$.

[61] A snapshot of a pendulum also gives no information about its direction of motion. But while
classically, a second picture taken a short time later can clarify this question (one then knows the
initial velocity in addition to the initial position), in quantum mechanics a second image would
produce (prepare) a new state.

In this way, we obtain the hopping equation

$$|nl, t\rangle = \alpha \, |nl, t + \tau\rangle + \beta \, |nl + l, t + \tau\rangle + \beta \, |nl - l, t + \tau\rangle. \quad (J.5)$$

This equation can of course only 'work' if the sum of states is in fact defined—in other words, if the *superposition principle* holds true; this is here the basic assumption. It leads necessarily to the appearance of *probabilities*. For if we assume that, at a fixed time, our quantum object can be measured by only *one* detector, the numbers α and β are related to the probabilities of finding the quantum object at the position n or $n + 1$ or $n - 1$.

Properties of the Coefficients

The coefficients of the states in (J.5) do not depend on t or n, since we are considering free quantum objects. The coefficients of the two states $|nl \pm l, t + \tau\rangle$ *must* be the same, since there is no preferred direction for a free quantum object.[62] For $\tau \to 0$, $\alpha \to 1$ and $\beta \to 0$ must hold, i.e. the quantum object remains at its initial position. Rearranging and dividing by τ, we find

$$\frac{-|nl, t+\tau\rangle + |nl, t\rangle}{\tau} = \frac{(\alpha - 1)}{\tau} |nl, t + \tau\rangle + \frac{\beta}{\tau} |nl + l, t + \tau\rangle + \frac{\beta}{\tau} |nl - l, t + \tau\rangle. \quad (J.6)$$

In order for this formulation to make sense, the limit $\tau \to 0$ must be defined. We set

$$\alpha - 1 = \hat{\alpha}\tau; \; \beta = \hat{\beta}\tau \quad (J.7)$$

where $\hat{\alpha}$ and $\hat{\beta}$ are complex numbers yet to be determined. Then we can take the limit $\tau \to 0$:

$$-\frac{d}{dt} |nl, t\rangle = \hat{\alpha} \, |nl, t\rangle + \hat{\beta} \, |nl + l, t\rangle + \hat{\beta} \, |nl - l, t\rangle. \quad (J.8)$$

We take the derivative of (J.3) with respect to t

$$0 = \frac{d}{dt} \langle nl, t| nl, t\rangle = \left(\frac{d}{dt} \langle nl, t|\right) |nl, t\rangle + \langle nl, t| \left(\frac{d}{dt} |nl, t\rangle\right) \quad (J.9)$$

and insert (J.8):

$$0 = \left[\hat{\alpha}^* \langle nl, t| + \hat{\beta}^* \langle nl + l, t| + \hat{\beta}^* \langle |nl - l, t\rangle|\right] |nl, t\rangle$$
$$+ \langle nl, t| \left[\hat{\alpha} \, |nl, t\rangle + \hat{\beta} \, |nl + l, t\rangle + \hat{\beta} \, |nl - l, t\rangle\right]. \quad (J.10)$$

With (J.3) and (J.4), it follows that

---

[62] Since objects with greater mass are less mobile, β must become smaller with increasing mass. Below, we show that $\beta \sim 1/m$.

$$\hat{\alpha}^* + \hat{\alpha} = 0 \tag{J.11}$$

This implies

$$\hat{\alpha} \in \mathbb{I} \text{ or } \hat{\alpha} = ia; \ a \in \mathbb{R} \tag{J.12}$$

We obtain analogously

$$\hat{\beta} = ib; \ b \in \mathbb{R} \tag{J.13}$$

Thus, we have for the coefficients in (J.5)

$$\begin{aligned} \alpha &= 1 + ia\tau; \ a \in \mathbb{R} \\ \beta &= ib\tau; \ b \in \mathbb{R}. \end{aligned} \tag{J.14}$$

Schrödinger Equation

We now consider a superposition of all states of the form

$$|\Phi\rangle = \sum_{n=-\infty}^{\infty} \Psi(nl, t) |nl, t\rangle \tag{J.15}$$

where the coefficients $\Psi(nl, t)$ are the 'weights' of the individual positions. Since we sum from $-\infty$ to ∞, we have

$$\sum_{n=-\infty}^{\infty} \Psi(nl, t) |nl, t\rangle = \sum_{n=-\infty}^{\infty} \Psi(nl, t + \tau) |nl, t + \tau\rangle. \tag{J.16}$$

We insert the hopping equation (J.5) into this equation:

$$\sum_{n=-\infty}^{\infty} \Psi(nl, t) [\alpha |nl, t + \tau\rangle + \beta |nl + l, t + \tau\rangle + \beta |nl - l, t + \tau\rangle]$$
$$= \sum_{n=-\infty}^{\infty} \Psi(nl, t + \tau) |nl, t + \tau\rangle. \tag{J.17}$$

Due to the orthonormality of the states $|nl, t + \tau\rangle$, it follows directly that:

$$\Psi(nl, t + \tau) = \alpha \Psi(nl, t) + \beta \Psi(nl - l, t) + \beta \Psi(nl + l, t). \tag{J.18}$$

We want to transform this expression into the SEq. First we rearrange:

$$\frac{\Psi(nl, t + \tau) - \Psi(nl, t)}{\tau} = \frac{(\alpha - 1)}{\tau} \Psi(nl, t) + \frac{\beta}{\tau} \Psi(nl - l, t) + \frac{\beta}{\tau} \Psi(nl + l, t). \tag{J.19}$$

Substituting (J.14) leads in the limit $\tau \to 0$ to[63]

$$\frac{\partial}{\partial t}\Psi(nl, t) = ia\Psi(nl, t) + ib\Psi(nl - l, t) + ib\Psi(nl + l, t). \tag{J.20}$$

Now we have to include the spatial dependence. It is

$$\begin{aligned}\Psi(nl + l, t) + \Psi(nl - l, t) \\ = [\Psi(nl+l, t) - \Psi(nl, t)] - [\Psi(nl, t) - \Psi(nl - l, t)] + 2\Psi(nl, t),\end{aligned} \tag{J.21}$$

and from this, it follows that:

$$\frac{\partial}{\partial t}\Psi(nl, t)$$
$$= (ia + 2ib)\Psi(nl, t) + ib\{[\Psi(nl + l, t) - \Psi(nl, t)] - [\Psi(nl, t) - \Psi(nl - l, t)]\}. \tag{J.22}$$

This implies

$$\frac{\partial}{\partial t}\Psi(nl, t)$$
$$= ibl^2 \frac{\{[\Psi(nl + l, t) - \Psi(nl, t)] - [\Psi(nl, t) - \Psi(nl - l, t)]\}}{l^2} + (ia + 2ib)\Psi(nl, t). \tag{J.23}$$

In the following, we set $a = -2b$, which is not necessary, but just serves to simplify the discussion.[64] We require $b = Bl^{-2}$ so that it makes sense to take the limit $l \to 0$, where B is a constant independent of l. Then it follows that

$$\frac{\partial}{\partial t}\Psi(nl, t) = iB \frac{\{[\Psi(nl + l, t) - \Psi(nl, t)] - [\Psi(nl, t) - \Psi(nl - l, t)]\}}{l^2}, \tag{J.24}$$

which we may write, for $l \to 0$ with $x \hat{=} nl$, as the second derivative w.r.t the spatial coordinate,

$$\frac{\partial}{\partial t}\Psi(x, t) = iB \frac{\partial^2}{\partial x^2}\Psi(x, t). \tag{J.25}$$

The precise value of B cannot be specified uniquely here; but at least we know that the equation of motion for a free quantum object *must* have the form

$$i\frac{\partial}{\partial t}\Psi(x, t) = -B \frac{\partial^2}{\partial x^2}\Psi(x, t). \tag{J.26}$$

To obtain the usual form of the SEq, we multiply by $\hbar$. Then we have for the units of $B\hbar$

---

[63] We use ∂ from the start, because we subsequently consider the spatial coordinates.

[64] For $a \neq 2b$, one obtains a contribution in the SEq which corresponds to a constant potential.

$$[B\hbar] = \frac{m^2}{s} Js = \frac{Js}{kg} Js = \frac{(Js)^2}{kg} = \frac{[\hbar^2]}{[m]}, \tag{J.27}$$

and the final result reads

$$i\hbar \frac{\partial}{\partial t} \Psi(x,t) = -\hat{B} \frac{\hbar^2}{2m} \frac{\partial^2}{\partial x^2} \Psi(x,t). \tag{J.28}$$

The number $\hat{B}$, which depends on the system of units chosen, cannot be determined here without additional information.

A possible item of additional information would be for example that plane waves $e^{i(kx-\omega t)}$ must be solutions of the last equation (under nonrelativistic conditions). It follows that

$$i\hbar(-i\omega) = -\hat{B} \frac{\hbar^2}{2m} (-k^2) \text{ or } \hat{B} = \frac{2m\hbar\omega}{\hbar^2 k^2} = \frac{2mE}{p^2} = 1. \tag{J.29}$$

Appendix K
The Phase Shift at a Beam Splitter

In Chap. 6, we used the fact that the *relative* phase shift between transmitted and reflected waves at a beam splitter is 90°. This will be demonstrated in detail here.[65]

We consider in Fig. K.1 a plane wave which is incident on a beam splitter with amplitude 1 and is split into a reflected wave with complex amplitude $R = \alpha e^{i\varphi}$ and a transmitted wave with complex amplitude $T = \beta e^{i\psi}$. The refractive index n is assumed to be constant throughout the beam splitter. Since the beam splitter is symmetric, the same amplitude ratios would occur if the plane wave were incident from the right instead of from the left.

The intensity is proportional to the absolute square of the amplitude. We assume that there are no absorption processes, and that the medium outside of the beam splitter is homogeneous. Then, due to energy conservation, it follows that:

$$1 = R^*R + T^*T. \tag{K.1}$$

Now we consider a superposition of two incoming waves with amplitudes R^* and T^*; cf. Fig. K.2. These waves are split as shown in Fig. K.1, and superpose to give two outgoing total waves. According to (K.1), the amplitude of the top right outgoing wave is 1. Thus, this wave already transports all of the incoming energy. Consequently, the amplitude of the top left outgoing wave must vanish:

$$R^*T + T^*R = 0. \tag{K.2}$$

Since T^*R is the complex conjugate of R^*T, it follows from (K.2) that T^*R is purely imaginary.[66] With $R = \alpha e^{i\varphi}$ and $T = \beta e^{i\psi}$, this leads to

$$\cos(\psi - \varphi) = 0. \tag{K.3}$$

---

[65] See also: J. Pade and L. Polley, Phasenverschiebung am Strahlteiler, PhyDid 1/3 (2004) 39, (Phase shift at a beam splitter, PhyDid 1/3 (2004) 39).

[66] Or it is real and zero, but this case is obviously of no interest.

© Springer Nature Switzerland AG 2018
J. Pade, *Quantum Mechanics for Pedestrians 1*, Undergraduate Lecture Notes in Physics, https://doi.org/10.1007/978-3-030-00464-4

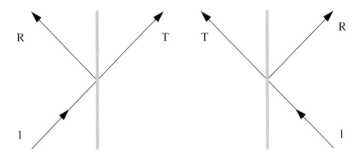

Fig. K.1 Amplitudes at the beam splitter, laterally reversed on the right side

Fig. K.2 Superposition of
two incoming waves

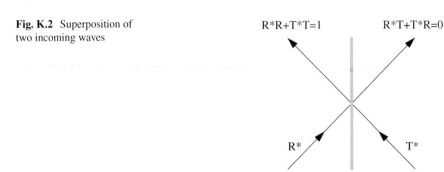

This corresponds to a relative phase of 90° between the reflected and transmitted
waves (The choice of −90°, at this point possible in principle, can be excluded
because of other considerations. It is important above all that the amplitudes be
perpendicular to each other.)

We note without proof (see textbooks on experimental physics) that the phase
shift between the incident and the reflected wave at a mirror is 180°.

Appendix L
The Quantum Zeno Effect

Reducing the time between successive measurements further and further, one ideally approaches *continuous* measurements. In this context, new quantum-mechanical phenomena such as the *quantum Zeno effect* [67] (QZE) can occur. This effect exhibits the fact that the more often one measures an unstable system, the more one prevents its decay. The effect has been known experimentally for about 20 years; a catchy formulation is "a watched pot never boils".

For some years now, the opposite effect has been discovered—the *anti-quantum Zeno effect*. Here, a more frequent observation of an unstable quantum does not have a stabilizing action (as in the Zeno effect), but rather a destabilizing effect. The more often one raises the lid, the faster the water is boiling: "boiling the pot by watching". And, recently, a third related phenomenon has been under discussion, the *Hamlet effect* . Here, measurements of a quantum system destroy the prognosis options to the point that no prediction is possible at all.[68] "Boiling or not boiling, that is the question".[69]

In the following, we want to make some illustrative comments on the QZE and the anti-QZE in unstable systems, before we present a simple calculation for the QZE.

---

[67]Zenon (or Zeno) of Elea (490–430 BC), Greek philosopher, was mainly concerned with the problem of the continuum. Perhaps best known is his paradox of the swift-footed Achilles and the tortoise: In a race, Achilles gives the tortoise a head start and therefore can never overtake it. Because, to achieve this, he must first catch up its lead. But during this time, the tortoise has gained a new (smaller) lead, which Achilles also has to catch up to. When he arrives at that point, the tortoise has again gained a (still smaller) lead and so on. From a present-day perspective, the argument misses the fact, among other things, that an infinite series can still have a finite sum.

[68]V. Pankovic, 'Quantum Hamlet effect—a new example', http://arxiv.org/PS_cache/arxiv/pdf/0908/0908.1301v2.pdf, (2009).

[69]Apropos Hamlet: "A somewhat saucy philosopher, I think Hamlet, Prince of Denmark, said that there were many things in heaven and earth not dreamt of in our Compendiis. If that simple-minded man, who was not in his right mind as is well known, made digs at our Compendia of physics, so one can confidently answer him, 'Well, but instead there are also many things in our Compendiis which do not exist, neither in heaven nor on earth'". Georg Christoph Lichtenberg, *Scrap Books*, Vol. L (155).

© Springer Nature Switzerland AG 2018
J. Pade, *Quantum Mechanics for Pedestrians 1*, Undergraduate Lecture
Notes in Physics, https://doi.org/10.1007/978-3-030-00464-4

Then we consider how we can improve the efficiency of an interaction-free quantum measurement using the QZE.[70]

L.1   Unstable Systems

We want to provide a conceptual idea here, without going into formal details.

An unstable state evolves eventually into a linear superposition of states, one of which can be observed in a measurement. The decay rate depends on several factors, among others on the energy spectrum of the final states (also called reservoir states) to which the unstable state is coupled.

Measurements which are performed at the frequency ν cause an energy uncertainty $\sim h\nu$, according to the energy-time uncertainty principle, which affects the range of accessible reservoir states and thus the decay rate. If the energy uncertainty due to successive measurements is large compared with both the width of the reservoir spectrum and the energy separation between the unstable state and the average energy of the reservoir, then the QZE should occur. If, on the other hand, the energy spread is initially comparatively small, it increases with ν, and therefore the number of attainable reservoir states into which transitions can occur also increases. In this case, the anti-QZE should occur first.

Indeed, this has been observed, for example in an experiment[71] in which sodium atoms are trapped in an optical standing wave. The atoms can escape this potential by the tunneling effect. The experimental result was that measurement intervals of 1 μs reduced the tunneling (i.e. the decay), while measurement intervals of 5 μs enhanced the tunneling.

L.2   Simple Model Calculation

We want to illustrate the basic idea of the QZE by a simple calculation. We start from the SEq in the form:

$$|\psi(t)\rangle = e^{-iHt/\hbar} |\psi(0)\rangle, \tag{L.1}$$

as well with an observable A whose spectrum is discrete and not degenerate for simplicity. Thus we may write

$$A = \sum_m a_m |\varphi_m\rangle \langle\varphi_m|. \tag{L.2}$$

---

[70]We mention that the quantum Zeno effect may also be used to generate entanglement as well to suppress decoherence (cf. Chaps. 20 and 24, Vol. 2).

[71] M.C. Fischer et al., Observation of the Quantum Zeno and Anti-Zeno Effects in an Unstable System, Phys. Rev. Lett. 87(4) (2001), 040402.

The scenario is now that we carry out repeated measurements of A at fixed time intervals τ; between the measurements, the SEq determines the evolution of the state.

The initial state is $|\psi(0)\rangle = |\varphi_n\rangle$. We ask for the probability that after N measurements, the system is still in the initial state. For sufficiently small t, it holds because of (L.1) that

$$|\psi(t)\rangle = \left[1 - \frac{iHt}{\hbar} - \frac{H^2 t^2}{2\hbar^2} + O\left(t^3\right)\right]|\varphi_n\rangle. \tag{L.3}$$

The first measurement takes place at $t = \tau$. The probability p_n of measuring the value a_n is given by

$$p_n(\tau) = |\langle\varphi_n|\psi(\tau)\rangle|^2 = \left|\langle\varphi_n|\left[1 - \frac{iH\tau}{\hbar} - \frac{H^2\tau^2}{2\hbar^2} + O\left(\tau^3\right)\right]|\varphi_n\rangle\right|^2. \tag{L.4}$$

Solving the brackets and rearranging yields

$$p_n(\tau) = 1 - \frac{\tau^2}{\hbar^2}(\Delta H)_n^2 + O\left(\tau^3\right) \tag{L.5}$$

with the energy uncertainty

$$(\Delta H)_n^2 = \langle\varphi_n|H^2|\varphi_n\rangle - \langle\varphi_n|H|\varphi_n\rangle^2. \tag{L.6}$$

In the context of these considerations, the time $t_Z = \frac{\hbar}{(\Delta H)_n}$ is called the Zeno time.

The quantity $p_n(\tau)$ is the probability that the system is still in the initial state $|\varphi_n\rangle$ after the time τ. After N measurements, the total time is $T = N\tau$; it follows for the probability (henceforth, we omit the terms of higher order):

$$p_n(T) \approx \left[1 - \frac{\tau^2}{\hbar^2}(\Delta H)_n^2\right]^N = \left[1 - \frac{\tau T}{\hbar^2 N}(\Delta H)_n^2\right]^N. \tag{L.7}$$

If we now fix T and let N become very large (i.e. the measurement intervals become shorter and approximate more and more closely a continuous measurement[72]), then we can use the definition of the exponential function $\left[1 + \frac{x}{N}\right]^N \underset{N\to\infty}{\to} e^x$ and obtain

$$p_n(T) \approx \exp\left(-\frac{(\Delta H)_n^2}{\hbar^2}\tau T\right) \underset{\tau\to 0}{\to} 1. \tag{L.8}$$

Hence, the system stays in the initial state in the limit of a continuous measurement: a watched pot never boils.

---

[72]This is of course an idealization. The measurement process always has a certain finite duration, even if it can possibly be made very short compared to the relevant time constants of the system.

The formal reason for this is that, according to (L.5), the probability of leaving the initial state is given by

$$1 - p_n(\tau) \sim \tau^2, \qquad (L.9)$$

while the number of measurements increases $\sim \frac{1}{\tau}$. Consequently, the state reduction caused by the successive measurements is faster than possible transitions into other states, provided that τ is sufficiently small.

L.3 Interaction-Free Quantum Measurement

We consider an interaction-free quantum measurement making use of the quantum Zeno effect. Here, the scenario is somewhat different, for we do not consider unstable states, but rather we want to force a system from an initial state into a different state by repeated measurements, and this should be all the 'smoother', the more often one repeats the measurements.[73]

The basic idea[74] is quite simple: We let light pass through N polarization rotators, each of which rotates the plane of polarization of the incident state by $\frac{\pi}{2N}$. Added together, the N rotators, connected in series, turn the state by $\frac{\pi}{2}$, so that e.g. an initially horizontally polarized state becomes vertically polarized. Now we add a horizontal analyzer behind each rotator. The probability that a photon passes one of these polarizers is then given by $p = \cos^2\left(\frac{\pi}{2N}\right)$, and the probability of passing all N polarizers is $p_N = \cos^{2N}\left(\frac{\pi}{2N}\right)$; for sufficiently large N, we thus have $p_N \approx 1 - \frac{\pi^2}{4N}$. Accordingly, the absorption probability is given by $\frac{\pi^2}{4N}$.

As to the interaction-free quantum measurement, we have seen in Chap. 6 that in one-fourth of the trials, the 'bomb test' works without the bomb blowing up. This percentage can be increased considerably by using the setup shown schematically in Fig. L.1. The 'inner' part, i.e. the arrangement of two mirrors and polarizing beam splitters (PBS), is called a polarization Mach–Zehnder interferometer (PMZI). The setup (L.1) allows, in principle, the detection of an object in the beam path with the probability 1 in an 'interaction-free' manner.

The basis states are not, as in Chap. 6, the horizontal and vertical directions of propagation (i.e. $|H\rangle$ and $|V\rangle$), but instead the horizontal and vertical polarization states of the photons, $|h\rangle$ and $|v\rangle$; the propagation direction does not matter.

At the beginning, the lower left mirror is switched open. It is closed after the photon has entered the setup. The photon can then make N rounds, after which the lower-right mirror is opened and the photon is directed out to further analysis.

On each iteration, first the polarizer is passed, whereby the plane of polarization is rotated in each passage by $\frac{\pi}{2N}$. In the PBS, the horizontally-polarized component is transmitted and the vertically-polarized component is reflected.

---

[73] Actually it is therefore more like the anti-Zeno effect, but the name 'Zeno effect' has been adopted in this context.

[74] This effect can be detected also in classical optics; the quantum-mechanical aspect lies in the fact that it is considered below for a *single* photon.

Fig. L.1 Setup for the zeno effect. mM = switchable mirror, M = mirror, P = polarization rotator, pBS = polarizing beam splitter, mB = movable blocker; red = vertically polarized component, blue = horizontally polarized component

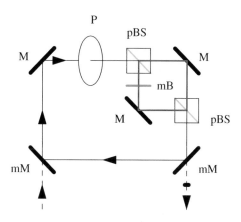

We start from a purely horizontally-polarized initial state. If no object is in the PMZI, the plane of polarization is finally vertical due to N rotations of the polarization plane by $\frac{\pi}{2N}$; if there is a obstacle, the interference is disturbed and the final state has e.g. only a horizontal component.

In this case also, the formal description is quite simple. The basis vectors (linear horizontal and vertical polarization), are as usual[75]

$$|h\rangle = \begin{pmatrix} 1 \\ 0 \end{pmatrix}, \ |v\rangle = \begin{pmatrix} 0 \\ 1 \end{pmatrix} \tag{L.10}$$

The polarizer (rotation of the polarization plane by the angle α) can be represented by

$$\begin{pmatrix} a \\ b \end{pmatrix} \rightarrow \begin{pmatrix} \cos\alpha & -\sin\alpha \\ \sin\alpha & \cos\alpha \end{pmatrix} \begin{pmatrix} a \\ b \end{pmatrix} \tag{L.11}$$

and the combined action of PBS and obstacle which blocks the vertical component in the PMZI is given by

$$\begin{pmatrix} a \\ b \end{pmatrix} \rightarrow \begin{pmatrix} 1 & 0 \\ 0 & \delta \end{pmatrix} \begin{pmatrix} a \\ b \end{pmatrix}, \ \delta \in C \tag{L.12}$$

with

$$\begin{aligned} \delta = 1 &: \text{ without obstacle} \\ \delta = 0 &: \text{ with obstacle.} \end{aligned} \tag{L.13}$$

We summarize the effects of polarizer, PBS and obstacle (the mirrors need not be considered because they produce the same phase shift for both polarization components), and obtain for one iteration

---

[75]For simplicity we dispense here with the distinction between $\cong$ and $=$.

$$\begin{pmatrix} a \\ b \end{pmatrix} \rightarrow \begin{pmatrix} 1 & 0 \\ 0 & \delta \end{pmatrix} \begin{pmatrix} \cos\alpha & -\sin\alpha \\ \sin\alpha & \cos\alpha \end{pmatrix} \begin{pmatrix} a \\ b \end{pmatrix}$$
$$= \begin{pmatrix} \cos\alpha & -\sin\alpha \\ \delta\sin\alpha & \delta\cos\alpha \end{pmatrix} \begin{pmatrix} a \\ b \end{pmatrix}. \tag{L.14}$$

For N iterations, each with the angle $\alpha = \frac{\pi}{2N}$, we find

$$\begin{pmatrix} a \\ b \end{pmatrix} \rightarrow \begin{pmatrix} \cos\frac{\pi}{2N} & -\sin\frac{\pi}{2N} \\ \delta\sin\frac{\pi}{2N} & \delta\cos\frac{\pi}{2N} \end{pmatrix}^N \begin{pmatrix} a \\ b \end{pmatrix} = M(N,\delta)\begin{pmatrix} a \\ b \end{pmatrix}. \tag{L.15}$$

The matrix $M(N,\delta)$ can readily be calculated for the special cases of $\delta = 1$ and $\delta = 0$.[76] It follows that

$$M(N,\delta = 1) = \begin{pmatrix} \cos\frac{\pi}{2} & -\sin\frac{\pi}{2} \\ \sin\frac{\pi}{2} & \cos\frac{\pi}{2} \end{pmatrix} = \begin{pmatrix} 0 & -1 \\ 1 & 0 \end{pmatrix} \tag{L.16}$$

and

$$M(N,\delta = 0) = \left(\cos\frac{\pi}{2N}\right)^{N-1}\begin{pmatrix} \cos\frac{\pi}{2N} & -\sin\frac{\pi}{2N} \\ 0 & 0 \end{pmatrix}. \tag{L.17}$$

For a purely horizontally-polarized initial state, we have in the absence of an obstacle ($\delta = 1$)

$$\begin{pmatrix} 1 \\ 0 \end{pmatrix} \rightarrow M(N,\delta = 1)\begin{pmatrix} 1 \\ 0 \end{pmatrix} = \begin{pmatrix} 0 \\ 1 \end{pmatrix}. \tag{L.18}$$

Thus, the original horizontal polarization is completely converted into *vertical* polarization.

With an obstacle ($\delta = 0$), we have instead:

$$\begin{pmatrix} 1 \\ 0 \end{pmatrix} \rightarrow M(N,\delta = 0)\begin{pmatrix} 1 \\ 0 \end{pmatrix} = \left(\cos\frac{\pi}{2N}\right)^N\begin{pmatrix} 1 \\ 0 \end{pmatrix}. \tag{L.19}$$

Thus, the original horizontal polarization is *completely conserved*. For sufficiently large N, we have $\left(\cos\frac{\pi}{2N}\right)^{2N} \approx 1 - \frac{\pi^2}{8N}$. The term $\frac{\pi^2}{8N}$ describes the 'loss', i.e. the absorption by the obstacle; this part can in principle be made arbitrarily small for $N \rightarrow \infty$.

In summary: The experimental arrangement makes it possible to determine the presence of an obstacle in an 'interaction-free' manner, substantially more efficiently than with the Mach–Zehnder setup from Chap. 6. There, the 'bomb test' worked in 25% of the cases; here, the percentage is $\left(\cos\frac{\pi}{2N}\right)^N \approx 1 - \frac{\pi^2}{8N}$.

---

[76]Partly transparent obstacles ($\delta \neq 0, 1$) are discussed in J. Pade and L. Polley, 'Wechselwirkungs-freie Quantenmessung', Physik in der Schule 38/5 (2000) 343, ('interaction-free quantum measurement', Physics in School 38/5 (2000) 343).

In practice, of course, N cannot be made arbitrarily large due to various experimental difficulties (e.g. the components are not ideal, there is some absorption, etc.). However, numbers such as $N \approx 15$ can be attained.[77]

[77]P.G. Kwiat et al., High-efficiency quantum interrogation measurements via the quantum Zeno effect, http://de.arxiv.org/abs/quant-ph/9909083.

Appendix M
Delayed Choice and the Quantum Eraser

The experiments discussed in this appendix are all based on the Mach–Zehnder interferometer (MZI). They show that the experimental setup or the observation, respectively, will decide whether a quantum object will behave (mainly) as a particle or (mainly) as a wave. Here, the so-called *which-way information* is crucial: if one can distinguish and identify the paths taken, then the photons behave like particles (no interference); if the paths are indistinguishable, then their behavior is like that of waves (interference). The consequences of the experiments lead right up to the question of whether we must also take into account a time-reversed effect of events. Since the experiments are relatively simple, they are found more and more frequently in textbooks and curricula for physics at the school level.[78]

M.1 Delayed Choice Experiments

The term 'delayed-choice experiment' (or 'experiment with delayed decision') denotes an experimental setup where it is decided whether one will allow (self-)interference or not, or which variables are measured, only *in the course of* the experiment. Proposed in 1978 as a thought experiment by John Archibald Wheeler, the effect has been confirmed experimentally in the meantime, including a measurement based on a setup similar to a Mach–Zehnder interferometer.[79]

---

[78]See e.g. the many hits of an internet search using the keywords 'quantum eraser' and 'school'.

[79]See V. Jacques et al., 'Experimental Realization of Wheeler's Delayed-choice Gedanken Experiment?, Science 315, 966 (2007), and references therein. In the experiment cited, polarization beam splitters are used instead of simple beam splitters. Recently, a delayed-choice experiment was performed with single photons, where the counters are also quantum objects and not (as usual) classical detecting devices; see Jian-Shun Tang et al., 'Realization of Wheeler's delayed-choice quantum experiment', *Nature Photonics* 6, 600–604 (2012), https://doi.org/10.1038/nphoton.2012.179.

© Springer Nature Switzerland AG 2018
J. Pade, *Quantum Mechanics for Pedestrians 1*, Undergraduate Lecture Notes in Physics, https://doi.org/10.1007/978-3-030-00464-4

Fig. M.1 Delayed choice—removing the second beam splitter

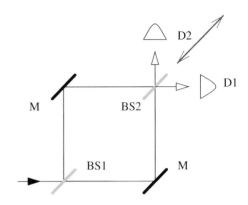

M.1.1 Setup 1

The basic idea: As in Chap. 6, we have a MZI setup through which a single photon passes. While the photon is in the apparatus, the second beam splitter BS2 can be removed or inserted, as shown in Fig. M.1, and this in such a way that an 'information' to the photon would have to be sent with a superluminal velocity.[80]

There are four ways to perform the experiment, which we denote by M1-M4. In the first two, the beam splitter BS2 is inserted or removed at the start and remains so during the entire experiment

M1: The photon is incident; BS2 remains inserted. Then, because of (self-) interference, only detector D1 is activated (D2 remains silent), and we cannot say which one of the two paths the photon has taken (wave nature).

M2: The photon is incident; BS2 remains removed. Then, with 50% probability, either D1 or D2 is activated. There is no interference and we can say clearly which path the photon has taken (particle nature).

The next two methods involve removing or inserting the second beam splitter BS2, *after* the photon has passed the first beam splitter (and possibly the mirror); these are the delayed decisions.[81]

M3: The photon is incident; BS2 is inserted. After the photon has passed BS1 and M, one removes BS2. With 50% probability, D1 or D2 is activated (particle nature).

M4: The photon is incident; BS2 is removed. After the photon has passed BS1 and M, one inserts BS2. Only D1 is activated, D2 remains silent (wave nature).

---

[80] In theory, we can assume identical optical paths, perfect 90° angles, etc.; as shown in Chap. 6, this leads to the result that with BS2 inserted, detector D1 is always activated and D2 never. In a real experiment, these ideal conditions are not found; the path difference therefore depends e.g. on the angle. When using laser light, with BS2 inserted we in fact obtain interference fringes on screens which are placed at the positions of the detectors; if BS2 is removed, we see only a 'bright spot' on the screen.

[81] The removal and insertion of BS2 can be delayed arbitrarily—it just has to occur before the photon arrives at the position of BS2.

It is interesting to consider M3 (or M4) more closely. The photon enters the MZI and passes BS1 and M, then BS2 is inserted. Consequently, as in M1, the photon has to explore *both* paths, since otherwise interference could not occur in principle (the photon cannot 'know' that we will remove BS2 a moment later). Therefore, it has to be in a coherent superposition state (wave nature).

Now we remove BS2, immediately before the photon passes this point. It will then end up in one of the two detectors, and we can tell which way it has gone (particle nature). Thus, the photon cannot be in a coherent superposition state—contrary to what we just said. From a classical point of view, we can resolve the conflict only if we assume that the photon, when entering BS1, already knows whether BS2 will remain or will be removed—i.e. it had to know the future. The delayed choice seems to cause an effect on the events in the past. If this interpretation is wrong, where should we look for the error?

The usual answer is that we cannot say anything about how the photon propagates in the MZI ('one path' or 'two-path') before an appropriate measurement is undertaken. Prior to the measurement, there is nothing that can be associated with such a which-way statement. Thus, the question of which path the photon takes is not meaningful before a measurement (and this also applies to the above argument, insofar as it is based on the observation of the paths taken).[82] There are questions that simply are not meaningful, just for the reason that we cannot answer them *in principle*.

On the other hand, there are voices that propose considering a time-reversed effect of events.[83] Actually, the fundamental laws of physics are all symmetric with respect to time reversal, and do not reflect the time-asymmetric notion of cause and effect.

However this discussion may turn out, we see that a photon is not just a particle or a wave, but something else (i.e. a quantum object), which can be forced to behave like a particle or like a wave only by performing a measurement.

M.1.2   Setup 2

A variant of the experimental setup, as shown in Fig. M.2, leaves both beam splitters in place, but one can introduce additional detectors (D3 and D4) into the paths of the photon streams. For instance, if we insert D3 (D4 remains outside), we have the information about which path the photon has taken.

The delayed choice here is to bring D3 and/or D4 into the paths (or to remove them) after the photon has passed the first beam splitter and the mirrors. The argument is analogous to that used in setup 1.

---

[82]"The past has no existence except as it is recorded in the present." (J.A. Wheeler, in Mathematical Foundations of Quantum Theory (ed A.R. Marlow), 9–48 (Academic, New York, 1978).

[83]See also the discussion in Chap. 27, Vol. 2 on locality and reality in quantum mechanics.

Fig. M.2 Delayed
Choice—inserting additional
detectors

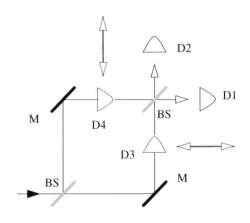

M.2 The Quantum Eraser

A quantum eraser can generally be understood as an experimental setup with which
one can delete information (i.e. 'erase' it) about the course of an experiment. Specifically, this usually concerns the possibility of restoring the ability to interfere (in
retrospect, so to speak) by the destruction of information. A quite simple example
is an MZI with a beam splitter (for example only BS1) as shown in Fig. M.1, which
delivers the information about which path the photon has taken through the apparatus. If we insert a second beam splitter BS2, we lose this information—it will be
rubbed out, so to speak.

   A slightly more elaborate setup is shown in Fig. M.3. Is an ideal MZI with fixed
beam splitters into which adjustable polarizers are inserted. Initially, the polarizers
P3 and P4 are not in the paths. The state entering the MZI is horizontally polarized.
If the polarizers are all set to zero, we have the usual finding that only D1 and not D2
is activated (interference, wave character, no path information). If we rotate P1 to

Fig. M.3 Quantum eraser

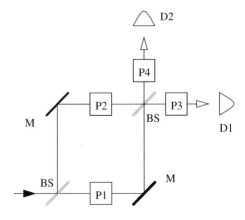

$+45°$ and P2 by $-45°$, we imprint on the photons a which-way information; there is no more interference and D1 and D2 are activated with 50% probabilities (whereby, of course, due to the setting of the polarizers, the number of transmitted photons or the intensity decreases to half its original value). Now we insert P3 and P4 into the paths, say with the setting $0°$. Hence, only one half of the $\pm45°$-polarized photons can pass detectors D3 and D4—but these photons are now capable of interference. Accordingly, we have again the finding that D1 is activated, while D2 is silent. In effect, from a classical point of view, we have therefore deleted the which-way information by means of the settings of P3 and P4, even though it was already present.

One can make a delayed choice, of course, by setting P3 and P4 only if the photon has passed the first beam splitter. Here also, one can delete the path information. Thus, this experiment again indicates that it is not meaningful to speak about physical reality without performing a measurement.

Appendix N
The Equation of Continuity

In the following, the derivation of the equation of continuity is briefly recapitulated, using as an example the mass density.

Mass and mass density are connected by the differential relation $dm = \rho dV$ or its integral formulation

$$M = \int_G \rho dV, \tag{N.1}$$

where the integration is over a certain closed (fixed) volume G. We want to exclude all processes in the following which destroy or create mass; the total mass in G can thus change only by mass transport through the surface of G. This formulation relates to the region G as a whole, and is therefore a global or integral principle. The local (that is, for a particular space-time point) or differential formulation is given by the *continuity equation*, which we will now derive.

The change of the mass with time is given by

$$\frac{d}{dt} M = \frac{d}{dt} \int_G \rho dV = \int_G \frac{\partial \rho}{\partial t} dV. \tag{N.2}$$

According to our assumption, it can occur only by means of mass transport through the surface ∂G of the volume G. With the usual definition of the current density $\mathbf{j}$ as $\mathbf{j} = \rho \mathbf{v}$ (magnitude of current density = mass flow per unit time through a unit area), we obtain

$$\int_G \frac{\partial \rho}{\partial t} dV = - \int_{\partial G} \mathbf{j} \cdot d\mathbf{A}. \tag{N.3}$$

Here, $d\mathbf{A}$ is an oriented surface element (oriented to the outside in the case of a closed volume); the minus sign indicates that the mass within the volume G decreases if there is an outward flow from G. This equation can be regarded as an integral formulation of the continuity equation. To arrive at the differential formulation, we transform

© Springer Nature Switzerland AG 2018
J. Pade, *Quantum Mechanics for Pedestrians 1*, Undergraduate Lecture
Notes in Physics, https://doi.org/10.1007/978-3-030-00464-4

the surface integral into a volume integral by using the Gaussian integral theorem (Appendix D, Vol. 1):

$$\int_G \frac{\partial \rho}{\partial t} dV = -\int_{\partial G} \mathbf{j} \cdot d\mathbf{A} = -\int_G \text{div} \mathbf{j} \, dV \equiv -\int_G \nabla \cdot \mathbf{j} \, dV$$

$$\text{or } \int_G \frac{\partial \rho}{\partial t} dV = -\int_G \nabla \cdot \mathbf{j} \, dV \qquad (N.4)$$

Since the last equation holds for any *arbitrary* volume G, the integrands must be equal and it follows that:

$$\frac{\partial \rho}{\partial t} + \nabla \cdot \mathbf{j} = 0 \qquad (N.5)$$

with $\rho = \rho(\mathbf{r}, t)$ and $\mathbf{j} = \mathbf{j}(\mathbf{r}, t)$. This is the differential formulation of the conservation of mass, called the continuity equation. Moreover, this equation is not only valid for the mass density, but also e.g. for the charge density and any other density subject to an integral conservation law.

Appendix O
Variance, Expectation Values

O.1 Variance, Moments

If one measures a quantity x several times (e.g. the duration of an oscillation, the lifetime of a radioactive nucleus, etc.), one generally obtains different values x_n with a relative frequency of occurrence[84] f_n. The *mean value* is then given by

$$\bar{x} = \langle x \rangle = \sum_n f_n x_n. \tag{O.1}$$

The mean value tells us nothing about how much the data spread; very different sets of data can yield the same mean value; cf. Fig. O.1.

One might think of taking the sum of the deviations of the data points from the mean value as a measure of the dispersion, i.e. $\sum_n f_n |x_n - \langle x \rangle|$. This idea is quite correct in itself, but has certain disadvantages. Thus, one sums instead first the squares of the deviations and takes the square root afterwards:

$$\sigma^2 = \sum_n f_n (x_n - \langle x \rangle)^2 \; ; \; \sigma = \sqrt{\sum_n f_n (x_n - \langle x \rangle)^2}. \tag{O.2}$$

We cast this in a more pleasing form:

$$
\begin{aligned}
\sigma^2 &= \sum_n f_n (x_n - \langle x \rangle)^2 = \sum_n f_n \left[x_n^2 - 2x_n \langle x \rangle + \langle x \rangle^2 \right] \\
&= \sum_n f_n x_n^2 - 2 \langle x \rangle \sum_n f_n x_n + \langle x \rangle^2 \sum_n f_n \\
&= \langle x^2 \rangle - 2 \langle x \rangle^2 + \langle x \rangle^2 = \langle x^2 \rangle - \langle x \rangle^2.
\end{aligned}
\tag{O.3}
$$

---

[84] Also called weight. The reliability of data can be characterized by the weights; a lower weight is allocated to less reliable data than to reliable data. It must hold that $\sum_n f_n = 1$.

© Springer Nature Switzerland AG 2018
J. Pade, *Quantum Mechanics for Pedestrians 1*, Undergraduate Lecture
Notes in Physics, https://doi.org/10.1007/978-3-030-00464-4

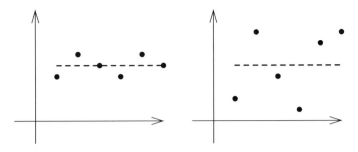

Fig. O.1 Data with different scatter but with the same mean value

The quantity σ^2 is called the *variance*; σ is the *standard deviation* (or mean square deviation, root deviation, dispersion).

Generalizing these concepts, one defines the *Nth moment* by

$$\langle x^N \rangle = \sum_n f_n x_n^N, \tag{O.4}$$

and the *central Nth moment* as $\langle (x - \langle x \rangle)^N \rangle$.

O.2 Expectation Value, Mean Value

These two terms are often used synonymously (not only in quantum mechanics), but strictly speaking, we should note that the mean value relates to a record of past data and is formulated in terms of relative frequencies, while the expectation value is meant to predict future occurrences in terms of probabilities. As an illustrative example, consider random dice throws, repeated 18 times in a (hypothetical) test series.:

Spots a_n	1	2	3	4	5	6
Probability w_n	$\frac{1}{6}$	$\frac{1}{6}$	$\frac{1}{6}$	$\frac{1}{6}$	$\frac{1}{6}$	$\frac{1}{6}$
Number of throws	3	4	2	2	3	4
Relative frequency f_n	$\frac{1}{6}$	$\frac{2}{9}$	$\frac{1}{9}$	$\frac{1}{9}$	$\frac{1}{6}$	$\frac{2}{9}$

Thus we obtain the expectation value $E = \sum_{n=1}^{6} w_n a_n = 3.5$ and the mean value $M = \sum_{n=1}^{6} f_n a_n = 3.56$.

O.3 Discrete and Continuous[85]

In the discrete case, the mean value is given by

$$\langle x \rangle = \sum_n f_n x_n. \tag{O.5}$$

In the continuous case, the summation becomes as usual an integration, and we find

$$\langle x \rangle = \int \rho(x) x \, dx \tag{O.6}$$

with the density function $\rho(x)$. For the variance, it follows accordingly:

$$\sigma^2 = \langle x^2 \rangle - \langle x \rangle^2 = \int \rho(x) x^2 \, dx - \left(\int \rho(x) . x \, dx \right)^2 \tag{O.7}$$

In a physical terms, this concept is familiar from the first semester on. The center of mass[86] $\mathbf{R}$ of a set of point masses, each with mass m_i at the position $\mathbf{r}_i$, is defined by

$$\mathbf{R} = \frac{\sum m_i \mathbf{r}_i}{\sum m_i} = \frac{\sum m_i \mathbf{r}_i}{M} = \sum \frac{m_i}{M} \mathbf{r}_i. \tag{O.8}$$

For a continuous mass distribution or a mass density $\rho(\mathbf{r})$, it follows with $dm = \rho dV$ (as always the integrals are over the entire domain of definition) that:

$$\mathbf{R} = \frac{\int \rho(\mathbf{r}) \mathbf{r} \, dV}{\int \rho(\mathbf{r}) \, dV} = \frac{\int \rho(\mathbf{r}) \mathbf{r} \, dV}{M} = \int \frac{\rho(\mathbf{r})}{M} \mathbf{r} \, dV. \tag{O.9}$$

O.4 Standard Deviation in Quantum Mechanics

The following are some remarks about the standard deviation in quantum mechanics.

O.4.1 Example: Two-State System

We calculate the dispersion for the example of the Pauli matrix σ_z with eigenvalues $\lambda_{1,2}$ and eigenvectors $v_{1,2}$:

---

[85] See also the chapter 'Discrete-continuous' in Appendix T, Vol. 1.
[86] Conventionally, one writes $\mathbf{R}$ instead of $\langle \mathbf{r} \rangle$.

$$\sigma_z = \begin{pmatrix} 1 & 0 \\ 0 & -1 \end{pmatrix}; \quad \lambda_{1,2} = \pm 1; \quad v_1 = \begin{pmatrix} 1 \\ 0 \end{pmatrix}; \quad v_2 = \begin{pmatrix} 0 \\ 1 \end{pmatrix}. \tag{O.10}$$

If the system is in the state v_1, its spin component has the value $+1$, and in the state v_2, it has the value -1. For all other states, a definite value cannot be assigned.

We now calculate the dispersion. Because of $\sigma_z^2 = 1$, we find that

$$(\Delta\sigma_z)^2 = \langle \sigma_z^2 \rangle - \langle \sigma_z \rangle^2 = 1 - \langle \sigma_z \rangle^2. \tag{O.11}$$

For a normalized state, it then holds that

$$(\Delta\sigma_z)^2 = 1 - \left[(a^* b^*) \begin{pmatrix} 1 & 0 \\ 0 & -1 \end{pmatrix} \begin{pmatrix} a \\ b \end{pmatrix} \right]^2 = 1 - \left(|a|^2 - |b|^2 \right)^2 = 4|a|^2 \left(1 - |a|^2 \right). \tag{O.12}$$

Hence, the standard deviation vanishes for the eigenvectors $v_{1,2}$; for all other states, it is in principle not zero. It is therefore a measure of the extent to which a system does *not* have a value for σ_z (i.e. one of the two allowed values ± 1).

O.4.2 General Case

We assume a Hermitian operator A with eigenvectors $|a_n\rangle$ and eigenvalues a_n:

$$A |a_n\rangle = a_n |a_n\rangle. \tag{O.13}$$

The variance for a normalized state $|\psi\rangle$ is then given by

$$\left(\Delta_\psi A \right)^2 = \langle A^2 \rangle_\psi - \langle A \rangle_\psi^2 = \langle \psi | A^2 |\psi\rangle - \langle \psi | A |\psi\rangle^2. \tag{O.14}$$

We want to show that it vanishes iff $|\psi\rangle$ is an eigenstate of A. To this end we use the Schwarz inequality in the form:

$$|\langle a | b \rangle|^2 \leq \langle a | a \rangle \langle b | b \rangle, \tag{O.15}$$

where the equality holds iff $|a\rangle$ and $|b\rangle$ are collinear, i.e. $|a\rangle \sim |b\rangle$.

Because of the Hermiticity of A, we have $\langle \psi | A^2 |\psi\rangle = \langle A\psi | A\psi \rangle$; in addition, $\langle \psi | A |\psi\rangle^* = \langle \psi | A |\psi\rangle$, and consequently $\langle \psi | A |\psi\rangle^2 = |\langle \psi | A\psi\rangle|^2$. We identify $|a\rangle$ with $|\psi\rangle$ and $|b\rangle$ with $|A\psi\rangle$; then the Schwarz inequality reads

$$|\langle \psi | A\psi\rangle|^2 = \langle \psi | A |\psi\rangle^2 \leq \langle \psi | \psi \rangle \langle A\psi | A\psi \rangle = \langle \psi | A^2 |\psi\rangle. \tag{O.16}$$

The equals sign applies iff $|\psi\rangle \sim A |\psi\rangle$ is valid—in other words, if $|\psi\rangle$ is an eigenfunction of A.

Appendix P
On Quantum Cryptography

P.1 Verification of the Key

After the procedure was performed as described in Chap. 10, two questions still need to be answered: (1) How can Alice and Bob ascertain that they have the same key, and how they can reliably eliminate discrepancies that may occur? (2) How can Alice and Bob ensure that no one but themselves, especially not Eve, possesses this improved key? These two problems are solved using classical (i.e. non-quantum-mechanical) multistage processes, which offer again not absolute, but in a practical sense adequate security. In each stage, the length of the key is reduced. It should again be emphasized that the following exchange of information between Alice and Bob is in every phase *public*. This can, in principle, of course, pose a problem if Eve exercises absolute control over the public channel. She could then pick up every message from Alice and Bob, modify it suitably and finally retransmit it. For example, she could lead them to believe that the error rate is zero. However, if Alice and Bob have remained watchful and adhere to the procedures described below, they can avoid this trap.

We consider once more the whole process. The transfer of N_A photons with fixed time intervals from Alice to Bob has ended. The first phase is then that Bob tells Alice at which times he did not receive a signal, although it was supposed to arrive. Both eliminate these dark counts and now each has a key of the same length N, the *raw key*. Now they compare publicly the settings of their polarizers, thereby excluding all measurements where the basis systems do not match. This key is often called the *sifted key*; it has a length of $n \approx N/2$.

The next step is the estimation of the rate of eavesdropping e. Suppose Eve spied on each $q-th$ bit. Then the two keys of Alice and Bob differ at approximately $n/(4q)$ sites; the interception rate is $e = 1/(4q)$. To estimate this number, Alice and Bob compare publicly t individual bits of their two keys and then delete them. Of course, the eavesdropping rate should not be too high; if it is above an agreed threshold, the whole key is discarded and the process is restarted again. With a threshold of e.g. 12.5%, it is guaranteed that Eve has at most spied on every second photon. Alice and Bob are now in possession of keys (also called the *plain key*) of length $n - t$ and

© Springer Nature Switzerland AG 2018
J. Pade, *Quantum Mechanics for Pedestrians 1*, Undergraduate Lecture
Notes in Physics, https://doi.org/10.1007/978-3-030-00464-4

have at their disposal an estimate of the rate of interception, e. Eve knows $(n - t) 3e$ bits of Alice's key.

In the next stage, Alice and Bob make sure that they eliminate the bits distorted by Eve, and will therefore obtain the same key, called the *reconciled key*. This can be done by various methods, which can also be carried out in series. The common feature is that not all individual bits are compared, but only some properties of subsets of the key, e.g. the parity of these subsets.

For example, Alice and Bob can choose publicly a random permutation of their key. This series is then cut into blocks of length l, where l is selected in such a way that the probability of the occurrence of two or more errors per block is sufficiently small. Another method is quite similar; here Alice and Bob publicly extract random series from their keys and use them to make blocks of length l. The parity of these blocks of length l is then compared publicly. If it does not coincide, a binary search is started for the wrong bit, i.e. halving the block, comparing the parities of the halves, again bisecting the block with the different parities, etc. up to a certain minimum length of the sub-blocks of different parities; this piece of the key is then eliminated. In this way, all l blocks are worked through. Subsequently, the next permutation or the next random set is selected and the comparison is started again; the entire process is performed several times. In all, d bits will be removed in this procedure. By the way, a nice side effect is that Alice and Bob get a confirmation of the error rate estimated in the first phase—if not, then something is suspect and the key will be discarded.

Once these procedures have been run through, one can assume, finally, that it is highly probable that the remaining key is without error, i.e. that the keys of Alice and Bob match in every position. Due to its construction, this improved or reconciled key has the length $n_v = n - t - d$. We note that it is only a partially secret key, because Eve knows $(n - t - d) 3e$ bits.

In the last stage, Alice and Bob make sure that this 'flaw' is corrected and the partially secret key becomes a totally secret one. This process is called *privacy amplification*. It can work like this: Alice and Bob have the reconciled key of length n_v and know that Eve knows about t_v bits of this key. Let $0 < s_v < n_v - t_v$ be a safety parameter, and let $r_v = n_v - t_v - s_v$. Then, Alice and Bob may chose e.g. r_v subsets of the reconciled key. The parities of these subsets are determined, and this is *not done publicly*. They then make up the final secret key. One can show that Eve's information about this key is at most $2^{-s_v} / \ln 2$.

We suppose, for example, that Alice sends 500 photons to Bob. After the elimination of the dark counts and the comparison of the basis systems, a key of say $n = 233$ positions remains. To estimate the interception rate, $t = 48$ bits are used. We assume that Eve's interception rate e is $e = 0.02$ or 2%. The key now has a length $n - t = 185$ bits. To construct the reconciled key by comparing the parities, we have to drop say $d = 67$ bits. The key now has a length of $n_v = n - t - d = 118$; Eve knows $118 \cdot 0.06 \approx 7$ bits. We finally choose, for example, $s_v = 10$ and make up the final key using the parities of 108 random subsets. An explicitly specified short example illustrates the situation.

P.2 An Example

Alice sends 64 photons to Bob. Obviously, this is a very small number for such purposes, as becomes apparent, *inter alia*, from the fact that the statistical errors are quite large. But for a short and concise toy example we accept this disadvantage.

Eve spies on every second photon; thus, $q = 2$ and $e = \frac{1}{4q} = \frac{1}{8}$. As an example, this might appear as shown in Table P.1.

The first and second columns indicate the polarization direction and polarization value of the photons sent by Alice. The third column denotes the polarization direction chosen by Eve. If two numbers appear in the fourth column, the result of measurement by Eve is 0 or 1. The fifth column shows one of Eve's possible measurement series. Analogous remarks apply to Bob's columns. Since there are no erroneous readings, Alice and Bob are now in possession of their raw keys.

Alice and Bob publicly compare their polarization directions and delete all the results with different settings. The result (sifted key) is given in Table P.2.

A glance at the table shows that Alice and Bob have different entries in three places; the predicted value is (see above) $\frac{n}{4q} = 31 \cdot e = \frac{31}{8} = 3.9$. In addition, we see that the keys of Alice and Eve match in 12 locations; the predicted value is $31 \cdot 3e = \frac{93}{8} = 11.6$. Alice and Bob cannot, of course, look at this table, but they need to estimate the error rate. To this end, they compare publicly e.g. 7 out of these 31 bits, at the positions 4, 8, 12, 16, 20, 24, 28. There is one deviation (bit 4); the error rate can therefore be estimated to be $e \approx \frac{1}{7}$. Accordingly, Eve has eavesdropped on about every second bit ($\frac{7}{4} \approx 2$). The checked bits are deleted,[87] and we obtain Table P.3 (plain key).

Alice and Bob have different entries in two places; the predicted value is $24 \cdot e = \frac{24}{8} = 3$. The keys of Alice and Eve agree in 10 positions; the predicted value is $24 \cdot 3e = \frac{72}{8} = 9$. In order to eliminate the influence of Eve, Alice and Bob now compare the parities of subsets. For the sake of simplicity, we choose in our toy example the consecutive blocks of length 4 as subsets. The two 4-blocks (5–8) and (9–12) have different parities, whereas all other blocks match. Halving the 'wrong' 4-blocks shows that the two 2-blocks 7, 8 and 9, 10 have different parities; they are deleted. Table P.4 shows the reconciled key.

Alice and Bob now have identical keys. Eve's key agrees with theirs at seven positions; the predicted value is $20 \cdot 3e = \frac{60}{8} = 7.5$, while at 11 positions (predicted value 10), Eve has no information.

The method outlined above for privacy amplification starts from $n_v = 20$ and $t_v = 7.5$. For the security parameter, we have $0 < s_v < 12.5$. We choose $s_v = 3.5$ and obtain $r_v = n_v - t_v - s_v = 9$. Alice and Bob choose $r_v = 9$ subsets of the reconciled key (there are $\binom{20}{9} = 167,960$ of length 9); the parities (not made public) of these subsets are then the final secret key. Eve's information about this key is not more than 13%.

---

[87]We see that the key must not be too short.

Table P.1 Initial data

	A	A	E	E	E	B	B	B		A	A	E	E	E	B	B	B
1	⊞	1				⊠	01	0	33	⊠	0				⊠	0	0
2	⊠	0	⊠	0	0	⊠	0	0	34	⊞	1	⊞	1	1	⊠	01	1
3	⊞	0				⊠	01	0	35	⊞	1				⊠	01	1
4	⊞	1	⊠	01	0	⊞	01	1	36	⊞	0	⊠	01	1	⊞	01	0
5	⊠	1				⊠	1	1	37	⊞	0				⊞	0	0
6	⊠	0	⊞	01	1	⊠	01	1	38	⊞	0	⊞	0	0	⊞	0	0
7	⊠	1				⊞	01	0	39	⊞	0				⊠	01	1
8	⊞	0	⊞	0	0	⊞	0	0	40	⊞	1	⊞	1	1	⊠	01	1
9	⊞	1				⊞	1	1	41	⊞	0				⊠	01	0
10	⊞	1	⊞	1	1	⊠	01	1	42	⊞	1	⊞	1	1	⊠	01	1
11	⊠	0				⊠	0	0	43	⊠	0				⊞	01	0
12	⊞	1	⊠	01	0	⊞	01	1	44	⊞	0	⊠	01	1	⊠	1	1
13	⊠	0				⊞	01	1	45	⊠	1				⊠	1	1
14	⊠	1	⊠	1	1	⊠	1	1	46	⊞	1	⊞	1	1	⊞	1	1
15	⊞	0				⊠	01	0	47	⊞	0				⊞	0	0
16	⊠	1	⊞	01	1	⊠	01	0	48	⊠	0	⊠	0	0	⊞	01	0
17	⊞	0				⊠	01	0	49	⊠	0				⊞	01	1
18	⊞	0	⊞	0	0	⊞	0	0	50	⊞	1	⊞	1	1	⊠	01	1
19	⊠	0				⊞	01	1	51	⊠	1				⊞	01	0
20	⊠	0	⊠	0	0	⊞	01	0	52	⊠	0	⊞	01	0	⊠	01	0
21	⊠	0				⊠	0	0	53	⊞	0				⊠	01	0
22	⊞	0	⊠	01	1	⊞	01	1	54	⊞	1	⊞	1	1	⊠	01	0
23	⊞	1				⊠	01	1	55	⊞	1				⊞	1	1
24	⊞	0	⊞	0	0	⊞	0	0	56	⊞	1	⊞	1	1	⊠	01	1
25	⊠	0				⊠	0	0	57	⊞	1				⊞	1	1
26	⊠	1	⊞	01	0	⊞	0	0	58	⊠	0	⊠	0	0	⊠	0	0
27	⊠	0				⊞	01	1	59	⊞	1				⊞	1	1
28	⊞	1	⊞	1	1	⊞	1	1	60	⊠	1	⊠	1	1	⊞	01	1
29	⊞	1				⊠	01	1	61	⊠	1				⊠	1	1
30	⊞	1	⊠	01	0	⊠	0	0	62	⊞	1	⊠	01	0	⊠	0	0
31	⊠	0				⊠	0	0	63	⊞	0				⊠	01	0
32	⊠	0	⊠	0	0	⊞	01	0	64	⊞	1	⊞	1	1	⊞	1	1

Table P.2 Sifted key

	A	E	B		A	E	B
1	0	0	0	17	0		0
2	1	0	1	18	0		0
3	1		1	19	0	1	0
4	0	1	1	20	0		0
5	0	0	0	21	0	0	0
6	1		1	22	1		1
7	0		0	23	1	1	1
8	1	0	1	24	0		0
9	1	1	1	25	0	0	0
10	1	1	0	26	1		1
11	0	0	0	27	1		1
12	0		0	28	0	0	0
13	0	1	1	29	1		1
14	0	0	0	30	1		1
15	0		0	31	1	1	1
16	1	1	1				

Table P.3 Plain key

	A	E	B		A	E	B
1	0	0	0	13	0		0
2	1	0	1	14	0		0
3	1		1	15	0	1	0
4	0	0	0	16	0	0	0
5	1		1	17	1		1
6	0		0	18	1	1	1
7	1	1	1	19	0	0	0
8	1	1	0	20	1		1
9	0	0	0	21	1		1
10	0	1	1	22	1		1
11	0	0	0	23	1		1
12	0		0	24	1	1	1

Table P.4 Reconciled key

	A	E	B		A	E	B
1	0	0	0	11	0	1	0
2	1	0	1	12	0	0	0
3	1		1	13	1		1
4	0	0	0	14	1	1	1
5	1		1	15	0	0	0
6	0		0	16	1		1
7	0	0	0	17	1		1
8	0		0	18	1		1
9	0		0	19	1		1
10	0		0	20	1	1	1

Table P.5 Final key

	1	2	3	4	5	6	7	8	9	10	11	12	13	14	15	16	17
A, B	0	1	0	0	0	1	0	1	0	1	0	1	0	0	1	1	0

This is a conservative estimate. If we proceed less stringently, we can choose, for example, $20 - 3 = 17$ random subsets (of which there are $\binom{20}{17} = 1,140$ of length 17), and determine their parities. In principle, one can also choose smaller subsets; it is important only that the subsets are large enough that Eve has, on average, always at least one 'flaw'. To make life easy here, we assume e.g. that the first six are from neighboring triplets 1–3, etc., the next five from groups of four, the next four from groups of five, the last two from the first two groups of six (a schematic approach like this is all right for our toy example, but not of course for serious cryptography). We see this final key in Table P.5.

There is not a single parity which Eve can determine exactly, since she always misses the information by at least one bit. Of course she can guess—but in doing so, she is in the same situation as if she had not tried to spy on the key. Thus, in our toy example, Alice and Bob have a common secret key. Quantum mechanics makes it possible.

Appendix Q
Schrödinger Picture, Heisenberg Picture, Interaction Picture

Q.1 Schrödinger and Heisenberg Picture

The Schrödinger equation in the form that we have used has time-dependent solutions or states, while operators such as the angular momentum do not depend on time. This type of description is called the *Schrödinger picture*. But there are also other forms for the state description, for example the *Heisenberg picture*, in which the states are constant and the operators change with time. In the Schrödinger picture, the time-variable state is given by

$$|\Psi(t)\rangle = e^{-i\frac{Ht}{\hbar}}|\Psi(0)\rangle. \tag{Q.1}$$

In the Heisenberg picture, the same state is defined as

$$|\Psi\rangle_H = e^{i\frac{Ht}{\hbar}}|\Psi(t)\rangle. \tag{Q.2}$$

Accordingly, the operator A, which is time independent in the Schrödinger picture, becomes the time-dependent operator A_H in the Heisenberg picture, with:

$$A_H = e^{i\frac{Ht}{\hbar}}Ae^{-i\frac{Ht}{\hbar}}. \tag{Q.3}$$

For the time evolution, we have

$$i\hbar\frac{d}{dt}A_H = [A_H, H] + i\hbar\frac{\partial}{\partial t}A_H \tag{Q.4}$$

with

$$\frac{\partial}{\partial t}A_H = e^{i\frac{Ht}{\hbar}}\left(\frac{\partial}{\partial t}A\right)e^{-i\frac{Ht}{\hbar}}. \tag{Q.5}$$

© Springer Nature Switzerland AG 2018
J. Pade, *Quantum Mechanics for Pedestrians 1*, Undergraduate Lecture Notes in Physics, https://doi.org/10.1007/978-3-030-00464-4

Intuitively, the difference between the two representations corresponds to the representation with a fixed coordinate system[88] and a moving vector (Schrödinger image), as compared to a representation with a fixed vector and a coordinate system moving in a corresponding manner (Heisenberg picture).

We use the Schrödinger picture almost exclusively in this book.

Q.2 Interaction Picture

The interaction picture (or interaction representation) is a third way in addition to the Schrödinger and the Heisenberg picture. It is central in quantum field theory.

In the Schrödinger picture, the operators are time-independent and the states time-dependent, in the Heisenberg picture the operators are time-dependent and the states time-independent. A certain 'division' of the time dependence is obtained in the *interaction picture*, where both the operators and the states can be (and are usually) time-dependent. It is especially useful if the Hamiltonian can be written as the sum $H = H_0 + H_1$, where H_1 is a small term[89] compared to H_0. H_0 is called free part and H_1 interaction part. Usually, H_0 is time-independent and allows for analytical solutions.

We start in the Schrödinger picture

$$i\hbar \frac{\mathrm{d}}{\mathrm{d}t} |\psi(t)\rangle = (H_0 + H_1) |\psi(t)\rangle = H |\psi(t)\rangle . \qquad (Q.6)$$

We define states $|\psi_I(t)\rangle$ and operators $A_I(t)$ in the interaction picture by

$$|\psi_I(t)\rangle = e^{i\frac{H_0 t}{\hbar}} |\psi(t)\rangle \; ; \; A_I(t) = e^{i\frac{H_0 t}{\hbar}} A e^{-i\frac{H_0 t}{\hbar}} . \qquad (Q.7)$$

Note that in these expressions only the free Hamiltonian H_0 occurs.

To calculate the equation of motion for $|\psi_I(t)\rangle$, we start from the Schrödinger equation. With $|\psi(t)\rangle = e^{-i\frac{H_0 t}{\hbar}} |\psi_I(t)\rangle$ we obtain

$$i\hbar \frac{\mathrm{d}}{\mathrm{d}t} e^{-i\frac{H_0 t}{\hbar}} |\psi_I(t)\rangle = (H_0 + H_1) e^{-i\frac{H_0 t}{\hbar}} |\psi_I(t)\rangle . \qquad (Q.8)$$

This gives for time-independent H_0

$$i\hbar \left(-i\frac{H_0}{\hbar} e^{-i\frac{H_0 t}{\hbar}} |\psi_I(t)\rangle + e^{-i\frac{H_0 t}{\hbar}} \frac{\mathrm{d}}{\mathrm{d}t} |\psi_I(t)\rangle \right) = (H_0 + H_1) e^{-i\frac{H_0 t}{\hbar}} |\psi_I(t)\rangle \quad (Q.9)$$

---

[88]Or measuring apparatus.

[89]Hence, the interaction picture is particularly suitable for perturbation theory.

or

$$i\hbar \frac{d}{dt} |\psi_I(t)\rangle = H_I(t) |\psi_I(t)\rangle \qquad (Q.10)$$

with

$$H_I(t) = e^{i\frac{H_0 t}{\hbar}} H_1 e^{-i\frac{H_0 t}{\hbar}} |\psi_I(t)\rangle . \qquad (Q.11)$$

Vividly, the state $|\psi_I(t)\rangle$ changes much more slowly than $|\psi(t)\rangle$, since the energy associated with H_1 is small compared to that of H_0, in general.

Next, we want to determine the time evolution operator in the interaction picture. In the Schrödinger picture, see (Q.6),

$$|\psi(t)\rangle = e^{-i\frac{H(t-t_0)}{\hbar}} |\psi(t_0)\rangle = U_S(t, t_0) |\psi(t_0)\rangle . \qquad (Q.12)$$

With the definition of $|\psi_I(t)\rangle$ follows

$$e^{i\frac{H_0 t}{\hbar}} |\psi(t)\rangle = |\psi_I(t)\rangle = e^{i\frac{H_0 t}{\hbar}} e^{-i\frac{H(t-t_0)}{\hbar}} e^{-i\frac{H_0 t_0}{\hbar}} |\psi_I(t_0)\rangle \qquad (Q.13)$$

which means that the time evolution operator in the interaction picture has the form

$$U_I(t, t_0) = e^{i\frac{H_0 t}{\hbar}} U_S(t, t_0) e^{-i\frac{H_0 t_0}{\hbar}} = e^{i\frac{H_0 t}{\hbar}} e^{-i\frac{H(t-t_0)}{\hbar}} e^{-i\frac{H_0 t_0}{\hbar}} \qquad (Q.14)$$

and we have

$$|\psi_I(t)\rangle = U_I(t, t_0) |\psi_I(t_0)\rangle . \qquad (Q.15)$$

As is seen, the knowledge of $U_I(t, t_0)$ enables us to calculate $|\psi_I(t)\rangle$ for a given $|\psi_I(t_0)\rangle$.

Note that transition probabilities are independent from the picture chosen. Assume that in a certain process the system is at time t_0 in the initial state $|\psi_I(t_0)\rangle = |i\rangle$. Then we have $|\psi_I(t)\rangle = U_I(t, t_0) |\psi_I(t_0)\rangle$, see (Q.15), and the probability P_{fi} to find it at time t in a final state $|\psi_I(t)\rangle = |f\rangle$ is given by

$$P_{fi} = |\langle f| U_I(t, t_0) |i\rangle|^2 . \qquad (Q.16)$$

As one can show, the transition amplitude is equal in the Schrödinger and the interaction picture, see the exercises.

Due to (Q.14), the equation of motion for $U_I(t, t_0)$ is given by

$$i\hbar \frac{d}{dt} U_I(t, t_0) = H_I(t) U_I(t, t_0) . \qquad (Q.17)$$

A formal solution of equation (Q.17) with the initial condition $U_I(t_0, t_0) = 1$ reads

$$U_I(t, t_0) = 1 - \frac{i}{\hbar} \int_{t_0}^{t} dt_1 \, H_I(t_1) U_I(t_1, t_0) . \qquad (Q.18)$$

Iterating this solution gives

$$
\begin{aligned}
U_I(t, t_0) &= 1 - \tfrac{i}{\hbar} \int_{t_0}^{t} dt_1 \, H_I(t_1) \, U_I(t_1, t_0) = \\
&= 1 - \tfrac{i}{\hbar} \int_{t_0}^{t} dt_1 \, H_I(t_1) \left[1 - \tfrac{i}{\hbar} \int_{t_0}^{t_1} dt_2 \, H_I(t_2) \, U_I(t_2, t_0) \right] = \\
&= 1 + \left(-\tfrac{i}{\hbar}\right) \int_{t_0}^{t} dt_1 \, H_I(t_1) + \left(-\tfrac{i}{\hbar}\right)^2 \int_{t_0}^{t} \int_{t_0}^{t_1} dt_2 \, H_I(t_2) \, U_I(t_2, t_0)
\end{aligned}
\tag{Q.19}
$$

and further iteration results in

$$
\begin{aligned}
U_I(t, t_0) &= 1 + \left(-\tfrac{i}{\hbar}\right) \int_{t_0}^{t} dt_1 \, H_I(t_1) + \left(-\tfrac{i}{\hbar}\right)^2 \int_{t_0}^{t} dt_1 \int_{t_0}^{t_1} dt_2 \, H_I(t_1) \, H_I(t_2) + \\
&\quad + \left(-\tfrac{i}{\hbar}\right)^3 \int_{t_0}^{t} dt_1 \int_{t_0}^{t_1} dt_2 \int_{t_0}^{t_2} dt_3 \, H_I(t_1) \, H_I(t_2) \, H_I(t_3) + \cdots
\end{aligned}
\tag{Q.20}
$$

Thus, we can write the (formal) solution in form of an infinite series

$$
U_I(t, t_0) = \sum_{n=0}^{\infty} \left(-\frac{i}{\hbar}\right)^n \int_{t_0}^{t} dt_1 \int_{t_0}^{t_1} dt_2 \ldots \int_{t_0}^{t_{n-1}} dt_n \, H_I(t_1) \, H_I(t_2) \ldots H_I(t_n).
\tag{Q.21}
$$

This series is called *Dyson series*.[90] Note that the $H_I(t_m)$ at different times will not commute, $[H_I(t_1), H_I(t_2)] \neq 0$, in general. Thus, the order of time is of great importance. Note furthermore that the upper limits of the integrals are all different and are ordered, $t_0 \leq t_n \leq t_{n-1} < \cdots \leq t_2 \leq t_1 \leq t$.

The Dyson series plays a important role in quantum field theory, among others. There, one applies an operation called time ordering to the series (Q.21) in order to get rid of the problem of different upper limits of the integrals, see Appendix W, Vol. 2.

Q.2.1 Exercises and Solutions

1. Given an operator A in the Schrödinger picture, show that the time evolution of the operator A_H in the Heisenberg picture is given by

$$
i\hbar \frac{d}{dt} A_H = [A_H, H] + i\hbar \frac{\partial}{\partial t} A_H \quad \text{with} \quad \frac{\partial}{\partial t} A_H = e^{i\frac{Ht}{\hbar}} \left(\frac{\partial}{\partial t} A\right) e^{-i\frac{Ht}{\hbar}}.
\tag{Q.22}
$$

Solution: We have

$$
\begin{aligned}
i\hbar \frac{d}{dt} A_H &= -H e^{i\frac{Ht}{\hbar}} A e^{-i\frac{Ht}{\hbar}} + i\hbar e^{i\frac{Ht}{\hbar}} \left(\frac{\partial}{\partial t} A\right) e^{-i\frac{Ht}{\hbar}} + e^{i\frac{Ht}{\hbar}} A e^{-i\frac{Ht}{\hbar}} H \\
&= A_H H - H A_H + i\hbar e^{i\frac{Ht}{\hbar}} \left(\frac{\partial}{\partial t} A\right) e^{-i\frac{Ht}{\hbar}} = [A_H, H] + i\hbar \frac{\partial}{\partial t} A_H.
\end{aligned}
\tag{Q.23}
$$

2. Given $|\psi_I(t_0)\rangle = |i_I\rangle$ and $|\psi_I(t)\rangle = |f_H\rangle$. The corresponding states in the Schrödinger picture are $|i_S\rangle$ and $|f_S\rangle$. Show $\langle f_I | i_I\rangle = \langle f_S | i_S\rangle$.

---

[90]It is assumed that the series is enough good-natured and will converge.

Solution: States in the Schrödinger and the interaction picture are related by
$|\psi_I(t)\rangle = e^{i\frac{H_0 t}{\hbar}} |\psi_S(t)\rangle$. It follows

$$\langle f_I | i_I \rangle = \langle f_s | e^{-i\frac{H_0 t}{\hbar}} e^{i\frac{H_0 t}{\hbar}} | i_S \rangle = \langle f_S | i_S \rangle. \qquad (Q.24)$$

3. Show $\langle f_I | U_I(t, t_0) | i_I \rangle = \langle f_S | U_S(t, t_0) | i_S \rangle$.
   Solution: We have

$$\langle f_I | U_I(t, t_0) | i_I \rangle = \langle f_S | e^{-i\frac{H_0 t}{\hbar}} U_I(t, t_0) e^{i\frac{H_0 t}{\hbar}} | i_S \rangle = \langle f_S | U_S(t, t_0) | i_S \rangle$$
$$\qquad (Q.25)$$

   due to the definition $H_I(t) = e^{i\frac{H_0 t}{\hbar}} H_S e^{-i\frac{H_0 t}{\hbar}}$.
4. Prove (Q.17).
   Solution: Due to (Q.14) we have

$$i\hbar \frac{d}{dt} U_I(t, t_0) = i\hbar \frac{d}{dt} e^{i\frac{H_0 t}{\hbar}} e^{-i\frac{H(t-t_0)}{\hbar}} e^{-i\frac{H_0 t_0}{\hbar}} =$$
$$= i\hbar \left(e^{i\frac{H_0 t}{\hbar}} \frac{i}{\hbar} H_0 e^{-i\frac{H(t-t_0)}{\hbar}} e^{-i\frac{H_0 t_0}{\hbar}} - e^{i\frac{H_0 t}{\hbar}} \frac{i}{\hbar} H_0 e^{-i\frac{H(t-t_0)}{\hbar}} e^{-i\frac{H_0 t_0}{\hbar}} \right) =$$
$$= e^{i\frac{H_0 t}{\hbar}} H e^{-i\frac{H(t-t_0)}{\hbar}} e^{-i\frac{H_0 t_0}{\hbar}} = e^{i\frac{H_0 t}{\hbar}} H e^{-i\frac{H_0 t}{\hbar}} e^{i\frac{H_0 t}{\hbar}} e^{-i\frac{H(t-t_0)}{\hbar}} e^{-i\frac{H_0 t_0}{\hbar}} =$$
$$= H_I(t) U_I(t, t_0). \qquad (Q.26)$$

Appendix R
The Postulates of Quantum Mechanics

After a few remarks on the term 'postulate', we present in this chapter various versions of the postulates as given in current textbooks.

R.1 Postulate, Axiom, Rule?

In Chap. 14, we formulated the 'rules of the game' of quantum mechanics by introducing several postulates. This term (from the Latin *postulatum*, demand) is widespread, but there are also other terms such as axiom, rule or principle.[91]

The variety of names alone is enough to show that here, the formal term 'axiom' is not appropriate; it is defined as a fundamental statement about a system S that is assumed without proof. All axioms together form a consistent (i.e. contradiction-free), minimal, independent system (independence: no axiom can be derived from another), from which all statements about S are logically derivable.

In our context, 'postulate' stands in contrast for a basic rule that is plausible in terms of physical considerations and has been very well confirmed experimentally. The postulates all together must be consistent, of course; but the question of whether they actually are all mutually independent is considered secondary. It is rather important that they are few and concise formulations; they are, so to speak, the load-bearing structure of quantum mechanics.

Currently, the postulates of quantum mechanics are not deducible in a strictly logically manner from a broader theory—which does not mean that this will not be possible someday. But even then, the postulates of quantum mechanics would retain their value because they describe the relevant phenomena very well. This is quite similar to the Newtonian axioms: Their limits were shown by quantum mechanics and special relativity; but within this framework, the Newtonian axioms are still useful because they are simple and their predictions accurate enough for many purposes.

---

[91]"I am a quantum engineer, but on Sundays I have principles." John Stewart Bell, Irish physicist, 'inventor' of Bell's inequality.

© Springer Nature Switzerland AG 2018
J. Pade, *Quantum Mechanics for Pedestrians 1*, Undergraduate Lecture
Notes in Physics, https://doi.org/10.1007/978-3-030-00464-4

Which facts of quantum mechanics have to be formulated as a postulate/axiom/rule /principle and how, is to some extent a matter of personal taste, as we shall see in the next section.

R.2 Formulations of Some Authors

We wish to get to know different representations of the basic rules of quantum mechanics. We do not aim at an exhaustive survey; rather, we give an illustration of how different and how similar the approaches can be.

We cite only the postulates themselves, giving for better comparability all the texts in English, and quoting them verbatim (apart from translation). Therefore, some of the following quotes are very sparse. To save space, we dispense with further explanations. These are found *in extenso* in the relevant sources, where they can be consulted if desired.

First, some remarks:

(a) Some books do not refer explicitly to 'rules' under any name. The subject matter itself is treated, of course—in any quantum mechanics textbook, e.g. the Hilbert space and time evolution are presented, but these issues are not always explicitly listed as postulates, for example in A. Messiah, *Quantum Mechanics* (1964) or T. Fließbach. *Quantum Mechanics* (2000). This can apply even if there is a separate chapter such as 'The principles of quantum dynamics' in E. Merzbacher, *Quantum Mechanics* (1998).

(b) Some authors summarize only kinematics, but not dynamics in the form of postulates (although dynamics is treated in great detail in their texts), e.g. W. Nolting, *Quantum Mechanics* (1992), or K. Gottfried, T.-M. Yan, *Quantum Mechanics* (2006).

(c) The formulations of many quantum mechanics books are more or less uniform and differ substantially only in details of the wording or the order of the postulates. It is striking that often the indistinguishability of identical quantum objects or the Pauli principle is not formulated as a postulate.

(d) In some quantum-mechanics books, such a peculiar terminology is used that simply citing the postulates without further comments would be incomprehensible, e.g. in A. Peres, *Quantum Theory* (1995). For space reasons, we refrain from citing such works.

(e) The placement of the postulates varies considerably; sometimes they are summarized compactly in a few pages, either at the beginning of the book or more often in the middle. In other cases, they are scattered throughout the book and are not recognizable at first glance as a coherent system.

R.2.1 J. Audretsch

Entangled Systems (2007), p. 32 ff, **postulates**

Postulate 1 (pure state) An isolated quantum system which is in a pure state is described by its state vector $|\psi\rangle$. This is a normalised vector in a Hilbert space $\mathcal{H}_d$ which is associated with the quantum system.

Postulate 2 (projective measurements, non-deterministic dynamic evolution)

(a) A projective measurement of a physical quantity (e.g. of the energy, angular momentum, etc.) carried out on a quantum system is described by an Hermitian operator which can be time dependent and acts on the vectors of $\mathcal{H}_d$. We speak of a measurement of the observable A and denote the operator with the same symbol A.

(b) The possible measured values which can occur as a result of a measurement of the observable A are the eigenvalues a_n of the associated operator A. For simplicity, we assume that its spectrum is discrete:

$$A \left| u_n^i \right\rangle = a_n \left| u_n^i \right\rangle; \quad i = 1, \ldots, g_n. \tag{R.1}$$

The eigenvectors $\left| u_n^i \right\rangle$ form an orthonormal basis or can, in the case of degeneracy, be correspondingly chosen. The g_n give the degree of degeneracy of the eigenvalues a_n.

(c) When a selective measurement of the observables A of a system with a normalised state vector $|\psi\rangle$ leads to the result a_n, then the non-normalised state vector $\left| \tilde{\psi}_n' \right\rangle$ immediately following the measurement is given by the projection of $|\psi\rangle$

$$|\psi\rangle \rightarrow \left| \tilde{\psi}_n' \right\rangle = P_n |\psi\rangle \tag{R.2}$$

with the projection operator

$$P_n = \sum_{i=1}^{g_n} \left| u_n^i \right\rangle \left\langle u_n^i \right| \tag{R.3}$$

which projects onto the space of the eigenvectors corresponding to a_n. Through normalisation of $\left| \tilde{\psi}_n' \right\rangle$, the state vector $\left| \psi_n' \right\rangle$ after the measurement is obtained.

(d) We denote by $N(a_n)$ the frequency with which a measured value a_n is obtained when the measurement is carried out on N identically prepared systems in the state $|\psi\rangle$. The relative frequencies $\frac{N(a_n)}{N}$ for all these ensembles approach the probability $p(a_n)$ as a limiting value in the limit $N \rightarrow \infty$:

$$\frac{N(a_n)}{N} \overset{N \rightarrow \infty}{\rightarrow} p(a_n). \tag{R.4}$$

(e) The probability $p(a_n)$ of obtaining a particular measured value a_n at a certain time is equal to the expectation value of the projection operator P_n computed with the state $|\psi\rangle$ prior to the measurement. Equivalently, it is equal to the square of the norm of the non-normalised state vector $\left| \tilde{\psi}_n' \right\rangle$ after the measurement:

$$p\,(a_n) = \langle \psi |\, P_n\, | \psi \rangle = \left\| \left| \tilde{\psi}'_n \right\rangle \right\|^2 . \tag{R.5}$$

Postulate 3 (deterministic dynamic evolution between preparation and measurement)

(a) For isolated systems, the probability distribution $p\,(a_n)$ evolves in a deterministic and reversible manner between the preparation and the measurement. Its time development between two times t_0 and t_1 is described by a unitary time-development operator $U\,(t_1, t_0)$:

$$U^\dagger(t_1, t_0) = U^{-1}(t_1, t_0). \tag{R.6}$$

This operator fulfills the conditions $U\,(t_0, t_0) = 1$ and

$$U\,(t_2, t_1) U\,(t_1, t_0) = U\,(t_2, t_0) \tag{R.7}$$

for arbitrary times t_0, t_1, t_2.

(b) The dynamic equation for $U\,(t, t_0)$ is

$$i\hbar \frac{\mathrm{d}}{\mathrm{d}t} U(t, t_0) = H(t) U(t, t_0). \tag{R.8}$$

(c) The Schrödinger representation is one of the many possible formulations of this time development. In this representation, the dynamic evolution of the state is given by the state vector alone, according to

$$|\psi\,(t)\rangle = U(t, t_0)\, |\psi\,(t_0)\rangle . \tag{R.9}$$

Observables can be only explicitly time dependent. At each time t, there is a corresponding probability distribution $p(a_n, t)$ for the results of a measurement of A given by (R.5). From (R.8) and (R.9), the Schrödinger equation follows

$$i\hbar \frac{\mathrm{d}}{\mathrm{d}t}\, |\psi\,(t)\rangle = H\, |\psi\,(t)\rangle . \tag{R.10}$$

R.2.2 J.-L. Basdevant and J. Dalibard

Quantum Mechanics (2002), p. 100 ff; **Principles**

First Principle: The Superposition Principle With each physical system one can associate an appropriate Hilbert space $\mathcal{E}_H$. At each time t, the state of the system is completely determined by a normalized vector $|\psi\,(t)\rangle$ of $\mathcal{E}_H$.

Second Principle: Measurements of Physical Quantities

(a) With each physical quantity A one can associate a linear Hermitian operator $\hat{A}$ acting in $\mathcal{E}_H$: $\hat{A}$ is the observable which represents the quantity A.

(b) We denote by $|\psi\rangle$ the state of the system before the measurement of A is performed. Whatever $|\psi\rangle$ may be, the only possible results of the measurement are the eigenvalues a_α of $\hat{A}$.

(c) We denote by $\hat{P}_\alpha$ the projector onto the subspace associated with the eigenvalue a_α. The probability of finding the value a_α in a measurement of A is

$$\mathcal{P}(a_\alpha) = \|\psi_\alpha\|^2, \; where \; |\psi_\alpha\rangle = \hat{P}_\alpha |\psi\rangle. \tag{R.11}$$

(d) Immediately after the measurement of A has been performed and has given the result a_α, the new state $|\psi'\rangle$ of the system is

$$|\psi'\rangle = \frac{|\psi_\alpha\rangle}{\|\psi_\alpha\|}. \tag{R.12}$$

Third Principle: Time Evolution   We denote by $|\psi(t)\rangle$ the state of the system at time t. As long as the system does not undergo any observation, its time evolution is given by the Schrödinger equation:

$$i\hbar \frac{\mathrm{d}}{\mathrm{d}t} |\psi(t)\rangle = \hat{H} |\psi(t)\rangle \tag{R.13}$$

where $\hat{H}$ is the energy observable, or Hamiltonian, of the system.

R.2.3   D.R. Bes

Quantum Mechanics (2004), p. 9 (1–3), p. 96 (4), p. 137 (5); Basic **principles**

Principle 1.   The state of a system is completely described by a vector Ψ—the state vector or state function—belonging to a Hilbert space.

Principle 2.   To every physical quantity there corresponds a single operator. In particular, the operators $\hat{x}$ and $\hat{p}$, corresponding to the coordinate and momentum of a particle, satisfy the commutation relation

$$[\hat{x}, \hat{p}] = i\hbar. \tag{R.14}$$

Principle 3.   The eigenvalues q_i of an operator $\hat{Q}$ constitute the possible results of a measurement of the physical quantity Q . The probability of obtaining the eigenvalue q_i is the modulus squared $|c_i|^2$ of the amplitude of the eigenvector φ_i in the state vector Ψ representing the state of the system.

Principle 4. There are only two kinds of particles in nature: bosons described by symmetric state vectors, and fermions described by antisymmetric state vectors.

Principle 5. The operator yielding the change of a state vector over time is proportional to the Hamiltonian

$$\frac{\partial}{\partial t'} U\left(t', t\right)\bigg|_{t'=t} = -\frac{i}{\hbar}\hat{H}\left(t\right). \tag{R.15}$$

R.2.4   B.H. Bransden and C.J. Joachain

Quantum Mechanics (2000), p. 194 to p. 231; **postulates**

Postulate 1   To an ensemble of physical systems one can, in certain cases, associate a wave function or state function which contains all the information that can be known about the ensemble. This function is in general complex; it may be multiplied by an arbitrary complex number without altering its physical significance.

Postulate 2   The superposition principle.[92]

Postulate 3   With every dynamical variable is associated a linear operator.

Postulate 4   The only result of a precise measurement of the dynamical variable $\mathcal{A}$ is one of the eigenvalues a_n of the linear operator A associated with $\mathcal{A}$.

Postulate 5   If a series of measurements is made of the dynamical variable $\mathcal{A}$ on an ensemble of systems, described by the wave function Ψ, the expectation or average value of this dynamical variable is

$$\langle A \rangle = \frac{\langle \Psi | A | \Psi \rangle}{\langle \Psi | \Psi \rangle}. \tag{R.16}$$

Postulate 6   A wave function representing any dynamical state can be expressed as a linear combination of the eigenfunctions of A, where A is the operator associated with a dynamical variable.

Postulate 7   The time evolution of the wave function of a system is determined by the time-dependent Schrödinger equation

$$i\hbar\frac{\partial}{\partial t}\Psi(t) = H(t)\Psi(t) \tag{R.17}$$

where H is the Hamiltonian, or total energy operator of the system.

R.2.5   C. Cohen-Tannoudji, B. Diu, and F. Laloë

Quantum mechanics, Volume 1 (1977), p. 215 ff; **postulates**

---

[92]It stands there really so short and crisp.

First Postulate: At a fixed time t_0, the state of a physical system is defined by specifying a ket $|\psi(t_0)\rangle$ belonging to the state space $\mathcal{E}$.

Second Postulate: Every measurable physical quantity $\mathcal{A}$ is described by an operator A acting in $\mathcal{E}$; this operator is an observable.

Third Postulate: The only possible result of the measurement of a physical quantity $\mathcal{A}$ is one of the eigenvalues of the corresponding observable A.

Fourth Postulate (**case of a discrete non-degenerate spectrum**): When the physical quantity $\mathcal{A}$ is measured on an system in the normalized state $|\psi\rangle$, the probability $\mathcal{P}(a_n)$ of obtaining the non-degenerate eigenvalue a_n of the corresponding variable A is:

$$\mathcal{P}(a_n) = |\langle u_n| \psi\rangle|^2 \tag{R.18}$$

where $|u_n\rangle$ is the normalized eigenvector of A associated with the eigenvalue a_n.

Fourth Postulate (**case of a discrete spectrum**): When the physical quantity $\mathcal{A}$ is measured on an system in the normalized state $|\psi\rangle$, the probability $\mathcal{P}(a_n)$ of obtaining the eigenvalue a_n of the corresponding variable A is:

$$\mathcal{P}(a_n) = \sum_{i=1}^{g_n} |\langle u_n^i| \psi\rangle|^2 \tag{R.19}$$

where g_n is the degree of degeneracy of a_n and $\{|u_n^i\rangle\}$ $\{i = 1, 2, \ldots, g_n\}$ is an orthonormal set of vectors which forms a basis in the eigensubspace $\mathcal{E}_n$ associated with the eigenvalue a_n of A.

Fourth Postulate (**case of a continuous non-degenerate spectrum**): When the physical quantity $\mathcal{A}$ is measured on an system in the normalized state $|\psi\rangle$, the probability $d\mathcal{P}(a_n)$ of obtaining a result included between α and $\alpha + d\alpha$ is equal to:

$$d\mathcal{P}(a) = |\langle v_\alpha| \psi\rangle|^2 \, d\alpha \tag{R.20}$$

where $|v_\alpha\rangle$ is the eigenvector corresponding to the eigenvalue α of the observable A associated with $\mathcal{A}$.

Fifth Postulate: If the measurement of the physical quantity $\mathcal{A}$ on the system in the state $|\psi\rangle$ gives the result a_n, the state of the system immediately after the measurement is the normalized projection, $\frac{P_n|\psi\rangle}{\sqrt{\langle\psi|P_n|\psi\rangle}}$, of $|\psi\rangle$ onto the eigensubspace associated with a_n.

Sixth Postulate: The time evolution of the state vector $|\psi(t)\rangle$ is governed by the Schrödinger equation

$$i\hbar\frac{d}{dt}|\psi(t)\rangle = H(t)|\psi(t)\rangle \qquad (R.21)$$

where $H(t)$ is the observable associated with the total energy of the system.

R.2.6 K. Gottfried and T.-M. Yan

Quantum Mechanics: Fundamentals (2006), p. 40ff; **postulates**
1. The most complete possible description of the state of any physical system S at any instant is provided by some particular vector $|\psi\rangle$ in the Hilbert space $\mathcal{H}$ appropriate to the system. Every linear combination of such state vectors represents a possible physical state of S.
2. The physically meaningful entities of classical mechanics, such as momentum, energy, position and the like, are represented by Hermitian operators.
3. A set of N identically prepared replicas of a system S described by the pure state $|\psi\rangle$, when subjected to a measurement designed to display the physical quantity represented by the observable A, will in each individual case display one of the values[93] $(a, a', \ldots)$, and as $N \to \infty$ will do so with the probabilities $p_\psi(a)$, $p_\psi(a')$, ..., where

$$p_\psi(a) = |\langle a|\psi\rangle|^2. \qquad (R.22)$$

(The dynamics is not given in the form of postulates.)

R.2.7 C.J. Isham

Lectures on Quantum Theory (2008), p. 84ff; **rules**
   Rule 1: The predictions of results of measurements made on an otherwise isolated system are probabilistic in nature. In situations where the maximum amount of information is available, this probabilistic information is represented mathematically by a vector in a complex Hilbert space $\mathcal{H}$ that forms the state space of the quantum theory. In so far as it gives the most precise predictions that are possible, the vector is to be thought of as the mathematical representative of the physical notion of 'state' of the system.
   Rule 2: The observables of the system are represented mathematically by self-adjoint operators that act on the Hilbert space $\mathcal{H}$.
   Rule 3: If an observable quantity A and a state are represented respectively by the self-adjoint operator $\hat{A}$ and the normalised vector $\psi \in \mathcal{H}$, then the expected result $\langle A\rangle_\psi$ of measuring A is

---

[93]These are the eigenvalues of A.

$$\langle A \rangle_\psi = \langle \psi | \hat{A} | \psi \rangle . \tag{R.23}$$

Rule 4: In the absence of any external influence (i.e., in a closed system), the state vector ψ changes smoothly in time t according to the time-dependent Schrödinger equation

$$i\hbar \frac{\mathrm{d}}{\mathrm{d}t} \psi = H\psi \tag{R.24}$$

where H is a special operator known as the Hamiltonian.

Another formulation of Rule 3 (p. 99 ff): (1) The only possible result of a measurement of A is one of the eigenvalues of the operator $\hat{A}$ that represents it. (2) If the state vector is $|\psi\rangle$ and a measurement of A is made, the probability that the result will be the particular eigenvalue a_n is $Prob\, (A = a_n; |\psi\rangle) = \langle \psi | \hat{P}_n | \psi \rangle$ where $\hat{P}_n = \sum_{j=1}^{d(n)} |a_{n,j}\rangle\langle a_{n,j}|$ is the projector onto the eigenspace of vectors with eigenvalue a_n. The rule $\langle A \rangle_\psi = \langle \psi | \hat{A} | \psi \rangle$ is entirely equivalent to this pair of rules.

R.2.8 M. LeBellac

Quantum Physics (2006), pp. 96–108; **postulates**

Postulate I: The space of states The properties of a quantum system are completely defined by specification of its state vector $|\varphi\rangle$, which fixes the mathematical representation of the physical state of the system. The state vector is an element of a complex Hilbert space $\mathcal{H}$ called the space of states. It will be convenient to choose $|\varphi\rangle$ to be normalized that is, to have unit norm: $\|\varphi\|^2 = \langle \varphi | \varphi \rangle = 1$.

Postulate II: Probability amplitudes and probabilities If $|\varphi\rangle$ is the vector representing the state of a system and if $|\chi\rangle$ represents another physical state, there exists a probability amplitude $a\,(\varphi \to \chi)$ of finding $|\varphi\rangle$ in state $|\chi\rangle$, which is given by a scalar product on $\mathcal{H}$: $a\,(\varphi \to \chi) = \langle \chi | \varphi \rangle$. The probability $p\,(\varphi \to \chi)$ for the state $|\varphi\rangle$ to pass the test $|\chi\rangle$ is obtained by taking the squared modulus $|\langle \chi | \varphi \rangle|^2$ of this amplitude

$$p\,(\varphi \to \chi) = |a\,(\varphi \to \chi)|^2 = |\langle \chi | \varphi \rangle|^2 . \tag{R.25}$$

Postulate III: Physical properties and operators With every physical property $\mathcal{A}$ (energy, position, momentum, angular momentum, and so on) there exists an associated Hermitian operator A which acts in the space of states $\mathcal{H}$: A fixes the mathematical representation of $\mathcal{A}$.

The WFC Postulate(**w**ave **f**unction **c**ollapse), complement to postulate II If a system is initially in a state $|\varphi\rangle$, and if the result of an ideal measurement of $\mathcal{A}$ is a_n, then immediately after this measurement the system is in the state projected on the subspace of the eigenvalue a_n:

$$|\varphi\rangle \to |\psi\rangle = \frac{\mathcal{P}_n |\varphi\rangle}{\sqrt{\langle \varphi | \mathcal{P}_n | \varphi \rangle}} . \tag{R.26}$$

Postulate IV: the evolution equation The time evolution of the state vector $|\varphi(t)\rangle$ of a quantum system is governed by the evolution equation

$$i\hbar\frac{d|\varphi(t)\rangle}{dt} = H(t)|\varphi(t)\rangle.$$ (R.27)

The Hermitian operator H is called the Hamiltonian.

Postulate IV': the evolution operator, alternative to postulate IV The state vector $|\varphi(t)\rangle$ at time t is derived from the state vector $|\varphi(t_0)\rangle$ at time t_0 by applying an unitary operator $U(t, t_0)$, called the evolution operator:

$$|\varphi(t)\rangle = U(t, t_0)|\varphi(t_0)\rangle.$$ (R.28)

R.2.9 G. Münster

Quantentheorie (Quantum Theory) (2006), p. 84; **postulates** (translated from the German)

I Pure states are represented by normalized vectors (or rays) of a complex Hilbert space.

Superposition principle: Each vector corresponds to a possible pure state.

II Self-adjoint operators are associated with the observables of a system. The possible measurement values are the eigenvalues of the operator.

III The expectation value of an observable A in the state $|\psi\rangle$ is given by

$$\langle A \rangle = \langle\psi|A|\psi\rangle.$$ (R.29)

IV The time evolution of states is determined by the Schrödinger equation:

$$i\hbar\frac{\partial}{\partial t}|\psi\rangle = H|\psi\rangle$$ (R.30)

where H is the Hamiltonian.

V If the observable A is measured on a system in the state $|\psi\rangle$ and the measured value a is found, the system changes by the measurement to the corresponding eigenstate $|a\rangle$ (state reduction).

R.2.10 W. Nolting

Quantenmechanik, Teil 1: Grundlagen (Quantum Mechanics, Part 1: Fundamentals) (1992), p. 181ff; **postulates** (translated from the German)

1. Postulate: Measuring device for a certain physical quantity (observable) ⇔ linear Hermitian operator.

2. Postulate: *Pure* state of the quantum system ⇔ Hilbert vector.

3. Postulate: Measurement $\hat{=}$ Interaction between system and apparatus ⇔ application of the operator A onto the state $|\psi\rangle$: $A|\psi\rangle = \sum \int a_i |a_i\rangle \langle a_i |\psi\rangle \overset{\text{filter}}{\to} |a_j\rangle\langle a_j |\psi\rangle$.

4. Postulate: Measurement results ⇔ eigenvalues a_i of the operator A.

5. Postulate: Measurement probability for a_i ⇔ $w(a_i |\psi\rangle) = |\langle a_j |\psi\rangle|^2$.

(The dynamics is not given in the form of postulates.)

R.2.11   A.I.M. Rae

Quantum Mechanics (2008), p. 68ff; **postulates**

Postulate 1   For every dynamical system there exists a wave function that is a continuous, square-integrable, single-valued function of the parameters of the system and of time, and from which all possible predictions about the physical properties of the system can be obtained.

Postulate 2   Every dynamical variable may be represented by a Hermitian operator whose eigenvalues represent the possible results of carrying out a measurement of the value of the dynamical variable. Immediately after such a measurement, the wave function of the system is identical to the eigenfunction corresponding to the eigenvalue obtained as a result of measurement.

Postulate 3   The operators representing the position and momentum of a particle are $\mathbf{r}$ and $-i\hbar\nabla$, respectively. Operators representing other dynamical quantities bear the same functional relation to these, as do the corresponding classical quantities to the classical position and momentum variables.

Postulate 4   When a measurement of a dynamic variable represented by a Hermitian operator $\hat{Q}$, is carried out on a system whose wave function is ψ, then the probability of the result being equal to a particular eigenvalue q_m will be $|a_m|^2$, where $\psi = \sum_n a_n \phi_n$ and the ϕ_n are the eigenfunctions of $\hat{Q}$ corresponding to the eigenvalues q_n.

R.2.12   H. Rollnik

Quantentheorie I (Quantum Theory I) (2003), p. 212ff; **axioms** (translated from the German)

• **State axiom**: Physical states are described by the vectors of a Hilbert space $\mathcal{H}$. More precisely: Physical states are mapped injectively onto the rays of $\mathcal{H}$.

• **Observable axiom 1**: Each physical observable $\mathcal{A}$ is represented by a linear Hermitian operator A of the state space $\mathcal{H}$ (p. 224).

• **Observable axiom 2**: The expectation value $\langle \mathcal{A} \rangle_\psi$ of $\mathcal{A}$ in the state ψ is given by

$$\langle \mathcal{A} \rangle_\psi = \sum_i a_i w_i = \frac{\langle \psi | A | \psi \rangle}{\langle \psi | \psi \rangle}. \tag{R.31}$$

For $\|\psi\| = 1$, it holds that:

$$\langle \mathcal{A} \rangle_\psi = \langle \psi | A | \psi \rangle. \tag{R.32}$$

Later on, Rollnik invokes: (1) Symmetry axiom: Physical symmetry groups are represented by unitary or anti-unitary operators, (2) Axiom of nonrelativistic quantum mechanics: For an N-particle system, the position operators $Q_i(t), i = 1, \ldots, 3N$ form a complete set of commuting observables. The same applies for the momentum operators $P_i(t)$. The commutation relation $\left[P_j(t), Q_k(t) \right] = \frac{\hbar}{2} \delta_{jk}$ holds.

R.2.13 H. Schulz

Physik mit Bleistift (Physics with a pencil) (2001), p. 302ff; **postulates** (translated from the German)

I. The complete information about a quantum system is contained in a one-valued function $\psi(x, t) \in \mathbb{C}$ (the information carrier). x is a set of variables, one for each degree of freedom. In general, one can write $x = 1, 2, \ldots$ with $1 :=$ set of variables for particle 1 and so on.

II. A linear Hermitian operator A is associated with each observable. A table of such associations is a constituent of the postulate:

Class. quantity	Name in quantum m.	Letter	Space	Action
Position (1D)	Position	X	$\psi(x)$	$X = x$
Momentum (1D)	Momentum	p	$\psi(x)$	$p = \frac{\hbar}{i}\partial_x$
Momentum (3D)	Momentum	$\vec{p}$	$\psi(\vec{r})$	$\vec{p} = \frac{\hbar}{i}\nabla$
Angular momentum	Angular momentum	$\vec{L}$	$\psi(\vec{r})$	$\vec{L} = \vec{r} \times \vec{p} = \vec{r} \times \frac{\hbar}{i}\nabla$
	Parity (3D)	P	$\psi(\vec{r})$	$P\psi(\vec{r}) = \psi(-\vec{r})$
	Spin (1. comp.)	σ^x	Two-component	$\sigma^x = \begin{pmatrix} 0 & 1 \\ 1 & 0 \end{pmatrix}$
Energy	Hamiltonian	H	E.g. $\psi(\vec{r})$	E.g. $H = -\frac{\hbar^2}{2m}\nabla^2 + V(\vec{r}, t)$
⋮	⋮	⋮	⋮	⋮

III. Possible measurement values are the eigenvalues of A, obtainable by solving $A\varphi_{av} = a\varphi_{av}$ and requiring univalence and normalizability.

IV. The eigenstates of A are to be normalized according to

$$\int dx\, \varphi_{a\nu}^* \varphi_{b\mu} = \delta_{ab}\delta_{\mu\nu}\, ,\ \delta\,(a-b)\,\delta_{\mu\nu}\, ,\ \delta_{ab}\delta\,(\mu-\nu)\, ,\ \delta\,(a-b)\,\delta\,(\mu-\nu) \quad (R.33)$$

depending on whether the index is in a discrete or continuous region of the spectrum of A. The actual state ψ of the system must always be normalized to one:

$$\int dx\, |\psi\,(x)|^2 = 1. \tag{R.34}$$

V. The probability of obtaining a discrete measured value a and the probability density for continuous measured values a follow the same formula:

$$P\,(a,t) = \sum_{\nu} |c_{a\nu}|^2 = \sum_{\nu} \left| \int dx\, \varphi_{a\nu}^* \psi\,(x,t) \right|^2 . \tag{R.35}$$

If ν is continuous, then $\sum_{\nu}$ is to be replaced by $\int d\nu$.

VI. The equation of motion of quantum mechanics is

$$i\hbar\dot{\psi} = H\psi, \tag{R.36}$$

where the operator H is given in the table in II. The equation also applies if H is time dependent (for example due to $V\,(\vec{r},t)$; see table).

VII. Pauli exclusion principle: Under permutation of the sets of variables of two identical particles, one must require

$$\psi\,(1,2,\ldots) = \mp\psi\,(2,1,\ldots)\,; \tag{R.37}$$

negative sign for fermions, positive sign for bosons.

R.2.14 F. Schwabl

Quantum Mechanics, 3. ed. (2002), p. 40; **axioms**

I. The state is described by the wavefunction $\psi\,(x)$.

II. The observables are represented by Hermitian operators $A\ldots$, with functions of observables being represented by the corresponding functions of the operators.

III. The expectation value of the observable represented by the operator A is given in the state ψ by $\langle A\rangle = (\psi, A\psi)$.

IV. The time evolution of the states is given by the Schrödinger equation

$$i\hbar\frac{\partial}{\partial t}\psi = H\psi;\ \ H = -\frac{\hbar^2}{2m}\nabla^2 + V\,(\mathbf{x})\,. \tag{R.38}$$

V. If in a measurement of A the value a_n is found, the wavefunction changes to the corresponding eigenfunction ψ_n.[94]

From axioms II and III, it follows that the only possible results of a measurement of an observable are the eigenvalues of the corresponding operator A, and the probabilities are given by $|c_n|^2$, where c_n are the expansion coefficients of $\psi(\mathbf{x})$ in the eigenfunctions of A. In particular, it follows that $|\psi(\mathbf{x})|^2$ is the probability density for the position.

R.2.15 N. Zettili

Quantum Mechanics, Concepts and Applications (2009), p. 165ff; **postulates**

Postulate 1: State of a system The state of any physical system is specified, at each time t, by a state vector $|\psi(t)\rangle$ in a Hilbert space $\mathcal{H}$; $|\psi(t)\rangle$ contains (and serves as the basis to extract) all the needed information about the system. Any superposition of state vectors as also a state vector.

Postulate 2: Observables and operators To every physically measurable quantity A, called an observable or dynamical variable, there corresponds a linear Hermitian operator $\hat{A}$ whose eigenfunctions form a complete basis.

Postulate 3: Measurements and eigenvalues of operators The measurement of an observable A may be represented formally by the action of $\hat{A}$ on a state vector $|\psi(t)\rangle$. The only possible result of such a measurement is one of the eigenvalues a_n (which are real) of the operator $\hat{A}$. If the result of a measurement of A on a state $|\psi(t)\rangle$ is a_n, the state of the system immediately after the measurement changes to $|\psi_n\rangle$:

$$\hat{A}|\psi(t)\rangle = a_n|\psi(t)\rangle$$

where $a_n = \langle\psi_n|\psi(t)\rangle$. Note: a_n is the component of $|\psi(t)\rangle$ when projected onto the eigenvector $|\psi_n\rangle$.

Postulate 4: Probabilistic outcome of measurements

Discrete spectra: When measuring an observable A of a system in a state $|\psi\rangle$, the probability of obtaining one of the nondegenerate eigenvalues a_n of the corresponding operator $\hat{A}$ is given by

$$P_n(a_n) = \frac{|\langle\psi_n|\psi\rangle|^2}{\langle\psi|\psi\rangle} = \frac{|a_n|^2}{\langle\psi|\psi\rangle} \tag{R.39}$$

where $|\psi_n\rangle$ is the eigenstate of $\hat{A}$ with eigenvalue a_n. If the eigenvalue a_n is m-degenerate, P_n becomes

---

[94]Should perhaps preferably be formulated as: The wavefunction ψ has changed to the corresponding eigenfunction ψ_n.

$$P_n(a_n) = \frac{\sum\limits_{j=1}^{m} \left| \left\langle \psi_n^j \mid \psi \right\rangle \right|^2}{\langle \psi \mid \psi \rangle} = \frac{\sum\limits_{j=1}^{m} \left| a_n^{(j)} \right|^2}{\langle \psi \mid \psi \rangle}. \tag{R.40}$$

The act of measurement changes the state of the system from $|\psi\rangle$ to $|\psi_n\rangle$. If the system is already in an eigenstate $|\psi_n\rangle$ of $\hat{A}$, a measurement of A yields with certainty the corresponding eigenvalue $a_n : \hat{A} |\psi_n\rangle = a_n |\psi_n\rangle$.

Continuous spectra: The relation (R.39), which is valid for discrete spectra, can be extended to determine the probability density that a measurement of $\hat{A}$ yields a value between a and $a + da$ on a system which is initially in a state $|\psi\rangle$:

$$\frac{dP(a)}{da} = \frac{|\psi(a)|^2}{\langle \psi \mid \psi \rangle} = \frac{|\psi(a)|^2}{\int\limits_{-\infty}^{\infty} |\psi(a')|^2 \, da'}. \tag{R.41}$$

For instance, the probability density for finding a particle between x and $x + dx$ is given by $dP(x)/dx = |\psi(x)|^2 / \langle \psi \mid \psi \rangle$.

Postulate 5: Time evolution of a system The time evolution of the state vector $|\psi(t)\rangle$ of a system is governed by the time-dependent Schrödinger equation

$$i\hbar \frac{\partial |\psi(t)\rangle}{\partial t} = \hat{H}(t) |\psi(t)\rangle \tag{R.42}$$

where $\hat{H}$ is the Hamiltonian operator corresponding to the total energy of the system.

Appendix S
System and Measurement: Some Concepts

We compile in this appendix some common notations concerning quantum systems and measurements.

S.1   System: Isolated, Closed, Open

In the following, we consider the relationship between a system S and its environment $\mathcal{U}$.

Here, a *system* is that part of the universe which we are examining. It must not necessarily be separated from the rest of the real universe in reality; there can also be imagined boundaries. Everything that lies outside these borders is called the system's surroundings or *environment*.[95] The individual parts of a system have to interact somehow with each other. Examples of systems are an atom, a pendulum or the Earth's ecosystem. Specifically, a quantum system is everything that allows for a consistent dynamic description in terms of quantum mechanics.

There are different types of interactions between system and environment. The nomenclature is derived from thermodynamics, where it has a well-defined meaning. In quantum mechanics, the situation is somewhat less consistent.

Thermodynamics

In an *isolated* system, there is no exchange of matter and energy (work, heat) with the surroundings. Hence, the total energy and mass stay constant. In a *closed* system, there is no exchange of matter with the environment, but only of energy. Accordingly, the total mass is constant, but not the total energy. In an *open* system, the system boundaries are permeable to matter and energy exchange; neither energy nor mass are constant.

Of course, the terms 'isolated' and 'closed' are nearly always approximations. In particular, there is no 'real' isolated system (perhaps apart from the entire universe);

---

[95] If necessary, the 'measuring apparatus' can be introduced as mediator between system and environment.

© Springer Nature Switzerland AG 2018

J. Pade, *Quantum Mechanics for Pedestrians 1*, Undergraduate Lecture Notes in Physics, https://doi.org/10.1007/978-3-030-00464-4

this is prevented in any case by the ubiquitous gravitational field. As an approximation, we can of course consider specific systems for some period of time to be isolated; for example, a Thermos bottle or its contents.

An example of a closed system is the earth, at least to a good approximation: There is an energy exchange with its environment (incoming solar radiation, radiation from the Earth into space), but no significant transfer of matter.

Examples of open systems are the ecological and the economic systems of a region. In this sense, also a human being or any living being is an open system; continually, matter and energy are taken up and given off.

For the environment, in thermodynamics there exists the term *reservoir* (an environment with an infinite number of degrees of freedom) and (heat or thermal) *bath* (a reservoir which is in thermal equilibrium).

Quantum Mechanics

Even in thermodynamics, where the terms 'isolated' and 'closed' are properly defined, one occasionally finds blurred formulations. A quote from the internet: "A closed system is a system in the state of being isolated from its surrounding environment."[96] In quantum mechanics, these two terms are often even used interchangeably. On the other hand, one can find also the distinction based on thermodynamics.

An *isolated* system (sometimes called a totally isolated system) is completely decoupled from its environment. In particular, its total energy is constant, which means that the Hamiltonian H is not time dependent. The complete separation from the environment also means that there must be no entanglement between system and environment.

If the environment is acting through external forces on the system, and one can formulate the dynamics of the system in terms of a possibly time-dependent Hamiltonian, the system is called *closed*. It is commonly assumed that there is no feedback from the system onto the environment, i.e. the interactions environment—system are a one-way street.

A system is called *open* if interactions and entanglements between system and environment are allowed (to and from). Usually, it is assumed that the aggregate (system + environment) is isolated or at least closed, and that its dynamics is described by a Hamiltonian.

S.2   Measurement

Measurement

By *measurement*, we understand an operation on a system which determines the values of one or more physical variables of the system immediately before the measurement in the form of distinct and storable numbers (see also Chap. 14).

---

[96]See e.g. http://en.wikidoc.org/index.php/Closed_system (November 2011) as one of many references.

Fig. S.1 Ideal measurement of a right circular-polarized photon with regard to linear polarization

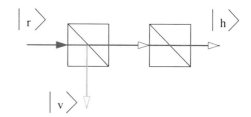

Classically, a value of a physical quantity A is measured which already exists before the measurement is made (pre-existing value). In quantum mechanics, this is the case only if the system is initially in an eigenstate of the measured observable; otherwise there is no unique observable value before the measurement.[97] The transition from a superposition to a single state is called *state reduction* or *collapse of the wavefunction*. It is an *irreversible* evolution that characterizes a direction in time (except for the case that the initial state is already an eigenstate of the operator, i.e. there is no initial superposition).

Ideal Measurement, QND

Actual measurements on quantum objects often destroy them or make them disappear for the observer. They can therefore be carried out only once.[98] Other measurements influence the objects so strongly that after the measurement or through the measurement, they take on a different value of the measured physical quantity.[99]

These types of measurements (also known as *real* measurements) are common in practice,[100] but for theoretical considerations it is useful to consider *ideal* measurements.[101] An ideal measurement is non-destructive and recoilless. In other words, an ideal measurement affects the system so little that the repetition of such a measurement within a short time interval [102] gives the same result.

As an example, we consider in Fig. S.1 a circular-polarized photon $|r\rangle$ incident on a polarizing beam splitter. The photon (irreversibly) changes to a linear-polarized state, say $|h\rangle$. A further measurement of this state again yields the same result. Therefore, measurements of this kind are also called quantum non-demolition measurements, QND.[103]

---

[97] We assume here that quantum mechanics is complete and there are no hidden variables.

[98] Example 1: A photon triggers a photomultiplier. It is absorbed and its energy is converted into an electrical signal. Example 2: An electron falls on a photographic plate and disappears among all the other electrons.

[99] Example: Measurement of the momentum of a neutron by observation of a recoil proton, which changes the momentum of the neutron during the interaction.

[100] Moreover, in these cases the probability statements concerning the measurement results also apply.

[101] Older designation by Pauli: measurement of the first kind (ideal) and the second kind (real).

[102] This means in such a short time interval that external influences cannot make themselves felt.

[103] The term 'non demolition' does not mean, however, that the wavefunction does not collapse.

We speak of an ideal measurement of a physical quantity A when the system is transferred by the measurement to an eigenstate of A.[104] If the system, for example, is in a superposition of energy eigenstates $|E_n\rangle$, then an ideal measurement of the energy will yield one of the the values E_N and the system will be transferred into the eigenstate $|E_N\rangle$. With a continuous quantity, of course, we can carry out only more or less ideal measurements (loosely formulated). Moreover, in the discrete case, also, one will not always be able to measure the exact spectrum in practice, because of the limited resolution of the detector or other instrumental limitations; here also, one often has to make do with only approximately ideal measurements.

Preparation

The term *preparation* refers to an operation which is intended to impose a given (initial) state on the system. The system (or the ensemble,[105] if one prefers) is thus forced into a certain state *after* the operation, while by a measurement the state of the system immediately *before* the measurement is probed. These different objectives are reflected in the fact that a preparation does not yield unique and storable numbers, in contrast to a measurement. But otherwise, ideal measurements and preparations share many properties, including the fact that both operations are nondestructive. Therefore, there are different opinions in the literature about the relationship between preparation and ideal measurement, ranging from "It is important to distinguish between measurement and preparation." through "Not all the processes of preparation are measurements in the traditional sense." to "Preparation is of course only one form of non-destructive measurement." (all quotes from the literature).

For example, consider the setup shown in Fig. S.1. Each measurement of the polarization by the second polarizing beam splitter necessarily gives $|h\rangle$. So one can understand this as indicating that the first polarizing beam splitter has prepared the system. This particular preparation is not a measurement—at least not in the sense that the result will be recorded before the following polarization measurement is made.

The following definition summarizes the situation: Let A be an observable with eigenvalues a_i (discrete spectrum). An operation on a set $\mathcal{E}$ of physical systems is called preparation (state preparation) for A if it leads to a division of $\mathcal{E}$ into subsets $\mathcal{E}_m$ such that for each m, a measurement of A immediately following the preparation is guaranteed to yield the result a_m for each system in the subset $\mathcal{E}_m$. If the operation is also a true measurement,[106] it is called an *ideal measurement* of A.

---

[104]For simplicity we assume that there is no degeneracy.

[105]The use of the word 'ensemble' does not mean that physical variables 'have' values that are distributed in an unknown way among the members of the ensemble. It is more of a code word that reminds us that in the pragmatic or instrumentalist approach, the predictions of the theory concern only the dispersion of the results of repeated measurements.

[106]That is, if a number is determined and stored.

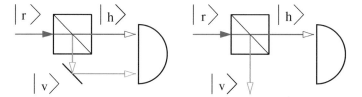

Fig. S.2 Nonselective (*left*) and selective (*right*) measurements

Indirect Quantum Measurement

To measure properties of a quantum object Q, one may allow another quantum object S (called the quantum probe) to interact with Q in a suitable way. S is then measured by a usual measuring apparatus.

Continuous Measurement

If one measures a system repeatedly, thereby letting the intervals between the times of measurement approach zero, one speaks (in the limiting case[107]) of a *continuous measurement*. Since in an ideal measurement, repeating the measurement within a sufficiently short time interval guarantees the same result, a continuous measurement may inhibit (under appropriate circumstances) the system from taking a different state. This is the quantum Zeno effect; the topic is formulated compactly in the sentence, 'a watched pot never boils'. More details are given in Appendix L, Vol. 1.

Selective Measurement

Given an observable A with a discrete non-degenerate spectrum; a measurement that selects only one of the eigenstates and does not register any other states is called a *selective measurement* (also known as filtering). Generalized, this means that if we have an initial ensemble and can split it by means of the measurement into different sub-ensembles, each of which would yield a different measurement result, then we speak of a selective measurement. If we mix the subsets after the measurement and then process them further (or do not select from the beginning), we refer to a non-selective measurement; cf. Fig. S.2.

---

[107]As an idealization, this limit is achievable, but not in reality, since every measurement takes a certain time.

Appendix T
Recaps and Outlines

In this chapter, we compile some material which is fundamental for the discussion of relativistic quantum theory and of relativistic quantum field theory in Vols. 1 and 2. The topics are special relativity, classical field theory and electrodynamics. We limit ourselves to issues actually needed for further discussion. We begin with a short comparison of the discrete and continuous description of functions.

T.1 Discrete - Continuous

There are two ways to describe a system. One can assume that it lives in either a finite or an infinite domain. Both versions occur, and to avoid mistakes one has to take care not to mingle them. This is made a little bit harder by a certain nonchalant way of notation. Therefore, here some words about the issue.

We start with the case that the system is confined to a finite volume (which may be arbitrarily large) with impermeable walls (infinite potential). Then we have something like the potential well (see Vol. 1 Chap. 5), and we know that the values of the momentum are not arbitrary but quantized, i.e., multiples of a basic wave length (in fact, of the half wave length). So we can count the allowed momenta and write them as e.g. $\mathbf{k}_n$, $n = 0, \pm 1, \pm 2, \ldots$, hence the name 'discrete case'. In addition, certain properties of the system are typically given by sums.

In contrast, if the system lives in an infinite volume, there is no basic wave length, i.e., every momentum is allowed; this is the continuous case. Thus, certain properties of the system are typically given by integrals, not by sums.

Now here enters the said nonchalant way of notation. In the discrete case, the sum over all possible plane waves should read $\sum_{n=-\infty}^{\infty} e^{-i\mathbf{k}_n \mathbf{x}}$ or something like that, but what one finds in literally all textbooks is a sort of shorthand notation, namely $\sum_{\mathbf{k}} e^{-i\mathbf{k}\mathbf{x}}$. The summation 'index' $\mathbf{k}$ indicates that we have the discrete case and means summation over all allowed momenta.

© Springer Nature Switzerland AG 2018
J. Pade, *Quantum Mechanics for Pedestrians 1*, Undergraduate Lecture
Notes in Physics, https://doi.org/10.1007/978-3-030-00464-4

Sometimes it is convenient to switch between the two ways of description. We compile here a few formulas. Essentially, they are based on the fact that each allowed **k**-value occupies a volume $(2\pi)^3 / V$ in reciprocal space.

1. The free solutions of the Klein–Gordon equation read in the case of a finite volume V

$$\phi(x) = \sum_{\mathbf{k}} \frac{1}{\sqrt{2V\omega_{\mathbf{k}}}} \left(a_{\mathbf{k}} e^{i(\mathbf{kr}-\omega_{\mathbf{k}}t)} + a_{\mathbf{k}}^{\dagger} e^{-i(\mathbf{kr}-\omega_{\mathbf{k}}t)} \right) \qquad (T.1)$$

where the sum runs over all allowed discrete values of **k**. The continuous variant reads :

$$\phi(x) = \frac{1}{(2\pi)^{3/2}} \int \frac{d^3 k}{\sqrt{2\omega_{\mathbf{k}}}} \left(a_{\mathbf{k}} e^{i(\mathbf{kr}-\omega_{\mathbf{k}}t)} + a_{\mathbf{k}}^{\dagger} e^{-i(\mathbf{kr}-\omega_{\mathbf{k}}t)} \right). \qquad (T.2)$$

2. It may happen that one has to calculate an integral like $\int d^3 x \, e^{ikx}$. Here one has to distinguish if it is about the continuous case or the discrete case. Usually, this is not noted by writing $\mathbf{k}_n$ or something like that, but is determined by the context. The result reads[108]

$$\int d^3 x \, e^{ikx} = \begin{cases} \delta_{\mathbf{k},0} \cdot V \\ \delta(\mathbf{k}) \cdot (2\pi)^3 \end{cases} \text{ for the } \begin{array}{c} \text{discrete} \\ \text{continuous} \end{array} \text{ case.} \qquad (T.4)$$

3. In the discrete case holds

$$\sum_{\mathbf{k}} e^{i\mathbf{k}(\mathbf{x}-\mathbf{y})} = V\delta(\mathbf{x}-\mathbf{y}) \qquad (T.5)$$

and in the continuous case

$$\int d^3 k \, e^{i\mathbf{k}(\mathbf{x}-\mathbf{y})} = (2\pi)^3 \, \delta(\mathbf{x}-\mathbf{y}). \qquad (T.6)$$

4. As the examples show, one can skip back and forth between 'discrete' and 'continuous' by certain replacements. Cum grano salis holds

$$\sum_{\mathbf{k}} \Leftrightarrow \int d^3 k \; ; \; V \Leftrightarrow (2\pi)^3 \; ; \; \delta_{\mathbf{k},0} \Leftrightarrow \delta(\mathbf{k}). \qquad (T.7)$$

---

[108] Due to the different factors V and $(2\pi)^3$, the use of the generalized Kronecker symbol $\delta(a,b)$ introduced in Vol. 1, Chap. 12:

$$\delta(a,b) = \begin{cases} \delta_{ab} \\ \delta(a-b) \end{cases} \text{ for } \begin{array}{c} a,b \text{ discrete} \\ a,b \text{ continuous} \end{array} \qquad (T.3)$$

does not facilitate the formulation considerably.

T.1.1 Exercises and Solutions

1. Prove (T.4).

   Solution: In the discrete case, the integration volume V is finite. The momentum **k** has only discrete values. We first consider the one dimensional case where V is the distance between say 0 and L_x (see also Vol. 1 Chap. 5). It follows

$$\int_0^{L_x} dx\, e^{ik_x x} = \left[\frac{e^{ik_x x}}{ik_x}\right]_0^{L_x} = \frac{e^{ik_x L_x}-1}{ik_x} \text{ for } k_x \neq 0 \; ; \; \int_0^{L_x} dx\, e^{ik_x x} = L_x \text{ for } k_x = 0.$$

$$(\text{T.8})$$

The allowed wave lengths are given by $n\frac{\lambda}{2} = L_x$. With $\lambda = \frac{2\pi}{k}$ follows $n\frac{\pi}{k_x} = L_x$ or $k_x = \frac{n\pi}{L_x}$. This yields

$$\int_0^{L_x} dx\, e^{ik_x x} = \frac{e^{i\frac{n\pi}{L_x}L_x}-1}{i\frac{n\pi}{L_x}} = \frac{e^{in\pi}-1}{i\frac{n\pi}{L_x}} = \frac{1-1}{i\frac{n\pi}{L_x}} = 0 \text{ for } k_x \neq 0 \text{ or } n \neq 0$$

$$\int_0^{L_x} dx\, e^{ik_x x} = L_x \text{ for } n = 0$$

$$(\text{T.9})$$

or

$$\int_0^{L_x} dx\, e^{ik_x x} = \delta_{k_x,0} \cdot L_x \qquad (\text{T.10})$$

Thus, we have in three dimensions

$$\int_V dx\, e^{i\mathbf{kx}} = \int_0^{L_x} dx\, e^{ik_x x} \cdot \int_0^{L_y} dx\, e^{ik_y y} \cdot \int_0^{L_z} dx\, e^{ik_z z} = \delta_{k_x,0}\cdot L_x \cdot \delta_{k_y,0}\cdot L_y \cdot \delta_{k_z,0}\cdot L_z = \delta_{\mathbf{k},\mathbf{0}}\cdot V. \; (\text{T.11})$$

In the continuous case, we use the definition of the delta function and obtain immediately

$$\int d^3x\, e^{i\mathbf{kx}} = (2\pi)^3\, \delta(\mathbf{k}). \qquad (\text{T.12})$$

T.2 Special Relativity

To define the notation and for the sake of completeness, we compile here some elements of special relativity (SR).[109]

---

[109] SR uses special notations and conventions which on the first view perhaps seem to be a little bit strange. But actually, they are sophisticated, perfectly adapted to the purpose and indispensable for topics like Quantum Field Theory.

T.2.1 Lorentz Boost and Four-Vectors

SR describes how the coordinates of an event in two inertial systems I and $\tilde{I}$ are related. If $\tilde{I}$ is moving relative to the frame I with velocity v along the x-axis, then the relation is given by the following *Lorentz boost* (in x-direction)

$$
\begin{aligned}
\tilde{t} &= \gamma \left(t - \tfrac{vx}{c^2} \right) \\
\tilde{x} &= \gamma \left(x - vt \right) \\
\tilde{y} &= y ; \ \tilde{z} = z
\end{aligned}
\tag{T.13}
$$

where $\gamma = \left(1 - \beta^2 \right)^{-1/2}$ and $\beta = v/c$.

In SR, time and space are on an equal footing. Thus, it is plausible to extend the notion of the three-dimensional position vector to a four-dimensional vector which also includes the time. These *four-vectors* (4-vectors) which are essential ingredients of SR have one time-like and three space-like components. We write the coordinates of an event in space and time as

$$
x^0 = ct ; \ x^1 = x ; \ x^2 = y ; \ , x^3 = z
\tag{T.14}
$$

or

$$
x = \left(x^0, x^1, x^2, x^3 \right).
\tag{T.15}
$$

Note that these are upper indices and not powers. x is the prototype of a 4-vector.

A remark on notation: In general, 4-vectors are displayed in italic script and 3-vectors in bold italic. Thus

$$
x = \left(x^0, x^1, x^2, x^3 \right) = \left(x^0, \mathbf{x} \right).
\tag{T.16}
$$

x^0 is always the time-component ($x^0 = ct$). In addition, the components of a 4-vector are listed with a *Greek index* like λ, μ or ν; the components of a 3-vector are usually labelled with a *Roman index* from i to n.

As in (T.15), 4-vectors are often written as row vectors which improves the readability of texts. However, if one wants to perform matrix calculations, it is better to think of x as a column vector. This is clearly seen by writing the Lorentz transformation (T.13) in matrix form. On the left and the right there are the components of the four-vector $\tilde{x}$ and x. Thus we have

$$
\begin{pmatrix} \tilde{x}^0 \\ \tilde{x}^1 \\ \tilde{x}^2 \\ \tilde{x}^3 \end{pmatrix} = \begin{pmatrix} \gamma & -\beta\gamma & 0 & 0 \\ -\beta\gamma & \gamma & 0 & 0 \\ 0 & 0 & 1 & 0 \\ 0 & 0 & 0 & 1 \end{pmatrix} \begin{pmatrix} x^0 \\ x^1 \\ x^2 \\ x^3 \end{pmatrix}.
\tag{T.17}
$$

We denote the transformation matrix by Λ with elements $\Lambda^{\mu}{}_{\nu}$ ($\mu = 0, 1, 2, 3$ labels the rows, $\nu = 0, 1, 2, 3$ the columns). Apparently, Λ is symmetrical. With the usual notation for matrix multiplication, we can write the Lorentz transformation (T.17) as

$$\tilde{x}^\mu = \sum_{\nu=0}^{3} \Lambda^\mu{}_\nu \, x^\nu \; ; \mu = 0, 1, 2, 3. \tag{T.18}$$

In SR, there is second kind of vector whose prototype is the vector of derivatives

$$\left(\frac{\partial}{c\partial t}, \nabla \right) = \left(\frac{\partial}{c\partial t}, \frac{\partial}{\partial x}, \frac{\partial}{\partial y}, \frac{\partial}{\partial z} \right) = \left(\frac{\partial}{\partial x^0}, \frac{\partial}{\partial x^1}, \frac{\partial}{\partial x^2}, \frac{\partial}{\partial x^3} \right). \tag{T.19}$$

Under a Lorentz-transformation, this vector does not transform as given in (T.18), but as

$$\frac{\partial}{\partial \tilde{x}^\mu} = \sum_{\nu=0}^{3} \left(\frac{\partial x^\nu}{\partial \tilde{x}^\mu} \right) \frac{\partial}{\partial x^\nu} \tag{T.20}$$

which expression may be written using the common short-hand notation $\frac{\partial}{\partial x^\mu} = \partial_\mu$ as

$$\tilde{\partial}_\mu = \sum_{\nu=0}^{3} \left(\frac{\partial x^\nu}{\partial \tilde{x}^\mu} \right) \partial_\nu = \sum_{\nu=0}^{3} \Lambda_\mu{}^\nu \, \partial_\nu \tag{T.21}$$

where $\Lambda_\mu{}^\nu = \frac{\partial x^\nu}{\partial \tilde{x}^\mu}$ is the inverse of $\Lambda^\mu{}_\nu = \frac{\partial \tilde{x}^\mu}{\partial x^\nu}$ or in matrix notation:

$$\Lambda = (\Lambda^\mu{}_\nu) = \begin{pmatrix} \gamma & -\beta\gamma & 0 & 0 \\ -\beta\gamma & \gamma & 0 & 0 \\ 0 & 0 & 1 & 0 \\ 0 & 0 & 0 & 1 \end{pmatrix} \; ; \; \Lambda^{-1} = (\Lambda_\mu{}^\nu) = \begin{pmatrix} \gamma & \beta\gamma & 0 & 0 \\ \beta\gamma & \gamma & 0 & 0 \\ 0 & 0 & 1 & 0 \\ 0 & 0 & 0 & 1 \end{pmatrix}. \tag{T.22}$$

Note that Λ and Λ^{-1} only differ in the sign of β which is essentially the relative velocity of the two reference frames.

We now generalize (T.18) and (T.21) and define general 4-vectors by means of their behavior under Lorentz transformations. A general 4-vector with components a^0, a^1, a^2, a^3 has to fulfill

$$\tilde{a}^\mu = \sum_{\nu=0}^{3} \left(\frac{\partial \tilde{x}^\mu}{\partial x^\nu} \right) a^\nu = \sum_{\nu=0}^{3} \Lambda^\mu{}_\nu \, a^\nu \tag{T.23}$$

and a general 4-vector with components a_0, a_1, a_2, a_3 has to fulfill

$$\tilde{a}_\mu = \sum_{\nu=0}^{3} \left(\frac{\partial x^\nu}{\partial \tilde{x}^\mu} \right) a_\nu = \sum_{\nu=0}^{3} \Lambda_\mu{}^\nu \, a_\nu. \tag{T.24}$$

As an example, the written-out (T.23) is found below in (T.25).

Note that the 4-vectors of SR are defined by their behavior under Lorentz transformations, and not, as the 'usual' 3-vectors, by the behavior under space transformations. This leads among others to a different definition of the inner product. For this and some examples for 4-vectors see below.

T.2.1.1   Contra- and Covariant Vectors

In SR, we have two ways to define the components of a vector a, namely (a^0, a^1, a^2, a^3) and (a_0, a_1, a_2, a_3). The names of the the two types are owed to the historical heritage: *contravariant vector* for (a^0, a^1, a^2, a^3) and *covariant vector* for (a_0, a_1, a_2, a_3). This naming is common, but a little bit unfortunate and a misnomer at least for two reasons.

(1) A covariant transformation is a properly defined transformation in the SR; in this sense, both types of vectors transform covariantly.

(2) Contravariant vector and covariant vector are not two different vectors, as maybe the names would suggest, but the very same vector a with different components, i.e., formulated in different coordinate systems. Take as an example the plane with a skewed coordinate system, i.e., the coordinate axes x^1 and x^2 enclose an angle $\neq 90°$. There are two ways to define the components of a vector a in such an oblique system, simply since the parallel to one axis is not perpendicular to the other axis.

– For the first way, one drops a line from the tip of the vector, parallel to the x^2 axis, onto the x^1 axis, and another line, parallel to the x^1 axis, onto the x^2 axis. The intercepts of these lines with the axes are called the contravariant components of the vector, a^1 and a^2.

– For the second way, one drops a perpendicular from the tip of the vector onto the x^1 axis and another perpendicular from the tip of a onto the x^2 axis. The intercepts are called covariant components of the vector, a_1 and a_2.

In a Cartesian coordinate system, the perpendiculars onto one axis are parallel to the other. Thus, contravariant and covariant components coincide, $a^1 = a_1$ and $a^2 = a_2$.

To circumvent the unfortunate naming contra- and covariant, some textbooks prefer terms like upstairs and downstairs vector or something like that.

T.2.1.2   Remarks on Notation

The way to index terms in SR has great advantages; it is excellently adapted to the questions of SR, simple, concise and elegant. However, for the beginner it offers perhaps some difficulties.

Labeling contra- and covariant 4-vectors Often it is not sufficient to label a vector simply a because it would be not clear if it is meant in its upstairs or downstairs formulation. It has become established to write a^μ for a contravariant and a_μ for a covariant vector. This could be confusing since μ is an index which takes values

0, 1, 2, 3. Thus, it is not clear whether a^μ means the vector with its 4 components or just a single component (as e.g. in (T.23) and (T.24)). However, in general, the context clearly defines the meaning. If necessary for the sake of clarity, one can use the notation a_{cov} and a^{con} instead of just a.

Labeling Lorentz transformations We are dealing with two types of vectors, co- and contravariant, distinguished by the position of the index. Thus, it is useful to have also upstairs and downstairs indices for operators (matrices, tensors) - the operators themselves should carry the information on which objects they act. The usual notation for a matrix like $\Lambda_{\mu\nu}$ would not provide this information. Thus, we write $\Lambda^\mu{}_\nu$ to make clear that this Lorentz transformation acts on a contravariant vector, and $\Lambda_\mu{}^\nu$ on a covariant vector. Note that the notation $\Lambda^\mu{}_\nu$ and $\Lambda_\mu{}^\nu$ comes from tensor calculus. Since we use this special way of indexing operators only for these two terms, we will not discuss it further. Just imagine that the purpose is to make clear from the notation what the terms are supposed to act on: $\Lambda^\mu{}_\nu$ on a vector of type a^ν, $\Lambda_\mu{}^\nu$ on a vector of type a_ν.

Labeling 3-vectors We repeat that for 3-vectors covariant and contravariant components are the same and therefore, the position of the indices is irrelevant. In the context of our discussion, we can write e.g. for the Pauli matrices $\boldsymbol{\sigma}$ either $(\sigma_1, \sigma_2, \sigma_3)$ or $(\sigma^1, \sigma^2, \sigma^3)$.

A certain caution may be required when dealing with the indices. For instance, consider the 4-momentum p with $p^\mu = (p^0, \mathbf{p})$ and $p_\mu = (p_0, -\mathbf{p})$. Written out, we have $p^\mu = (p^0, p^1, p^2, p^3)$ and $p_\mu = (p_0, p_1, p_2, p_3)$, i.e., $p^0 = p_0$ and $p_1 = -p^1$ and so on. Now take the 3-momentum $\mathbf{p}$. If we write the components as $\mathbf{p} = (p^1, p^2, p^3)$, we have $\mathbf{p} = (p_1, p_2, p_3)$ since for a 3-vector, contravariant and covariant components are equal, as said above. Thus, we would have for the 4-vector $p_1 = -p^1$ and for the 3-vector $p^1 = p_1$. One can avoid this ambiguity by writing $\mathbf{p} = (p_x, p_y. p_z)$, for instance.

T.2.2   Four-Vector Inner Product, Metric Tensor, Einstein Convention

T.2.2.1   Four-Vector Inner Product

Being a scalar, the inner product $a \cdot b$ of two 4-vectors should not depend on the reference frame, but should be invariant with respect to Lorentz transformations, i.e., $a \cdot b = \tilde{a} \cdot \tilde{b}$. With (T.23), i.e.,

$$\tilde{a}^0 = \gamma\left(a^0 - \beta a^1\right) \; ; \; \tilde{a}^1 = \gamma\left(-\beta a^0 + a^1\right) \; ; \; \tilde{a}^2 = a^2 \; ; \; \tilde{a}^3 = a^3 \qquad (T.25)$$

follows immediately that the 'familiar rule' does not work: $\sum_{\mu=0}^3 \tilde{a}^\mu \tilde{b}^\mu \neq \sum_{\mu=0}^3 a^\mu b^\mu$, since on the l.h.s there are mixed terms like $a^0 b^1$ which do not cancel and are not found on the r.h.s.. Thus, we make the ansatz

$$a \cdot b = p a^0 b^0 + q \mathbf{a} \cdot \mathbf{b} \tag{T.26}$$

with two yet to be determined constants p and q. We insert (T.25) into the equation

$$\tilde{a} \cdot \tilde{b} = p \tilde{a}^0 \tilde{b}^0 + q \tilde{\mathbf{a}} \cdot \tilde{\mathbf{b}} \overset{!}{=} a \cdot b = p a^0 b^0 + q \mathbf{a} \cdot \mathbf{b}. \tag{T.27}$$

This yields

$$a^0 b^0 \left[p\gamma^2 + q\gamma^2\beta^2 \right] + a^0 b^1 \left[-p\gamma^2\beta - q\gamma^2\beta \right] + a^1 b^0 \left[-p\gamma^2\beta - q\gamma^2\beta \right] + \\ a^1 b^1 \left[p\gamma^2\beta^2 + q\gamma^2 \right] + q a^2 b^2 + q a^3 b^3 \overset{!}{=} p a^0 b^0 + q a^1 b^1 + q a^2 b^2 + q a^3 b^3. \tag{T.28}$$

As mentioned before, the mixed terms have to vanish which leads to $q = -p$. Accordingly, since $\gamma^2 \left(1 - \beta^2 \right) = 1$, the prefactors of $a^0 b^0$ (and of $a^1 b^1$) are equal on both sides. Finally, to hold things simple, we choose $p = 1 = -q$ and obtain the expression [110]

$$a \cdot b = a^0 b^0 - \mathbf{a} \cdot \mathbf{b}. \tag{T.29}$$

With contra- and covariant vectors, we have

$$a \cdot b = a^\mu b_\mu = a_\mu b^\mu. \tag{T.30}$$

Inner products always involve a contra- and a covariant vector. It does not import which one of the two vectors is written as contravariant or as covariant.

Thus, the definition of the inner product for two four-vectors in SR differs obviously from that of Euclidean 3-vectors. Especially, this leads to another definition of the length of a 4-vector, namely

$$a \cdot a = a^0 a^0 - \mathbf{a} \cdot \mathbf{a} = a^0 a^0 - \left(a^1 a^1 + a^2 a^2 + a^3 a^3 \right). \tag{T.31}$$

Note that this term is not positive definite.

Let us stress once more that the inner product of two 4-vectors, defined in this way, does not depend on the reference frame, i.e., it is *invariant* under Lorentz transformations. As an example, we consider the length of the 4-vector $x = \left(x^0, x^1, x^2, x^3 \right) = \left(x^0, \mathbf{x} \right)$. It is given by

$$\|x\|^2 = \left(x^0 \right)^2 - \mathbf{x} \cdot \mathbf{x} = \left(x^0 \right)^2 - \left[\left(x^1 \right)^2 + \left(x^2 \right)^2 + \left(x^3 \right)^2 \right]. \tag{T.32}$$

---

[110]One can equally well choose $p = -1$ and $q = 1$ which is done in some books. Then, for time-like and space-like vectors a holds $\|a\|^2 < 0$ and $\|a\|^2 > 0$. See below the discussion about the metric tensor.

This expression is an invariant known as the *spacetime interval*. Four-vectors x with $\|x\|^2 > 0$ are called time-like vectors, with $\|x\|^2 < 0$ space-like vectors, and with $\|x\|^2 = 0$ light-like.[111]

T.2.2.2 Metric Tensor

We now look for the relation between the two types of four-vectors. We write (T.23) and (T.24) in the form $\tilde{a}^{con} = \Lambda a^{con}$ and $\tilde{a}_{cov} = \Lambda^{-1} a_{cov}$. Our ansatz reads $a_{cov} = G a^{con}$, where G is a 4×4-matrix yet to be determined. Since the transformation should not depend on the frame of reference, we have also $\tilde{a}_{cov} = G \tilde{a}^{con}$. Thus, starting with (T.23) yields

$$\tilde{a}^{con} = \Lambda a^{con} \rightarrow G \tilde{a}^{con} = G \Lambda G^{-1} G a^{con} \rightarrow \tilde{a}_{cov} = \Lambda^{-1} a_{cov} = G \Lambda G^{-1} a_{cov}.$$
(T.33)

In other words, G must fulfill the equation $G \Lambda G^{-1} = \Lambda^{-1}$ or $G \Lambda = \Lambda^{-1} G$. As it turns out (see exercise), the simplest solution is given by

$$G = \left(g_{\mu\nu} \right) = \begin{pmatrix} 1 & 0 & 0 & 0 \\ 0 & -1 & 0 & 0 \\ 0 & 0 & -1 & 0 \\ 0 & 0 & 0 & -1 \end{pmatrix}.$$
(T.34)

This object (also named $\eta_{\mu\nu}$) is called *metric tensor*; it is at the heart of SR. Obviously, the metric tensor is symmetrical.

Note that the other simplest and equivalent solution has reversed signs, $g_{00} = -1$ and $g_{11} = g_{22} = g_{33} = 1$ which choice is made in some textbooks. In this context one uses the term *metric signature* which gives the number of positive, negative and zero eigenvalues. For the metric tensor in SR, it is often denoted by $(+, -, -, -)$ or $(-, +, +, +)$. Of course, it reflects the choice of the sign for the inner product (T.29). The choice of the metric signature is theoretically inconsequential, but one has simply to choose one of the two alternatives for purposes of internal consistency.

For later purposes, it is useful to introduce $g^{\mu\nu} = g_{\mu\nu}$. With (T.34), the connection between the two types of four-vectors is given by

$$a_{\mu} = \sum_{\nu=0}^{3} g_{\mu\nu} a^{\nu}$$
(T.35)

or explicitly in matrix form

---

[111] The relativistic *line element* ds is given by $ds^2 = \left(dx^0 \right)^2 - (d\mathbf{x})^2$ and the *eigentime* by $d\tau = \frac{1}{c} ds$. We see, that by construction the line element and the eigentime are Lorentz invariant quantities.

$$
\begin{pmatrix} a_0 \\ a_1 \\ a_2 \\ a_3 \end{pmatrix} = \begin{pmatrix} 1 & 0 & 0 & 0 \\ 0 & -1 & 0 & 0 \\ 0 & 0 & -1 & 0 \\ 0 & 0 & 0 & -1 \end{pmatrix} \begin{pmatrix} a^0 \\ a^1 \\ a^2 \\ a^3 \end{pmatrix} = \begin{pmatrix} a^0 \\ -a^1 \\ -a^2 \\ -a^3 \end{pmatrix} \tag{T.36}
$$

or

$$
a^{con} = \left(a^0, \mathbf{a} \right) \ , \ a_{cov} = \left(a^0, -\mathbf{a} \right). \tag{T.37}
$$

T.2.2.3 Einstein Summation Convention

Since summations over the four indices $\mu = 0, 1, 2, 3$ as in (T.35) occur quite often in SR, we adopt the extremely useful *Einstein summation convention* by which twice appearing indices are to be summed up (provided, one is 'upstairs' and one 'downstairs'), thereby omitting the summation sign. Using this convention, the connection between co- and contravariant vectors (T.35) reads

$$
a_\mu = g_{\mu\nu} a^\nu \tag{T.38}
$$

and the inverse transformation is given by

$$
a^\mu = g^{\mu\nu} a_\nu \tag{T.39}
$$

with $g_{\mu\nu} = g^{\mu\nu}$. As is seen, we can lower or raise an index by inserting the metric tensor.[112]

With the summation convention, the inner product (T.29) is written as[113]

$$
a \cdot b = g_{\mu\nu} a^\mu b^\nu = a_\nu b^\nu = a_\mu b^\mu. \tag{T.40}
$$

Finally, the behavior of contra- and covariant vectors under Lorentz transformations (see (T.23) and (T.24)) is written as

$$
\tilde{a}^\mu = \Lambda^\mu{}_\nu \, a^\nu \ ; \ \tilde{a}_\mu = \Lambda_\mu{}^\nu \, a_\nu. \tag{T.41}
$$

T.2.2.4 Special Four-Vectors

As stated above, a four-vector in SR transforms in a specific way under Lorentz transformations. It has four components, but its length is determined differently from an Euclidean vector. In SR, there is a bunch of common 4-vectors, as for instance position, the momentum or the potential.

---

[112]This holds also for tensors, e.g. $g_{\mu\nu} \Lambda^\nu{}_\rho = \Lambda_{\mu\rho}$.

[113]Note that if the same index occurs upstairs and downstairs, it is used up by the summation; accordingly, it may be named arbitrarily (dummy index): $a_\mu b^\mu = a_\nu b^\nu = a_\rho b^\rho$.
The inner product always involves one contravariant and one covariant vector.

It is clear that all problems in SR could be solved without invoking 4-vectors.[114]
But they are a powerful tool which makes life very much easier. In fact, many
problems would be nearly impossible to treat without the use of 4-vectors. On the
one hand, equations between 4-vectors which hold in a particular inertial system
are automatically valid in all systems. On the other hand, the 4-vector inner product
is invariant and the same in all frames. So in treating a problem, one may choose
the frame in which the problem appears in its simplest form. In addition, the use of
4-vectors may give new insights. For example, the conservation of the 4-momentum
includes the conservation of energy.

Here are some examples for 4-vectors.

(1) The position 4-vector or 4-position could be denoted by $x = (ct, \mathbf{x})$ or $x = (x_0, \mathbf{x})$. But in this notation it is not clear if we mean the contravariant or the covariant
version. It is common practice to write $x^\mu = (ct, \mathbf{x})$ though it is, as stated above, a
certain misuse of notation; the context has to clear if x^μ means one component of
the 4-vector or the whole 4-vector. So we write

$$x^\mu = (x_0, \mathbf{x}) \tag{T.42}$$

and its inner product, called spacetime interval, is given by

$$x_\mu x^\mu = \left(x^0\right)^2 - \mathbf{x} \cdot \mathbf{x}. \tag{T.43}$$

(2) The energy-momentum 4-vector or 4-momentum is given by

$$p^\mu = \left(\frac{E}{c}, \mathbf{p}\right). \tag{T.44}$$

The inner product reads

$$p_\mu p^\mu = \frac{E^2}{c^2} - \mathbf{p}^2 = \frac{m^2 c^4 + c^2 \mathbf{p}^2}{c^2} - \mathbf{p}^2 = m^2 c^2 \tag{T.45}$$

where m is the rest mass. Note that mc^2 is an invariant. For objects with vanishing
rest mass like the photon, we have $p_\mu p^\mu = 0$.

(3) In connection with the 4-momentum we can also define the 4-wavenumber
$k^\mu = p^\mu / \hbar$:

$$k^\mu = \left(\frac{E}{\hbar c}, \frac{\mathbf{p}}{\hbar}\right) = \left(\frac{\omega}{c}, \mathbf{k}\right). \tag{T.46}$$

The inner product of the two 4-vectors k and x is given by

---

[114]The same holds for contra- and covariant vectors. Due to the relations (T.38) and (T.39), one
could formulate the SR with one sort of index only, e.g. contravariant vectors only. But this would
result in a quite cumbersome formalism without the transparency and elegance of the established
method.

$$kx = k^\mu x_\mu = k_\mu x^\mu = k^0 x_0 - \mathbf{k}\mathbf{x} = \omega t - \mathbf{k}\mathbf{x} \tag{T.47}$$

and it follows

$$e^{ikx} = e^{i(\omega t - \mathbf{k}\mathbf{x})}. \tag{T.48}$$

(4) The vector potential 4-vector is given by

$$A^\mu = \left(\frac{\Phi}{c}, \mathbf{A}\right). \tag{T.49}$$

(5) The four-dimensional derivative operator or 4-gradient is given by

$$\partial_\mu = \frac{\partial}{\partial x^\mu} = \left(\frac{1}{c}\frac{\partial}{\partial t}, \nabla\right) = (\partial_0, \nabla) \tag{T.50}$$

and the contravariant form by

$$\partial^\mu = \frac{\partial}{\partial x_\mu} = \left(\frac{1}{c}\frac{\partial}{\partial t}, -\nabla\right) = (\partial_0, -\nabla). \tag{T.51}$$

The inner product (also called d'Alembert operator) reads

$$\partial^2 = \partial_\mu \partial^\mu = \frac{1}{c^2}\frac{\partial^2}{\partial t^2} - \nabla^2. \tag{T.52}$$

Note that since ∂ is an operator, one can not attribute a length to it.

(6) The 4-current (4-current density, current density 4-vector) is defined as

$$j^\mu = (c\rho, \mathbf{j}) \tag{T.53}$$

where ρ and $\mathbf{j}$ are the charge and current density. With j^μ, the continuity equation $\frac{\partial \rho}{\partial t} + \nabla \mathbf{j} = 0$ reads

$$\partial_\mu j^\mu = 0. \tag{T.54}$$

(7) The sum of two 4-vectors is a 4-vector. An example is given by the difference of the 4-momentum and the 4-potential

$$p^\mu - qA^\mu = \left(\frac{E}{c}, \mathbf{p}\right) - q\left(\frac{\Phi}{c}, \mathbf{A}\right) = \left(\frac{E - q\Phi}{c}, \mathbf{p} - q\mathbf{A}\right). \tag{T.55}$$

T.2.3 Exercises and Solutions

1. Show $\Lambda \cdot \Lambda^{-1} = 1$.
   Solution:

$$\Lambda \cdot \Lambda^{-1} = \begin{pmatrix} \gamma & -\beta\gamma & 0 & 0 \\ -\beta\gamma & \gamma & 0 & 0 \\ 0 & 0 & 1 & 0 \\ 0 & 0 & 0 & 1 \end{pmatrix} \begin{pmatrix} \gamma & \beta\gamma & 0 & 0 \\ \beta\gamma & \gamma & 0 & 0 \\ 0 & 0 & 1 & 0 \\ 0 & 0 & 0 & 1 \end{pmatrix} = \begin{pmatrix} \gamma^2(1-\beta)^2 & 0 & 0 & 0 \\ 0 & \gamma^2(1-\beta)^2 & 0 & 0 \\ 0 & 0 & 1 & 0 \\ 0 & 0 & 0 & 1 \end{pmatrix}. \quad \text{(T.56)}$$

   It is $\gamma = \left(1 - \beta^2\right)^{-1/2}$ and therefore $\Lambda \cdot \Lambda^{-1} = 1$.
2. Show

$$\sum_{\mu=0}^{3} \Lambda^\mu{}_\nu \Lambda_\mu{}^\rho = \sum_{\mu=0}^{3} \left(\frac{\partial \tilde{x}^\mu}{\partial x^\nu}\right)\left(\frac{\partial x^\rho}{\partial \tilde{x}^\mu}\right) = \delta^\rho_\nu \quad \text{(T.57)}$$

   where δ^ρ_ν is the Kronecker symbol in the SR:

$$\delta^\rho_\nu = \begin{cases} 1 \\ 0 \end{cases} \text{ for } \begin{matrix} \nu = \rho \\ \nu \neq \rho \end{matrix}. \quad \text{(T.58)}$$

   Solution: This is the same question as in exercise 1, written out explicitly.
3. Solve the equation $G\Lambda = \Lambda^{-1}G$ and determine the two simplest solutions.
   Solution: From $G\Lambda = \Lambda^{-1}G$ we have

$$\begin{pmatrix} g_{00} & g_{01} & g_{02} & g_{03} \\ g_{10} & g_{11} & g_{12} & g_{13} \\ g_{20} & g_{21} & g_{22} & g_{23} \\ g_{30} & g_{31} & g_{32} & g_{33} \end{pmatrix} \begin{pmatrix} \gamma & -\beta\gamma & 0 & 0 \\ -\beta\gamma & \gamma & 0 & 0 \\ 0 & 0 & 1 & 0 \\ 0 & 0 & 0 & 1 \end{pmatrix} = \begin{pmatrix} \gamma & \beta\gamma & 0 & 0 \\ \beta\gamma & \gamma & 0 & 0 \\ 0 & 0 & 1 & 0 \\ 0 & 0 & 0 & 1 \end{pmatrix} \begin{pmatrix} g_{00} & g_{01} & g_{02} & g_{03} \\ g_{10} & g_{11} & g_{12} & g_{13} \\ g_{20} & g_{21} & g_{22} & g_{23} \\ g_{30} & g_{31} & g_{32} & g_{33} \end{pmatrix}. \quad \text{(T.59)}$$

   Due to the block structure of Λ, it is advantageous to define the matrices

$$A = \begin{pmatrix} g_{00} & g_{01} \\ g_{10} & g_{11} \end{pmatrix}, B = \begin{pmatrix} g_{02} & g_{03} \\ g_{12} & g_{13} \end{pmatrix}, C = \begin{pmatrix} g_{20} & g_{21} \\ g_{30} & g_{31} \end{pmatrix}, D = \begin{pmatrix} g_{22} & g_{23} \\ g_{32} & g_{33} \end{pmatrix}$$
$$\text{(T.60)}$$

   and

$$S(\beta) = \gamma \begin{pmatrix} 1 & \beta \\ \beta & 1 \end{pmatrix}. \quad \text{(T.61)}$$

   Then, (T.59) reads

$$\begin{pmatrix} A & B \\ C & D \end{pmatrix} \begin{pmatrix} S(-\beta) & 0 \\ 0 & 1 \end{pmatrix} = \begin{pmatrix} S(\beta) & 0 \\ 0 & 1 \end{pmatrix} \begin{pmatrix} A & B \\ C & D \end{pmatrix}. \quad \text{(T.62)}$$

   Multiplying the matrices yields

$$\begin{pmatrix} AS(-\beta) & B \\ CS(-\beta) & D \end{pmatrix} = \begin{pmatrix} S(\beta)A & S(\beta)B \\ C & D \end{pmatrix}. \tag{T.63}$$

One sees immediately that the block matrix D remains undetermined. For B we have

$$B = S(\beta)B \rightarrow \begin{pmatrix} g_{02} & g_{03} \\ g_{12} & g_{13} \end{pmatrix} = \gamma \begin{pmatrix} 1 & \beta \\ \beta & 1 \end{pmatrix} \begin{pmatrix} g_{02} & g_{03} \\ g_{12} & g_{13} \end{pmatrix} = \gamma \begin{pmatrix} g_{02} + \beta g_{12} & g_{03} + \beta g_{13} \\ \beta g_{02} + g_{12} & \beta g_{03} + g_{13} \end{pmatrix} \tag{T.64}$$

which leads to $g_{02} = g_{03} = g_{12} = g_{13} = 0$ or $B = 0$. Analogously we have $C = 0$.

The remaining equation $AS(-\beta) = S(\beta)A$ reads explicitly

$$\begin{pmatrix} g_{00} & g_{01} \\ g_{10} & g_{11} \end{pmatrix} \begin{pmatrix} 1 & -\beta \\ -\beta & 1 \end{pmatrix} = \begin{pmatrix} 1 & \beta \\ \beta & 1 \end{pmatrix} \begin{pmatrix} g_{00} & g_{01} \\ g_{10} & g_{11} \end{pmatrix} \tag{T.65}$$

or

$$\begin{pmatrix} g_{00} - \beta g_{01} & -\beta g_{00} + g_{01} \\ g_{10} - \beta g_{22} & -\beta g_{10} + g_{11} \end{pmatrix} = \begin{pmatrix} g_{00} + \beta g_{10} & g_{01} + \beta g_{11} \\ \beta g_{00} + g_{10} & \beta g_{01} + g_{11} \end{pmatrix}. \tag{T.66}$$

The equation is apparently fulfilled for $g_{11} = -g_{00}$ and $g_{10} = -g_{01}$. Thus, the result is

$$G = \begin{pmatrix} g_{00} & g_{01} & 0 & 0 \\ -g_{01} & -g_{00} & 0 & 0 \\ 0 & 0 & g_{22} & g_{23} \\ 0 & 0 & g_{32} & g_{33} \end{pmatrix}. \tag{T.67}$$

We can choose all these entries arbitrarily. First, we want the off-diagonal elements to be zero. Second, treating the space coordinates on a equal footing, we put $g_{22} = g_{33} = -g_{00}$. Finally, we choose as simplest case $g_{00} = 1$ and arrive at

$$G = (g_{\mu\nu}) = \begin{pmatrix} 1 & 0 & 0 & 0 \\ 0 & -1 & 0 & 0 \\ 0 & 0 & -1 & 0 \\ 0 & 0 & 0 & -1 \end{pmatrix}. \tag{T.68}$$

The other simplest solution is given by the choice $g_{00} = -1$ and $g_{11} = g_{22} = g_{33} = 1$.

4. We have $a_{cov} = Ga^{con}$ with G given in (T.34). Show that also holds $a^{con} = Ga_{cov}$.

   Solution: Obviously, it is $G^{-1} = G$. Thus, from $a_{cov} = Ga^{con}$ follows $a^{con} = G^{-1}a_{cov} = Ga_{cov}$.

5. Check that

$$\sum_{\nu=0}^{4} g_{\mu\nu}g^{\nu\rho} = \delta_\mu^\rho \tag{T.69}$$

6. Show that one can write

$$a \cdot b = g^{\mu\nu} a_\mu b_\nu = a^\mu b_\mu \tag{T.70}$$

Solution: It is[115]

$$a^\mu b_\mu = g^{\mu\nu} a_\mu b_\nu = a_\mu g^{\mu\nu} b_\nu = a_\mu b^\mu. \tag{T.71}$$

7. Show that $\partial_\mu x^\mu = 4$.
   Solution:

$$\partial_\mu x^\mu = 4 \quad \text{due to} \quad \partial_\mu x^\nu = \delta_\mu^\nu. \tag{T.72}$$

8. The current density 4-vector is given by

$$j = (\rho c, \mathbf{j}) . \tag{T.73}$$

   Calculate $\partial_\mu j^\mu$ in terms of ρ and $\mathbf{j}$.
   Solution:

$$\partial_\mu j^\mu = \partial_0 j^0 + \partial_k j^k = \partial_t \rho + \nabla \mathbf{j} = \dot{\rho} + \operatorname{div} \mathbf{j}. \tag{T.74}$$

9. Regarding E and $\mathbf{p}$ as operators $i\hbar\partial_t$ and $\frac{\hbar}{i}\nabla$, show that holds

$$p^\mu = i\hbar\partial^\mu \quad (\text{and, of course,} \quad p_\mu = i\hbar\partial_\mu). \tag{T.75}$$

   Solution:

$$p^\mu = \left(p^0, p^k\right) = \left(\frac{E}{c}, \mathbf{p}\right) = \left(\frac{1}{c}i\hbar\partial_t, \frac{\hbar}{i}\nabla\right) = i\hbar\left(\frac{1}{c}\partial_t, -\nabla\right) = i\hbar\left(\partial^0, -\partial^k\right) = i\hbar\partial^\mu. \tag{T.76}$$

10. The velocity 4-vector (4-velocity) is given by

$$v = \frac{dx}{d\tau} = \frac{dx}{dt}\frac{dt}{d\tau} = \gamma(c, \mathbf{v}) \tag{T.77}$$

    where τ is the proper time. Determine $v_\mu v^\mu$.
    Solution:[116]

$$v_\mu v^\mu = \gamma^2 \left(c^2 - \mathbf{v}^2\right) = \frac{1}{1 - \left(\frac{v}{c}\right)^2} c^2 \left(1 - \left(\frac{\mathbf{v}}{c}\right)^2\right) = c^2. \tag{T.78}$$

---

[115] Note that all terms are scalars and thus may be written in arbitrary order.
[116] Remind $\gamma = \left(1 - \beta^2\right)^{-1/2}$ and $\beta = v/c$.

T.3   Classical Field Theory

In order to quantize a classical system one is interested in a generally valid approach which has not to be tailored to the special system under consideration. As it is known, this universal method is the Lagrange-Hamilton formalism. It answers the relevant questions as, for instant, how to find those variables which in the process of quantizing will become non-commuting operators, how to find the energy density (Hamiltonian density) and so on.

After a compressed revision of the formalism for particles, we review the basics of classical field theory. The emphasis is on the presentation of the most important results, not on their derivation.

T.3.1   Particles

T.3.1.1   One Coordinate q

We consider the one-dimensional motion of a particle whose (generalized) coordinate q depends on time, $q = q(t)$. Let L be a given function, called *Lagrange function* or Lagrangian[117] which depends on q and its time derivative $\dot{q}$, i.e., $L = L(q, \dot{q})$. The action S is defined as the integral of L over time, $S = \int_{t_1}^{t_2} L(q, \dot{q})\, dt$. *Hamilton's principle of least action* states that the motion of the particle is determined by the condition that the variation of the action disappears[118], $\delta S = \delta \int_{t_1}^{t_2} L(q, \dot{q})\, dt = 0$. Thus, the orbit of the particle between (q_1, t_1) and (q_2, t_2) is that one for which the action is stationary. Performing the variation leads to the *Euler–Lagrange equation*

$$\frac{\partial L}{\partial q} - \frac{d}{dt}\frac{\partial L}{\partial \dot{q}} = 0. \tag{T.79}$$

The *conjugated momentum* p is defined by

$$p = \frac{\partial L}{\partial \dot{q}}. \tag{T.80}$$

The *Hamiltonian* H is a function of q and p and is given by[119]

$$H(p, q) = p\dot{q} - L(q, \dot{q}). \tag{T.81}$$

The equations of motion, known as *canonical equations of Hamilton* , are given by

---

[117]Often, L is given by the difference of kinetic and potential energy, i.e., $L = T - V$.

[118]One can imagine that the difference between kinetic and potential energy $T - V$ becomes minimal if it is averaged over the entire motion.

[119]If $L = T - V$, then $H = T + V$.

$$\dot{q} = \{H, q\}_{PB} = \frac{\partial H}{\partial p} \; ; \; \dot{p} = \{H, p\}_{PB} = -\frac{\partial H}{\partial q} \tag{T.82}$$

where the *Poisson bracket* is defined by

$$\{A, B\}_{PB} = \frac{\partial A}{\partial q} \frac{\partial B}{\partial p} - \frac{\partial A}{\partial p} \frac{\partial B}{\partial q}. \tag{T.83}$$

T.3.1.2 Several Coordinates q_k

If L is a function of several coordinates $q_k, k = 1, 2, \ldots$, i.e., $L = L(q_1, q_2, \ldots, \dot{q}_1, \dot{q}_2, \ldots)$, we have

$$\frac{\partial L}{\partial q_k} - \frac{d}{dt} \frac{\partial L}{\partial \dot{q}_k} = 0 \; ; \; k = 1, 2, \ldots \tag{T.84}$$

and, correspondingly,

$$p_k = \frac{\partial L}{\partial \dot{q}_k} \; ; \; H = \sum_k p_k \dot{q}_k - L. \tag{T.85}$$

The equations of motion read

$$\dot{q}_k = \frac{\partial H}{\partial p_k} \; ; \; \dot{p}_k = -\frac{\partial H}{\partial q_k} \tag{T.86}$$

and the Poisson bracket is given by

$$\{A, B\}_{PB} = \sum_k \frac{\partial A}{\partial q_k} \frac{\partial B}{\partial p_k} - \frac{\partial A}{\partial p_k} \frac{\partial B}{\partial q_k}. \tag{T.87}$$

Especially, we have

$$\{q_i, p_j\}_{PB} = \sum_k \frac{\partial q_i}{\partial q_k} \frac{\partial p_j}{\partial p_k} - \frac{\partial q_i}{\partial p_k} \frac{\partial p_j}{\partial q_k} = \sum_k \delta_{ik} \delta_{jk} - 0 = \delta_{ij} \tag{T.88}$$

and

$$\{q_i, q_j\}_{PB} = \{p_i, p_j\}_{PB} = 0. \tag{T.89}$$

These relations are the starting point for the canonical quantization, see below.

T.3.2 Fields

T.3.2.1 Lagrangian Density

The formalism developed so far is used for systems with a finite number of degrees of freedom. We now expand the formalism to cover continua and fields with an infinite number of degrees of freedom (called Lagrangian field theory). We consider three space dimensions $\mathbf{x} = (x, y, z)$ or the spacetime (t, x, y, z). The basic term is the *Lagrangian density* $\mathcal{L}$ which for one field $\varphi\,(t, x, y, z)$ is written as

$$\mathcal{L} = \mathcal{L}\left(\varphi, \frac{\partial \varphi}{\partial t}, \frac{\partial \varphi}{\partial x}, \frac{\partial \varphi}{\partial y}, \frac{\partial \varphi}{\partial z}\right) = \mathcal{L}\left(\varphi, \frac{\partial \varphi}{\partial t}, \nabla \varphi\right) = \mathcal{L}\left(\varphi, \partial_\mu \varphi\right). \quad \text{(T.90)}$$

Remember that the last expression means that $\mathcal{L}$ is a function of all derivatives, $\mu = 0, 1, 2, 3$. We point out that we assume that $\mathcal{L}$ is a function of the field and its first derivatives only and does not depend explicitly on the space-time coordinates. In other words, we consider only closed systems which do not exchange energy and momentum with the environment.[120]

Lagrangian L and Lagrangian density $\mathcal{L}$ are related by

$$L\,(t) = \int \mathrm{d}^3 x \; \mathcal{L}\left(\varphi, \partial_\mu \varphi\right) \quad \text{(T.91)}$$

where $\int \mathrm{d}^3 x$ means the integration over space. The action is given by

$$S = \int \mathrm{d}t \; L = \int \mathrm{d}t \; \mathrm{d}^3 x \; \mathcal{L} = \int \mathrm{d}^4 x \; \mathcal{L}. \quad \text{(T.92)}$$

The variation $\delta S = 0$ leads to the equations of motion (Euler-Lagrange equations). We have

$$\delta S = \int \mathrm{d}^4 x \left[\frac{\partial \mathcal{L}}{\partial \varphi} \delta \varphi + \frac{\partial \mathcal{L}}{\partial\left(\partial_\mu \varphi\right)} \delta\left(\partial_\mu \varphi\right)\right] \quad \text{(T.93)}$$

where in the second term we have adopted the summation convention (i.e., summation over μ). With

$$\frac{\partial \mathcal{L}}{\partial\left(\partial_\mu \varphi\right)} \delta\left(\partial_\mu \varphi\right) = \frac{\partial \mathcal{L}}{\partial\left(\partial_\mu \varphi\right)} \partial_\mu\left(\delta\varphi\right) = \partial_\mu \left\{\frac{\partial \mathcal{L}}{\partial\left(\partial_\mu \varphi\right)}\left(\delta\varphi\right)\right\} - \left(\partial_\mu \frac{\partial \mathcal{L}}{\partial\left(\partial_\mu \varphi\right)}\right)\left(\delta\varphi\right)$$
$$\text{(T.94)}$$

---

[120]Given a physical system, the Lagrangian $\mathcal{L}$ is the central expression from which 'everything' can be derived. However, there does not seem to be a unique way to identify $\mathcal{L}$. Indeed, to find the right expression seems to be more a matter of experience, based on trial and error. Ultimately, the exact form of the Lagrangians has to be confirmed by experiment. Of course, there are some guidelines in tayloring $\mathcal{L}$ which can be adressed, apart from general principles as symmetries and so on.

we arrive at

$$\delta S = \int d^4x \left[\left\{ \frac{\partial \mathcal{L}}{\partial \varphi} - \left(\partial_\mu \frac{\partial \mathcal{L}}{\partial (\partial_\mu \varphi)} \right) \right\} \delta \varphi + \partial_\mu \left\{ \frac{\partial \mathcal{L}}{\partial (\partial_\mu \varphi)} (\delta \varphi) \right\} \right]. \qquad \text{(T.95)}$$

By means of the divergence theorem, the last term may be written as a surface integral

$$\int d^4x \, \partial_\mu \left\{ \frac{\partial \mathcal{L}}{\partial (\partial_\mu \varphi)} (\delta \varphi) \right\} = \int_\Gamma d\sigma_\mu \, \frac{\partial \mathcal{L}}{\partial (\partial_\mu \varphi)} (\delta \varphi) \qquad \text{(T.96)}$$

where Γ denotes the surface of the (4-dimensional) integration volume and $d\sigma_\mu$ is the μ-component of the of the surface element. Assuming that $\frac{\partial \mathcal{L}}{\partial (\partial_\mu \varphi)}$ goes to zero sufficiently quickly at infinity, the surface integral vanishes. Finally, demanding again $\delta S = 0$ for arbitrary variations $\delta \varphi$ we get the Euler–Lagrange equation(s)

$$\frac{\partial \mathcal{L}}{\partial \varphi} - \partial_\mu \frac{\partial \mathcal{L}}{\partial (\partial_\mu \varphi)} = 0. \qquad \text{(T.97)}$$

Remind the summation convention. If the Lagrangian is a function of several fields $\varphi_i, i = 1, 2, \ldots$, the equation holds for each field separately.

Equation (T.97) can be written more compactly if one introduces the variational derivative.[121] For a function which depends as in our case on the field φ and its derivatives $\partial_\mu \varphi$, the variational derivative is defined by

$$\frac{\delta}{\delta \varphi} = \frac{\partial}{\partial \varphi} - \partial_\mu \frac{\partial}{\partial (\partial_\mu \varphi)}. \qquad \text{(T.98)}$$

Note the minus on the r.h.s.; in addition, the letter δ differs from the ∂ of the partial derivative ∂. Using this notation, (T.97) may be written $\frac{\delta \mathcal{L}}{\delta \varphi} = 0$.

T.3.2.2 Hamiltonian Density

With the abbreviation $\dot{\varphi} = \frac{\partial \varphi}{\partial t}$, the conjugated momentum (or conjugated momentum field) is defined by

$$\pi = \frac{\delta L}{\delta \dot{\varphi}} = \frac{\partial \mathcal{L}}{\partial \dot{\varphi}} \qquad \text{(T.99)}$$

and the Hamiltonian density is given by

$$\mathcal{H}(\pi, \varphi) = \pi \dot{\varphi} - \mathcal{L}. \qquad \text{(T.100)}$$

---

[121] The variational derivative can be seen as a generalization of the directional derivative, so to speak as the derivative 'in direction of a function'.

The Hamiltonian is given by

$$H(t) = \int d^3x \, \mathcal{H}(\pi, \varphi).$$

(T.101)

The equations of motion are given by

$$\dot{\varphi} = \frac{\delta \mathcal{H}}{\delta \pi} \; ; \; \dot{\pi} = -\frac{\delta \mathcal{H}}{\delta \varphi}.$$

(T.102)

This leads to

$$\dot{\varphi} = \frac{\partial \mathcal{H}}{\partial \pi} - \partial_\mu \frac{\partial \mathcal{H}}{\partial (\partial_\mu \pi)} \; ; \; \dot{\pi} = -\frac{\partial \mathcal{H}}{\partial \varphi} + \partial_\mu \frac{\partial \mathcal{H}}{\partial (\partial_\mu \varphi)}.$$

(T.103)

T.3.2.3   Poisson Brackets

In field theory, the derivation of the Poisson brackets is a little bit cumbersome and not as straightforward as for the discrete case. We report just the result:

$$\begin{aligned} \{\varphi(t, \mathbf{x}), \pi(t, \mathbf{x}')\}_{PB} &= \delta^{(3)}(\mathbf{x} - \mathbf{x}') \\ \{\varphi(t, \mathbf{x}), \varphi(t, \mathbf{x}')\}_{PB} &= 0 \\ \{\pi(t, \mathbf{x}), \pi(t, \mathbf{x}')\}_{PB} &= 0. \end{aligned}$$

(T.104)

Here, the Poisson bracket is defined by

$$\{A, B\}_{PB} = \int d^3x \, \sum_i \left(\frac{\delta A}{\delta \phi_i} \frac{\delta B}{\delta \pi_i} - \frac{\delta B}{\delta \phi_i} \frac{\delta A}{\delta \pi_i} \right).$$

(T.105)

T.3.2.4   Several Fields

If there are several fields φ_r, $r = 1, 2, \ldots$ we have

$$\pi_r = \frac{\partial \mathcal{L}}{\partial \dot{\varphi}_r} \quad \text{and} \quad \mathcal{H} = \sum_r \pi_r \dot{\varphi}_r - \mathcal{L}.$$

(T.106)

The equations of motion are given by

$$\dot{\varphi}_r = \frac{\partial \mathcal{H}}{\partial \pi_r} - \partial_\mu \frac{\partial \mathcal{H}}{\partial (\partial_\mu \pi_r)} \; ; \; \dot{\pi}_r = -\frac{\partial \mathcal{H}}{\partial \varphi_r} - \partial_\mu \frac{\partial \mathcal{H}}{\partial (\partial_\mu \varphi_r)}$$

(T.107)

and the Poisson brackets by

$$\left\{\varphi_r\left(t,\mathbf{x}\right),\pi_s\left(t,\mathbf{x}'\right)\right\}_{PB}=\delta^3\left(\mathbf{x}-\mathbf{x}'\right)\delta_{rs}$$
$$\left\{\varphi_r\left(t,\mathbf{x}\right),\varphi_s\left(t,\mathbf{x}'\right)\right\}_{PB}=0 \qquad\qquad (\text{T.108})$$
$$\left\{\pi_r\left(t,\mathbf{x}\right),\pi_s\left(t,\mathbf{x}'\right)\right\}_{PB}=0.$$

T.3.3 Canonical Quantization

The Lagrangian contains the complete information about the physical system. It enables us to derive the equations of motions, the conjugated momentum and the Hamiltonian.[122] In addition, there is another benefit: the knowledge of the (classical, i.e., non-quantum mechanical) Lagrangian offers the means to quantize this classical system.

We sketch the essential steps first for a Lagrange function L. The conjugated momenta p_k and the Hamiltonian H are given by

$$p_k=\frac{\partial L}{\partial \dot{q}_k}\ k=1,\dots N\ ;\ \ H=\sum_{k=1}^{N}p_k\dot{q}_k-L. \qquad (\text{T.109})$$

The Poisson brackets for two quantities q_i and p_j are given by

$$\left\{q_i,p_j\right\}_{PB}=\delta_{ij}\ ;\ \ \left\{q_i,q_j\right\}_{PB}=\left\{p_i,p_j\right\}_{PB}=0. \qquad (\text{T.110})$$

This relation can be considered as the key element of quantization. The method runs as follows: We replace (1) the variables $q_i,\ p_j$ by operators $\hat{q}_i,\ \hat{p}_j$; (2) the Poisson bracket $\{,\}$ by a commutator $[,]$, (3) the Kronecker symbol δ_{ij} by $i\hbar\delta_{ij}$. The well-known result reads

$$\left[\hat{q}_i,\hat{p}_j\right]=i\hbar\delta_{ij}\ ;\ \ \left[\hat{q}_i,\hat{q}_j\right]=\left[\hat{p}_i,\hat{p}_j\right]=0. \qquad (\text{T.111})$$

This three-step procedure is called *canonical quantization*.

We adopt this approach for fields. From the Lagrangian density $\mathcal{L}$ we can deduce the conjugated momentum fields and the Hamiltonian density $\mathcal{H}$ as

$$\pi_r=\frac{\partial\mathcal{L}}{\partial\left(\partial_0\varphi_r\right)}\ r=1,\dots N\ ;\ \ \mathcal{H}=\sum_{r=1}^{N}\pi_r\left(\partial_0\varphi_r\right)-\mathcal{L}. \qquad (\text{T.112})$$

We can introduce Poisson brackets; for $\varphi_r\left(t,\mathbf{x}\right)$ and $\pi_s\left(t,\mathbf{x}'\right)$ they read

---

[122]We point out that the information content of the Lagrangian, of the Hamiltonian and of the equations of motion is equivalent.

$$\{\varphi_r\,(t,\mathbf{x})\,,\,\pi_s\,(t,\mathbf{x}')\}_{PB} = \delta^3\,(\mathbf{x}-\mathbf{x}')\,\delta_{rs}$$
$$\{\varphi_r\,(t,\mathbf{x})\,,\,\varphi_s\,(t,\mathbf{x}')\}_{PB} = 0 \qquad (\text{T.113})$$
$$\{\pi_r\,(t,\mathbf{x})\,,\,\pi_s\,(t,\mathbf{x}')\}_{PB} = 0.$$

Again, we perform the canonical quantization by the above-mentioned three steps and arrive at

$$[\hat{\varphi}_r\,(t,\mathbf{x})\,,\,\hat{\pi}_s\,(t,\mathbf{x}')] = i\hbar\delta^3\,(\mathbf{x}-\mathbf{x}')\,\delta_{rs}$$
$$[\hat{\varphi}_r\,(t,\mathbf{x})\,,\,\hat{\varphi}_s\,(t,\mathbf{x}')] = 0\;;\;\;[\hat{\pi}_r\,(t,\mathbf{x})\,,\,\hat{\pi}_s\,(t,\mathbf{x}')] = 0 \qquad (\text{T.114})$$

where $\hat{\varphi}_r\,(t,\mathbf{x})$ and $\hat{\pi}_s\,(t,\mathbf{x})$ are now field operators.[123]

One may ask if the step from the Poisson bracket to the commutator is logically mandatory. The answer is 'no'. But the step is, in a certain sense, very plausible, and, most importantly, the method works, i.e., the resulting equations lead to outcomes which agree very well with the experiment. However, there is an important limitation of the method, since it requires the knowledge of the classical Lagrangian. In other words: If there is no macroscopic Lagrangian, the canonical quantization can not be applied.

T.3.4 Some Lagrangian Densities

Oscillating string

$$\mathcal{L} = \frac{1}{2}\left[\mu\left(\frac{\partial\varphi}{\partial t}\right)^2 - E\left(\frac{\partial\varphi}{\partial x}\right)^2\right] \qquad (\text{T.115})$$

Newtonian gravity

$$\mathcal{L} = -\rho\,(t,\mathbf{x})\,\varphi\,(t,\mathbf{x}) - \frac{1}{8\pi G}\,(\nabla\varphi\,(t,\mathbf{x}))^2 \qquad (\text{T.116})$$

where G is the gravitational constant.

Klein–Gordon Lagrangian

$$\mathcal{L} = \frac{1}{2}\left(\frac{\partial\varphi}{\partial x^\mu}\frac{\partial\varphi}{\partial x_\mu} - m^2\varphi^2\right) \qquad (\text{T.117})$$

---

[123]For the sake of clearness, we write here $\hat{\varphi}_r\,(t,\mathbf{x})$ and $\hat{\pi}_s\,(t,\mathbf{x})$ for the field operators. Otherwise, we will omit the hats and simply write $\varphi_r\,(t,\mathbf{x})$ and $\pi_s\,(t,\mathbf{x})$ for the operators (as is common in many textbooks).

Dirac Lagrangian

$$\mathcal{L} = i\hbar c \bar{\varphi} \partial\!\!\!/ \varphi - mc^2 \bar{\varphi}\varphi \tag{T.118}$$

where φ is a Dirac spinor, $\bar{\varphi} = \varphi^\dagger \gamma^0$ is its Dirac adjoint, and $\partial\!\!\!/$ is the Feynman slash notation for $\gamma^\mu \partial_\mu$.

For more Lagrangians see the next section.

T.3.5 Exercises and Solutions

1. Show (T.79).
   Solution: We have[124]

$$\delta S = \delta \int_{t_1}^{t_2} L\left(q, \dot{q}\right) dt = \int_{t_1}^{t_2} \left(\frac{\partial L}{\partial q} \delta q + \frac{\partial L}{\partial \dot{q}} \delta \dot{q} \right) dt. \tag{T.120}$$

Using integration by parts for the second term, we arrive at

$$\delta S = \int_{t_1}^{t_2} \frac{\partial L}{\partial q} \delta q \, dt + \left[\frac{\partial L}{\partial \dot{q}} \delta q \right]_{t_1}^{t_2} - \int_{t_1}^{t_2} \left(\frac{d}{dt} \frac{\partial L}{\partial \dot{q}} \right) \delta q \, dt. \tag{T.121}$$

The term in brackets vanishes due to $\delta q\left(t_1\right) = \delta q\left(t_2\right) = 0$. So we have

$$\delta S = \int_{t_1}^{t_2} \left[\frac{\partial L}{\partial q} - \frac{d}{dt} \frac{\partial L}{\partial \dot{q}} \right] \delta q \, dt. \tag{T.122}$$

Demanding $\delta S = 0$ for arbitrary variations δq yields

$$\frac{\partial L}{\partial q} - \frac{d}{dt} \frac{\partial L}{\partial \dot{q}} = 0. \tag{T.123}$$

2. Prove $\left\{ \varphi\left(t, \mathbf{x}\right), \pi\left(t, \mathbf{x}'\right) \right\}_{PB} = \delta^3\left(\mathbf{x} - \mathbf{x}'\right)$.

---

[124]Variation means in this context, that we change $q(t)$ by $q(t) \to q(t) + \delta q(t)$ with $\delta q(t_1) = \delta q(t_2) = 0$ which, of course, means that $\dot{q}$ is changed correspondingly. This induces a change of L with respect to its two arguments q and $\dot{q}$. It follows (we use Taylor expansion)

$$\begin{aligned} \delta S &= \delta \int_{t_1}^{t_2} L\left(q, \dot{q}\right) dt = \int_{t_1}^{t_2} \left[L(q + \delta q, \dot{q} + \delta \dot{q}) - L(q, \dot{q}) \right] dt = \\ &= \int_{t_1}^{t_2} \left[L\left(q, \dot{q}\right) + \frac{\partial L(q, \dot{q})}{\partial q} \delta q + \frac{\partial L(q, \dot{q})}{\partial \dot{q}} \delta \dot{q} - L\left(q, \dot{q}\right) \right] dt = \\ &= \int_{t_1}^{t_2} \left[\frac{\partial L(q, \dot{q})}{\partial q} \delta q + \frac{\partial L(q, \dot{q})}{\partial \dot{q}} \delta \dot{q} \right] dt \end{aligned} \tag{T.119}$$

Solution:

$$\{\varphi(t, \mathbf{x}), \pi(t, \mathbf{x}')\}_{PB} = \int d^3 x'' \left(\frac{\delta \varphi(t,\mathbf{x})}{\delta \varphi(t,\mathbf{x}'')} \frac{\delta \pi(t,\mathbf{x}')}{\delta \pi(t,\mathbf{x}'')} - \frac{\delta \pi(t,\mathbf{x}')}{\delta \varphi(t,\mathbf{x}'')} \frac{\delta \varphi(t,\mathbf{x})}{\delta \pi(t,\mathbf{x}'')} \right) =$$
$$= \int d^3 x'' \, \delta^{(3)}(\mathbf{x} - \mathbf{x}'') \, \delta^{(3)}(\mathbf{x} - \mathbf{x}') \left(\frac{\delta \varphi(t,\mathbf{x})}{\delta \varphi(t,\mathbf{x}'')} \frac{\delta \pi(t,\mathbf{x}')}{\delta \pi(t,\mathbf{x}'')} - \frac{\delta \pi(t,\mathbf{x}')}{\delta \varphi(t,\mathbf{x}'')} \frac{\delta \varphi(t,\mathbf{x})}{\delta \pi(t,\mathbf{x}'')} \right) =$$
$$= \int d^3 x'' \, \delta^{(3)}(\mathbf{x} - \mathbf{x}'') \, \delta^{(3)}(\mathbf{x} - \mathbf{x}') = \delta^{(3)}(\mathbf{x} - \mathbf{x}').$$
(T.124)

3. Consider

$$\mathcal{L} = \frac{1}{2} \left[\mu \left(\frac{\partial \varphi}{\partial t} \right)^2 - E \left(\frac{\partial \varphi}{\partial x} \right)^2 \right]. \tag{T.125}$$

This is the Lagrangian of a one-dimensional oscillating string; μ is the linear mass density and E the modulus of elasticity. Determine the Euler–Lagrange equation, the conjugated momentum, the Hamiltonian and Hamilton's equations.
Solution: We have

$$\frac{\partial \mathcal{L}}{\partial \varphi} = 0 \;; \quad \frac{\partial \mathcal{L}}{\partial \frac{\partial \varphi}{\partial t}} = \mu \frac{\partial \varphi}{\partial t} \;; \quad \frac{\partial \mathcal{L}}{\partial \frac{\partial \varphi}{\partial x}} = -E \frac{\partial \varphi}{\partial x}. \tag{T.126}$$

Thus, the Euler–Lagrange equation reads

$$0 - \frac{d}{dt} \mu \frac{\partial \varphi}{\partial t} - \frac{d}{dx} \left(-E \frac{\partial \varphi}{\partial x} \right) = 0 \rightarrow \mu \frac{\partial^2 \varphi}{\partial t^2} = E \frac{\partial^2 \varphi}{\partial x^2} \text{ or } \mu \ddot{\varphi} = E \varphi''. \tag{T.127}$$

The conjugated momentum is given by

$$\pi = \frac{\partial \mathcal{L}}{\partial \dot{\varphi}} = \mu \frac{\partial \varphi}{\partial t} \rightarrow \dot{\varphi} = \frac{\pi}{\mu} \tag{T.128}$$

and the Hamiltonian density reads

$$\mathcal{H} = \pi \dot{\varphi} - \frac{1}{2} \left[\mu \left(\frac{\partial \varphi}{\partial t} \right)^2 - E \left(\frac{\partial \varphi}{\partial x} \right)^2 \right] = \frac{\pi^2}{\mu} - \frac{\mu}{2} \left(\frac{\pi}{\mu} \right)^2 + \frac{E}{2} \left(\frac{\partial \varphi}{\partial x} \right)^2 = \frac{\pi^2}{2\mu} + \frac{E}{2} \left(\frac{\partial \varphi}{\partial x} \right)^2.$$
(T.129)

Hamilton's equations are given by

$$\dot{\varphi} = \frac{\partial \mathcal{H}}{\partial \pi} - \frac{\partial}{\partial x} \left(\frac{\partial \mathcal{H}}{\partial (\frac{\partial \pi}{\partial x})} \right) = \frac{\pi}{\mu}$$
$$\dot{\pi} = -\frac{\partial \mathcal{H}}{\partial \varphi} + \frac{\partial}{\partial x} \left(\frac{\partial \mathcal{H}}{\partial (\frac{\partial \varphi}{\partial x})} \right) = \frac{\partial}{\partial x} E \left(\frac{\partial \varphi}{\partial x} \right) = E \frac{\partial^2 \varphi}{\partial x^2}$$
(T.130)

and the equation of motion reads

$$\ddot{\varphi} = \frac{\dot{\pi}}{\mu} \rightarrow \ddot{\varphi} = \frac{E}{\mu} \varphi''. \tag{T.131}$$

4. Show

$$\frac{\partial}{\partial\left(\partial_\mu\phi\right)}\left(\partial_\alpha\phi\right)\left(\partial^\alpha\phi\right) = 2\left(\partial^\mu\phi\right). \tag{T.132}$$

Solution: We have

$$\frac{\partial}{\partial(\partial_\mu\phi)}\left(\partial_\alpha\phi\right)\left(\partial^\alpha\phi\right) = \delta_{\mu\alpha}\left(\partial^\alpha\phi\right) + \left(\partial_\alpha\phi\right)\frac{\partial}{\partial(\partial_\mu\phi)}\left(g^{\alpha\nu}\partial_\nu\phi\right) =$$
$$= \left(\partial^\mu\phi\right) + \left(\partial_\alpha\phi\right)g^{\alpha\nu}\delta_{\mu\nu}\phi = \left(\partial^\mu\phi\right) + \left(\partial^\nu\phi\right)\delta_{\mu\nu}\phi = \left(\partial^\mu\phi\right) + \left(\partial^\mu\phi\right) = 2\left(\partial^\mu\phi\right). \tag{T.133}$$

5. By adding appropriate terms to the Lagrangians (Hamiltonians) of free fields, we can model or describe interactions of fields. One of the simplest examples, by obvious reasons called φ^4-theory, is based on the Klein–Gordon field. Its Lagrangian is given by

$$\mathcal{L} = \frac{1}{2}\left(\partial_\mu\varphi\right)\left(\partial^\mu\varphi\right) - \frac{1}{2}m\varphi^2 - \frac{g}{4!}\varphi^4. \tag{T.134}$$

Determine the equations of motion.
Solution: The Euler–Lagrange equations are given by

$$\left(\partial_\mu\frac{\partial\mathcal{L}}{\partial\left(\partial_\mu\varphi\right)}\right) - \frac{\partial\mathcal{L}}{\partial\varphi} = 0. \tag{T.135}$$

With

$$\frac{\partial\mathcal{L}}{\partial\left(\partial_\mu\varphi\right)} = \partial^\mu\varphi \,; \quad \frac{\partial\mathcal{L}}{\partial\varphi} = -m\varphi - \frac{g}{3!}\varphi^3 \tag{T.136}$$

follows

$$\partial_\mu\partial^\mu\varphi + m\varphi + \frac{g}{3!}\varphi^3 = 0. \tag{T.137}$$

T.4 Electrodynamics

We repeat here some facts from electrodynamics as far as we need them in further chapters. Apart from providing the necessary formulas and expressions, we aim at a consistent notation.

T.4.1 Maxwell Equations, Potentials, Gauge

In SI-units, the maxwell equations are given by

$$\boldsymbol{\nabla} \cdot \vec{E}\left(\mathbf{r}, t\right) = \frac{1}{\varepsilon_0} \rho \; ; \; \boldsymbol{\nabla} \cdot \vec{B}\left(\mathbf{r}, t\right) = 0$$
$$\boldsymbol{\nabla} \times \mathbf{E}\left(\mathbf{r}, t\right) = -\frac{\partial}{\partial t} \mathbf{B}\left(\mathbf{r}, t\right) \; ; \; \boldsymbol{\nabla} \times \mathbf{B}\left(\mathbf{r}, t\right) = \frac{1}{c^2} \frac{\partial}{\partial t} \mathbf{E}\left(\mathbf{r}, t\right) + \mu_0 \mathbf{j} \tag{T.138}$$

with

$$c^2 \varepsilon_0 \mu_0 = 1. \tag{T.139}$$

Introducing the scalar and the vector potential Φ and $\mathbf{A}$

$$\mathbf{B} = \boldsymbol{\nabla} \times \mathbf{A} \; ; \; \mathbf{E} = -\boldsymbol{\nabla} \Phi - \frac{\partial \mathbf{A}}{\partial t} \tag{T.140}$$

transforms (T.138) into

$$\boldsymbol{\nabla} \cdot \left(-\boldsymbol{\nabla} \Phi - \frac{\partial \mathbf{A}}{\partial t}\right) = \frac{1}{\varepsilon_0} \rho \; ; \; \boldsymbol{\nabla} \cdot (\boldsymbol{\nabla} \times \mathbf{A}) = 0$$
$$\boldsymbol{\nabla} \times \left(-\boldsymbol{\nabla} \Phi - \frac{\partial \mathbf{A}}{\partial t}\right) = -\frac{\partial}{\partial t} \boldsymbol{\nabla} \times \mathbf{A} \; ; \; \boldsymbol{\nabla} \times (\boldsymbol{\nabla} \times \mathbf{A}) = \frac{1}{c^2} \frac{\partial}{\partial t} \left(-\boldsymbol{\nabla} \Phi - \frac{\partial \mathbf{A}}{\partial t}\right) + \mu_0 \mathbf{j}. \tag{T.141}$$

The source-free equations are automatically satisfied. The other two equations are

$$-\boldsymbol{\nabla}^2 \Phi - \frac{\partial}{\partial t} \boldsymbol{\nabla} A = \frac{1}{\varepsilon_0} \rho$$
$$\frac{1}{c^2} \frac{\partial^2 \mathbf{A}}{\partial t^2} - \boldsymbol{\nabla}^2 \mathbf{A} + \boldsymbol{\nabla} \left(\boldsymbol{\nabla} \vec{A}\right) + \frac{1}{c^2} \boldsymbol{\nabla} \frac{\partial}{\partial t} \Phi = \mu_0 \mathbf{j}. \tag{T.142}$$

Adding zero to the first equation gives

$$\frac{1}{c^2} \frac{\partial^2 \Phi}{\partial t^2} - \boldsymbol{\nabla}^2 \Phi - \frac{\partial}{\partial t} \left(\frac{1}{c^2} \frac{\partial \Phi}{\partial t} + \boldsymbol{\nabla} \vec{A}\right) = \frac{1}{\varepsilon_0} \rho$$
$$\frac{1}{c^2} \frac{\partial^2 \mathbf{A}}{\partial t^2} - \boldsymbol{\nabla}^2 \mathbf{A} + \boldsymbol{\nabla} \left(\frac{1}{c^2} \frac{\partial}{\partial t} \Phi + \boldsymbol{\nabla} \vec{A}\right) = \mu_0 \mathbf{j} \tag{T.143}$$

or

$$\left(\frac{1}{c^2} \frac{\partial^2}{\partial t^2} - \boldsymbol{\nabla}^2\right) \Phi - \frac{\partial}{\partial t} \left(\frac{1}{c^2} \frac{\partial \Phi}{\partial t} + \boldsymbol{\nabla} \vec{A}\right) = \frac{1}{\varepsilon_0} \rho$$
$$\left(\frac{1}{c^2} \frac{\partial^2}{\partial t^2} - \boldsymbol{\nabla}^2\right) \mathbf{A} + \boldsymbol{\nabla} \left(\frac{1}{c^2} \frac{\partial}{\partial t} \Phi + \boldsymbol{\nabla} \vec{A}\right) = \mu_0 \mathbf{j}. \tag{T.144}$$

Using the 4-vectors $j^{\mu} = (c\rho, \mathbf{j})$, $A^{\mu} = (\frac{\Phi}{c}, \mathbf{A})$, $\partial_{\mu} = \left(\frac{1}{c} \frac{\partial}{\partial t}, \boldsymbol{\nabla}\right)$ and $\partial^{\mu} = \left(\frac{1}{c} \frac{\partial}{\partial t}, -\boldsymbol{\nabla}\right)$, we can cast (T.144) into the form

$$\partial_{\nu} \partial^{\nu} A^0 - \partial_0 \left(\partial_0 A^0 + \partial_k A^k\right) = \frac{1}{c^2 \varepsilon_0} j^0$$
$$\partial_{\nu} \partial^{\nu} A^k - \partial_k \left(\partial_0 A^0 + \partial_k A^k\right) = \mu_0 j^k \tag{T.145}$$

which may be written in covariant manner as

$$\partial_{\nu} \partial^{\nu} A^{\mu} - \partial^{\mu} \left(\partial_{\nu} A^{\nu}\right) = \mu_0 j^{\mu}. \tag{T.146}$$

This equation, formulated in terms of the potentials A^μ, replaces the Maxwell equations,[125] formulated in terms of the fields $\mathbf{E}$ and $\mathbf{B}$. However, there is one distinctive difference: by equation (T.146), the potentials are not determined uniquely, in contrast to the fields $\mathbf{E}$ and $\mathbf{B}$. Indeed, a transformation (called *gauge transformation*) with an arbitrary Γ

$$A^\mu \to \tilde{A}^\mu = A^\mu + \partial^\mu \Gamma \tag{T.147}$$

leaves the fields $\mathbf{E}$ and $\mathbf{B}$ invariant as well as the equation (T.146). This fact is called *gauge invariance*.

One can exploit the freedom of the choice of the gauge to make things as simple as possible. Depending on the system under consideration, there are among others two common choices, namely the Lorenz[126] gauge $\partial_\nu A^\nu = 0$ and the Coulomb (or radiation) gauge $\partial_k A^k = \mathbf{V} \cdot \mathbf{A} = 0$. As an example, (T.146) reads in the Lorenz gauge simply

$$\partial_\nu \partial^\nu A^\mu = \mu_0 j^\mu. \tag{T.148}$$

T.4.2 Free Solutions

Without sources, the equations of motion read

$$\partial_\mu \partial^\mu A^\nu = 0. \tag{T.149}$$

This is essentially the Klein–Gordon equation for vanishing mass, apart from the fact that the potential is not a scalar, but a 4-vector. This means, we can immediately write down the solutions which read in the discrete and continuous case[127]

$$A^\mu(x) = \sum_{\mathbf{k},r} \sqrt{\frac{1}{2V\omega_\mathbf{k}}} \varepsilon_r^\mu(\mathbf{k}) \left[\alpha_r(\mathbf{k}) e^{-ikx} + \alpha_r^\dagger(\mathbf{k}) e^{ikx} \right]$$
$$A^\mu(x) = \sum_r \int \frac{d^3k}{\sqrt{2(2\pi)^3 \omega_\mathbf{k}}} \varepsilon_r^\mu(\mathbf{k}) \left[\alpha_r(\mathbf{k}) e^{-ikx} + \alpha_r^\dagger(\mathbf{k}) e^{ikx} \right]. \tag{T.150}$$

Obviously, the potential is real, as it should be. The 4-vectors $\varepsilon_r^\mu(\mathbf{k})$ are called *polarization vectors*. Their specific form depends on the chosen gauge. In the Coulomb gauge $\partial_k A^k = \mathbf{V} \cdot \mathbf{A} = 0$ we have for the polarization vectors

$$\varepsilon_r^0(\mathbf{k}) = 0 \, ; \; k_l \varepsilon_r^l(\mathbf{k}) = \mathbf{k} \cdot \varepsilon_r(\mathbf{k}) = 0 \, ; \; \varepsilon_r(\mathbf{k}) \cdot \varepsilon_{r'}(\mathbf{k}) = \delta_{rr'}. \tag{T.151}$$

In other words, in this gauge we need only two polarization vectors ε_r^μ, $r = 1, 2$. They are orthogonal to each other and orthogonal to $\mathbf{k}$ (i.e., transversal). Moreover, in the

---

[125]Note that the homogenous Maxwell equations in (T.138) are automatically fulfilled.

[126]It is indeed Lorenz (Ludvig Valentin Lorenz, Dane, 1829–1891), and not Lorentz (Hendrik Antoon Lorentz, Dutch, 1853–1928).

[127]The normalization is chosen with regard to the application in quantum field theory.

Lorenz gauge $\partial_\nu A^\nu = 0$ we have $k_\nu \varepsilon_r^\nu (\mathbf{k}) = 0$, i.e., three independent polarization vectors (see exercises).

T.4.3 Electromagnetic Field Tensor

A very compact and elegant description of electrodynamics is provided by the *electromagnetic field tensor* $F_{\mu\nu}$. It is defined by

$$F^{\mu\nu} = \frac{1}{c}\begin{pmatrix} 0 & -E^1 & -E^2 & -E^3 \\ E^1 & 0 & -cB^3 & cB^2 \\ E^2 & cB^3 & 0 & -cB^1 \\ E^3 & -cB^2 & cB^1 & 0 \end{pmatrix} \; ; \; F_{\mu\nu} = \frac{1}{c}\begin{pmatrix} 0 & E^1 & E^2 & E^3 \\ -E^1 & 0 & -cB^3 & cB^2 \\ -E^2 & cB^3 & 0 & -cB^1 \\ -E^3 & -cB^2 & cB^1 & 0 \end{pmatrix}$$

(T.152)

(due to $F^{\mu\nu} = g^{\mu\alpha} F_{\alpha\beta} g^{\beta\nu}$). With $F^{\mu\nu}$, the inhomogeneous and homogenous Maxwell equations (T.148) read

$$\partial_\nu F^{\mu\nu} = \mu_0 j^\mu \text{ and } \partial^\lambda F^{\mu\nu} + \partial^\mu F^{\nu\lambda} + \partial^\nu F^{\lambda\mu} = 0.$$

(T.153)

Expressing $F^{\mu\nu}$ by means of the 4-potential $A^\mu a$, i.e., by

$$F_{\mu\nu} = \partial_\mu A_\nu - \partial_\nu A_\mu \; ; \; F^{\mu\nu} = \partial^\nu A^\mu - \partial^\mu A^\nu$$

(T.154)

gives for the inhomogeneous equations

$$\partial_\nu \partial^\nu A^\mu - \partial^\mu (\partial_\nu A^\nu) = \mu_0 j^\mu$$

(T.155)

while the homogenous equations are automatically fulfilled.

Note that the electromagnetic field tensor is gauge invariant:

$$\partial^\nu \tilde{A}^\mu - \partial^\mu \tilde{A}^\nu = \partial^\nu [A^\mu + \partial^\mu \Gamma] - \partial^\mu [A^\nu + \partial^\nu \Gamma] = \partial^\nu A^\mu + \partial^\nu \partial^\mu \Gamma - \partial^\mu A^\nu - \partial^\mu \partial^\nu \Gamma =$$
$$= \partial^\nu A^\mu - \partial^\mu A^\nu + \partial^\nu \partial^\mu \Gamma - \partial^\mu \partial^\nu \Gamma = \partial^\nu A^\mu - \partial^\mu A^\nu.$$

(T.156)

A Lorentz scalar is given e.g. by

$$F_{\mu\nu} F^{\mu\nu} = 2 \left(\mathbf{B}^2 - \frac{\mathbf{E}^2}{c^2} \right)$$

(T.157)

which is, up to a constant factor, the energy density of the electromagnetic field. As is seen, the term is gauge invariant.

T.4.4 Lagrangian $\mathcal{L}$

Due to gauge invariance, there is no unique Lagrangian density $\mathcal{L}$ for the electromagnetic field. The only criterion is that $\mathcal{L}$ reproduces the correct equations of motion, i.e., the Maxwell equations. The criterion is fulfilled, for instance, by the choice

$$\mathcal{L} = -\frac{1}{4\mu_0} F_{\mu\nu} F^{\mu\nu} - j_\mu A^\mu \tag{T.158}$$

which leads to the Euler–Lagrange equations $\partial_\nu F^{\mu\nu} = \mu_0 j^\mu$.
Another common Lagrangian density $\mathcal{L}$ reads

$$\mathcal{L} = -\frac{1}{2\mu_0} \left(\partial_\nu A_\mu\right)\left(\partial^\nu A^\mu\right) - j_\mu A^\mu \tag{T.159}$$

which leads to $\partial_\nu \partial^\nu A^\mu = \mu_0 j^\mu$.

T.4.5 Some Lagrangian Densities

The Lagrangian for the electromagnetic field (T.158) is something like the starting point for advanced formulations. In quantum electrodynamics (QED), the Lagrangian $\mathcal{L}$ may be written

$$\mathcal{L} = i\hbar c \bar{\psi}\,\slashed{D}\psi - mc^2 \bar{\psi}\psi - \frac{1}{4\mu_0} F_{\mu\nu} F^{\mu\nu} \tag{T.160}$$

where $\slashed{D} = \gamma^\mu D_\mu$ is the QED gauge covariant derivative with $D_\mu = \partial_\mu - iq A_\mu$ and ψ the Dirac field. The Lagrangian for quantum chromodynamics (QCD) follows in a certain sense the same pattern. Of course it is more complex, but the structure inherited from QED is clearly apparent. We report the result without going into details:

$$\mathcal{L} = \sum_n \left(i\hbar c \bar{\psi}_n D/\psi_n - m_n c^2 \bar{\psi}_n \psi_n\right) - \frac{1}{4} G^\alpha_{\mu\nu} G^{\mu\nu}_\alpha. \tag{T.161}$$

Here, D is the QCD gauge covariant derivative, $n = 1, \ldots, 6$ counts the quark types, and $G^\alpha_{\mu\nu}$ is the gluon field strength tensor.

T.4.6 Exercises and Solutions

1. Starting from (T.145), derive (T.146).
   Solution: We insert

$$\partial_\nu \partial^\nu = \frac{1}{c^2}\frac{\partial^2}{\partial t^2} - \mathbf{V}^2 \; ; \; \Phi = cA^0 \; ; \; \mathbf{A} = A^k \; ; \; \frac{\partial}{\partial t} = c\partial^0 \; ; \; \mathbf{V} = \partial_k \quad \text{(T.162)}$$

and arrive at

$$\begin{aligned} \partial_\nu \partial^\nu cA^0 - c\partial^0 \left(\tfrac{1}{c^2}c\partial^0 cA^0 + \partial_k A^k \right) &= \tfrac{1}{\varepsilon_0}\rho \\ \partial_\nu \partial^\nu A^k - \partial^k \left(\tfrac{1}{c^2}c\partial^0 cA^0 + \partial_k A^k \right) &= \mu_0 \mathbf{j} \end{aligned} \quad \text{(T.163)}$$

or

$$\begin{aligned} \partial_\nu \partial^\nu A^0 - \partial^0 \left(\partial_0 A^0 + \partial_k A^k \right) &= \tfrac{1}{c\varepsilon_0}\rho = \tfrac{1}{c^2 \varepsilon_0} j^0 = \mu_0 j^0 \\ \partial_\nu \partial^\nu A^k - \partial^k \left(\partial_0 A^0 + \partial_k A^k \right) &= \mu_0 \mathbf{j} = \mu_0 j^k. \end{aligned} \quad \text{(T.164)}$$

Merging the two equations leads to

$$\partial_\nu \partial^\nu A^\mu - \partial^\mu \left(\partial_\nu A^\nu \right) = \mu_0 j^\mu \quad \text{(T.165)}$$

2. Show that (T.146) is invariant with respect to the transformation (T.147).
   Solution: We have

$$\begin{aligned} \partial_\nu \partial^\nu \tilde{A}^\mu - \partial^\mu \left(\partial_\nu \tilde{A}^\nu \right) &= \partial_\nu \partial^\nu [A^\mu + \partial^\mu \Lambda] - \partial^\mu \left(\partial_\nu [A^\nu + \partial^\nu \Lambda] \right) = \\ &= [\partial_\nu \partial^\nu A^\mu + \partial_\nu \partial^\nu \partial^\mu \Lambda] - [\partial^\mu \partial_\nu A^\nu + \partial^\mu \partial_\nu \partial^\nu \Lambda] = \\ &= \partial_\nu \partial^\nu A^\mu - \partial^\mu \partial_\nu A^\nu + \partial_\nu \partial^\nu \partial^\mu \Lambda - \partial^\mu \partial_\nu \partial^\nu \Lambda = \partial_\nu \partial^\nu A^\mu - \partial^\mu \partial_\nu A^\nu. \end{aligned} \quad \text{(T.166)}$$

3. Derive the Maxwell equations from the Lagrangian (T.159).
   Solution: The Euler–Lagrange equations for the electromagnetic field (in terms of A^μ and $\partial_\nu A^\mu$) are

$$\frac{\partial \mathcal{L}}{\partial A^\nu} - \partial^\mu \frac{\partial \mathcal{L}}{\partial(\partial^\mu A^\nu)} = 0. \quad \text{(T.167)}$$

This yields with (T.159)

$$\partial^\mu \frac{\partial}{\partial(\partial^\mu A^\nu)} \left(-\frac{1}{2\mu_0} \partial_\alpha A_\beta \partial^\alpha A^\beta - j_\alpha A^\alpha \right) = \frac{\partial}{\partial A^\nu} \left(-\frac{1}{2\mu_0} \partial_\alpha A_\beta \partial^\alpha A^\beta - j_\alpha A^\alpha \right). \quad \text{(T.168)}$$

From this follows step by step

$$\begin{aligned} \partial^\mu \frac{\partial}{\partial(\partial^\mu A^\nu)} \left(\tfrac{1}{2\mu_0} g_{\delta\alpha} g_{\epsilon\beta} \partial^\delta A^\epsilon \partial^\alpha A^\beta \right) &= \frac{\partial}{\partial A^\nu} j_\alpha A^\alpha \\ \tfrac{1}{2\mu_0} \partial^\mu g_{\delta\alpha} g_{\epsilon\beta} \left(\delta^\delta_\mu \delta^\epsilon_\nu \partial^\alpha A^\beta + \partial^\delta A^\epsilon \delta^\alpha_\mu \delta^\beta_\nu \right) &= j_\alpha \delta^\alpha_\nu \\ \tfrac{1}{2\mu_0} \partial^\mu \left(g_{\mu\alpha} g_{\nu\beta} \partial^\alpha A^\beta + g_{\delta\mu} g_{\epsilon\nu} \partial^\delta A^\epsilon \right) &= j_\nu \\ \tfrac{1}{2\mu_0} \partial^\mu \left(\partial_\mu A_\nu + \partial_\mu A_\nu \right) &= j_\nu \\ \tfrac{1}{\mu_0} \partial^\mu \partial_\mu A_\nu = j_\nu \quad \text{or} \quad \partial^\mu \partial_\mu A_\nu &= \mu_0 j_\nu. \end{aligned} \quad \text{(T.169)}$$

4. Determine the polarization vectors in (T.150) for the Coulomb and the Lorenz gauge.
   Solution: W.l.o.g we can identify the z-axis with the direction of propagation, $\mathbf{k} = (0, 0, k_3)$.

(a) Coulomb gauge $\partial_m A^m = \nabla \cdot \mathbf{A} = 0$. It follows

$$\partial_m A^m (x) = \sum_{\mathbf{k},r} \sqrt{\frac{1}{2V\omega_\mathbf{k}}} \varepsilon_r^m (\mathbf{k}) \, i k_m \left[\alpha_r (\mathbf{k}) \, e^{-ikx} - \alpha_r^\dagger (\mathbf{k}) \, e^{ikx} \right] =$$
$$= i \sum_{\mathbf{k},r} \sqrt{\frac{1}{2V\omega_\mathbf{k}}} \varepsilon_r (\mathbf{k}) \cdot \mathbf{k} \left[\alpha_r (\mathbf{k}) \, e^{-ikx} - \alpha_r^\dagger (\mathbf{k}) \, e^{ikx} \right] = 0 \rightarrow \varepsilon_r (\mathbf{k}) \cdot \mathbf{k} \overset{!}{=} 0.$$

$$(\text{T.170})$$

Thus, we can focus on the inner product $\varepsilon_r (\mathbf{k}) \cdot \mathbf{k}$.which tells that the polarizations vectors are orthogonal to $\mathbf{k}$. Due to $\mathbf{k} = (0, 0, k_3)$, we have $\varepsilon_r^3 (\mathbf{k}) = 0$ whereas ε_r^1 and ε_r^2 are undetermined. We choose the simplest solution,[128] $\varepsilon_1^1 (\mathbf{k}) = 1; \varepsilon_1^2 (\mathbf{k}) = 0$ and $\varepsilon_2^1 (\mathbf{k}) = 0; \varepsilon_2^2 (\mathbf{k}) = 1$. In addition, also $\varepsilon_r^0 (\mathbf{k})$ may be chosen freely; we set $\varepsilon_r^0 (\mathbf{k}) = 0$. Thus, we have two polarization vectors, namely $\varepsilon_1 (\mathbf{k}) = (0, 1, 0, 0)$ and $\varepsilon_2 (\mathbf{k}) = (0, 0, 1, 0)$.

(b) Lorenz gauge $\partial_\nu A^\nu = \partial_0 A^0 - \nabla \vec{A} = 0$ Here the defining inner product is $k \varepsilon_r (\mathbf{k}) = k_0 \varepsilon_r^0 (\mathbf{k}) - k_3 \varepsilon_r^3 (\mathbf{k}) = 0$. Again, ε_r^1 and ε_r^2 are undetermined; we choose them as in the Coulomb case to be $\varepsilon_1^1 (\mathbf{k}) = 1; \varepsilon_1^2 (\mathbf{k}) = 0$ and $\varepsilon_2^1 (\mathbf{k}) = 0; \varepsilon_2^2 (\mathbf{k}) = 1$. The remaining components are connected by $\varepsilon_r^0 (\mathbf{k}) = (k_3/k_0) \varepsilon_r^3 (\mathbf{k})$. Thus we have three pairwise orthogonal polarization vectors, e.g. $\varepsilon_1 (\mathbf{k}) = (0, 1, 0, 0)$ and $\varepsilon_2 (\mathbf{k}) = (0, 0, 1, 0)$ and $\varepsilon_3 (\mathbf{k}) = N \left(\frac{k_3}{k_0}, 0, 0, 1 \right)$ with $N^{-2} = 1 - \left(\frac{k_3}{k_0} \right)^2$.

---

[128]The two polarization vectors have to be orthogonal to the direction of propagation. Thus, other solutions, e.g. circular polarization and so on, are possible, of course.

Appendix U
Elements of Relativistic Quantum Mechanics

U.1 Introduction

The bulk of the two volumes of this book is devoted to non-relativistic quantum mechanics (NRQM). In the following, we will discuss *relativistic quantum mechanics* (RQM). This field of physics is important in itself and show issues that do not (and can not) emerge in non-relativistic theories - we mention just spin or antiparticles. In addition, it is an indispensable prerequisite for advanced theories as quantum field theory (QFT), elements of which are presented in Vol. 2. Hence, by providing some of the basics RQM we hope that this will help to a better understanding of some fundamentals of modern physics and an easier access to some of the latest topics in modern physics.

The content plan is as follows: We first derive the Klein–Gordon and the Dirac equation as basic elements of RQM. Then we construct plane wave solutions for both equations. As we will see, we encounter immediately the notorious problem of solutions with negative energy which will be solved in a satisfying manner only within the context of QFT. In case of the Dirac equation, spin 1/2 emerges quasi unexpectedly by our quantization procedure, without any requirement about angular momentum, let alone spin 1/2. After that, we prove that the Dirac equation is in accordance with special relativity (i.e., is covariant) and couple it to the electromagnetic field. This formulation allows among others for deriving the Pauli equation as the non-relativistic limit of the Dirac equation. Finally, we discuss the pros and cons of the Dirac equation and discuss how modern interpretations show a way out of the dilemma of negative energies.

In addition, there is a section about the relativistic description of the Hydrogen atom and its energy spectrum. Because of the thematic context, it is found in Vol. 2.

© Springer Nature Switzerland AG 2018
J. Pade, *Quantum Mechanics for Pedestrians 1*, Undergraduate Lecture Notes in Physics, https://doi.org/10.1007/978-3-030-00464-4

U.2   Constructing Relativistic Equations

In this section, we construct relativistic quantum-mechanical equations of motion. The guideline is our derivation of the free Schrödinger equation (see Vol. 1, Chap. 1). It was based on the non-relativistic dispersion relation

$$E = \frac{\mathbf{p}^2}{2m} \tag{U.1}$$

and the substitutions

$$E \to i\hbar \frac{\partial}{\partial t} \; ; \; \mathbf{p} \to \frac{\hbar}{i} \nabla. \tag{U.2}$$

Following these principles, we will derive in the following the Klein–Gordon equation and the Dirac equation.

U.2.1   Klein–Gordon Equation

The relativistic dispersion relation is given by

$$E^2 = m^2 c^4 + c^2 \mathbf{p}^2 \tag{U.3}$$

where m is the rest mass. The substitution (U.2) leads directly to the *Klein–Gordon equation* (cf. Vol. 1, Chap. 3)

$$-\hbar^2 \frac{\partial^2}{\partial t^2} \phi(\mathbf{r}, t) = -c^2 \hbar^2 \nabla^2 \phi(\mathbf{r}, t) + c^4 m^2 \phi(\mathbf{r}, t). \tag{U.4}$$

As it turns out, this equation is valid for quantum objects with spin zero. In the beginnings of Quantum Mechanics, it was Schrödinger who found this equation. However, he discarded it since it does not allow for a positive definite probability density.

U.2.1.1   Probability Density

Remember that for the Schrödinger equation, we have found that $\rho = \psi^* \psi$ can be regarded as probability density, since it is positive definite, $|\psi|^2 \geq 0$ (see Vol. 1, Chap. 7) and fulfills the continuity equation

$$\frac{\partial \rho}{\partial t} + \nabla \cdot \mathbf{j} = 0 \; ; \; \mathbf{j} = \frac{\hbar}{2im} \left(\psi^* \nabla \psi - \psi \nabla \psi^* \right). \tag{U.5}$$

In case of the Klein–Gordon equation (U.4), we multiply the equation and its complex conjugate by ϕ^* and ϕ and obtain the equations

$$\phi^* \frac{\partial^2}{\partial t^2} \phi = c^2 \phi^* \nabla^2 \phi - \frac{c^4 m^2}{\hbar^2} \phi^* \phi$$
$$\phi \frac{\partial^2}{\partial t^2} \phi^* = c^2 \phi \nabla^2 \phi^* - \frac{c^4 m^2}{\hbar^2} \phi \phi^*.$$
(U.6)

Subtracting the two equations gives

$$\phi^* \frac{\partial^2}{\partial t^2} \phi - \phi \frac{\partial^2}{\partial t^2} \phi^* = c^2 \phi^* \nabla^2 \phi - c^2 \phi \nabla^2 \phi^*$$
(U.7)

which can be written as

$$\frac{\partial}{\partial t} \left(\phi^* \frac{\partial}{\partial t} \phi - \phi \frac{\partial}{\partial t} \phi^* \right) = c^2 \nabla \left(\phi^* \nabla \phi - \phi \nabla \phi^* \right).$$
(U.8)

Again, we define the probability current density by $\mathbf{j} = \frac{\hbar}{2im} (\phi^* \nabla \phi - \phi \nabla \phi^*)$. Since the right hand side of (U.8) is given by $\frac{2im}{\hbar} c^2 \nabla \mathbf{j}$, comparison with the continuity equation leads to the conclusion that the probability density is defined by

$$\rho = \frac{i\hbar}{2mc^2} \left(\phi^* \frac{\partial}{\partial t} \phi - \phi \frac{\partial}{\partial t} \phi^* \right).$$

This expression is not positive definite which means that the concepts of probability cannot be applied for the Klein–Gordon equation - a sufficient argument for Erwin Schrödinger to reject this equation. We note that the problem stems from the fact that in the Klein–Gordon equation there occurs a second time derivative $\frac{\partial^2}{\partial t^2}$; a first derivative $\frac{\partial}{\partial t}$ would lead to $\rho = |\phi|^2$ as in the case of the Schrödinger equation.

U.2.1.2 Plane Waves and Negative Energies

Let us look for plane wave solutions of equation (U.4). The ansatz

$$\phi = \alpha (\mathbf{k}) e^{i\mathbf{kr} - i\omega t}$$
(U.9)

with an amplitude $\alpha (\mathbf{k})$ yields

$$\hbar^2 \omega^2 = c^2 \hbar^2 \mathbf{k}^2 + c^4 m^2.$$
(U.10)

As expected, by means of the deBroglie relations $E = \hbar\omega$ and $\mathbf{p} = \hbar\mathbf{k}$ we arrive at

$$E^2 = c^2 \mathbf{p}^2 + c^4 m^2.$$
(U.11)

Thus, the general plane wave solution is a linear combination of all partial solutions (U.9) and their complex conjugates, i.e.,

$$\phi = \int d^3k \left[\alpha\,(\mathbf{k})\, e^{i\mathbf{k}\mathbf{r}-i\omega t} + \alpha^*\,(\mathbf{k})\, e^{-i\mathbf{k}\mathbf{r}+i\omega t} \right]. \tag{U.12}$$

Now we ask which values of E (or ω) are allowed in (U.11) for a given momentum $\mathbf{p}$. The answer, of course, is that we have *two* solutions, namely

$$E = \pm\sqrt{c^2\mathbf{p}^2 + c^4 m^2}. \tag{U.13}$$

This means that we have also solutions with *negative* energy (or frequency). The problem is that nature does not know negative energies. So what to do with these solutions?

We remark that negative energies are not a speciality of the Klein–Gordon equation, but that this problem is in a certain sense common to all relativistic equations. We will meet and discuss it also when considering the Dirac equation, see below.

In classical physics, there occur negative energies, but this is due to a shifted energy zero point. As an example we consider the Hydrogen atom. The energies are given by (see Appendix F, Vol. 2)[129]

$$E_{nj} = m_0 c^2 \left[1 + \left(\frac{\alpha}{n - \left(j + \frac{1}{2}\right) + \sqrt{\left(j + \frac{1}{2}\right)^2 - \alpha^2}} \right)^2 \right]^{-1/2} \approx m_0 c^2 \left[1 - \frac{\alpha^2}{2n^2} + \cdots \right]$$

$$\tag{U.14}$$

where we have expanded the root with respect to the fine structure constant $\alpha \approx \frac{1}{137}$. Obviously, the exact expression E_{nj} is always positive. In order to obtain in classical physics expressions which are easy to handle, one subtracts the rest energy from the total energy and obtains $E_{nj,\,\text{classical}} = \left(E_{nj} - m_0 c^2\right) \approx -\frac{\alpha^2}{2n^2} m_0 c^2$. So these classical negative energies are not 'true' negative energies, in contrast to e.g. those with the lower sign given in (U.13).

U.2.2 Dirac Equation

Since the Klein–Gordon equation does not allow for the familiar probability interpretation, it was rejected. About the year 1928, Paul Dirac[130] found the equation which later was named after him. Since the problem of negative probabilities in the Klein–Gordon equation is connected with the second time derivative, he made the ansatz

---

[129] m_0 is the electron mass, α the fine structure constant, n the main quantum number and j the total angular momentum, $j = l \pm \frac{1}{2}$.

[130] Dirac, Paul Adrien Maurice, 1902–1984; British physicist, nobel prize 1933.

$$ i\hbar \frac{\partial}{\partial t} \psi = H\psi \tag{U.15} $$

with a yet to be determined operator H. As mentioned above in the introduction, it is perhaps surprising that this assumption together with the relativistic dispersion relation leads so to say automatically to an equation for particles with spin 1/2, without any additional requirements about angular momentum.

U.2.2.1 Statement of the Problem

We note that the dispersion relation (U.3) is not linear in E, as it is in the nonrelativistic case. To arrive at an equation of the form (U.15), i.e. a linear expression for E, one could take the root of (U.3)

$$ E = \pm\sqrt{c^2\mathbf{p}^2 + c^4 m^2} \tag{U.16} $$

which leads to

$$ i\hbar \frac{\partial}{\partial t} \psi = \pm\sqrt{-\hbar^2 c^2 \nabla^2 + c^4 m^2} \ \psi. \tag{U.17} $$

But this formulation is problematic. First, since it is a relativistic theory, time and space coordinates should be on equal footing. But in (U.17), spatial and temporal derivatives appear in an unsymmetrical manner. Second, how to cope with the square root operator? If we expand the square root in a power series by $\sqrt{-\hbar^2 c^2 \nabla^2 + c^4 m^2} = mc^2\sqrt{1 - \frac{\hbar^2}{c^2 m^2}\nabla^2} = mc^2\left[1 - \frac{\hbar^2}{2c^2 m^2}\nabla^2 + \frac{1}{8}\left(\frac{\hbar^2}{2c^2 m^2}\nabla^2\right)^2 \pm \ldots\right]$, we get an equation which contains all powers of the differential operator ∇^2, i.e. a nonlocal theory.

How to get rid of these problems? The basic idea is to use matrices in taking the square root. Consider the equation $x^2 = 1$. If x is an ordinary number, we have the two solutions $x = \pm 1$. But allowing for matrices and understanding the 1 as n-dimensional unit matrix 1_n, we can find other solutions (literally without the explicit use of the square root symbol $\sqrt{\ }$) as is seen in the next exercise. It was Paul Dirac who applied this basic idea in this context.

One problem remains, namely the meaning of the two signs on the right hand side of (U.16) and (U.17). This issue is not only a technical one, but is deeply connected with the structure of the world. We see that we have to deal not only with positive, but also with negative energies - if we have found a solution for positive energy, there exists a solution for negative energy, too. One could perhaps suppose that the solutions with negative energy are an artefact and could be neglected. But as it turns out, this is not true. Indeed, solutions with negative energy are a common feature of relativistic theories and are connected to the existence of *antiparticles*; the point is treated in further chapters in this and the second volume.

U.2.2.2   The Structure of the Dirac Equation

We want to find an expression which is linear in energy *and* momentum (considering relativistic time and space coordinates to be on an equal footing), whereby the squared expression must give the dispersion relation (U.3). The ansatz reads[131]

$$E = c \cdot \boldsymbol{\alpha} \cdot \mathbf{p} + mc^2 \beta \qquad \text{(U.18)}$$

where $\boldsymbol{\alpha} \neq 0$ and $\beta \neq 0$ are mathematical objects whose properties are to be determined. With the substitution (U.2), we arrive at the so-called *Dirac equation*, i.e. the quantum mechanical equation corresponding to (U.18):

$$i\hbar \frac{\partial}{\partial t} \psi = c \frac{\hbar}{i} \boldsymbol{\alpha} \cdot \boldsymbol{\nabla}\psi + mc^2 \beta \psi \qquad \text{(U.19)}$$

where $\psi = \psi(\mathbf{r}, t)$ is the wave function.

Before discussing this equation in more detail, we need to know more about $\boldsymbol{\alpha}$ and β. Like $\mathbf{p}$, the term $\boldsymbol{\alpha}$ has three components; β has one component. The four terms and β do not necessarily commute - we keep in mind the use of matrices.

Information about $\boldsymbol{\alpha}$ and β is obtained by comparison with (U.3). Squaring the ansatz (U.18) yields

$$E^2 = c^2 \cdot (\boldsymbol{\alpha} \cdot \mathbf{p})(\boldsymbol{\alpha} \cdot \mathbf{p}) + mc^3 (\boldsymbol{\alpha} \cdot \mathbf{p})\beta + mc^3 \beta(\boldsymbol{\alpha} \cdot \mathbf{p}) + m^2 c^4 \beta^2 \qquad \text{(U.20)}$$

and with $E^2 = c^2 \mathbf{p}^2 + m^2 c^4$, we get

$$\begin{aligned}
(\boldsymbol{\alpha} \cdot \mathbf{p})(\boldsymbol{\alpha} \cdot \mathbf{p}) &= \mathbf{p}^2 \\
(\boldsymbol{\alpha} \cdot \mathbf{p})\beta + \beta(\cdot \mathbf{p}) &= 0 \\
\beta^2 &= 1.
\end{aligned} \qquad \text{(U.21)}$$

By means of these equations, we have to determine β and the three components of $\boldsymbol{\alpha}$ as far as possible. We assume that the momentum $\mathbf{p}$ commutes with β and $\boldsymbol{\alpha}$.

To get an idea why we introduce matrices, let us first look at the second equation. Since it must hold for arbitrary momentum, we can write

$$(\alpha\beta + \beta\alpha) \cdot \mathbf{p} = 0 \rightarrow \alpha\beta + \beta\alpha = 0. \qquad \text{(U.22)}$$

Evidently, the last equation cannot be fulfilled if β and are 'ordinary' numbers - but with *matrices*, it works!

Rearranging the equations (U.21), we can write them as

---

[131] The factors c and c^2 ensure that the matrices α and β have no physical dimension.

Note that here β is *not* $\frac{v}{c}$. Moreover, the matrices α have nothing to do with the fine structure constant α. To label two totally different things by the same symbol is perhaps annoying and confusing especially for beginners in SR, but it is common practice.

$$\alpha_j \alpha_k + \alpha_k \alpha_j = 2\delta_{jk}$$
$$\alpha_j \beta + \beta \alpha_j = 0 \tag{U.23}$$
$$\beta^2 = 1$$

where all indices run from 1 to 3 and the unit matrix is abbreviated by 1 (yet we do not know the dimension of the matrices).

General properties of the matrices α and β With (U.23), we have ten equations for the four matrices α_1, α_2, α_3 and β. Firstly, we demand that the matrices are hermitian in order that the ansatz (U.18) gives an hermitian Hamilton operator (this and the required commutativity with the momentum are assumptions to make life easier - they do not follow per se). In addition, the matrices are unitary; this is due to $\beta^2 = 1 \rightarrow \beta = \beta^{-1}$ (analogously for the α_i). This means that the eigenvalues of the matrices are $+1$ or -1.

Thus, in (U.23) we have four unitary matrices which anticommute pairwise.[132] To get some information about the dimension of the matrices, we use the trace of the matrices. As an example, we consider the equation $\alpha_j \beta + \beta \alpha_j = 0$ which we multiply from the right by β to arrive at.

$$\alpha_j = -\beta \alpha_j \beta. \tag{U.24}$$

Taking the trace of both sides and making use of the cyclical commutativity under the trace,[133] we get

$$tr\ \alpha_j = -tr\ \beta \alpha_j \beta = -tr\ \alpha_j \beta^2 = -tr\ \alpha_j \tag{U.25}$$

from which follows $tr\ \alpha_j = 0$; analogously $tr\ \beta = 0$. The argument now runs as follows: since the matrices are unitary, they are diagonalizable. Thus, they can be represented by diagonal matrices, whereby the eigenvalues appear in the diagonal; all other entries vanish.[134] Since the eigenvalues are restricted to be $+1$ or -1, the condition $tr\ \alpha_j = 0$ can only be fulfilled if the matrices have even dimension - 2, 4, 6 and so on.

The question, which dimension it should be, cannot be answered unambiguously. But at least, we can exclude dimension 2. This is due to the fact that in the space of 2×2 matrices there are only three linearly independent anticommuting matrices, but in (U.23) we need four matrices.

Remembering Occham's razor, we attempt (as the next simplest case) to satisfy equations (U.23) by use of unitary 4×4 -matrices. Since these are ten equations for the four hermitian matrices α_1, α_2, α_3 and β (with altogether 64 complex entries), it comes as no surprise that the problem is underdetermined and, correspondingly, that no unique solution exists.

---

[132] A system like (U.23) is called a *Clifford algebra*.
[133] Remind $tr\ AB = tr\ BA$.
[134] Remember that eigenvalues and trace do not depend on the representation.

Standard representation of the matrices We summarize our results: under some weak assumptions, we arrived at a relativistic quantum mechanical description (the Dirac equation)

$$ i\hbar\frac{\partial}{\partial t}\psi = c\frac{\hbar}{i}\boldsymbol{\alpha}\cdot\nabla\psi + mc^2\beta\psi. \tag{U.26} $$

$\boldsymbol{\alpha}$ and β are 4×4-matrices which fulfill (U.23). These conditions do not give an unique solution; correspondingly, there are different representations of the Dirac equation. Some of them are common, and in the following, we want to make plausible one of them, the so-called *standard representation* of the Dirac equation.

We start with $\beta^2 = 1$, the third equation in (U.23), and *choose*

$$ \beta = \begin{pmatrix} 1 & 0 \\ 0 & -1 \end{pmatrix}. \tag{U.27} $$

The 1 in the diagonal is the 2×2 unit matrix (writing 1 instead of E is common in this context). We emphasize that (U.27) is not logically mandatory and every other choice is possible, provided it is an unitary matrix with vanishing trace.

To evaluate the second equation of (U.23), we write the hermitian matrices α_i as

$$ \alpha_i = \begin{pmatrix} A_i & B_i \\ B_i^\dagger & D_i \end{pmatrix}. \tag{U.28} $$

Note that A_i and the other entries in (U.28) are 2×2 matrices. Inserting (U.28) in $\alpha_i\beta + \beta\alpha_i = 0$, we obtain

$$ \alpha_i\beta + \beta\alpha_i = \begin{pmatrix} 2A_i & 0 \\ 0 & -2D_i \end{pmatrix} = 0. \tag{U.29} $$

It follows $A_i = D_i = 0$, i.e.

$$ \alpha_i = \begin{pmatrix} 0 & B_i \\ B_i^\dagger & 0 \end{pmatrix}. \tag{U.30} $$

Obviously, the condition $tr\ \alpha_i = 0$ is fulfilled.

Finally, we evaluate the first equation of (U.23), $\alpha_j\alpha_k + \alpha_j\alpha_k = 2\delta_{jk}$. With (U.30) it follows

$$ \alpha_j\alpha_k + \alpha_j\alpha_k = \begin{pmatrix} B_j B_k^\dagger + B_k B_j^\dagger & 0 \\ 0 & B_j^\dagger B_k + B_k^\dagger B_j \end{pmatrix} = 2\delta_{jk} \tag{U.31} $$

or in short

$$ B_j B_k^\dagger + B_k B_j^\dagger = 2\delta_{jk}. \tag{U.32} $$

We assume that the B_j are Hermitian matrices, $B_j = B_j^\dagger$ (again, this is not mandatory, but just convenient) and arrive at

$$B_j B_k + B_k B_j = 2\delta_{jk}.$$ (U.33)

Comparing this equation with the anticommutation rule (U.125) of the Pauli matrices shows that we can identify the B_j with the Pauli matrices σ_j, i.e.

$$\alpha = \begin{pmatrix} 0 & \sigma \\ \sigma & 0 \end{pmatrix} \; ; \; \beta = \begin{pmatrix} 1 & 0 \\ 0 & -1 \end{pmatrix}.$$ (U.34)

Note the block structure of the matrices α and β. This is the so-called *standard representation* of these matrices. Depending on the matter of question, other representations as the Weyl or the Majorana representation may be better suited, but we won't use them.

In this way we get as final result the standard representation of the Dirac equation

$$i\hbar \frac{\partial}{\partial t}\psi = c\frac{\hbar}{i}\alpha \cdot \nabla\psi + mc^2\beta\psi = c\alpha \cdot \mathbf{p}\psi + mc^2\beta\psi$$
with
$$\alpha = \begin{pmatrix} 0 & \sigma \\ \sigma & 0 \end{pmatrix} \; ; \; \beta = \begin{pmatrix} 1 & 0 \\ 0 & -1 \end{pmatrix}.$$ (U.35)

Remind that α and β are a 4×4-matrices.

U.3   Plane Wave Solutions

In this section, we consider the plane wave solutions of the Klein–Gordon and the Dirac equation with the focus on the results for the Dirac equation. We present solutions for the discrete case (finite Volume V) and the continuous case (infinite volume).

U.3.1   Klein–Gordon Equation

The free solutions in the case of a finite volume V read[135]

$$\phi(x) = \sum_{\mathbf{k}} \frac{1}{\sqrt{2V\omega_{\mathbf{k}}}} \left(a(\mathbf{k}) e^{i(\mathbf{kr}-\omega_{\mathbf{k}}t)} + a^\dagger(\mathbf{k}) e^{-i(\mathbf{kr}-\omega_{\mathbf{k}}t)} \right)$$ (U.36)

where the sum runs over all allowed discrete values of $\mathbf{k}$. The term $a(\mathbf{k})$ is here an arbitrary amplitude. We state in advance that in quantum field theory, it is this term which will be quantized.

---

[135]From now on, we write $E_{\mathbf{k}}$ and $\omega_{\mathbf{k}}$ instead of E and ω.

As stated above, energy and momentum are related by the relativistic dispersion relation

$$E_{\mathbf{k}} = \hbar\omega_{\mathbf{k}} = \sqrt{c^2\hbar^2\mathbf{k}^2 + c^4m^2}. \tag{U.37}$$

The continuous variant reads (cf. the preceding section) :

$$\phi(x) = \frac{1}{(2\pi)^{3/2}} \int \frac{d^3k}{\sqrt{2\omega_{\mathbf{k}}}} \left(a(\mathbf{k}) e^{i(\mathbf{kr}-\omega_{\mathbf{k}}t)} + a^\dagger(\mathbf{k}) e^{-i(\mathbf{kr}-\omega_{\mathbf{k}}t)}\right). \tag{U.38}$$

Notes: (1) There are different conventions for normalization. This one is chosen in view of later applications in quantum field theory; see the exercises. (2) With regard to considerations in quantum field theory, we write $a^\dagger(\mathbf{k})$ (Hermitian adjoint) and not simply $a^*(\mathbf{k})$ (complex conjugated) since later on, in quantum field theory, the amplitudes will be quantized, i.e., turned into operators. At this point, the notation $a^\dagger$ is is not per se evident. (3) Note that we use the same symbol $a(\mathbf{k})$ in the discrete and the continuous case. Strictly speaking one would have to make a distinction e.g. by different names. But using the same notation is quite common and functional, and confusion should be unlikely.

All three points apply the Dirac equation, too.

U.3.2   Dirac Equation

Next we want to find the free solutions of the Dirac equation

$$i\hbar\frac{\partial}{\partial t}\psi = c\frac{\hbar}{i}\boldsymbol{\alpha}\cdot\nabla\psi + mc^2\beta\psi = c\boldsymbol{\alpha}\cdot\mathbf{p}\psi + mc^2\beta\psi \tag{U.39}$$

where the matrices are given e.g. in standard representation

$$\alpha = \begin{pmatrix} 0 & \sigma \\ \sigma & 0 \end{pmatrix} \; ; \; \beta = \begin{pmatrix} 1 & 0 \\ 0 & -1 \end{pmatrix}. \tag{U.40}$$

In contrast to the Klein–Gordon case, the solutions are not scalar, since ψ has four components which complicates things a little bit. In addition, ψ is not a 'normal' 4-vector but a 4-*spinor*. The name has to do with the transformation behavior of ψ under Lorentz transformations (see Appendix T, Vol. 1) which in turn leads to another definition of the inner product of two spinors. We know that the inner product of two vectors a and b is defined by $a^\dagger b$ where $a^\dagger$ is the Hermitian adjoint of a. In contrast, the inner product of two 4-spinors ψ and φ is defined by $\bar{\psi}\varphi$ where the *adjungated spinor* or *Dirac adjoint* $\bar{\psi}$ is defined by $\bar{\psi} := \psi^\dagger\beta$ with $\psi^\dagger$ the Hermitian adjoint.[136] We remark that in the bulk of the book, treating nonrelativistic quantum mechanics,

---

[136]Note the differences in the definition of the inner product of 3-vectors, 4-vectors and 4-spinors.

it was not necessary to distinguish between 'adjungated' and 'hermitian adjungated'. Thus, these terms are often used synonymously. But here, in the context of the Dirac equation, we have to make precise distinctions.

U.3.2.1 Particle at Rest

Before we attempt the general plane wave solution, it is instructive to consider the case of a particle at rest, i.e., $\mathbf{p} = 0$. This means that the spatial derivatives vanish and we have to look for solutions of the simpler equation

$$i\hbar \frac{\partial}{\partial t}\psi_{\mathbf{p}=0} = mc^2 \beta \psi_{\mathbf{p}=0} = mc^2 \begin{pmatrix} 1 & 0 \\ 0 & -1 \end{pmatrix}\psi_{\mathbf{p}=0}. \tag{U.41}$$

The ansatz $\psi_{\mathbf{p}=0} = \psi_0 e^{-iEt/\hbar}$ leads to

$$E\psi_0 = mc^2 \beta \psi_0 = mc^2 \begin{pmatrix} 1 & 0 & 0 & 0 \\ 0 & 1 & 0 & 0 \\ 0 & 0 & -1 & 0 \\ 0 & 0 & 0 & -1 \end{pmatrix}\psi_0. \tag{U.42}$$

We see immediately that there are two solutions $\psi^{(+)}$ with positive energy, $E = mc^2$, and two solutions $\psi^{(-)}$ with negative energy, $E = -mc^2$. With $\omega = \frac{mc^2}{\hbar}$ they read explicitly

$$\psi_{1,\mathbf{p}=0}^{(+)} = \begin{pmatrix} 1 \\ 0 \\ 0 \\ 0 \end{pmatrix} e^{-i\omega t} \; ; \; \psi_{2,\mathbf{p}=0}^{(+)} = \begin{pmatrix} 0 \\ 1 \\ 0 \\ 0 \end{pmatrix} e^{-i\omega t}$$

$$\psi_{1,\mathbf{p}=0}^{(-)} = \begin{pmatrix} 0 \\ 0 \\ 1 \\ 0 \end{pmatrix} e^{i\omega t} \; ; \; \psi_{2,\mathbf{p}=0}^{(-)} = \begin{pmatrix} 0 \\ 0 \\ 0 \\ 1 \end{pmatrix} e^{i\omega t}. \tag{U.43}$$

Introducing the 2-spinors

$$\chi_1 = \begin{pmatrix} 1 \\ 0 \end{pmatrix} , \chi_2 = \begin{pmatrix} 0 \\ 1 \end{pmatrix} \tag{U.44}$$

we can write this more compactly as

$$\psi_{s,\mathbf{p}=0}^{(+)} = \begin{pmatrix} \chi_s \\ 0 \end{pmatrix} e^{-i\omega t} \; ; \; \psi_{s,\mathbf{p}=0}^{(-)} = \begin{pmatrix} 0 \\ \chi_s \end{pmatrix} e^{i\omega t} \; ; \; s = 1, 2. \tag{U.45}$$

U.3.2.2 Moving Particle

Guided by these results, we make for the general plane wave solution the ansatz for positive and negative energy

$$\psi^{(+)} = u_s \, e^{i(\mathbf{kr}-\omega t)} \text{ and } \psi^{(-)} = v_s \, e^{-i(\mathbf{kr}-\omega t)} \; ; s = 1, 2. \qquad \text{(U.46)}$$

Inserting (U.46) into the Dirac equation yields

$$\hbar\omega u_s = c\hbar\boldsymbol{\alpha} \cdot \mathbf{k}u_s + mc^2\beta u_s \text{ and } -\hbar\omega v_s = -c\hbar\boldsymbol{\alpha} \cdot \mathbf{k}v_s + mc^2\beta v_s. \qquad \text{(U.47)}$$

Using $E = \hbar\omega$ and $\mathbf{p} = \hbar\mathbf{k}$ brings $E\psi_0 = c\boldsymbol{\alpha} \cdot \mathbf{p}\psi_0 + mc^2\beta\psi_0$ or

$$\left(c\boldsymbol{\alpha} \cdot \mathbf{p} + mc^2\beta - E\right)u_s = 0 \text{ and } \left(c\boldsymbol{\alpha} \cdot \mathbf{p} - mc^2\beta - E\right)v_s = 0. \qquad \text{(U.48)}$$

With $E_\mathbf{p} = \sqrt{m^2c^4 + c^2\mathbf{p}^2}$ and the 2-spinors χ_s, the normalized solutions read

$$u_s(\mathbf{p}) = \sqrt{\frac{E_\mathbf{p} + mc^2}{2mc^2}}\left(\begin{array}{c}\chi_s \\ \frac{c\sigma\vec{p}}{E_\mathbf{p}+mc^2}\chi_s\end{array}\right) \; ; \; v_s(\mathbf{p}) = \sqrt{\frac{E_\mathbf{p} + mc^2}{2mc^2}}\left(\begin{array}{c}\frac{c\sigma\vec{p}}{E_\mathbf{p}+mc^2}\chi_s \\ \chi_s\end{array}\right) \; ; s = 1, 2. \text{ (U.49)}$$

The four spinors $u_s(\mathbf{p})$ and $v_s(\mathbf{p})$ as given in (U.49) are linearly independent. They satisfy the following relations:

$$\begin{array}{c}\bar{u}_r(\mathbf{p})\,u_s(\mathbf{p}) = \delta_{rs} \; ; \; \bar{v}_r(\mathbf{p})\,v_s(\mathbf{p}) = -\delta_{rs} \\ \bar{u}_r(\mathbf{p})\,v_s(\mathbf{p}) = 0 \; ; \; \bar{v}_r(\mathbf{p})\,u_s(\mathbf{p}) = 0\end{array} \; \forall r, s \qquad \text{(U.50)}$$

where $\bar{u}_r$ is the Dirac adjoint of u_r.

The general solution is a superposition of all allowed partial solutions .It reads

$$\begin{array}{ll}\psi(x) = \sum_{\mathbf{p},s}\sqrt{\frac{m}{V\omega_\mathbf{p}}}\left(b_s(\mathbf{p})\,u_s(\mathbf{p})\,e^{i(\mathbf{pr}-E_\mathbf{p}t)/\hbar} + d_s^\dagger(\mathbf{p})\,v_s(\mathbf{p})\,e^{-i(\mathbf{pr}-E_\mathbf{p}t)/\hbar}\right) & \text{discrete case} \\ \psi(x) = \sum_s\int d^3p\sqrt{\frac{m}{(2\pi)^3\omega_\mathbf{p}}}\left(b_s(\mathbf{p})\,u_s(\mathbf{p})\,e^{i(\mathbf{pr}-\omega_\mathbf{p}t)/\hbar} + d_s^\dagger(\mathbf{p})\,v_s(\mathbf{p})\,e^{-i(\mathbf{pr}-\omega_\mathbf{p}t)/\hbar}\right) & \text{continuous case}\end{array}$$
$$\text{(U.51)}$$

where $b_s(\mathbf{p})$ and $d_s(\mathbf{p})$ are arbitrary amplitudes. We write $d_s^\dagger(\mathbf{p})$ and not $d_s^*(\mathbf{p})$ in view of later applications in Quantum Field where the terms $b_s(\mathbf{p})$ and $d_s(\mathbf{p})$ are changed into operators.

U.3.2.3 Plane Waves, Limiting Cases

To shed some light on the physical meaning of the plane waves (U.51), we consider the basic spinors (U.49) for the two limiting cases $p \rightarrow 0$ (i.e., the nonrelativistic case) and $p \rightarrow \infty$. Remember that $u_s(\mathbf{p})$ describes solutions with positive energy and $v_s(\mathbf{p})$ solutions with negative energy.

Case $p \rightarrow 0$ In the limit $p \rightarrow 0$, we have from (U.49)

$$u_s\,(\mathbf{p}) \to \begin{pmatrix} \chi_s \\ 0 \end{pmatrix} \quad ; \quad v_s\,(\mathbf{p}) \to \begin{pmatrix} 0 \\ \chi_s \end{pmatrix} \quad ; \quad s = 1, 2 \qquad (\text{U.52})$$

which agrees with (U.43).

Considering the solution for positive energy, we see that there is a certain resemblance to the two basic states of a spin-1/2-particle, at least with respect to the first and second entry. This assumption is enhanced by the fact that the third and fourth entry vanish and the solutions for positive and negative energy are strictly separated. At this point, one could perhaps nurture the hope of just neglecting the 'lower' parts with negative energy, being something like an artefact of the theory.

Plane wave solutions, $p \to \infty$ Unfortunately (indeed rather fortunately) this faint hope is immediately destroyed by inspection of (U.49). To get the point clearer we consider the fully relativistic case $p \to \infty$ which means $E_\mathbf{p} \to c\,|\mathbf{p}| \gg mc^2$. This yields

$$u_s\,(\mathbf{p}) \to \sqrt{\frac{|\mathbf{p}|}{2mc}} \begin{pmatrix} \chi_s \\ \boldsymbol{\sigma}\frac{\mathbf{p}}{|\mathbf{p}|}\chi_s \end{pmatrix} \quad ; \quad v_s\,(\mathbf{p}) = \sqrt{\frac{|\mathbf{p}|}{2mc}} \begin{pmatrix} \boldsymbol{\sigma}\frac{\mathbf{p}}{|\mathbf{p}|}\chi_s \\ \chi_s \end{pmatrix} \quad ; \quad s = 1, 2. \quad (\text{U.53})$$

We see that all four entries in the solutions for positive and negative energy have the same order of magnitude and are inextricably coupled (note $\left\|\boldsymbol{\sigma}\frac{\mathbf{p}}{|\mathbf{p}|}\right\| = 1$).[137] This means we cannot neglect neither the third and fourth entry in u_s nor v_s as such. Thus, we have to discuss the question how to interpret the solutions for negative energy and, in addition, clarify the question why we have dimension 4 for spin 1/2 instead of the familiar dimension 2 of the state space.

U.3.2.4 Spin

The spinors (U.45) look very much like particles with spin 1/2, and this impression is confirmed by application of the spin operator Σ. Since we here have 4-spinors, the spin operator is not given just by $\boldsymbol{\sigma}$, but by

$$\Sigma = \frac{\hbar}{2} \begin{pmatrix} \boldsymbol{\sigma} & 0 \\ 0 & \boldsymbol{\sigma} \end{pmatrix}. \qquad (\text{U.54})$$

This relation can be shown formally, but the derivation is quite lengthy, so we just report the result.[138] Applying e.g. Σ_3, we have

$$\Sigma_3 \psi_{1,\mathbf{p}=0}^{(+)} = \frac{\hbar}{2} \begin{pmatrix} \sigma_3 & 0 \\ 0 & \sigma_3 \end{pmatrix} \begin{pmatrix} \chi_1 \\ 0 \end{pmatrix} e^{-i\omega t} = \frac{\hbar}{2} \begin{pmatrix} \sigma_3 \chi_1 \\ 0 \end{pmatrix} e^{-i\omega t} = \frac{\hbar}{2} \begin{pmatrix} \chi_1 \\ 0 \end{pmatrix} = \frac{\hbar}{2} \psi_{1,\mathbf{p}=0}^{(+)}$$

$$(\text{U.55})$$

---

[137] By the way, the operator $\boldsymbol{\sigma}\frac{\mathbf{p}}{|\mathbf{p}|}$ is called helicity operator.

[138] Note that Σ has 3 components (like $\boldsymbol{\sigma}$), but each of the components is a 4×4 matrix in spinor space.

and analogously

$$\Sigma_3 \psi_{2,\mathbf{p}=0}^{(+)} = -\frac{\hbar}{2} \psi_{1,\mathbf{p}=0}^{(+)} \ ; \ \ \Sigma_3 \psi_{1,\mathbf{p}=0}^{(-)} = \frac{\hbar}{2} \psi_{1,\mathbf{p}=0}^{(-)} \ ; \ \ \Sigma_3 \psi_{2,\mathbf{p}=0}^{(-)} = -\frac{\hbar}{2} \psi_{1,\mathbf{p}=0}^{(-)}. \quad \text{(U.56)}$$

Thus, we have two particles with spin 1/2, one for positive energy and one for negative energy. They have four components instead of two which fact is connected with the occurrence of negative energies. We postpone the interpretation of the negative energy solutions, but accept them temporarily as mathematical correct solutions (in fact, as we will see below, they are also physical correct solutions).

   If the particles are moving, the situation is not as simple. For faster and faster relativistic particles, the spin aligns more and more to the velocity vector. Indeed, for massless particles which travel at c, the spin is always directed parallel or antiparallel to the velocity vector. Thus, in general, things are more complicated than in the rest system (i.e., $\mathbf{p} = 0$).There is one exception, namely $p_1 = p_2 = 0$ and $p = p_3 > 0$, i.e., translation in the spin axis direction. We have for instance (with $\mathcal{E} = \sqrt{\frac{E_{\mathbf{p}}+mc^2}{2mc^2}}$)

$$\Sigma_3 u_1 (0,0,p_3) = \mathcal{E}\frac{\hbar}{2} \begin{pmatrix} \sigma_3 & 0 \\ 0 & \sigma_3 \end{pmatrix} \begin{pmatrix} \chi_1 \\ \frac{c\sigma_3 p_3}{E_{\mathbf{p}}+mc^2}\chi_1 \end{pmatrix} = \mathcal{E}\frac{\hbar}{2} \begin{pmatrix} \sigma_3\chi_1 \\ \frac{c\sigma_3 p_3 \sigma_3}{E_{\mathbf{p}}+mc^2}\chi_1 \end{pmatrix} = \frac{\hbar}{2} u_1 (0,0,p_3) \quad \text{(U.57)}$$

(due to $\sigma_3\chi_1 = \chi_1$) and analogously for the other spinors.

   Taken all together we can state the solutions (U.51) describe two types of particles with spin 1/2. One type is related to positive energy, the other to negative energy. In the frame of Dirac equation there is no really convincing explanation of the particle with negative energies but there is reason to believe that it is the antiparticle of the electron, i.e. the positron. This will be corroborated in quantum field theory.

   In any case, the important point here is that the Dirac equation allows for the description of relativistic particles with spin 1/2 whereby this fact follows without further assumptions from the ansatz itself.

U.4   Covariant Formulation of the Dirac Equation

We have derived the Dirac equation (DE)

$$i\hbar \frac{\partial}{\partial t} \psi = c\frac{\hbar}{i} \boldsymbol{\alpha} \cdot \nabla\psi + mc^2 \beta\psi. \quad \text{(U.58)}$$

α and β are 4×4 matrices, and the state ψ has four components.

   Deriving such an equation is only the first step and several questions are open at this stage. For instance, we have to verify that (U.58) is fully compatible with all requirements of special relativity. Since this is most easily done using a covariant notation of the equation, we will tackle this issue now. Thereafter, we will connect the DE to the electromagnetic field. In this way, we will obtain the Pauli equation as the non-relativistic approximation of the DE.

First a short comment on the term *covariance* (covariant). It has distinct, but related meanings which all have to do with the behavior under transformations.

(1) A physical quantity can be (Lorentz) covariant. Consider e.g. a 4-vector. Its length remains unchanged under Lorentz transformations (coordinate transformations), but its components change covariantly. Scalars are invariant under coordinate transformations, vector components are covariant. Other covariant objects are spinors and tensors.

(2) An equation or a theory is said to be covariant if it can be written in terms of covariant-only quantities. As a consequence, such a theory or equation has the same form in all reference frames (inertial systems) and is said to be form invariant (this is the main criterion). As an example, the Dirac equation must have the same form for all observers, independently from their inertial system.

(3) Do not confuse this meaning of 'covariant' with the use of the terms covariant and contravariant vectors, common in SR. These labels are established, but quite unfortunate. In any case, covariant and contravariant vectors are both transforming covariantly.

U.4.1 Introducing γ Matrices

To tackle the mentioned questions, we first introduce and discuss the so-called γ matrices. We start with the Dirac equation (U.58), i.e.,

$$i\hbar\frac{\partial}{\partial t}\psi = c\frac{\hbar}{i}\boldsymbol{\alpha}\cdot\boldsymbol{\nabla}\psi + mc^2\beta\psi \tag{U.59}$$

where $\boldsymbol{\alpha}$ has the components α_i, $i = 1, 2, 3$.[139] Dividing both sides by c and using $x^0 = ct$, we arrive at[140]

$$i\hbar\frac{\partial}{\partial x^0}\psi = \frac{\hbar}{i}\sum_{k=1}^{3}\alpha_k\frac{\partial}{\partial x^k}\psi + mc\beta\psi. \tag{U.60}$$

Note that we use the covariant form of the gradient (see Appendix T, Vol. 1). Multiplying both sides by β and writing $\frac{\partial}{\partial x^\mu} = \partial_\mu$ and dividing by $i\hbar$ yields (remember $\beta^2 = 1$)

$$\left(\beta\partial_0 + \sum_{k=1}^{3}\beta\alpha_k\partial_k\right)\psi = \frac{mc}{i\hbar}\psi. \tag{U.61}$$

---

[139]Note that in this context $\boldsymbol{\alpha}$ behaves like a 3-vector; hence, we can use upper or lower indices, $\alpha^i = \alpha_i$.

[140]$\partial_\mu = \frac{\partial}{\partial x^\mu} = \left(\frac{1}{c}\frac{\partial}{\partial t}, \boldsymbol{\nabla}\right) = (\partial_0, \boldsymbol{\nabla})$ and $\partial^\mu = \frac{\partial}{\partial x_\mu} = \left(\frac{1}{c}\frac{\partial}{\partial t}, -\boldsymbol{\nabla}\right) = (\partial_0, -\boldsymbol{\nabla})$

In order to simplify the notation, we define new matrices by[141]

$$\gamma^0 := \beta \; ; \; \gamma^k := \beta\alpha_k. \tag{U.62}$$

These matrices are called γ *matrices* or *Dirac matrices*; note that they are upstairs objects, i.e., contravariant. Invoking the summation rule, we can write

$$\left(\gamma^0\partial_0 + \gamma^k\partial_k\right)\psi = \gamma^\mu\partial_\mu\psi = \frac{mc}{i\hbar}\psi. \tag{U.63}$$

With $p_\mu = i\hbar\partial_\mu$ it follows

$$\gamma^\mu p_\mu\psi = mc\psi. \tag{U.64}$$

Thus, on the l.h.s, we have with $\gamma^\mu p_\mu$ somewhat like a inner product (somewhat, because the γ^μ are matrices; but this point can be cleared positively), and we know that inner products are invariant; the same holds for the scalar mc on the r.h.s. This means that (U.64) is a good candidate for Lorentz covariance.

Since inner products like $\gamma^\mu\partial_\mu$ occur often in relativistic theories, a special short-hand has been established for these objects, also called Feynman slash notation, namely

$$\not{a} = \gamma \cdot a = \gamma^\mu a_\mu = \gamma_\mu a^\mu = \gamma^0 a_0 - \gamma^k a_k = \gamma^0 a_0 - \boldsymbol{\gamma} \cdot \mathbf{a}. \tag{U.65}$$

With this notation, the Dirac equation reads

$$\not{\partial}\psi = \frac{mc}{i\hbar}\psi. \tag{U.66}$$

With $p = i\hbar\partial$, we arrive at the presumably most streamlined form of the Dirac equation

$$\not{p}\psi = mc\psi. \tag{U.67}$$

U.4.1.1 Properties of the γ Matrices

The new found γ matrices play a dominant role in 'higher' relativistic theories where they replace completely the matrices α and β. In the following, we will discuss some of their properties. Their covariant form reads $\gamma_\mu = g_{\mu\nu}\gamma^\nu = \left(\gamma^0, -\gamma^1, -\gamma^2, -\gamma^3\right)$. In most manipulations, it is easiest to think of the 4-tuple γ^μ as of a matrix-valued 4-vector, though it is a slight misnomer.

---

[141]Unfortunately, these matrices are named also by the letter γ^μ but should not confused with $\gamma = \frac{1}{\sqrt{1-(\frac{v}{c})^2}}$.

Explicit formulation in the standard representation In the standard representation (also called Dirac representation), i.e.,

$$\beta = \begin{pmatrix} 1 & 0 \\ 0 & -1 \end{pmatrix} \; ; \; \alpha = \begin{pmatrix} 0 & \sigma \\ \sigma & 0 \end{pmatrix} \tag{U.68}$$

the γ matrices are given by

$$\gamma^0 = \begin{pmatrix} 1 & 0 \\ 0 & -1 \end{pmatrix} \; ; \; \gamma = \begin{pmatrix} 0 & \sigma \\ -\sigma & 0 \end{pmatrix} \tag{U.69}$$

or explicitly by

$$\gamma^0 = \begin{pmatrix} 1 & 0 & 0 & 0 \\ 0 & 1 & 0 & 0 \\ 0 & 0 & -1 & 0 \\ 0 & 0 & 0 & -1 \end{pmatrix} \; ; \; \gamma^1 = \begin{pmatrix} 0 & 0 & 0 & 1 \\ 0 & 0 & 1 & 0 \\ 0 & -1 & 0 & 0 \\ -1 & 0 & 0 & 0 \end{pmatrix}$$

$$\gamma^2 = \begin{pmatrix} 0 & 0 & 0 & -i \\ 0 & 0 & i & 0 \\ 0 & i & 0 & 0 \\ -i & 0 & 0 & 0 \end{pmatrix} \; ; \; \gamma^3 = \begin{pmatrix} 0 & 0 & 1 & 0 \\ 0 & 0 & 0 & -1 \\ -1 & 0 & 0 & 0 \\ 0 & 1 & 0 & 0 \end{pmatrix} . \tag{U.70}$$

There are other representations of the gamma matrices like the Weyl or chiral representation or the Majorana representation. They have different useful properties for certain calculations, but we will not need them. For details see the literature.[142]

U.4.1.2 (Anti-) Commutation Rules for the γ Matrices

The matrices α and β obey the commutation rules

$$\alpha_j \alpha_k + \alpha_k \alpha_j = 2\delta_{jk}$$
$$\alpha_k \beta + \beta \alpha_k = 0 \text{ or } \alpha_k = -\beta \alpha_k \beta \tag{U.71}$$
$$\alpha_i^2 = \beta^2 = 1.$$

With

$$\gamma^0 := \beta \; ; \; \gamma^k := \beta \alpha^k \; \rightarrow \; \beta = \gamma^0 \; ; \; \alpha^k = \beta \gamma^k = -\gamma^k \beta \tag{U.72}$$

we arrive for the Dirac matrices at

---

[142]In addition to the four matrices γ^μ one defines a matrix γ^5 by $\gamma^5 = i\gamma^0 \gamma^1 \gamma^2 \gamma^3$. The index 5 stems from the former notation of γ^4 instead of today's γ^0. γ^5 is hermitian, its eigenvalues are ± 1, and it anticommutes with the four γ^μ.

$$-\gamma^j \beta\beta\gamma^k - \gamma^k \beta\beta\gamma^j = -\gamma^j\gamma^k - \gamma^k\gamma^j = 2\delta_{jk}$$
$$\beta\gamma^k + \gamma^k\beta = 0 \rightarrow \gamma^0\gamma^k + \gamma^k\gamma^0 = 0 \tag{U.73}$$
$$\left(\gamma^0\right)^2 = 1 \ ; \ \left(\gamma^i\right)^2 = -1.$$

This relations may be formulated compactly with the help of the elements $g^{\mu\nu}$ of the metric tensor:

$$\gamma^\mu\gamma^\nu + \gamma^\nu\gamma^\mu = \{\gamma^\mu, \gamma^\nu\} = 2g^{\mu\nu} \cdot 1 \tag{U.74}$$

where 1 is the 4×4 unit matrix.[143]

U.4.1.3   Adjoint Dirac Equation, Continuity Equation

We start from the DE in the form

$$i\gamma^\mu\partial_\mu\psi - \frac{mc}{\hbar}\psi = 0. \tag{U.75}$$

Taking the hermitian conjugate of this equation and multiplying it from the right by γ^0 gives the *adjoint Dirac equation* ($\bar\psi = \psi^\dagger\gamma^0$):

$$i\left(\partial_\mu\bar\psi\right)\gamma^\mu + \frac{mc}{\hbar}\bar\psi = 0. \tag{U.76}$$

We multiply the Dirac equation from left by $\bar\psi$ and the adjoint Dirac equation from right by ψ:

$$i\bar\psi\gamma^\mu\partial_\mu\psi - \frac{mc}{\hbar}\bar\psi\psi = 0 \ ; \ i\left(\partial_\mu\bar\psi\right)\gamma^\mu\psi + \frac{mc}{\hbar}\bar\psi\psi = 0. \tag{U.77}$$

Adding theses two equations gives

$$\bar\psi\gamma^\mu\partial_\mu\psi + \left(\partial_\mu\bar\psi\right)\gamma^\mu\psi = 0 \ \text{ or } \ \partial_\mu\left(\bar\psi\gamma^\mu\psi\right) = 0. \tag{U.78}$$

Reading this equation as continuity equation[144] $\partial_\mu j^\mu = 0$ defines the *Dirac 4-current* by

$$j^\mu = \bar\psi\gamma^\mu\psi. \tag{U.79}$$

Hence, the probability density $\rho = j^0$ is given by

$$\rho = j^0 = \bar\psi\gamma^0\psi = \psi^\dagger\psi. \tag{U.80}$$

Evidently, the probability density ρ is positive definite.

---

[143]Like the matrices α and β, also the γ matrices generate a Clifford algebra.
[144]See Appendix T, Vol. 1.

U.4.2 How to Show the Covariance of the Dirac Equation - A Short Outline

In theoretical physics, an important principle is that coordinates are manmade and do not exist in nature; hence, they should play no role in the formulation of physical laws. In our case this means that the Dirac equation has to be form invariant, i.e., has to have the same form in all inertial systems.

We assume two inertial systems I and $\tilde{I}$ with coordinates x and $\tilde{x} = \Lambda x$ and wave functions $\psi(x)$ and $\tilde{\psi}(\tilde{x})$.[145] There has to exist a unique relation between $\psi(x)$ and $\tilde{\psi}(\tilde{x})$. Since both the Dirac equation and the Lorentz transformation are linear, this relation has to be linear; furthermore, since the wave functions have 4 components, this relation is a 4×4-matrix $S(\Lambda)$, i.e.,

$$\tilde{\psi}(\tilde{x}) = S(\Lambda)\psi(x) = S(\Lambda)\psi\left(\Lambda^{-1}\tilde{x}\right). \tag{U.81}$$

Thus, Lorentz covariance means that, by use of $\tilde{x} = \Lambda x$ and $\tilde{\psi}(\tilde{x}) = S(\Lambda)\psi\left(\Lambda^{-1}\tilde{x}\right)$, the Dirac equation in I is transformed into a Dirac equation in $\tilde{I}$. In other words: in I and $\tilde{I}$, the Dirac equations read

$$\left(i\hbar\gamma^{\mu}\partial_{\mu} - mc\right)\psi(x) = 0 \text{ and } \left(i\hbar\gamma^{\mu}\tilde{\partial} - mc\right)\tilde{\psi}(\tilde{x}) = 0 \tag{U.82}$$

with $\tilde{\partial}_{\mu} = \frac{\partial}{\partial\tilde{x}^{\mu}}$. To arrive at an equation for $S(\Lambda)$, we use

$$\partial_{\mu} = \frac{\partial}{\partial x^{\mu}} = \frac{\partial\tilde{x}^{\nu}}{\partial x^{\mu}}\frac{\partial}{\partial\tilde{x}^{\nu}} = \Lambda^{\nu}{}_{\mu}\frac{\partial}{\partial\tilde{x}^{\nu}}. \tag{U.83}$$

We insert this and $\psi(x) = S^{-1}(\Lambda)\tilde{\psi}(\tilde{x})$ into the Dirac equation of I. This yields

$$\left(i\hbar\gamma^{\mu}\Lambda^{\nu}{}_{\mu}\frac{\partial}{\partial\tilde{x}^{\nu}} - mc\right)S^{-1}(\Lambda)\tilde{\psi}(\tilde{x}) = 0. \tag{U.84}$$

We multiply from the left with S:

$$i\hbar S(\Lambda)\gamma^{\mu}\Lambda^{\nu}{}_{\mu}\tilde{\partial}_{\nu}S^{-1}(\Lambda)\tilde{\psi}(\tilde{x}) - mc\tilde{\psi}(\tilde{x}) = 0. \tag{U.85}$$

Comparison with the Dirac equation in $\tilde{I}$ shows that we have to look for a solution S of the equation

$$S(\Lambda)\gamma^{\mu}\Lambda^{\nu}{}_{\mu}\tilde{\partial}_{\nu}S^{-1}(\Lambda) = \gamma^{\mu}\tilde{\partial}_{\mu} = \gamma^{\nu}\tilde{\partial}_{\nu}. \tag{U.86}$$

---

[145]We here can assume that Λ encompasses all Lorentz transformations, i.e., apart from the boost also rotations, space reflections etc. In advanced theories there occurs e.g. parity violation (in weak interactions), but we are not concerned with this.

One also can perform the considerations for the case of the more general Poincare transformation $\tilde{x} = \Lambda x + a$, of course with the same result.

It follows

$$\gamma^{\mu}\Lambda^{\nu}{}_{\mu}\tilde{\partial}_{\nu} = S^{-1}(\Lambda)\,\gamma^{\nu}\tilde{\partial}_{\nu}S(\Lambda) \tag{U.87}$$

or[146]

$$S^{-1}(\Lambda)\,\gamma^{\nu}S(\Lambda) = \Lambda^{\nu}{}_{\mu}\gamma^{\mu}. \tag{U.88}$$

This is the fundamental equation to determine $S(\Lambda)$. Finding a solution for all Λ proves the covariance of the Dirac equation. As an example, consider space reflection which is given by

$$\Lambda^{\nu}{}_{\mu} = \begin{pmatrix} 1 & 0 & 0 & 0 \\ 0 & -1 & 0 & 0 \\ 0 & 0 & -1 & 0 \\ 0 & 0 & 0 & -1 \end{pmatrix} = g^{\nu\mu}. \tag{U.89}$$

Equation (U.88) reads in this case

$$S^{-1}\gamma^{\nu}S = \Lambda^{\nu}{}_{\mu}\gamma^{\mu} = g^{\nu\mu}\gamma^{\mu} = g^{\nu\nu}\gamma^{\nu} \tag{U.90}$$

without summation over ν in the last expression. It follows

$$S^{-1}\gamma^{0}S = \gamma^{0}\ ;\ S^{-1}\gamma^{k}S = -\gamma^{k} \rightarrow S = \gamma^{0} \tag{U.91}$$

where we have taken into account (U.74).[147] In this way, the transformation (U.81) reads

$$\tilde{\psi}(\tilde{x}) = \tilde{\psi}(t,\tilde{\mathbf{x}}) = \tilde{\psi}(t,-\mathbf{x}) = \gamma^{0}\psi(t,\mathbf{x})\,. \tag{U.92}$$

In total, the parity transformation for spinors can be written as

$$P = \gamma^{0}P^{(\mathbf{x})} \tag{U.93}$$

where $P^{(\mathbf{x})}$ causes the space reflection $\mathbf{x} \to -\mathbf{x}$.

In a similar way one can show that for all Lorentz transformations there is a solution $S(\Lambda)$ of equation (U.88) which fact states the covariance of the Dirac equation. The calculations are a little bit lengthy and cumbersome and we omit them. For the extensive details see the literature.

A remark on nomenclature: A wave function ψ which transforms corresponding to (U.81) with S given by (U.88) is called (4-component) *Lorentz spinor* or 4-spinor. Note that ψ it is not a 4-vector which is defined by its behavior under Lorentz transformation as $\tilde{a}^{\mu} = \Lambda^{\mu}{}_{\nu}\,a^{\nu}$.

---

[146]Note that $\Lambda^{\nu}{}_{\mu}$ are matrix *elements* which, as being scalars, commute with the Dirac matrices γ^{μ}, of course.

[147]One can allow for an arbitrary phase factor $e^{i\varphi}$, in addition.

U.4.3   Coupling to the Electromagnetic Field

Finally, we want to formulate the Dirac equation in an electromagnetic field. To this end, we start with the free Dirac equation in the form

$$\gamma^\mu p_\mu \psi = \not{p}\psi = mc\psi. \tag{U.94}$$

As in nonrelativistic quantum mechanics, we invoke the principle of *minimal coupling*,[148] i.e., replace the 4-momentum p by[149]

$$p \rightarrow p - qA. \tag{U.95}$$

A is the 4-potential, $A^\mu = \left(\frac{\Phi}{c}, \mathbf{A}\right)$ with the scalar potential Φ and the three components of the vector potential $\mathbf{A}$. The resulting equation reads

$$\gamma^\mu \left(p_\mu - qA_\mu\right)\psi = (\not{p} - q\not{A})\,\psi = mc\psi. \tag{U.96}$$

The substitution does not affect the considerations on the covariance of the Dirac equation. This argumentation was based on the fact that p is a 4-vector, but A is a 4-vector, too, and of course also the difference $p - qA$. Thus, (U.96) is invariant in the sense that every observer would find exactly this form of the equation in his reference frame, i.e., inertial system.

To make contact with the Dirac equation in the formulation (U.58), i.e. $i\hbar\frac{\partial}{\partial t}\psi = c\frac{\hbar}{i}\boldsymbol{\alpha} \cdot \boldsymbol{\nabla}\psi + mc^2\beta\psi$, we use $p^\mu = i\hbar\partial^\mu$ (cf. Appendix T, Vol. 1). This means explicitly

$$\begin{aligned} p^0 &\rightarrow p^0 - qA^0 \Longrightarrow i\hbar\partial^0 \rightarrow i\hbar\partial^0 - \tfrac{q}{c}\Phi \Longrightarrow i\hbar\tfrac{\partial}{\partial t} \rightarrow i\hbar\tfrac{\partial}{\partial t} - q\Phi \\ p^k &\rightarrow p^k - qA^k \Longrightarrow \tfrac{\hbar}{i}\partial^k \rightarrow \tfrac{\hbar}{i}\partial^k - qA^k \Longrightarrow \tfrac{\hbar}{i}\boldsymbol{\nabla} \rightarrow \tfrac{\hbar}{i}\boldsymbol{\nabla} - q\mathbf{A}. \end{aligned} \tag{U.97}$$

Thus, the Dirac equation in an electromagnetic field reads

$$i\hbar\frac{\partial}{\partial t}\psi = c\boldsymbol{\alpha}\left(\frac{\hbar}{i}\boldsymbol{\nabla} - q\mathbf{A}\right)\psi + q\Phi\psi + \beta mc^2\psi. \tag{U.98}$$

U.4.4   Nonrelativistic Limit of the Dirac Equation

Since the Dirac equation is formulated for 4-spinors and the nonrelativistic Pauli equation for 2-spinors, it seems advantageous to consider the 4-spinor ψ as composed of two 2-spinors, i.e.,

---

[148]Called minimal coupling, because it is the simplest nontrivial coupling compatible with gauge invariance. As far as we know, it is also the possibility which is realized by nature.

[149]q is the charge of the particle under consideration, e.g. $q = -e_0$ for an electron.

$$\psi = \begin{pmatrix} \varphi \\ \chi \end{pmatrix}. \tag{U.99}$$

With $\alpha = \begin{pmatrix} 0 & \sigma \\ \sigma & 0 \end{pmatrix}$ and $\beta = \begin{pmatrix} 1 & 0 \\ 0 & -1 \end{pmatrix}$ and the abbreviation

$$\boldsymbol{\pi} = \mathbf{p} - q\mathbf{A} \tag{U.100}$$

we can write the Dirac equation (U.98) in the form

$$
\begin{aligned}
i\hbar \tfrac{\partial}{\partial t}\varphi &= c\boldsymbol{\sigma} \cdot \boldsymbol{\pi}\chi + q\Phi\varphi + mc^2\varphi \\
i\hbar \tfrac{\partial}{\partial t}\chi &= c\boldsymbol{\sigma} \cdot \boldsymbol{\pi}\varphi + q\Phi\chi - mc^2\chi.
\end{aligned}
\tag{U.101}
$$

In the nonrelativistic limit, the rest mass mc^2 is the by far the largest energy in the expression $E = \sqrt{m^2c^4 + c^2p^2}$. Therefore, for $p \to 0$ the ansatz

$$\begin{pmatrix} \varphi \\ \chi \end{pmatrix} = e^{-imc^2t/\hbar} \begin{pmatrix} \varphi_{nr} \\ \chi_{nr} \end{pmatrix} \tag{U.102}$$

is appropriate to describe the solution for positive energy. It follows

$$
\begin{aligned}
i\hbar \tfrac{\partial}{\partial t}\varphi_{nr} &= c\boldsymbol{\sigma} \cdot \boldsymbol{\pi}\chi_{nr} + q\Phi\varphi_{nr} \\
i\hbar \tfrac{\partial}{\partial t}\chi_{nr} &= c\boldsymbol{\sigma} \cdot \boldsymbol{\pi}\varphi_{nr} + q\Phi\chi_{nr} - 2mc^2\chi_{nr}.
\end{aligned}
\tag{U.103}
$$

In the non-relativistic limit, we have $|q\Phi| \ll 2mc^2$. In addition, the functions φ_{nr} and χ_{nr} vary only very slowly with respect to time. Thus, we have in the second equation $\left| i\hbar \tfrac{\partial}{\partial t}\chi_{nr}\right| \ll \left|2mc^2\chi_{nr}\right|$, from which follows $c\boldsymbol{\sigma} \cdot \boldsymbol{\pi}\varphi_{nr} \approx 2mc^2\chi_{nr}$ or, as an estimation of the order of magnitude,

$$\left|\frac{\chi_{nr}}{\varphi_{nr}}\right| \approx \left|\frac{c\boldsymbol{\sigma} \cdot \boldsymbol{\pi}}{2mc^2}\right| \approx \left|\frac{\mathbf{p}}{2mc}\right| \approx \left|\frac{\mathbf{v}}{2c}\right|. \tag{U.104}$$

We see that in the nonrelativistic limit, χ_{nr} is smaller than φ_{nr} by a factor $\sim v/c$. Therefore, for $v \to 0$, one often calls φ and χ large and small component of the spinor ψ.

Using this result, we can write the first equation in (U.103) in a good approximation as

$$i\hbar\frac{\partial}{\partial t}\varphi_{nr} = \boldsymbol{\sigma} \cdot \boldsymbol{\pi}\frac{\boldsymbol{\sigma} \cdot \boldsymbol{\pi}}{2m}\varphi_{nr} + q\Phi\varphi_{nr}. \tag{U.105}$$

To evaluate the term $\boldsymbol{\sigma} \cdot \boldsymbol{\pi}\,\boldsymbol{\sigma} \cdot \boldsymbol{\pi}$, we use $\boldsymbol{\sigma} \cdot \mathbf{a}\,\boldsymbol{\sigma} \cdot \mathbf{b} = \mathbf{a} \cdot \mathbf{b} + i\boldsymbol{\sigma} \cdot (\mathbf{a} \times \mathbf{b})$. Thus, we have

$$\boldsymbol{\sigma} \cdot \boldsymbol{\pi}\,\boldsymbol{\sigma} \cdot \boldsymbol{\pi} = \pi^2 + i\boldsymbol{\sigma} \cdot (\boldsymbol{\pi} \times \boldsymbol{\pi}). \tag{U.106}$$

Note that $\boldsymbol{\pi} \times \boldsymbol{\pi}$ does not vanish, since $\boldsymbol{\pi}$ is an operator. In fact, we have

$$\boldsymbol{\pi} \times \boldsymbol{\pi} = (\mathbf{p} - q\mathbf{A}) \times (\mathbf{p} - q\mathbf{A}) = -q\,(\mathbf{p} \times \mathbf{A} + \mathbf{A} \times \mathbf{p}) \qquad \text{(U.107)}$$

from which follows[150]

$$(\boldsymbol{\pi} \times \boldsymbol{\pi})\,\varphi_{nr} = -\frac{\hbar}{i}q\left(\boldsymbol{\nabla} \times \vec{A} + \mathbf{A} \times \boldsymbol{\nabla}\right)\varphi_{nr} = -\frac{\hbar}{i}q\,[(\boldsymbol{\nabla} \times A)\,\varphi_{nr} + (\boldsymbol{\nabla}\varphi_{nr}) \times \mathbf{A} + \mathbf{A} \times \boldsymbol{\nabla}\varphi_{nr}]$$
$$\text{(U.108)}$$

or with $\mathbf{B} = rot\,\mathbf{A}$

$$(\boldsymbol{\pi} \times \boldsymbol{\pi})\,\varphi_{nr} = -\frac{\hbar}{i}q\left(\boldsymbol{\nabla} \times \vec{A}\right)\varphi_{nr} = -\frac{\hbar}{i}q\mathbf{B}\varphi_{nr}. \qquad \text{(U.109)}$$

Thus, we can write (U.105) in the form

$$i\hbar\frac{\partial}{\partial t}\varphi_{nr} = \frac{\boldsymbol{\pi}^2}{2m}\varphi_{nr} - \frac{q\hbar}{2m}\boldsymbol{\sigma} \cdot \vec{B}\varphi_{nr} + q\Phi\varphi_{nr} \qquad \text{(U.110)}$$

which is the nonrelativistic *Pauli equation* for the Pauli spinor φ_{nr}.

A note concerning the interaction of spin and magnetic field. Formulated by means of the spin vector $\mathbf{s} = \boldsymbol{\sigma}/2$, it is given by

$$-\frac{q\hbar}{2m}\boldsymbol{\sigma} \cdot \mathbf{B} = -2\frac{q\hbar}{2m}\mathbf{s} \cdot \mathbf{B} = -g\frac{q\hbar}{2m}\mathbf{s} \cdot \mathbf{B}. \qquad \text{(U.111)}$$

The factor $g = 2$ is called g-*factor* or Landé factor (or, more precisely, electron g-factor g_e). It relates the magnetic moment of the electron to its spin, $\boldsymbol{\mu} = g_e\frac{q}{2m}\mathbf{s}$. Remind that a classical consideration (e.g., for the orbital angular momentum) leads to $g = 1$. High precision experiments show that g_e is somewhat greater than given by the Dirac equation, namely roughly equal to $2,002319$. The reason is explained by quantum field theory (see Appendix W, Vol. 2). Indeed, g_e is known with a striking precision, both theoretically and experimentally.

U.5 Dirac Equation and the Hydrogen Atom

Due to reasons of content, this section is found in Appendix F, Vol. 2.

---

[150]Remind $\boldsymbol{\nabla} \times (f\mathbf{F}) = f \cdot (\boldsymbol{\nabla} \times \mathbf{F}) + (\boldsymbol{\nabla} f) \times \mathbf{F}$ and $(\boldsymbol{\nabla}\varphi) \times \mathbf{A} + \mathbf{A} \times (\boldsymbol{\nabla}\varphi) = 0$.

U.6   Discussion of the Dirac Equation

In this section, we compile the pros and cons of the Dirac equation (DE).[151] Despite all
the convincing properties of the equation, there is a major problem, namely negative
energies. This point remains without a convincing solution within the framework
of the equation. A way out of this dilemma towards an advanced theory offers the
Feynman–Stückelberg interpretation which paves the way for quantum field theory.

U.6.1   Pros and Cons of the Dirac Equation

We briefly mention some advantages and disadvantages of the DE.
   The list of advantages is impressive:

1. The Dirac equation allows for the description of relativistic particles with spin
   $1/2$. The fact is all the more remarkable since in the derivation of the DE there
   is no assumption spin $1/2$ or any angular momentum. The DE follows from
   two basic ingredients, namely (1) the relativistic dispersion relation, and (2) the
   assumption that there exists a Hamilton function H so that $i\hbar\dot\psi = H\psi$.
2. The equation provides the hydrogen spectrum with high accuracy, far better than
   the Schrödinger equation.
3. The DE provides the g-factor[152] of the electron ($g = 2$); its non-relativistic limit
   leads correctly to the Pauli equation.
4. The DE allows for an explanation of the Zitterbewegung (see below).
5. Historically, relativistic particles with spin $1/2$ could not be described before
   1928 when the DE appeared. In addition, the DE was the first one which suggested the existence of antiparticles. Indeed, shortly after the publication of the
   DE, the positron e^+ was discovered experimentally, i.e., the antiparticle of the
   electron e^-.

But there are also disadvantages, among them:

1. There are solutions with negative energies. From a mathematical point of view,
   these solutions are perfectly correct, but the problem is that negative energies
   are not realized in nature and would lead to paradoxical consequences. So how
   to deal with the negative energy solutions?
2. In deriving the DE, we started with the aim of finding a single-particle theory.
   But as we have seen, it is impossible to achieve this objective. We always have

---

[151] For each spin there exists a special equation, e.g. Klein–Gordon for $s = 0$, Dirac for $s = 1/2$,
Proca for $s = 1$, Rarita–Schwinger for $s = 3/2$.

[152] The electron g-factor is one of the most precisely measured values in physics. The DE says
$g = 2$, and quantum electrodynamics corrects this value to $g = 2.00231930436182$ with a relative
standard uncertainty of 10^{-13} (see Appendix W, Vol. 2).

a 4-spinor, i.e., vividly the simultaneous occurrence of two particles with spin 1/2, particle and antiparticle. Only in the ultra-nonrelativistic case, they are decoupled.

U.6.2 Antiparticles

As far as is known, for each elementary particle there exists an *antiparticle*. Mass, lifetime and spin of particle and antiparticle are the same, likewise the nature and strength of their interactions. Some neutral particles are their own antiparticle (e.g. the photon), but others are not (e.g. the neutron). Electrical charge, magnetic moment, and all charge-like quantum numbers are opposite. Antiparticles are produced naturally in different processes, e.g. beta decay (the electron antineutrino) or interactions of cosmic rays in the Earth's atmosphere. Antiparticles can build up antimatter, just as particles can build up matter. If a particle and an antiparticle are brought into contact, they annihilate eventually. Electron and positron decay into two or three photons, proton and antiproton into several pions. Vice versa, a photon can be converted into an electron and a positron, provided that the energy of the photon is sufficiently high. By the way, electron-positron annihilation is exploited in positron emission tomography.

U.6.3 Negative Energies

A main drawback of the DE is certainly the appearance of negative energies, not existing in nature. In this section we present the hole theory as a former way to handle this puzzling problem.

U.6.3.1 Do We Need Negative Energy Solutions?

The simplest approach would be to ignore these solutions. But we cannot do so since they are inextricably coupled to the solutions with positive energy. In addition, there are also physical phenomena which can be ascribed to these 'negative' solutions. For a simple example, consider an one-dimensional plane wave

$$\psi = Ae^{i(pz-Et)/\hbar} + Be^{-i(pz-Et)/\hbar} \tag{U.112}$$

where A and B are the amplitudes of the parts with positive and with negative energy. Taking the inner product leads to

$$\bar{\psi}\psi = \left[\bar{A}e^{-i(pz-Et)/\hbar} + \bar{B}e^{i(pz-Et)/\hbar}\right]\left[Ae^{i(pz-Et)/\hbar} + Be^{-i(pz-Et)/\hbar}\right] =$$
$$= \bar{A}A + \bar{B}B + \bar{A}Be^{-2i(pz-Et)/\hbar} + \bar{B}Ae^{2i(pz-Et)/\hbar}. \qquad \text{(U.113)}$$

Thus, if only negative or only positive energies exist, $\bar{\psi}\psi$ is constant, as we know it from nonrelativistic plane waves. However, if both positive and negative energies occur, an interference term appears in form of a high-frequency oscillation $e^{2iEt/\hbar}$. In a similar way, this also applies to other systems in which both negative and positive energy solutions are superimposed.

This effect is called *Zitterbewegung* (German, means 'trembling motion'). For a free relativistic electron, the effect has never been observed; it is very tiny with a frequency of about $2mc^2/\hbar = 1.6 \times 10^{21}\text{s}^{-1}$ and an amplitude of about 10^{-13}m. But it was simulated in two different experimental situations, firstly in 2010 with a trapped ion in an appropriate environment and secondly in 2013 in a setup with Bose–Einstein condensates.[153]

Thus, for several reasons, negative energy solutions cannot be simply neglected. We have to look for an physically reasonable interpretation.

U.6.3.2   Why Does the World Still Exist?

We know that in a Hydrogen atom there are certain discrete energy levels with positive energy. Excited states disintegrate eventually, i.e., the electron falls down into states of lower energy, provided these are not occupied, thereby emitting radiation. The ground state with the lowest energy $E_> = E_{\text{rest mass}} - E_0$ is stable and does not disintegrate. Now let us assume there are states with negative energy. If these states would be empty, the electron could fall into the highest negative state with $E_< = -E_{\text{rest mass}} + E_0$, thereby emitting radiation of energy $2(E_{\text{rest mass}} - E_0)$. Since there are infinitely many negative energy states available, the electron would keep falling 'down' and radiating, in the end, infinite amounts of energy. In other words, in this scenario all matter would be unstable and disintegrating eventually, until there is nothing left than radiation. Apparently, this is not observed at all.

U.6.3.3   Hole Theory

The *hole theory*[154] attempted to eliminate the problem of radiation disintegration and thus to 'save' the Dirac equation. Hardly surprising, it was suggested by Paul Dirac himself (first in 1928, and in an improved version in 1931). It is assumed that *all* states of negative energy are occupied, in accordance with the Pauli exclusion principle (note that we describe electrons, i.e., fermions). Thus an electron with positive energy remains in the ground state and can not 'fall down' into the range

---

[153]I. Stepanov et al., 'Coherent Electron Zitterbewegung', arXiv:1612.06190v1, [cond-mat.mes-hall], 19.12.2016; and references therein.

[154]In fact, it is not a theory, but more an interpretation.

of negative energies since all those states are occupied and a further occupation is forbidden according to the Pauli principle. The *vacuum state* is thus the state in which all states of negative energy are occupied and all states of positive energy are empty.

Accordingly, the negative states form something like an 'underworld' which also is called *Dirac sea*. One can argue that we do not notice any dynamics of the Dirac sea because if a particle of this sea would change its state, it would have to assume another, already occupied state, and this is forbidden due to the Pauli principle.

Actually, we can imagine to see something of this underworld - keyword *pair production*. We assume the vacuum state and consider the particle with the highest negative energy of about $-mc^2$. If this particle absorbs radiation with an energy $> 2mc^2$, it can leave the Dirac sea and become an electron with *positive* energy which has the charge $-e$ and the energy mc^2. Simultaneously, a *hole* emerges in the Dirac sea (hence the name hole theory) which shows the absence of the charge $-e$ with the energy $-E$. With respect to the vacuum, this corresponds to a charge $+e$ with the energy $+E$, and accordingly, this hole (object) is interpreted by an observer as a *positron*, the antiparticle of the electron. The process is called *pair production*, since a electron–positron pair is produced by radiation with an energy $> 2mc^2$.

The reverse process exists also: an electron falls into a free hole, whereby radiation is emitted. This looks like that an electron and a positron collide and dissolve in radiation. The process is called *electron-positron annihilation* or *pair annihilation*.

U.6.3.4 Hole Theory From Today's View

It is certainly a merit of this interpretation to predict the existence of positrons. The hole theory was set up by Paul Dirac in 1928. Initially, he assumed protons as 'holes', but in 1931 he changed over to positrons. Indeed, Carl David Anderson[155] detected the positron experimentally in 1932.

On the other hand, the hole theory has some severe flaws; we mention three of them.

(1) In order to maintain stable ground states, the Dirac sea has to consist of an infinite number of electrons with negative energy. This means an infinite mass and infinite negative charge, not interacting with the environment. In addition, one has to assume that these electrons do not interact with each other. Dirac was aware of this problem; he tried to argue that for us this situation would be the 'normal' state of charge zero.

(2) Also bosons have antiparticles. But since they are not subjected to the Pauli principle, no sort of hole theory would work for them.

(3) The hole theory is unsymmetrical with respect to the role of electrons and positrons - one electron hovers over an 'sea' of infinitively many positrons. The same holds true with interchanged roles if we start from the Dirac equation for positrons.

---

[155]Anderson, Carl David, 1905–1991, US-American physicist, nobel prize 1936.

Today, the hole theory is considered obsolete. A consistent description of particles and antiparticles is provided by quantum field theory which solves the addressed problems and makes the interpretation of antiparticles as holes unnecessary. Note that this does not hold for the Dirac equation itself which is still the basic equation from which e.g. quantum electrodynamics emerges.

Thus, the hole theory is no longer up-to-date, but it is interesting from the point of view of science history and/or sociology. Despite all contradictions, it was accepted as a working hypothesis for quite some time. This shows that scientists, if necessary, bite the bullet to retain a theory which they instinctively are convinced of.

U.6.4 Feynman–Stückelberg Interpretation

As mentioned above, quantum field theory is the established theory to describe particles and antiparticles. To make clear one of its basic ideas, we briefly outline the so-called Feynman–Stückelberg interpretation.[156] Here, the antiparticles are considered as the corresponding particles traveling *backwards* in time. Perhaps this seems at the first sight to be a weird idea from science fiction, and replacing the ominous infinite Dirac sea by a motion backwards in time may give the impression of jumping out of a frying pan into the fire. But the approach is compatible with the fundamental symmetry principles of physics in contrast to the Dirac sea.

To motivate this somewhat surprising approach in a simple manner, we consider a one-dimensional plane wave $\psi = Ae^{i(pz-Et)/\hbar} + Be^{-i(pz-Et)/\hbar}$, and especially the part with negative energy:

$$\psi_{neg} = Be^{-i(pz-Et)/\hbar}. \tag{U.114}$$

Now assume that we change the sign of the time t. Replacing t by $-t$ results in an exponent $pz - E(-t)$. This means an object travelling backward in time, like playing a film backwards. In addition, reversing the time reverses all momenta so we also need to change p by $p \to -p$ for consistency. In order that the term pz does not change its sign, this has to be compensated by changing the sign of the position $z \to -z$, i.e., we change the parity. In short, we start with an expression $Et - pz$ where the energy is negative, $E < 0$. We replace this term simply by the equivalent expression $(-E)(-t) - (-p)(-z)$. In this way, we have a particle with positive energy which moves backwards in time.

We have to take into account another point, namely the coupling to the electromagnetic field. Denote the charge of the electron or positron by q. For the sake of simplicity, we consider only the Lorentz force $\mathbf{F} = q(\mathbf{v} \times \mathbf{B})$. As it is seen, changing the direction of motion has the same effect as changing the sign of the charge q (which corresponds to the transition from particle to antiparticle).

---

[156]Feynman, Richard Phillips, 1918–1988; American physicist, Nobel prize 1965. Stückelberg, Ernst Carl Gerlach, 1905–1984; Swiss physicist.

Thus, we have changed the sign of the time (time reversal, T), of the position (parity transformation, P), and of the charge (charge conjugation, C). In other words, we have performed a *CPT-transformation*. In this way, we get rid of negative energies and turn them into positive energies.

To summarize: The relativistic dispersion relation $E = \pm\sqrt{\mathbf{p}^2 + m^2}$ allows for solutions with negative energy, but nature does not know those energies. The way out of this dilemma consists in interpreting the formerly negative solutions as antiparticles with positive energy whereas their momenta point in the opposite direction of the corresponding particle; in addition, particle and antiparticle have the same mass and spin, but opposite charges.

In quantum field theory, these ideas lead to diagrams, in which lines or arrows represent particles propagating for- or backwards in time. These diagrams are precise graphical realizations of scattering processes; they are called Feynman diagrams, see Appendix W, Vol. 2.

Supplement: CPT theorem The *CPT symmetry* is a fundamental symmetry of physics. It states the following: For a given physical process, exchanging matter by antimatter (which covers changing the sign of the charge) and performing in addition a reflection of the space and a reversal of the time direction yields again an allowed physical process. This is also called CPT invariance of the physical laws. The CPT theorem says that CPT symmetry holds for all physical phenomena, or to put it more technically that any Lorentz invariant local quantum field theory with a Hermitian Hamiltonian must have CPT symmetry.

Vividly, the CPT theorem states that a 'mirror universe' of our universe is possible. We can build it by replacing all matter by antimatter (charge inversion C), reversing all momenta (time inversion T) and reflecting all positions (parity inversion P). This mirror universe evolves exactly according to our known physical laws. The CPT transformation changes our universe to its mirror image and vice versa.

The Dirac equation has the individual symmetries C, P and T and their combinations as, for example, CP. This is in contrast to other fundamental equations (e.g. weak interaction) where CPT is the only combination of the three transformations C, P and T that is observed to be an exact symmetry of nature.

A violation of CPT symmetry would automatically indicate a violation of the special relativity. Within today's limits of accuracy, the CPT theorem is experimentally confirmed. However, it is an open question, if there are violations below these limits which are predicted by some modern theories, e.g. quantum gravitation or string theories.

U.7 Exercises and Solutions

1. Determine all 2-dimensional matrices M with $M^2 = 1$ (or E_2).
   Solution: We have

$$M = \begin{pmatrix} a & b \\ c & d \end{pmatrix} \rightarrow M^2 = \begin{pmatrix} a^2 + bc & b(a+d) \\ c(a+d) & d^2 + bc \end{pmatrix} \overset{!}{=} \begin{pmatrix} 1 & 0 \\ 0 & 1 \end{pmatrix}. \tag{U.115}$$

This leads to the equations

$$\begin{aligned} a^2 + bc = 1 \; ; \; d^2 + bc = 1 \\ b(a+d) = 0 \; ; \; c(a+d) = 0. \end{aligned} \tag{U.116}$$

Case (1) $a + d = 0$. This gives $d = -a$ and we have

$$M_1 = \begin{pmatrix} a & b \\ c & -a \end{pmatrix} \text{ with } bc = 1 - a^2. \tag{U.117}$$

Case (2) $a + d \neq 0$. It follows $b = c = 0$ and $a^2 = d^2 = 1$. Since a and d both have the values ± 1 and due to $a + d \neq 0$, we have $a = d$:

$$M_2 = \begin{pmatrix} a & 0 \\ 0 & a \end{pmatrix} \text{ with } a^2 = 1 \rightarrow M_2 = \pm E_2. \tag{U.118}$$

As is seen, we have not only the two solutions ± 1 (or $\pm E_2$), but with M_1 an additional infinity of solutions.

2. Show that the equations (U.21) may be written in the form (U.23).
   Solution: In a first step we write $c^2 (\boldsymbol{\alpha} \cdot \mathbf{p}) (\boldsymbol{\alpha} \cdot \mathbf{p}) = c^2 \mathbf{p}^2$ as

$$\sum_{j=1}^{3} \alpha_j p_j \sum_{k=1}^{3} \alpha_k p_k = \sum_{j,k=1}^{3} \alpha_j \alpha_k p_j p_k = \sum_{j=1}^{3} p_j^2. \tag{U.119}$$

Since the α_i are matrices, we have to take into account their order, i.e. $\alpha_j \alpha_k \neq \alpha_j \alpha_k$. To arrive at a compact formulation, we use therefore a small trick and add the left side with reversed indices:

$$\sum_{j,k=1}^{3} \alpha_j \alpha_k p_j p_k + \sum_{j,k=1}^{3} \alpha_k \alpha_j p_k p_j = 2 \sum_{j=1}^{3} p_j^2. \tag{U.120}$$

Since $p_j p_k = p_k p_j$, this may be written as

$$\sum_{j,k=1}^{3} \left(\alpha_j \alpha_k + \alpha_k \alpha_j \right) p_j p_k = 2 \sum_{j=1}^{3} p_j^2. \tag{U.121}$$

For $j = k$ on the left hand side, we have $\alpha_j \alpha_j + \alpha_j \alpha_j = 2$, whereas for $j \neq k$ holds $\alpha_j \alpha_k + \alpha_k \alpha_j = 0$. Combining these results, we can write $\alpha_j \alpha_k + \alpha_j \alpha_k = 2\delta_{jk}$.

The second equation, namely $(\boldsymbol{\alpha} \cdot \mathbf{p}) \beta + \beta (\boldsymbol{\alpha} \cdot \mathbf{p}) = 0$, can be written as

$$\sum_{j=1}^{3} \alpha_j p_j \beta + \beta \sum_{j=1}^{3} \alpha_j p_j = 0. \tag{U.122}$$

Since p_j is a scalar, it commutes with the matrices α_j and β, and we can write

$$\sum_{j=1}^{3} \left(\alpha_j p_j \beta + \beta \alpha_j p_j \right) = \sum_{j=1}^{3} \left(\alpha_j \beta + \beta \alpha_j \right) p_j. \tag{U.123}$$

Since the three components of the momentum are independent, the bracketed term must vanish, i.e., $\alpha_j \beta + \beta \alpha_j = 0$.

3. Show that in the space of 2×2 matrices there are not four linearly independent anticommuting matrices.

Solution: The space of 2×2 matrices is spanned, for instance, by the unit matrix E_2 and the three Pauli matrices σ_i, $i = 1, 2, 3$. (In this exercise, we note the unit matrix not by 1, but explicitly by E_2.) Each other matrix A in this space is a linear combination of these four linear independent matrices (of the form)

$$A = a_0 E_2 + \sum_{k=1}^{3} a_k \sigma_k. \tag{U.124}$$

The three Pauli matrices anticommute pairwise

$$\sigma_i \sigma_j + \sigma_j \sigma_i = 2 \delta_{ij} E_2 \tag{U.125}$$

with scalar coefficients a_i. We have to look for a fourth matrix A (i.e., to determine the coefficients a_j) which anticommutes with all Pauli matrices, i.e. which fulfills

$$A \sigma_l + \sigma_l A = 0 \text{ for } l = 1, 2, 3. \tag{U.126}$$

Inserting (U.124) into (U.126) gives

$$a_0 \left(E_2 \sigma_l + \sigma_l E_2 \right) + \sum_{k=1}^{3} a_k \left(\sigma_k \sigma_l + \sigma_l \sigma_k \right) = 0 \text{ for } l = 1, 2, 3. \tag{U.127}$$

With (U.125) and the fact that the Pauli matrices commute with the unit matrix, $E_2 \sigma_l = \sigma_l E_2 = \sigma_l$, we get

$$2 a_0 \sigma_l + \sum_{k=1}^{3} a_k 2 \delta_{kl} E_2 = 0 \text{ for } l = 1, 2, 3 \tag{U.128}$$

which may be written as

$$a_0\sigma_l + a_l E_2 = 0 \text{ for } l = 1, 2, 3. \tag{U.129}$$

But since the four matrices E_2 and σ_l are linearly independent, this equation can only be satisfied by $a_0 = a_1 = a_2 = a_3 = 0$.

4. Write down explicitly the matrices α and $\alpha \cdot \nabla$.

Solution:

$$\alpha_1 = \begin{pmatrix} 0 & \sigma_1 \\ \sigma_1 & 0 \end{pmatrix} = \begin{pmatrix} 0 & 0 & 0 & 1 \\ 0 & 0 & 1 & 0 \\ 0 & 1 & 0 & 0 \\ 1 & 0 & 0 & 0 \end{pmatrix}$$

$$\alpha_2 = \begin{pmatrix} 0 & \sigma_2 \\ \sigma_2 & 0 \end{pmatrix} = \begin{pmatrix} 0 & 0 & 0 & -i \\ 0 & 0 & i & 0 \\ 0 & -i & 0 & 0 \\ i & 0 & 0 & 0 \end{pmatrix} \tag{U.130}$$

$$\alpha_3 = \begin{pmatrix} 0 & \sigma_3 \\ \sigma_3 & 0 \end{pmatrix} = \begin{pmatrix} 0 & 0 & 1 & 0 \\ 0 & 0 & 0 & -1 \\ 1 & 0 & 0 & 0 \\ 0 & -1 & 0 & 0 \end{pmatrix}$$

and

$$\alpha \cdot \nabla = \alpha_1 \partial_x + \alpha_2 \partial_y + \alpha_3 \partial_z = \begin{pmatrix} 0 & 0 & \partial_z & \partial_x - i\partial_y \\ 0 & 0 & \partial_x + i\partial_y & -\partial_z \\ \partial_z & \partial_x - i\partial_y & 0 & 0 \\ \partial_x + i\partial_y & -\partial_z & 0 & 0 \end{pmatrix}. \tag{U.131}$$

5. Show that the matrices α_i are unitary.

Solution: We have to show that $\alpha_i\alpha_i^\dagger = \alpha_i^\dagger\alpha_i = 1$ We have (remind $\alpha_i = \alpha_i^\dagger$)

$$\alpha_i\alpha_i^\dagger = \begin{pmatrix} 0 & \sigma_i \\ \sigma_i & 0 \end{pmatrix}\begin{pmatrix} 0 & \sigma_i \\ \sigma_i & 0 \end{pmatrix} = \begin{pmatrix} \sigma_i^2 & 0 \\ 0 & \sigma_i^2 \end{pmatrix} = \begin{pmatrix} 1 & 0 \\ 0 & 1 \end{pmatrix} = 1 \tag{U.132}$$

6. Write the Lorentz matrix Λ and the matrices $\Lambda\alpha_i$ in form of block matrices.

Solution:

$$\Lambda = \begin{pmatrix} \gamma I_2 - \beta\gamma\sigma_1 & 0 \\ 0 & I_2 \end{pmatrix} ; \Lambda\alpha_i = \begin{pmatrix} 0 & \gamma\sigma_i - \beta\gamma\sigma_1\sigma_i \\ \sigma_i & 0 \end{pmatrix}. \tag{U.133}$$

7. Show that $(c\alpha \cdot \mathbf{p} + mc^2\beta - E)(c\alpha \cdot \mathbf{p} + mc^2\beta + E) = 0$. Remind that $c\alpha \cdot \mathbf{p} + mc^2\beta \pm E$ is a short-hand notation for 4×4 matrices, so $(c\alpha \cdot \mathbf{p} + mc^2\beta - E)(c\alpha \cdot \mathbf{p} + mc^2\beta + E) = 0$ does not necessarily imply that at least one of the two factors (brackets) vanishes.

Solution: Performing the multiplication leads to

$$\left(c\boldsymbol{\alpha}\cdot\mathbf{p}\right)\left(c\boldsymbol{\alpha}\cdot\mathbf{p}+mc^2\beta+E\right)+mc^2\beta\left(c\boldsymbol{\alpha}\cdot\mathbf{p}+mc^2\beta+E\right)-E\left(c\boldsymbol{\alpha}\cdot\mathbf{p}+mc^2\beta+E\right)=$$
$$=c^2\left(\boldsymbol{\alpha}\cdot\mathbf{p}\right)\left(\boldsymbol{\alpha}\cdot\mathbf{p}\right)+mc^3\left(\boldsymbol{\alpha}\cdot\mathbf{p}\right)\beta+c\left(\boldsymbol{\alpha}\cdot\mathbf{p}\right)E+$$
$$+mc^3\beta\left(\boldsymbol{\alpha}\cdot\mathbf{p}\right)+m^2c^4\beta^2+Emc^2\beta-Ec\left(\boldsymbol{\alpha}\cdot\mathbf{p}\right)-Emc^2\beta-E^2=$$
$$=c^2\left(\boldsymbol{\alpha}\cdot\mathbf{p}\right)\left(\boldsymbol{\alpha}\cdot\mathbf{p}\right)+mc^3\left(\boldsymbol{\alpha}\cdot\mathbf{p}\right)\beta+mc^3\beta\left(\boldsymbol{\alpha}\cdot\mathbf{p}\right)+m^2c^4\beta^2-E^2.$$

(U.134)

Due to $(\boldsymbol{\alpha}\cdot\mathbf{p})(\boldsymbol{\alpha}\cdot\mathbf{p})=\mathbf{p}^2$, $\beta^2=1$ and $\beta+\beta\boldsymbol{\alpha}\vec{=}0$ (see (U.23)) follows

$$\left(c\boldsymbol{\alpha}\cdot\mathbf{p}+mc^2\beta-E\right)\left(c\boldsymbol{\alpha}\cdot\mathbf{p}+mc^2\beta+E\right)=c^2\mathbf{p}^2+m^2c^4-E^2=0.$$

(U.135)

8. Given the spinor ψ with the components ψ_α, $\alpha=1,\ldots,4$. Write down explicitly $\bar{\psi}$.
Solution:

$$\psi=\begin{pmatrix}\psi_1\\\psi_2\\\psi_3\\\psi_4\end{pmatrix}\ ;\ \bar{\psi}=\left(\psi_1^*\ \psi_2^*\ \psi_3^*\ \psi_4^*\right)\begin{pmatrix}1&0&0&0\\0&1&0&0\\0&0&-1&0\\0&0&0&-1\end{pmatrix}=\left(\psi_1^*\ \psi_2^*\ -\psi_3^*\ -\psi_4^*\right).$$

(U.136)

9. Given a solution of the free Klein–Gordon equation with positive energy only, i.e., $\phi(x)=\int d^3k\,N(\mathbf{k})\,a(\mathbf{k})\,e^{-ikx}$. Determine the normalization constant $N(\mathbf{k})$ under the assumptions $\int d^3x\,\rho(x)=1$ and $\int d^3k\,|a(\mathbf{k})|^2=1$.
Solution: The probability density ρ is given by

$$\rho(x)=i\left(\dot{\phi}\phi^*-\dot{\phi}^*\phi\right).$$

(U.137)

With

$$\phi(x)=\int d^3k\,N(\mathbf{k})\,a(\mathbf{k})\,e^{-ikx}\ ;\ \dot{\phi}(x)=-i\int d^3k\,ck_0N(\mathbf{k})\,a(\mathbf{k})\,e^{-ikx}$$

(U.138)

follows

$$\rho(x)=\frac{\int d^3k\,ck_0N(\mathbf{k})\,a(\mathbf{k})\,e^{-ikx}\cdot\int d^3k'\,N^*(\mathbf{k}')\,a^*(\mathbf{k}')\,e^{ik'x}+}{+\int d^3k'\,ck_0'N^*(\mathbf{k}')\,a^*(\mathbf{k}')\,e^{ik'x}\cdot\int d^3k\,A(\mathbf{k})\,e^{-ikx}}=$$
$$=\int d^3k\,\int d^3k'\,\left[ck_0N(\mathbf{k})\,a(\mathbf{k})\,N^*(\mathbf{k}')\,a^*(\mathbf{k}')+ck_0'N^*(\mathbf{k}')\,a^*(\mathbf{k}')\,N(\mathbf{k})\,a(\mathbf{k})\right]e^{-ikx}e^{ik'x}=$$
$$=\int d^3k\,\int d^3k'\,c\left(k_0+k_0'\right)N(\mathbf{k})\,a(\mathbf{k})\,N^*(\mathbf{k}')\,a^*(\mathbf{k}')\,e^{-ikx}e^{ik'x}.$$

(U.139)

Integration brings

$$\int d^3x\,\rho(x)=\int d^3k\,\int d^3k'\,c\left(k_0+k_0'\right)N(\mathbf{k})\,a(\mathbf{k})\,N^*(\mathbf{k}')\,a^*(\mathbf{k}')\int d^3x\,e^{-ikx}e^{ik'x}=$$
$$=\int d^3k\,\int d^3k'\,c\left(k_0+k_0'\right)N(\mathbf{k})\,a(\mathbf{k})\,N^*(\mathbf{k}')\,a^*(\mathbf{k}')\,\delta(\mathbf{k}-\mathbf{k}')(2\pi)^3=$$
$$=(2\pi)^3\int d^3k\,2ck_0N(\mathbf{k})\,a(\mathbf{k})\,N^*(\mathbf{k})\,a^*(\mathbf{k})=(2\pi)^3\int d^3k\,2ck_0\,|N(\mathbf{k})\,a(\mathbf{k})|^2.$$

(U.140)

Choosing

$$N(\mathbf{k})=\frac{1}{(2\pi)^{3/2}}\frac{1}{\sqrt{2ck_0}}=\frac{1}{(2\pi)^{3/2}}\frac{1}{\sqrt{2\omega_\mathbf{k}}}$$

(U.141)

brings

$$\int d^3x\, \rho\,(x) = (2\pi)^3 \int d^3k\, 2\omega_\mathbf{k} \frac{1}{(2\pi)^3}\frac{1}{2\omega_\mathbf{k}}\, |a\,(\mathbf{k})|^2 = \int d^3k\, |a\,(\mathbf{k})|^2 = 1.$$
(U.142)

10. Show (U.49).

Solution: From (U.48), we take $\left(c\boldsymbol{\alpha}\cdot\mathbf{p} + mc^2\beta - E\right)u_s = 0$ and write it in the form

$$\begin{pmatrix} 0 & c\boldsymbol{\sigma}\mathbf{p} \\ c\boldsymbol{\sigma}\mathbf{p} & 0 \end{pmatrix}\begin{pmatrix} u_{s,u} \\ u_{s,l} \end{pmatrix} + \begin{pmatrix} mc^2 & 0 \\ 0 & -mc^2 \end{pmatrix}\begin{pmatrix} u_{s,u} \\ u_{s,l} \end{pmatrix} - \\ -\begin{pmatrix} E & 0 \\ 0 & E \end{pmatrix}\begin{pmatrix} u_{s,u} \\ u_{s,l} \end{pmatrix} = 0 \qquad ;\ u_s = \begin{pmatrix} u_{s,u} \\ u_{s,l} \end{pmatrix}. \quad \text{(U.143)}$$

This gives the equations

$$c\boldsymbol{\sigma}\mathbf{p}u_{s,l} + mc^2 u_{s,u} - Eu_{s,u} = 0 \text{ and } c\boldsymbol{\sigma}\mathbf{p}u_{s,u} - mc^2 u_{s,l} - Eu_{s,l} = 0. \quad \text{(U.144)}$$

Solving the second equation for $u_{s,l}$ yields

$$u_{s,l} = \frac{c\boldsymbol{\sigma}\mathbf{p}}{E + mc^2}u_{s,u} \qquad \qquad \text{(U.145)}$$

and hence

$$u_s = \begin{pmatrix} u_{s,u} \\ \frac{c\boldsymbol{\sigma}\mathbf{p}}{E_\mathbf{p}+mc^2}u_{s,u} \end{pmatrix}. \qquad \qquad \text{(U.146)}$$

Note that u_s is a 4-spinor. Hence, its norm is determined by $\bar{u}_s u_s = |u_s|^2 = |u_{s,u}|^2 - |u_{s,l}|^2$ (note the sign on the r.h.s; see Appendix T, Vol. 1). Thus, to normalize u_s in (U.146) means to determine $|u_{s,u}|^2$ by the equation

$$|u_{s,u}|^2 - \left|\frac{c\boldsymbol{\sigma}\mathbf{p}}{E_\mathbf{p} + mc^2}u_{s,u}\right|^2 = 1. \qquad \qquad \text{(U.147)}$$

This leads to[157]

$$1 = \left(1 - \left|\frac{c\boldsymbol{\sigma}\mathbf{p}}{E_\mathbf{p}+mc^2}\right|^2\right)|u_{s,u}|^2 = \frac{(E_\mathbf{p}+mc^2)^2 - |c\boldsymbol{\sigma}\mathbf{p}|^2}{(E_\mathbf{p}+mc^2)^2}|u_{s,u}|^2 = \\ = \frac{E_\mathbf{p}^2 + 2E_\mathbf{p}mc^2 + m^2c^4 - c^2|\mathbf{p}|^2}{(E_\mathbf{p}+mc^2)^2}|u_{s,u}|^2 = \frac{2mc^2}{E_\mathbf{p}+mc^2}|u_{s,u}|^2 \quad \text{(U.148)}$$

or

$$|u_{s,u}| = \sqrt{\frac{E_\mathbf{p} + mc^2}{2mc^2}}. \qquad \qquad \text{(U.149)}$$

---

[157] We use $(\boldsymbol{\sigma}\cdot\mathbf{a})\,(\boldsymbol{\sigma}\cdot\mathbf{b}) = \mathbf{a}\cdot\mathbf{b} + i\boldsymbol{\sigma}\cdot(\mathbf{a}\times\mathbf{b})$.

Thus, we can write $u_{s,u}$ in the form $\left| u_{s,u} \right| \cdot \hat{e}$ where $\hat{e}$ is an arbitrary unit 2-vector. Choosing $\hat{e} = \chi_1$ and $\hat{e} = \chi_2$ brings the desired result.

The calculation of $u_{s,l}$ runs analogously.

11. Prove (U.50).

Solution: From (U.49) we have

$$u_s(\mathbf{p}) = \sqrt{\frac{E_\mathbf{p} + mc^2}{2mc^2}} \begin{pmatrix} \chi_s \\ \frac{c\boldsymbol{\sigma}\mathbf{p}}{E_\mathbf{p}+mc^2}\chi_s \end{pmatrix} ; \ v_s(\mathbf{p}) = \sqrt{\frac{E_\mathbf{p} + mc^2}{2mc^2}} \begin{pmatrix} \frac{c\boldsymbol{\sigma}\mathbf{p}}{E_\mathbf{p}+mc^2}\chi_s \\ \chi_s \end{pmatrix} ; \ s = 1, 2.$$
(U.150)

The adjoints are given by

$$\bar{u}_s(\mathbf{p}) = \sqrt{\frac{E_\mathbf{p} + mc^2}{2mc^2}} \begin{pmatrix} \chi_s^\dagger & -\chi_s^\dagger \frac{c\boldsymbol{\sigma}\mathbf{p}}{E_\mathbf{p}+mc^2} \end{pmatrix} ; \ \bar{v}_s(\mathbf{p}) = \sqrt{\frac{E_\mathbf{p} + mc^2}{2mc^2}} \begin{pmatrix} \chi_s^\dagger \frac{c\boldsymbol{\sigma}\mathbf{p}}{E_\mathbf{p}+mc^2} & -\chi_s^\dagger \end{pmatrix} ; \ s = 1, 2.$$
(U.151)

Exemplarily, we consider $\bar{v}_r(\mathbf{p}) v_s(\mathbf{p})$. We have

$$\bar{v}_r(\mathbf{p}) v_s(\mathbf{p}) = \sqrt{\frac{E_\mathbf{p}+mc^2}{2mc^2}} \begin{pmatrix} \chi_r^\dagger \frac{c\boldsymbol{\sigma}\mathbf{p}}{E_\mathbf{p}+mc^2} & -\chi_r^\dagger \end{pmatrix} \sqrt{\frac{E_\mathbf{p}+mc^2}{2mc^2}} \begin{pmatrix} \frac{c\boldsymbol{\sigma}\mathbf{p}}{E_\mathbf{p}+mc^2}\chi_s \\ \chi_s \end{pmatrix} =$$

$$= \frac{E_\mathbf{p}+mc^2}{2mc^2} \left[\chi_r^\dagger \frac{c\boldsymbol{\sigma}\mathbf{p}}{E_\mathbf{p}+mc^2} \frac{c\boldsymbol{\sigma}\mathbf{p}}{E_\mathbf{p}+mc^2}\chi_s - \chi_r^\dagger \chi_s \right] = \frac{E_\mathbf{p}+mc^2}{2mc^2} \left[\left(\frac{c\mathbf{p}}{E_\mathbf{p}+mc^2}\right)^2 - 1 \right] \delta_{rs} =$$

$$= \frac{E_\mathbf{p}+mc^2}{2mc^2} \left[\left(\frac{c\mathbf{p}}{E_\mathbf{p}+mc^2}\right)^2 - \frac{E_\mathbf{p}^2+2E_\mathbf{p}mc^2+m^2c^4}{(E_\mathbf{p}+mc^2)^2} \right] \delta_{rs} =$$

$$= \frac{E_\mathbf{p}+mc^2}{2mc^2} \left[\left(\frac{c\mathbf{p}}{E_\mathbf{p}+mc^2}\right)^2 - \frac{m^2c^4+c^2\mathbf{p}^2+2E_\mathbf{p}mc^2+m^2c^4}{(E_\mathbf{p}+mc^2)^2} \right] \delta_{rs} =$$

$$= \frac{E_\mathbf{p}+mc^2}{2mc^2} \left[-\frac{2m^2c^4+2E_\mathbf{p}mc^2}{(E_\mathbf{p}+mc^2)^2} \right] \delta_{rs} = -\frac{1}{2mc^2} \left[\frac{2m^2c^4+2E_\mathbf{p}mc^2}{E_\mathbf{p}+mc^2} \right] \delta_{rs} = -\left[\frac{mc^2+E_\mathbf{p}}{E_\mathbf{p}+mc^2} \right] \delta_{rs} = -\delta_{rs}.$$
(U.152)

The other relations analogously.

12. Show that γ^0 is hermitian, γ^k is anti-hermitian.

Solution: We have

$$\gamma^{0\dagger} = \begin{pmatrix} 1 & 0 \\ 0 & -1 \end{pmatrix}^\dagger = \begin{pmatrix} 1 & 0 \\ 0 & -1 \end{pmatrix} = \gamma^0$$
(U.153)

and

$$\gamma^{k\dagger} = \begin{pmatrix} 0 & \sigma_k \\ -\sigma_k & 0 \end{pmatrix}^\dagger = \begin{pmatrix} 0 & -\sigma_k^\dagger \\ \sigma_k^\dagger & 0 \end{pmatrix} = \begin{pmatrix} 0 & -\sigma_k \\ \sigma_k & 0 \end{pmatrix} = -\gamma^k.$$
(U.154)

Appendix V
Exercises and Solutions to Chaps. 1–14

V.1 Exercises, Chap. 1

1. Consider the relativistic energy-momentum relation

$$E^2 = m_0^2 c^4 + p^2 c^2. \tag{V.1}$$

Show that in the nonrelativistic limit $v \ll c$, it gives approximately (up to an additive positive constant)

$$E = \frac{p^2}{2m_0}. \tag{V.2}$$

Solution: In the nonrelativistic limiting case, we find that

$$E = \sqrt{m_0^2 c^4 + p^2 c^2} = m_0 c^2 \sqrt{1 + \frac{p^2}{m_0^2 c^2}} \approx m_0 c^2 \left(1 + \frac{p^2}{2m_0^2 c^2}\right) \tag{V.3}$$

where we have used $\sqrt{1 + \varepsilon} \approx 1 + \varepsilon/2$ (see Appendix D, Vol. 1; Taylor expansion). It follows that

$$E \approx m_0 c^2 + \frac{p^2}{2m_0}, \tag{V.4}$$

where $m_0 c^2$ is the above-mentioned positive constant. Since one can choose the zero point of classical energies arbitrarily, we choose it in such a way that this term vanishes. Incidentally, one usually writes simply m instead of m_0 because the velocity dependence of the mass is negligible for $v \ll c$.

2. Show that the relation $E = p \cdot c$ (c is the speed of light) holds only for objects with zero rest mass.
   Solution: The result follows directly from $E^2 = m_0^2 c^4 + p^2 c^2$.

© Springer Nature Switzerland AG 2018
J. Pade, *Quantum Mechanics for Pedestrians 1*, Undergraduate Lecture
Notes in Physics, https://doi.org/10.1007/978-3-030-00464-4

3. A (relativistic) object has zero rest mass. Show that in this case the dispersion relation reads $\omega^2 = c^2 \mathbf{k}^2$.
   Solution: The result follows directly from exercise 2 and the de Broglie relations.

4. Let $k < 0, \omega > 0$. Is $e^{i(kx-\omega t)}$ a right- or left-moving plane wave?
   Solution: If we set the exponent equal to zero, we find from $kx - \omega t = 0$ the inequality:

$$\frac{x}{t} = v = \frac{\omega}{k} < 0. \tag{V.5}$$

   Because of $v < 0$, the wave runs from the right to the left.

5. Solve the three-dimensional wave equation

$$\frac{\partial^2 \Psi(\mathbf{r}, t)}{\partial t^2} = c^2 \nabla^2 \Psi(\mathbf{r}, t) \tag{V.6}$$

   explicitly by using the separation of variables.

6. Given the three-dimensional wave equation for a vector field $\mathbf{A}(\mathbf{r}, t)$,

$$\frac{\partial^2 \mathbf{A}(\mathbf{r}, t)}{\partial t^2} = c^2 \nabla^2 \mathbf{A}(\mathbf{r}, t). \tag{V.7}$$

   (a) What is a solution in the form of a plane wave?
   Solution:
$$\mathbf{A}(\mathbf{r}, t) = \mathbf{A}_0 e^{i(\mathbf{k}\mathbf{r}-\omega t)}; \quad \omega^2 = c^2 \mathbf{k}^2. \tag{V.8}$$

   (b) Which condition must $\mathbf{A}_0$ satisfy if $\mathbf{A}$ is (a) a longitudinal, (b) a transverse wave?
   Solution: For a longitudinal wave, the amplitude vector is parallel to the propagation direction, while for a transverse wave it is perpendicular to it. For a longitudinal wave, therefore, $\mathbf{A}_0 \sim \mathbf{k}$; for a transverse wave, $\mathbf{A}_0 \cdot \mathbf{k} = 0$.

7. Given the SEq

$$i\hbar \frac{\partial}{\partial t} \Psi(\mathbf{r}, t) = -\frac{\hbar^2}{2m} \nabla^2 \Psi(\mathbf{r}, t) + V(\mathbf{r}, t) \Psi(\mathbf{r}, t) \tag{V.9}$$

   and two solutions $\psi_1(\mathbf{r}, t)$ and $\psi_2(\mathbf{r}, t)$. Show explicitly that any linear combination of these solutions is again a solution.
   Solution: Since ψ_1 and ψ_2 are solutions of the SEq, we have

$$i\hbar \frac{\partial}{\partial t} \psi_i(\mathbf{r}, t) = -\frac{\hbar^2}{2m} \nabla^2 \psi_i(\mathbf{r}, t) + V(\mathbf{r}, t) \psi_i(\mathbf{r}, t); \quad i = 1, 2. \tag{V.10}$$

   We have to show that for a linear combination $\Phi(\mathbf{r}, t) = a\psi_1(\mathbf{r}, t) + b\psi_2(\mathbf{r}, t)$ with $a, b \in \mathbb{C}$, we obtain

$$i\hbar \frac{\partial}{\partial t} \Phi(\mathbf{r}, t) = -\frac{\hbar^2}{2m} \nabla^2 \Phi(\mathbf{r}, t) + V(\mathbf{r}, t) \Phi(\mathbf{r}, t). \tag{V.11}$$

This holds true because of (for brevity, we omit the arguments $\mathbf{r}, t$):

$$i\hbar \frac{\partial}{\partial t} \Phi = i\hbar \frac{\partial}{\partial t} (a\psi_1 + b\psi_2) = ai\hbar \frac{\partial}{\partial t} \psi_1 + bi\hbar \frac{\partial}{\partial t} \psi_2$$

$$= a \left[-\frac{\hbar^2}{2m} \nabla^2 \psi_1 + V\psi_1 \right] + b \left[-\frac{\hbar^2}{2m} \nabla^2 \psi_2 + V\psi_2 \right] \tag{V.12}$$

$$= -\frac{\hbar^2}{2m} \nabla^2 (a\psi_1 + b\psi_2) + V(a\psi_1 + b\psi_2) = -\frac{\hbar^2}{2m} \nabla^2 \Phi + V\Phi.$$

8. The wavefunction of a quantum object of mass m is given by

$$\psi(x, t) = \psi_0 \exp\left(-\frac{x^2}{2b^2} - i\frac{\hbar}{2mb^2} t \right), \tag{V.13}$$

b is a fixed length. Determine the potential energy $V(x)$ of the quantum object.
Solution: We determine $V(x)$ by inserting $\psi(x, t)$ into the time-dependent SEq.
With

$$\frac{\partial}{\partial t} \psi(x, t) = -i\frac{\hbar}{2mb^2} \psi_0 \exp\left(-\frac{x^2}{2b^2} - i\frac{\hbar}{2mb^2} t \right)$$

$$\frac{\partial}{\partial x} \psi(x, t) = -\frac{x}{b^2} \psi_0 \exp\left(-\frac{x^2}{2b^2} - i\frac{\hbar}{2mb^2} t \right)$$

$$\frac{\partial^2}{\partial x^2} \psi(x, t) = -\frac{1}{b^2} \psi_0 \exp\left(-\frac{x^2}{2b^2} - i\frac{\hbar}{2mb^2} t \right) + \frac{x^2}{b^4} \psi_0 \exp\left(-\frac{x^2}{2b^2} - i\frac{\hbar}{2mb^2} t \right) \tag{V.14}$$

it follows that

$$i\hbar \left(-i\frac{\hbar}{2mb^2} \right) = -\frac{\hbar^2}{2m} \left(-\frac{1}{b^2} + \frac{x^2}{b^4} \right) + V \tag{V.15}$$

or

$$V = \frac{\hbar^2}{2m} \frac{x^2}{b^4} \tag{V.16}$$

i.e. a harmonic-oscillator potential.
9. Given the plane waves

$$\Phi_1(x, t) = \Phi_{01} e^{\pm i(kx - \omega t)}; \quad \Phi_2(x, t) = \Phi_{02} e^{\pm i(kx + \omega t)}; \quad k, \omega > 0; \quad \Phi_{0i} \in \mathbb{R}. \tag{V.17}$$

Explain in a visual way that $\Phi_1(x, t)$ is a right- and $\Phi_2(x, t)$ a left-moving plane wave.
Solution: For heuristic reasoning we have to consider the real and imaginary parts of the functions (otherwise we would have to operate in a four-dimensional

Fig. V.1 Plane wave
$\cos(kx - \omega t)$ with $k > 0$,
$\omega > 0$, travelling to the right.
Blue for $t = 0$, *red* for $t > 0$

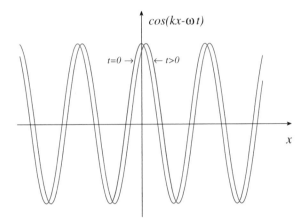

space, which would not be intuitively accessible). We restrict ourselves to Φ_1; the argument is analogous for Φ_2 . We have

$$\Phi_1(x, t) = \Phi_{01} \cos(kx - \omega t) \pm i\Phi_{01} \sin(kx - \omega t). \qquad (V.18)$$

We now consider the real part $\Phi_{01} \cos(kx - \omega t)$. At the time $t = 0$, we have $\Phi_{01} \cos(kx)$; one of the maxima of the function is where the argument of the cosine disappears, i.e. at $x = 0$. After a short period of time τ, the function reads $\Phi_{01} \cos(kx - \omega\tau)$; the maximum is now at the point $kx - \omega\tau = 0$, i.e. at $x = \frac{\omega\tau}{k}$, see Fig. V.1. In other words, the maximum, and thus the entire curve moves to the right. The same result is obtained by considering $\Phi_{01} \sin(kx - \omega t)$. Hence, we can regard $\Phi_1(x, t)$ in sum as a plane wave, travelling to the right.

V.2   Exercises, Chap. 2

1. Given an electromagnetic wave $\mathbf{E}(\mathbf{r}, t) = \mathbf{E}_0 e^{i(\mathbf{kr} - \omega t)}$ in a charge-free region of space (we consider only the electric field); show that the wave is transverse, i.e. that $\mathbf{k} \cdot \mathbf{E}_0 = 0$ holds (Hint: cf. the Maxwell equation $\nabla \vec{E} = 0$). Specialize to $\mathbf{k} = (0, 0, k)$.
   Solution:

$$0 = \nabla\mathbf{E} = \nabla\mathbf{E}_0 e^{i(\mathbf{kr} - \omega t)} = \partial_x E_{0x} e^{i(\mathbf{kr} - \omega t)} + \partial_y \cdots$$

$$= E_{0x} \partial_x e^{i(\mathbf{kr} - \omega t)} + \partial_y \cdots = E_{0x} \frac{\partial i(\mathbf{kr} - \omega t)}{\partial x} e^{i(\mathbf{kr} - \omega t)} + \partial_y \cdots \qquad (V.19)$$

Because of

$$\frac{\partial \,(\mathbf{kr} - \omega t)}{\partial x} = \frac{\partial \mathbf{kr}}{\partial x} = \frac{\partial \left(k_x x + k_y y + k_z z\right)}{\partial x} = k_x, \qquad \text{(V.20)}$$

it follows that

$$0 = i\,E_{0x}k_x e^{i(\mathbf{kr} - \omega t)} + i\,E_{0y}k_y e^{i(\mathbf{kr} - \omega t)} + i\,E_{0z}k_z e^{i(\mathbf{kr} - \omega t)}$$
$$= i\,\mathbf{E}_0 \cdot \mathbf{k} e^{i(\mathbf{kr} - \omega t)} \qquad \text{(V.21)}$$

and hence the assertion is demonstrated directly. For $\mathbf{k} = (0, 0, k)$, we have $\mathbf{k} \cdot \mathbf{E}_0 = k E_{0z} = 0$, which leads to $E_{0z} = 0$ due to $k \neq 0$.

2. Linear combinations

   (a) Express $|r\rangle$ as a linear combination of $|h\rangle$ and $|v\rangle$. Do the same for $|l\rangle$.

   (b) Express $|h\rangle$ as a linear combination of $|r\rangle$ and $|l\rangle$. Do the same for $|v\rangle$.

3. A phase shift of $90°$ is described by $e^{i\pi/2} = i$. What follows for a phase shift of $180°$?

Solution

$$180° \,\hat{=}\, e^{i\pi} = -1. \qquad \text{(V.22)}$$

4. Elliptical polarization: Given the state $|z\rangle = \alpha\,|h\rangle + \beta\,|v\rangle$ with $|\alpha|^2 + |\beta|^2 = 1$; express $|z\rangle$ as superposition of $|r\rangle$ and $|l\rangle$.

V.3   Exercises, Chap. 3

1. Show explicitly that the solutions of the Schrödinger equation (3.1) span a vector space.

2. Calculate $\left[x, \frac{\partial^2}{\partial x^2}\right]$.

Solution:

$$\left(x\frac{\partial^2}{\partial x^2} - \frac{\partial^2}{\partial x^2}x\right) f = x\frac{\partial^2 f}{\partial x^2} - \frac{\partial}{\partial x}\left(x\frac{\partial f}{\partial x} + f\,(x)\right)$$
$$= x\frac{\partial^2 f}{\partial x^2} - \left(x\frac{\partial^2 f}{\partial x^2} + \frac{\partial f}{\partial x} + \frac{\partial f}{\partial x}\right) = -2\frac{\partial f}{\partial x} \qquad \text{(V.23)}$$

i.e. more compactly:

$$\left[x, \frac{\partial^2}{\partial x^2}\right] = \left(x\frac{\partial^2}{\partial x^2} - \frac{\partial^2}{\partial x^2}x\right) = -2\frac{\partial}{\partial x}. \qquad \text{(V.24)}$$

3. Given the relativistic energy-momentum relation $E^2 = m_0^2 c^4 + c^2 p^2$; from this dispersion relation, deduce a differential equation.

Solution: With $E \leftrightarrow i\hbar\frac{\partial}{\partial t}$ and $\mathbf{p} \leftrightarrow \frac{\hbar}{i}\nabla$, it follows that:

$$\left(i\hbar \frac{\partial}{\partial t}\right)^2 = m_0^2 c^4 + c^2 \left(\frac{\hbar}{i}\nabla\right)^2 \quad \text{or} \quad \frac{\partial^2}{\partial t^2} = c^2 \nabla^2 - \frac{m_0^2 c^4}{\hbar^2}. \tag{V.25}$$

4. Separation: Deduce the time-independent Schrödinger equation from the time-dependent Schrödinger equation by means of the separation of variables.
5. Given the eigenvalue problem

$$\frac{\partial}{\partial x} f(x) = \gamma f(x); \ \gamma \in \mathbb{C} \tag{V.26}$$

with $f(x)$ satisfying the boundary conditions $f(0) = 1$ and $f(1) = 2$, calculate eigenfunction and eigenvalue.
Solution: The general solution of the differential equation reads $f(x) = f_0 e^{\gamma x}$. The boundary conditions lead to

$$f(0) = f_0 = 1 \text{ and } f(1) = e^\gamma = 2. \tag{V.27}$$

Hence, $\gamma = \ln 2$ is the only eigenvalue.
6. Given the eigenvalue problem

$$\frac{\partial^2}{\partial x^2} f = \delta^2 f; \ \delta \in \mathbb{C} \tag{V.28}$$

with $f(x)$ satisfying the boundary conditions $f(0) = f(L) = 0; L \neq 0, \delta \neq 0$, calculate eigenfunctions and eigenvalues.
Solution: The general solution of the differential equation reads $f(x) = f_+ e^{\delta x} + f_- e^{-\delta x}$, with the integration constants f_+ and f_-. Inserting the boundary conditions leads to

$$f(0) = f_+ + f_- = 0$$
$$f(L) = f_+ e^{\delta L} + f_- e^{-\delta L} = 0. \tag{V.29}$$

From this, it follows that:

$$f_- = -f_+ \tag{V.30}$$

and thus

$$f_+ e^{\delta L} - f_+ e^{-\delta L} = 0. \tag{V.31}$$

This equation has nontrivial solutions only if

$$e^{\delta L} - e^{-\delta L} = 0 \text{ or } e^{2\delta L} = 1. \tag{V.32}$$

It follows then

$$\delta = \frac{i m \pi}{L}; \ m = 0, \pm 1, \pm 2, \ldots \tag{V.33}$$

The quantity δ *must* therefore be imaginary, in order that the eigenvalue problem has a solution. The spectrum is discrete, whereby the eigenvalues δ^2 are always negative numbers, $\delta^2 = -\left(\frac{m\pi}{L}\right)^2$.
The eigenfunctions are given by

$$f(x) = f_+ e^{\delta x} - f_+ e^{-\delta x} = f_+\left(e^{\frac{im\pi}{L}x} - e^{-\frac{im\pi}{L}x}\right) = 2if_+ \sin\frac{m\pi}{L}x. \quad \text{(V.34)}$$

The constant f_+ remains undetermined, since the differential equation $\frac{\partial^2}{\partial x^2}f = \delta^2 f$ is linear (hence, for each solution, a multiple is again a solution); to determine it, an additional condition is required; see Chap. 5.

7. Given the nonlinear differential equation

$$y'(x) = \frac{dy(x)}{dx} = y^2(x). \quad \text{(V.35)}$$

$y_1(x)$ and $y_2(x)$ are two different nontrivial solutions of (V.35), i.e. $y_1 \neq const \cdot y_2$ and $y_1 y_2 \neq 0$.

(a) Show that a multiple of a solution, i.e. $f(x) = cy_1(x)$ with $c \neq 0, c \neq 1$, is not a solution of (V.35).
Solution: If $f(x)$ is a solution of (V.35), then $f' = f^2$ must be fulfilled. Because of

$$\begin{aligned} f' &= cy_1' = cy_1^2 \\ f^2 &= c^2 y_1^2 \end{aligned}, \quad \text{(V.36)}$$

we obtain immediately $c^2 = c$ with the solutions $c = 0$ and $c = 1$, which contradicts the preconditions.

(b) Show that a linear combination of two solutions, i.e. $g(x) = ay_1(x) + by_2(x)$ with $ab \neq 0$, but otherwise arbitrary, is not a solution of (V.35).
Solution: If $g(x)$ is a solution of (V.35), then $g' = g^2$ must hold. Because of

$$\begin{aligned} g' &= ay_1' + by_2' = ay_1^2 + by_2^2 \\ g^2 &= a^2 y_1^2 + 2aby_1 y_2 + b^2 y_2^2 \end{aligned}, \quad \text{(V.37)}$$

we obtain

$$\left(a^2 - a\right) y_1^2 + 2aby_1 y_2 + \left(b^2 - b\right) y_2^2 = 0. \quad \text{(V.38)}$$

This equation may be solved e.g. for y_1, and the result has the form $y_1 = const \cdot y_2$ or $y_1 = 0$, which contradicts the preconditions.

Explicitly, we have

$$y_1 = \frac{-ab \pm \sqrt{ab\,(a+b-1)}}{a\,(a-1)} y_2 \text{ for } a \neq 1; \quad y_1 = \frac{1-b}{2} y_2 \text{ for } a = 1.$$

$$(V.39)$$

(c) Find the general solution of (V.35).

Solution: Invoking the separation of variables, we can write (V.35) as

$$\frac{dy}{y^2} = dx; \quad y \neq 0.$$

$$(V.40)$$

Integrating both sides yields

$$-\frac{1}{y} = x - C,$$

$$(V.41)$$

where C is an arbitrary integration constant. Solving for y leads to

$$y = \frac{1}{C-x}.$$

$$(V.42)$$

8. Radial momentum

(a) Show that for the classical momentum $\mathbf{p}$ obeys

$$\mathbf{p}^2 = \left(\mathbf{p}\hat{\mathbf{r}}\right)^2 + \left(\mathbf{p} \times \hat{\mathbf{r}}\right)^2.$$

$$(V.43)$$

Solution:

$$\left(\mathbf{p}\hat{\mathbf{r}}\right)^2 = \mathbf{p}^2 \cdot \hat{\mathbf{r}}^2 \cdot \cos^2 \theta; \quad \left(\mathbf{p} \times \hat{\mathbf{r}}\right)^2 = \mathbf{p}^2 \cdot \hat{\mathbf{r}}^2 \cdot \sin^2 \theta.$$

$$(V.44)$$

The proposition follows because of $\hat{\mathbf{r}}^2 = 1$.

(b) Deduce the quantum-mechanical expression p_r for the classical radial momentum $\hat{\mathbf{r}}p \left(= p\hat{\mathbf{r}}\right)$.

Solution: For the translation into quantum mechanics, we have:

$$\mathbf{p}\hat{\mathbf{r}}f = \frac{\hbar}{i}\nabla\hat{\mathbf{r}}f = \frac{\hbar}{i}\left(\nabla\hat{\mathbf{r}}\right)f + \frac{\hbar}{i}\hat{\mathbf{r}}\nabla f.$$

$$(V.45)$$

With

$$\nabla\hat{\mathbf{r}} = \nabla\frac{\mathbf{r}}{r} = \frac{1}{r}\nabla\mathbf{r} + \mathbf{r}\nabla\frac{1}{r} = \frac{3}{r} - \frac{r^2}{r^3} = \frac{2}{r}$$

$$(V.46)$$

and $\hat{\mathbf{r}}\nabla f = \frac{\partial}{\partial r} f$, it follows that:

$$\mathbf{p}\hat{\mathbf{r}}f = \frac{\hbar}{i}\nabla \hat{\mathbf{r}}f = \frac{\hbar}{i}\frac{2}{r}f + \frac{\hbar}{i}\frac{\partial}{\partial r}f. \tag{V.47}$$

On the other hand, we can write

$$\hat{\mathbf{r}}pf = \frac{\hbar}{i}\hat{\mathbf{r}}\nabla f = \frac{\hbar}{i}\frac{\partial}{\partial r}f, \tag{V.48}$$

obtaining

$$(\hat{\mathbf{r}}p)_{qm} f = p_r f = \frac{\mathbf{p}\hat{\mathbf{r}}f + \hat{\mathbf{r}}pf}{2} = \frac{\hbar}{i}\left(\frac{1}{r}f + \frac{\partial}{\partial r}f\right) = \frac{\hbar}{i}\frac{1}{r}\frac{\partial}{\partial r}rf \tag{V.49}$$

or, written as an operator,

$$p_r = \frac{\hbar}{i}\frac{1}{r}\frac{\partial}{\partial r}r. \tag{V.50}$$

For the square, we have

$$p_r^2 = -\hbar^2\frac{1}{r}\frac{\partial^2}{\partial r^2}r = -\hbar^2\frac{1}{r^2}\frac{\partial}{\partial r}r^2\frac{\partial}{\partial r} = -\hbar^2\left(\frac{\partial^2}{\partial r^2} + \frac{2}{r}\frac{\partial}{\partial r}\right). \tag{V.51}$$

Compare this expression with the representation of the Laplacian in spherical coordinates.

9. Show explicitly that the classical expression $\mathbf{l} = \mathbf{r}\times\mathbf{p}$ needs not be symmetrized for the translation into quantum mechanics.
Solution: Here we have a product of operators, and we need to check whether the translation into quantum mechanics of the classical expressions $\mathbf{r}\times\mathbf{p}$ and $-\mathbf{p}\times\mathbf{r}$, which are the same in the classical view, yields the same result. If not, we have to symmetrize. We first consider only the x components. We have:

$$(\mathbf{r}\times\mathbf{p})_x = yp_z - zp_y; \quad -(\mathbf{p}\times\mathbf{r})_x = -p_y z + p_z y. \tag{V.52}$$

Since y commutes with p_z and z with p_y (analogously to the components l_y and l_z), then clearly $\mathbf{r}\times\mathbf{p} = -\mathbf{p}\times\mathbf{r}$ holds true also in quantum mechanics; hence, we need not symmetrize.

10. Given the operators $A = x\frac{d}{dx}$, $B = \frac{d}{dx}x$ and $C = \frac{d}{dx}$:

(a) Calculate $Af_i(x)$ for the functions $f_1(x) = x^2$, $f_2(x) = e^{ikx}$ and $f_3(x) = \ln x$.
Solution:

$$Af_1 = x\frac{d}{dx}x^2 = 2x^2$$

$$Af_2 = x\frac{d}{dx}e^{ikx} = ikxe^{ikx} \qquad (V.53)$$

$$Af_3 = x\frac{d}{dx}\ln x = 1.$$

(b) Calculate $A^2 f(x)$ for arbitrary $f(x)$.
   Solution:

$$A^2 f(x) = x\frac{d}{dx}x\frac{d}{dx}f(x) = x\frac{d}{dx}xf' = x\left(xf'' + f'\right) = x^2 f'' + xf'. $$
$$\qquad (V.54)$$

(c) Calculate the commutators $[A, B]$ and $[B, C]$.
   Solution:

$$[A, B]f = x\frac{d}{dx}\frac{d}{dx}xf - \frac{d}{dx}xx\frac{d}{dx}f = x\frac{d}{dx}\left(xf' + f\right) - \frac{d}{dx}x^2 f'$$
$$= x\left(xf'' + 2f'\right) - \left(x^2 f'' + 2xf'\right) = 0,$$
$$\qquad (V.55)$$

or, in compact form,

$$[A, B] = 0. \qquad (V.56)$$

For the second commutator, we have

$$[B, C]f = \frac{d}{dx}x\frac{d}{dx}f - \frac{d}{dx}\frac{d}{dx}xf = \frac{d}{dx}xf' - \frac{d}{dx}\left(xf' + f\right) = -\frac{d}{dx}f,$$
$$\qquad (V.57)$$

or, in compact form,

$$[B, C] = -\frac{d}{dx}. \qquad (V.58)$$

(d) Calculate $e^{iC}x^2 - (x + i)^2$. Prove the equation $e^{iC}e^{ikx} = e^{-k}e^{ikx}$.
   Solution: For e^{iC} we use the power series expansion of the e-function:

$$e^{iC} = e^{i\frac{d}{dx}} = \sum_{n=0}^{\infty} \frac{i^n}{n!}\frac{d^n}{dx^n}. \qquad (V.59)$$

It follows that

$$e^{iC}x^2 = \sum_{n=0}^{\infty} \frac{i^n}{n!}\frac{d^n}{dx^n}x^2 = \left(1 + \frac{i}{1!}\frac{d}{dx} + \frac{i^2}{2!}\frac{d^2}{dx^2}\right)x^2$$
$$= x^2 + 2ix - 1 = (x + i)^2 \qquad (V.60)$$

or

$$e^{iC}x^2-(x+i)^2=0. \tag{V.61}$$

For $e^{iC}e^{ikx}$, we find

$$e^{iC}e^{ikx} = \sum_{n=0}^{\infty}\frac{i^n}{n!}\frac{d^n}{dx^n}e^{ikx} = \left(1+\frac{i}{1!}\frac{d}{dx}+\frac{i^2}{2!}\frac{d^2}{dx^2}+\cdots+\frac{i^n}{n!}\frac{d^n}{dx^n}+\cdots\right)$$

$$e^{ikx} = \left(1+\frac{i}{1!}(ik)+\frac{i^2}{2!}(ik)^2+\cdots+\frac{i^n}{n!}(ik)^n+\cdots\right)e^{ikx}$$

$$= \left(1-k+\frac{k^2}{2!}+\cdots+\frac{(-1)^{2n}}{n!}k^n+\cdots\right)e^{ikx} = e^{-k}e^{ikx}. \tag{V.62}$$

V.4 Exercises, Chap. 4

1. Find examples for state spaces which

   (a) have the structure of a vector space,
   Solution: States of light waves, acoustical waves, water waves (insofar as they can be considered as linear phenomena), continuous functions on an interval, $n \times n$ matrices, $\mathbb{R}^n$, polynomials of degree $n \le 8$ etc.
   (b) do not have the structure of a vector space.
   Solution: States of a coin (heads or tails), of dice (1, 2, 3, 4, 5, 6), of a ball in a roulette wheel, cruising altitudes of an airplane, number of fish in an aquarium, blood pressure or temperature of a patient etc.

2. Polarization: Determine the length of the vector $\frac{1}{\sqrt{2}}\binom{1}{i}$.
3. Given $\langle y| = i\begin{pmatrix}1 & -2\end{pmatrix}$ and $\langle z| = \begin{pmatrix}2 & i\end{pmatrix}$, determine $\langle y|z\rangle$.
4. The Pauli matrices are

$$\sigma_x = \begin{pmatrix}0 & 1\\ 1 & 0\end{pmatrix} : \sigma_y = \begin{pmatrix}0 & -i\\ i & 0\end{pmatrix}; \ \sigma_z = \begin{pmatrix}1 & 0\\ 0 & -1\end{pmatrix}. \tag{V.63}$$

In addition to $\sigma_x, \sigma_y, \sigma_z$, the notation $\sigma_1, \sigma_2, \sigma_3$ is also common.

   (a) Show that $\sigma_i^2 = 1, i = x, y, z$.
   Solution:

$$\sigma_x^2 = \begin{pmatrix}0 & 1\\ 1 & 0\end{pmatrix}\begin{pmatrix}0 & 1\\ 1 & 0\end{pmatrix} = \begin{pmatrix}1 & 0\\ 0 & 1\end{pmatrix} \tag{V.64}$$

   and analogously for σ_y^2 and σ_z^2 with the same result.
   (b) Determine the commutators $[\sigma_i, \sigma_j] = \sigma_i\sigma_j - \sigma_j\sigma_i$ and the anticommutators $\{\sigma_i, \sigma_j\} = \sigma_i\sigma_j + \sigma_j\sigma_i$ $(i \ne j)$.
   Solution: For the commutator $[\sigma_x, \sigma_y]$, it holds that:

$$[\sigma_x, \sigma_y] = \begin{pmatrix} 0 & 1 \\ 1 & 0 \end{pmatrix} \begin{pmatrix} 0 & -i \\ i & 0 \end{pmatrix} - \begin{pmatrix} 0 & -i \\ i & 0 \end{pmatrix} \begin{pmatrix} 0 & 1 \\ 1 & 0 \end{pmatrix} = 2i\sigma_z, \qquad (V.65)$$

and similarly for the other indices.
For the anticommutator $\{\sigma_x, \sigma_y\}$, we have

$$\{\sigma_x, \sigma_y\} = \begin{pmatrix} 0 & 1 \\ 1 & 0 \end{pmatrix} \begin{pmatrix} 0 & -i \\ i & 0 \end{pmatrix} + \begin{pmatrix} 0 & -i \\ i & 0 \end{pmatrix} \begin{pmatrix} 0 & 1 \\ 1 & 0 \end{pmatrix} = 0 \qquad (V.66)$$

and similarly for the other indices.

(c) Calculate the eigenvalues and eigenvectors for each Pauli matrix.
Solution: The eigenvalue equations read $\sigma_i v_i = \lambda_i v_i$. Because of $\sigma_i^2 = 1$, we find

$$\sigma_i^2 v_i = \begin{cases} v_i \\ \sigma_i \lambda_i v_i = \lambda_i^2 v_i \end{cases} \rightarrow \lambda_i = \pm 1. \qquad (V.67)$$

Hence, all three Pauli matrices have the eigenvalues ± 1. The (normalized) eigenvectors are given by

$$v_{x,\pm 1} = \frac{1}{\sqrt{2}} \begin{pmatrix} 1 \\ \pm 1 \end{pmatrix}; \quad v_{y,\pm 1} = \frac{1}{\sqrt{2}} \begin{pmatrix} 1 \\ \pm i \end{pmatrix}$$

$$v_{z,+1} = \begin{pmatrix} 1 \\ 0 \end{pmatrix}; \quad v_{z,-1} = \begin{pmatrix} 0 \\ 1 \end{pmatrix}. \qquad (V.68)$$

5. Determine the eigenvalues and eigenvectors of the matrix

$$M = \begin{pmatrix} 1 & 4 \\ 2 & -1 \end{pmatrix}. \qquad (V.69)$$

Normalize the eigenvectors. Are they orthogonal?

6. Given the CONS $\{|a_1\rangle, |a_2\rangle\}$, determine the eigenvalues and eigenvectors of the operator

$$M = |a_1\rangle \langle a_1| - |a_2\rangle \langle a_2|. \qquad (V.70)$$

Solution: We have

$$M |a_1\rangle = (|a_1\rangle \langle a_1| - |a_2\rangle \langle a_2|) |a_1\rangle = |a_1\rangle$$
$$M |a_2\rangle = (|a_1\rangle \langle a_1| - |a_2\rangle \langle a_2|) |a_2\rangle = -|a_2\rangle. \qquad (V.71)$$

Thus, the eigenvalues are 1 and -1; the associated eigenvectors are $|a_1\rangle$ and $|a_2\rangle$.

7. Given a CONS $\{|\varphi_n\rangle\}$ and a state $|\psi\rangle = \sum_n c_n |\varphi_n\rangle$, $c_n \in \mathbb{C}$, calculate the coefficients c_n.

Solution:

$$\langle \varphi_i | \, \psi \rangle = \langle \varphi_i | \sum_n c_n \, | \varphi_n \rangle = \sum_n c_n \, \langle \varphi_i | \, \varphi_n \rangle = \sum_n c_n \delta_{in} = c_i, \quad \text{(V.72)}$$

or, in compact form,

$$c_n = \langle \varphi_n | \, \psi \rangle . \quad \text{(V.73)}$$

8. Show in bra-ket notation: The system $\{|r\rangle, |l\rangle\}$ is a CONS. Use the fact that $\{|h\rangle, |v\rangle\}$ is a CONS.
   Solution: We start from

$$|r\rangle = \frac{1}{\sqrt{2}} |h\rangle + \frac{i}{\sqrt{2}} |v\rangle ; \quad |l\rangle = \frac{1}{\sqrt{2}} |h\rangle - \frac{i}{\sqrt{2}} |v\rangle . \quad \text{(V.74)}$$

It follows that

$$\langle r | \, r \rangle = \left[\frac{1}{\sqrt{2}} \langle h | - \frac{i}{\sqrt{2}} \langle v | \right] \left[\frac{1}{\sqrt{2}} |h\rangle + \frac{i}{\sqrt{2}} |v\rangle \right] = \frac{1}{2} - \frac{i^2}{2} = 1, \quad \text{(V.75)}$$

and analogously for $\langle l | \, l \rangle$. Furthermore, we have:

$$\langle r | \, l \rangle = \left[\frac{1}{\sqrt{2}} \langle h | - \frac{i}{\sqrt{2}} \langle v | \right] \left[\frac{1}{\sqrt{2}} |h\rangle - \frac{i}{\sqrt{2}} |v\rangle \right] = \frac{1}{2} + \frac{i^2}{2} = 0. \quad \text{(V.76)}$$

Hence, the orthonormality is proved. Completeness follows from

$$|r\rangle \langle r | + |l\rangle \langle l |$$
$$= \left[\frac{1}{\sqrt{2}} |h\rangle + \frac{i}{\sqrt{2}} |v\rangle \right] \left[\frac{1}{\sqrt{2}} \langle h | - \frac{i}{\sqrt{2}} \langle v | \right] + \left[\frac{1}{\sqrt{2}} |h\rangle - \frac{i}{\sqrt{2}} |v\rangle \right] \left[\frac{1}{\sqrt{2}} \langle h | + \frac{i}{\sqrt{2}} \langle v | \right]$$
$$= \frac{1}{2} |h\rangle \langle h | - \frac{i}{2} |h\rangle \langle v | + \frac{i}{2} |v\rangle \langle h | - \frac{i^2}{2} |v\rangle \langle v | + c.c$$
$$= |h\rangle \langle h | + |v\rangle \langle v | = 1 \quad \text{(V.77)}$$

where $c.c.$ means the complex conjugate of the preceding expression. If we use alternatively the representation

$$|r\rangle \cong \frac{1}{\sqrt{2}} \begin{pmatrix} 1 \\ i \end{pmatrix} ; \quad |l\rangle \cong \frac{1}{\sqrt{2}} \begin{pmatrix} 1 \\ -i \end{pmatrix} , \quad \text{(V.78)}$$

the same result follows:

$$\langle l | \, r \rangle = 0 \leftrightarrow |r\rangle \perp |l\rangle$$
$$\langle l | \, l \rangle = \langle r | \, r \rangle = 1 \quad \text{(V.79)}$$
$$|r\rangle \langle r | + |l\rangle \langle l | = 1.$$

So $\{|r\rangle, |l\rangle\}$ is also a CONS. Accordingly, $\{|h\rangle, |v\rangle\}$ as well as $\{|r\rangle, |l\rangle\}$ each form a basis in $\mathcal{V}$; every vector $|z\rangle \in \mathcal{V}$ can be written as $|z\rangle = c_1 |h\rangle + c_2 |v\rangle$ or $|z\rangle = d_1 |r\rangle + d_2 |l\rangle$ with $c_i, d_i \in \mathbb{C}$.

9. Given the operator $|h\rangle \langle r|$:

(a) Is it a projection operator?
Solution: No, because of

$$|h\rangle \langle r |h\rangle \langle r| = \frac{1}{\sqrt{2}} |h\rangle \langle r| \neq |h\rangle \langle r| \text{ due to } \langle r |h\rangle = \frac{1}{\sqrt{2}}. \qquad \text{(V.80)}$$

(b) How does the operator appear in the representation (4.1)?
Solution:

$$|h\rangle \langle r| \cong \frac{1}{\sqrt{2}} \begin{pmatrix} 1 \\ 0 \end{pmatrix} \begin{pmatrix} 1 & -i \end{pmatrix} = \frac{1}{\sqrt{2}} \begin{pmatrix} 1 & -i \\ 0 & 0 \end{pmatrix}. \qquad \text{(V.81)}$$

(c) Given the state $|z\rangle$ with the representation $|z\rangle \cong \begin{pmatrix} z_1 \\ z_2 \end{pmatrix}$, apply the operator $|h\rangle \langle r|$ to this state (calculation making use of the representation).
Solution:

$$|h\rangle \langle r| z\rangle \cong \frac{1}{\sqrt{2}} \begin{pmatrix} 1 & -i \\ 0 & 0 \end{pmatrix} \begin{pmatrix} z_1 \\ z_2 \end{pmatrix} = \frac{1}{\sqrt{2}} \begin{pmatrix} z_1 - i z_2 \\ 0 \end{pmatrix}. \qquad \text{(V.82)}$$

(d) Use the concrete representation to prove the equality

$$(|h\rangle \langle r| z\rangle)^\dagger = \langle z| r\rangle \langle h|. \qquad \text{(V.83)}$$

Solution:

$$\text{(1)} \quad (|h\rangle \langle r| z\rangle)^\dagger \cong \left(\frac{1}{\sqrt{2}} \begin{pmatrix} z_1 - i z_2 \\ 0 \end{pmatrix} \right)^\dagger = \tfrac{1}{\sqrt{2}} \begin{pmatrix} z_1^* + i z_2^* & 0 \end{pmatrix}$$

$$\text{(2)} \quad \langle z| r\rangle \langle h| \cong \tfrac{1}{\sqrt{2}} \begin{pmatrix} z_1^* & z_2^* \end{pmatrix} \begin{pmatrix} 1 \\ i \end{pmatrix} \begin{pmatrix} 1 & 0 \end{pmatrix} \qquad \text{(V.84)}$$

$$= \tfrac{z_1^* + i z_2^*}{\sqrt{2}} \begin{pmatrix} 1 & 0 \end{pmatrix} = \tfrac{1}{\sqrt{2}} \begin{pmatrix} z_1^* + i z_2^* & 0 \end{pmatrix}.$$

10. We choose the following representation for the states $|h\rangle$ and $|v\rangle$:

$$|h\rangle \cong \frac{1}{\sqrt{2}} \begin{pmatrix} i \\ 1 \end{pmatrix}; \quad |v\rangle \cong \frac{a}{\sqrt{2}|a|} \begin{pmatrix} 1 \\ i \end{pmatrix}. \qquad \text{(V.85)}$$

(a) Show that the representing vectors form a CONS.

Solution:

$$N: \langle h \, | h \rangle = \frac{1}{2} \left(-i \;\; 1 \right) \begin{pmatrix} i \\ 1 \end{pmatrix} = \frac{1}{2} \left(-i^2 + 1 \right) = 1; \quad \text{for } |v\rangle \text{ analogously;}$$

$$O: \langle v \, | h \rangle = \frac{a^+}{2|a|} \left(1 \;\; -i \right) \begin{pmatrix} i \\ 1 \end{pmatrix} = \frac{a^+}{2|a|} \left(i - i \right) = 0;$$

$$C: |h\rangle \langle h| + |v\rangle \langle v| \cong \frac{1}{2} \begin{pmatrix} i \\ 1 \end{pmatrix} \left(-i \;\; 1 \right) + \frac{1}{2} \begin{pmatrix} 1 \\ i \end{pmatrix} \left(1 \;\; -i \right)$$

$$= \frac{1}{2} \begin{pmatrix} -i^2 & i \\ -i & 1 \end{pmatrix} + \frac{1}{2} \begin{pmatrix} 1 & -i \\ i & -i^2 \end{pmatrix} = \begin{pmatrix} 1 & 0 \\ 0 & 1 \end{pmatrix}.$$

$$(V.86)$$

(b) Determine $|r\rangle$ and $|l\rangle$ in this representation. Specialize to the cases of $a = 1, -1, i, -i$.

Solution: With

$$|r\rangle = \frac{|h\rangle + i \, |v\rangle}{\sqrt{2}}; \quad |r\rangle = \frac{|h\rangle - i \, |v\rangle}{\sqrt{2}}, \quad (V.87)$$

it follows that:

$$\frac{|h\rangle \pm i \, |v\rangle}{\sqrt{2}} \cong \frac{1}{2} \begin{pmatrix} i \\ 1 \end{pmatrix} \pm \frac{ia}{2 \, |a|} \begin{pmatrix} 1 \\ i \end{pmatrix} = \frac{1}{2} \begin{pmatrix} i \left(1 \pm \frac{a}{|a|} \right) \\ 1 \mp \frac{a}{|a|} \end{pmatrix}. \quad (V.88)$$

11. Show that the three vectors

$$\mathbf{a} = \frac{1}{\sqrt{2}} \begin{pmatrix} 1 \\ i \\ 0 \end{pmatrix}; \quad \mathbf{b} = \begin{pmatrix} 0 \\ 0 \\ 1 \end{pmatrix}; \quad \mathbf{c} = -\frac{1}{\sqrt{2}} \begin{pmatrix} 1 \\ -i \\ 0 \end{pmatrix} \quad (V.89)$$

form a CONS. Do the same for

$$\mathbf{a} = \frac{1}{\sqrt{2}} \begin{pmatrix} 1 \\ 0 \\ -1 \end{pmatrix}; \quad \mathbf{b} = \frac{1}{2} \begin{pmatrix} 1 \\ \sqrt{2} \\ 1 \end{pmatrix}; \quad \mathbf{c} = \frac{1}{2} \begin{pmatrix} 1 \\ -\sqrt{2} \\ 1 \end{pmatrix}. \quad (V.90)$$

12. A three-dimensional problem: Given the CONS $\{|u\rangle, |v\rangle, |w\rangle\}$ and the operator

$$L = |v\rangle \langle u| + (|u\rangle + |w\rangle) \langle v| + |v\rangle \langle w|. \quad (V.91)$$

(a) Determine the eigenvalues and eigenvectors of L.

Solution: The eigenvalue problem reads

$$L \, |\psi\rangle = l \, |\psi\rangle. \quad (V.92)$$

Since $\{|u\rangle, |v\rangle, |w\rangle\}$ are a CONS, we can represent $|\psi\rangle$ as

$$|\psi\rangle = a\,|u\rangle + b\,|v\rangle + c\,|w\rangle. \tag{V.93}$$

We insert (V.91) and (V.93) into (V.92) and initially obtain

$$[|v\rangle\langle u| + (|u\rangle + |w\rangle)\langle v| + |v\rangle\langle w|][a|u\rangle + b|v\rangle + c|w\rangle]$$
$$= l[a|u\rangle + b|v\rangle + c|w\rangle]. \tag{V.94}$$

The multiplication of the left side, due to the orthonormality of the states $|u\rangle, |v\rangle, |w\rangle$, yields:

$$a\,|v\rangle + b\,(|u\rangle + |w\rangle) + c\,|v\rangle = l\,[a\,|u\rangle + b\,|v\rangle + c\,|w\rangle] \tag{V.95}$$

and from this follow the equations

$$\begin{array}{ll} |u\rangle: & b = la \\ |v\rangle: & a + c = lb; \text{ with } b = lc \text{ follows} \\ |w\rangle: & b = lc \end{array} \qquad \begin{array}{l} lc = la \\ a + c = l^2 c. \end{array} \tag{V.96}$$

We can now have either $l = 0$, from which follows $b = 0, c = -a$; or $l \neq 0$. In the latter case, $c = a$ and $2a = l^2 a$, i.e. $l = \pm\sqrt{2}$.
To summarize: The three eigenvalues are $= 0$ and $l = \pm\sqrt{2}$. The corresponding eigenvectors are given initially by

$$\begin{array}{l} l = 0: \quad |\psi\rangle_0 = a\,|u\rangle - a\,|w\rangle \\ l = \pm\sqrt{2}: \quad |\psi\rangle_{\pm\sqrt{2}} = a\,|u\rangle \pm \sqrt{2}a\,|v\rangle + a\,|w\rangle. \end{array} \tag{V.97}$$

Normalizing theses states yields the final result:

$$\begin{array}{l} l = 0: \quad |\psi\rangle_0 = \frac{|u\rangle - |w\rangle}{\sqrt{2}} \\ l = \pm\sqrt{2}: \quad |\psi\rangle_{\pm\sqrt{2}} = \frac{|u\rangle \pm \sqrt{2}|v\rangle + |w\rangle}{2}. \end{array} \tag{V.98}$$

(b) Show that the three eigenvectors (V.98) form a CONS.

V.5 Exercises, Chap. 5

1. Given the free stationary SEq

$$E\Phi(x) = -\frac{\hbar^2}{2m}\Phi''(x), \tag{V.99}$$

formulate the corresponding equation for the Fourier transform of Φ.
Solution: $\Phi(x)$ is connected with its Fourier transform $\Theta(k)$ by

$$\Phi\left(x\right)=\frac{1}{\sqrt{2\pi}}\int_{-\infty}^{\infty}\Theta\left(k\right)e^{ikx}\mathrm{d}k;\quad\Theta\left(k\right)=\frac{1}{\sqrt{2\pi}}\int_{-\infty}^{\infty}\Phi\left(x\right)e^{-ikx}\mathrm{d}x.\quad\text{(V.100)}$$

By inserting into the free SEq, we obtain

$$E\frac{1}{\sqrt{2\pi}}\int_{-\infty}^{\infty}\Theta\left(k\right)e^{ikx}\mathrm{d}k=-\frac{\hbar^2}{2m}\frac{1}{\sqrt{2\pi}}\int_{-\infty}^{\infty}\left(-k^2\right)\Theta\left(k\right)e^{ikx}\mathrm{d}k \qquad\text{(V.101)}$$

$$\text{or } E\Theta\left(k\right)=\frac{\hbar^2k^2}{2m}\Theta\left(k\right).$$

2. Given the stationary SEq

$$E\Phi\left(x\right)=-\frac{\hbar^2}{2m}\Phi^{''}\left(x\right)+V\left(x\right)\Phi\left(x\right),\qquad\text{(V.102)}$$

formulate the corresponding equation for the Fourier transform of Φ.
Solution: Inserting into the free SEq initially gives

$$E\int_{-\infty}^{\infty}\Theta\left(k\right)e^{ikx}\mathrm{d}k=-\frac{\hbar^2}{2m}\int_{-\infty}^{\infty}\left(-k^2\right)\Theta\left(k\right)e^{ikx}\mathrm{d}k+V\left(x\right)\int_{-\infty}^{\infty}\Theta\left(k\right)e^{ikx}\mathrm{d}k.$$

$$\text{(V.103)}$$

To eliminate the variable x, we consider the Fourier transform $W\left(k\right)$ of $V\left(x\right)$:

$$V\left(x\right)=\frac{1}{\sqrt{2\pi}}\int_{-\infty}^{\infty}W\left(k\right)e^{ikx}\mathrm{d}k;\quad W\left(k\right)=\frac{1}{\sqrt{2\pi}}\int_{-\infty}^{\infty}V\left(x\right)e^{-ikx}\mathrm{d}x.\quad\text{(V.104)}$$

With this, we have

$$\int_{-\infty}^{\infty}\left[E-\frac{\hbar^2k^2}{2m}\right]\Theta\left(k\right)e^{ikx}\mathrm{d}k=\frac{1}{\sqrt{2\pi}}\int_{-\infty}^{\infty}W\left(k_1\right)e^{ik_1x}\mathrm{d}k_1\int_{-\infty}^{\infty}\Theta\left(k_2\right)e^{ik_2x}\mathrm{d}k_2.$$

$$\text{(V.105)}$$

We multiply by e^{-iKx} and integrate with respect to x:

$$\int_{-\infty}^{\infty}\left[E-\frac{\hbar^2k^2}{2m}\right]\Theta\left(k\right)\left[\int_{-\infty}^{\infty}e^{i(k-K)x}\mathrm{d}x\right]\mathrm{d}k$$

$$\text{(V.106)}$$

$$=\frac{1}{\sqrt{2\pi}}\int_{-\infty}^{\infty}W\left(k_1\right)\left[\int_{-\infty}^{\infty}e^{i(k_1+k_2-K)x}\mathrm{d}x\right]\mathrm{d}k_1\int_{-\infty}^{\infty}\Theta\left(k_2\right)\mathrm{d}k_2.$$

The square brackets are essentially the delta function,[158] and thus it follows that

$$\int_{-\infty}^{\infty} \left[E - \frac{\hbar^2 k^2}{2m} \right] \Theta(k) \, \delta(k - K) \, dk$$

$$= \frac{1}{\sqrt{2\pi}} \int_{-\infty}^{\infty} W(k_1) \, \delta(k_1 + k_2 - K) \, dk_1 \int_{-\infty}^{\infty} \Theta(k_2) \, dk_2 \qquad \text{(V.107)}$$

or

$$\left[E - \frac{\hbar^2 K^2}{2m} \right] \Theta(K) = \frac{1}{\sqrt{2\pi}} \int_{-\infty}^{\infty} W(k_1) \, \Theta(K - k_1) \, dk_1. \qquad \text{(V.108)}$$

The final result reads

$$\left[E - \frac{\hbar^2 k^2}{2m} \right] \Theta(k) = \frac{1}{\sqrt{2\pi}} \int_{-\infty}^{\infty} W(k') \, \Theta(k - k') \, dk'. \qquad \text{(V.109)}$$

This integral equation for $\Theta(k)$ replaces the SEq (V.102) in momentum space, as originally formulated in the position representation. Since the two equations yield the same information in the end, it is more a matter of taste which one will be applied. We use the 'usual' SEq (V.102), since the corresponding concepts, methods of solution etc. are for most people more familiar than those of integral equations.

3. The Hamiltonian has discrete nondegenerate eigenvalues $E_n, n = 1, 2, \ldots$. What is the general solution of the time-dependent SEq?

4. Infinite potential well: Show that the eigenfunctions in the form $\varphi_n(x) = \sqrt{\frac{2}{a}} e^{i\delta_n} \sin(k_n x)$ constitute an orthonormal system of functions ($\int_0^a \varphi_m^*(x)\varphi_n(x) = \delta_{mn}$). Hint: The integrals can be calculated for example by means of $\sin x \sin y = \frac{\cos(x-y)-\cos(x+y)}{2}$ or the exponential representation of the sine functions.

5. Infinite potential well: Formulate the general solution of the time-dependent SEq and verify that the specification of the initial condition determines the wavefunction. Concretize the consideration to the special cases ($C \in \mathbb{C}$ is an arbitrary complex constant):

(a) $\Psi(x, t = 0) = C\delta(x - \frac{a}{2})$;
(b) $\Psi(x, t = 0) = C$;
(c) $\Psi(x, t = 0) = Ce^{iKx}$.

Solution: As stated in the text, the general solution is (for simplicity, we

---

[158]It is $\delta(k - k') = \frac{1}{2\pi} \int_{-\infty}^{\infty} dx \, e^{ix(k-k')}$. Some remarks concerning the delta function and its properties are found in Appendix H, Vol. 1.

have set the phases δ_n equal to zero):

$$\Psi(x,t) = \sqrt{\frac{2}{a}} \sum_{n=1}^{\infty} c_n \sin k_n x e^{-i\omega_n t}; \quad c_n \in \mathbb{C}; \quad \omega_n = \frac{E_n}{\hbar} = \frac{\hbar k_n^2}{2m}. \quad (V.110)$$

It follows that

$$\Psi(x,0) = \sqrt{\frac{2}{a}} \sum_{n=1}^{\infty} c_n \sin k_n x \text{ or}$$

$$\sqrt{\frac{2}{a}} \int_0^a \sin k_m x \Psi(x,0) \, dx = \frac{2}{a} \sum_{n=1}^{\infty} c_n \int_0^a \sin k_m x \sin k_n dx. \quad (V.111)$$

The last integral equals $\frac{a}{2}\delta_{mn}$, and we obtain

$$c_m = \sqrt{\frac{2}{a}} \int_0^a \sin k_m x \Psi(x,0) \, dx. \quad (V.112)$$

(a) $\Psi(x, t=0) = C\delta(x - \frac{a}{2})$.
    Solution:

$$c_m = \sqrt{\frac{2}{a}} \int_0^a \sin k_m x \cdot C\delta(x - \frac{a}{2}) dx$$

$$= \sqrt{\frac{2}{a}} C \sin \frac{k_m a}{2} = \sqrt{\frac{2}{a}} C \sin \frac{m\pi}{2} = \frac{1-(-1)^m}{2}(-1)^{\frac{m-1}{2}}. \quad (V.113)$$

6. Given the three-dimensional SEq $E\psi(\mathbf{r}) = -\frac{\hbar^2}{2m}\nabla^2\psi(\mathbf{r})$, which energy eigen-values are allowed if one requires the following periodic boundary conditions:
$\psi(x, y, z) = \psi(x + L_x, y, z) = \psi(x, y + L_y, z) = \psi(x, y, z + L_z)$?
Note: with such periodic boundary conditions one can model, among other things, three-dimensional periodic structures, such as solid lattices. In two dimensions, one can also imagine that these conditions define a torus on whose surface the quantum system is located; see Fig. V.2.

Fig. V.2 Torus
(two-dimensional surface)

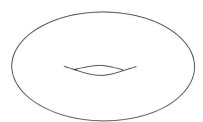

Solution: We use again the separation *ansatz*, $\psi(\mathbf{r}) = f(x)g(y)h(z)$. It leads in the usual manner to $f(x) = A_x e^{ik_x x} + B_x e^{-ik_x x}$ and corresponding expressions for $g(y)$ and $h(z)$. Insertion into the SEq yields:

$$E = \frac{\hbar^2 \mathbf{k}^2}{2m} \tag{V.114}$$

with $\mathbf{k} = (k_x, k_y, k_z)$. The periodic boundary condition, e.g. for x, leads to

$$\psi(x, y, z) = \psi(x + L_x, y, z) \rightarrow$$
$$\rightarrow A_x e^{ik_x x} + B_x e^{-ik_x x} = A_x e^{ik_x(x+L_x)} + B_x e^{-ik_x(x+L_x)}. \tag{V.115}$$

Hence, it follows that $e^{\pm ik_x L_x} = 1$. For integers n, we have $1 = e^{2\pi i n}$ and it follows that

$$k_x L_x = 2\pi n_x \text{ or } k_x = \frac{2\pi n_x}{L_x}, \quad n_x \in \mathbb{N} \tag{V.116}$$

With analogous results for y, z we obtain the following energy levels:

$$E = \frac{2\pi^2 \hbar^2}{m} \left[\left(\frac{n_x}{L_x} \right)^2 + \left(\frac{n_y}{L_y} \right)^2 + \left(\frac{n_z}{L_z} \right)^2 \right]. \tag{V.117}$$

We see nicely that the degree of degeneracy increases with increasing symmetry (e.g. $L_x = L_y$ or $L_x = L_y = L_z$).

Remark: In this case, one can work from the outset with the *ansatz* $\psi(\mathbf{r}) = A e^{i\mathbf{kr}}$.

7. An electron is located between the two walls of an infinite potential well, which are one light year apart. Calculate roughly the magnitude of the difference between two adjacent energy levels.

Solution: $1 ly \approx 9.5 \cdot 10^{15} m \approx 10^{16}$ m; $\hbar \approx 10^{-34}$ Js; $m_e \approx 10^{-30}$ kg; $1J \approx 6 \cdot 10^{18}$ eV. We assume $k_n \approx \frac{n\pi}{L}$. It follows that

$$E_n \approx \frac{\hbar^2 \pi^2}{2mL^2} n^2 \approx 10^{-69} \frac{n^2}{2} J \approx 3 \cdot 10^{-51} n^2 eV. \tag{V.118}$$

8. Find examples for functions which

   (a) are integrable, but not square-integrable;

   Solution: $f(x) = \frac{1}{\sqrt{x}}$ is integrable in the interval $[0, 1]$, but not square

   integrable: $\int_0^1 \frac{1}{\sqrt{x}} dx = 2$.

(b) square-integrable, but not integrable.

Solution: $f(x) = \frac{1}{x}$ is not integrable in the interval $[1, \infty]$, but is square integrable: $\int_{1}^{\infty} \frac{1}{x^2} dx = 1$.

9. Given the stationary SEq

$$E\varphi(x) = -\frac{\hbar^2}{2m}\varphi''(x) + V(x)\varphi(x), \tag{V.119}$$

rewrite this equation for a dimensionless independent variable.

Solution: We choose

$$z = Kx; \quad \varphi(x) = \psi(Kx) = \psi(z) \tag{V.120}$$

with the yet undetermined constant K (unit 1/m) and the dimensionless variable z. Insertion yields initially

$$E\psi(z) = -\frac{\hbar^2}{2m}\frac{d^2}{dx^2}\psi(z) + V(x)\psi(z). \tag{V.121}$$

Changing variables in the derivative yields

$$E\psi(z) = -\frac{\hbar^2 K^2}{2m}\frac{d^2}{dz^2}\psi(z) + V\left(\frac{z}{K}\right)\psi(z). \tag{V.122}$$

One should now choose K in such a manner that the prefactors of the functions are as simple as possible. If we use for example $K^2 = \frac{2mE}{\hbar^2}$ (which need not necessarily be the cleverest choice), we obtain

$$\psi(z) = -\frac{d^2}{dz^2}\psi(z) + \tilde{V}(z)\psi(z) \tag{V.123}$$

with the dimensionless potential $\tilde{V}(z) = \frac{1}{E}V\left(\frac{z}{K}\right)$.

10. A short outlook into string theory (compactified or rolled-up dimensions): String theory assumes that the elementary building blocks are not point objects, but rather one-dimensional objects (strings) with a certain energy—comparable to an object in a one-dimensional potential well. Strings have a spatial extension of order of the Planck length and live in higher-dimensional spaces (e.g. dim $= 10$ or dim $= 26$), where only four dimensions are not rolled up (compactified)—quite similar to our following simple example.

For the formal treatment, we take the two-dimensional SEq

$$-\frac{\hbar^2}{2m}\left(\frac{\partial^2\psi}{\partial x^2} + \frac{\partial^2\psi}{\partial y^2}\right) = E\psi \tag{V.124}$$

Fig. V.3 The 'cylinder
world' of our toy string

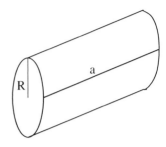

as starting point. In x direction, we have an infinite potential well

$$V = \begin{cases} 0 & \text{for } 0 < x < a \\ \infty & \text{otherwise} \end{cases} \qquad (V.125)$$

and for the y coordinate we postulate

$$\psi(x, y) = \psi(x, y + 2\pi R). \qquad (V.126)$$

So we have a combination of two different boundary conditions: In the x direction, $\psi(0, y) = \psi(a, y) = 0$ applies, while in the y direction the periodic boundary condition $\psi(x, y) = \psi(x, y + 2\pi R)$ is valid. In other words, the quantum object 'lives' on the surface of a cylinder of length a and of radius R (see Fig. V.3). The problem reads is now to calculate the possible energy levels. Discuss in particular the situation when $R \ll a$.

Solution: For the solution of the SEq, we use the separation *ansatz*:

$$\psi(x, y) = \Phi(x) \Psi(y) \qquad (V.127)$$

and obtain

$$-\frac{\hbar^2}{2m} \frac{1}{\Phi(x)} \frac{d^2\Phi(x)}{dx^2} - \frac{\hbar^2}{2m} \frac{1}{\Psi(y)} \frac{d^2\Psi(y)}{dy^2} = E. \qquad (V.128)$$

The terms which depend on x or y have to be constant:

$$
\begin{array}{c}
-\dfrac{\hbar^2}{2m} \dfrac{1}{\Phi(x)} \dfrac{d^2\Phi(x)}{dx^2} = E_x \\[2ex]
-\dfrac{\hbar^2}{2m} \dfrac{1}{\Psi(y)} \dfrac{d^2\Psi(y)}{dy^2} = E - E_x
\end{array}
\quad \text{or} \quad
\begin{array}{c}
\dfrac{d^2\Phi(x)}{dx^2} = -\dfrac{2m}{\hbar^2} E_x \Phi(x) = -k_x^2 \Phi(x) \\[2ex]
\dfrac{d^2\Psi(y)}{dy^2} = -\dfrac{2m}{\hbar^2}(E - E_x)\Psi(y) = -k_y^2 \Psi(y).
\end{array}
$$

$$(V.129)$$

As usual, we obtain as solutions (real form)

$$
\begin{aligned}
\Phi(x) &= A \sin k_x x + B \cos k_x x \\
\Psi(y) &= C \sin k_y y + D \cos k_y y.
\end{aligned}
\qquad (V.130)
$$

With the boundary condition $\Phi(0) = \Phi(a) = 0$, the first equation gives

$$B = A \text{ and } \sin k_x a = 0 \rightarrow k_x = \frac{N\pi}{a}, \; n = 1, 2, \ldots \quad \text{(V.131)}$$

For $\Psi(y)$, we obtain with $\Psi(y) = \Psi(y + 2\pi R)$:

$$C \sin k_y y + D \cos k_y y = C \sin k_y (y + 2\pi R) + D \cos k_y (y + 2\pi R). \quad \text{(V.132)}$$

Since C and D are independent integration constants, their coefficients must be equal on both sides. This leads to

$$k_y 2\pi R = 2M\pi \text{ or } k_y = \frac{M}{R}, \; M = 0, 1, 2, \ldots \quad \text{(V.133)}$$

The range of values of M also includes zero, since in this case the trivial solution does not occur, because we have $\Psi(y) = D$ for $k_y = 0$. This fact is especially important for the discussion in the case $R \ll a$, as we shall see shortly. For the energy, it follows because of $E = E_x + E_y = \frac{\hbar^2}{2m}\left(k_x^2 + k_y^2\right)$ that:

$$E_{N,M} = \frac{\hbar^2}{2m}\left[\left(\frac{N\pi}{a}\right)^2 + \left(\frac{M}{R}\right)^2\right]; \; N = 1, 2, \ldots; \; M = 0, 1, 2, \ldots \quad \text{(V.134)}$$

where N and M assume values independently of each other.

Due to the second dimension, the energy spectrum has changed significantly overall. In particular, it can now be degenerate; that is the case for $R = \frac{a}{\pi}\sqrt{\frac{p}{q}}$, where p and q are differences of squares of natural numbers.

For $M = 0$, the energy levels $E_{N,0}$ are those of the one-dimensional infinite potential well. Where is the lowest new energy level? We evidently find it for $N = 1$ ($N = 0$ is not allowed) and $M = 1$, i.e.

$$E_{1,1} = \frac{\hbar^2}{2m}\left[\left(\frac{\pi}{a}\right)^2 + \left(\frac{1}{R}\right)^2\right]. \quad \text{(V.135)}$$

If we now consider a very 'thin' cylinder, i.e. $R \ll a$, it follows that

$$E_{1,1} \approx \frac{\hbar^2}{2m}\left(\frac{1}{R}\right)^2 \text{ for } R \ll a. \quad \text{(V.136)}$$

In comparison to the 'unperturbed' energy levels $E_{N,0}$, this means that

$$E_{1,1} \approx E_{K,0} \text{ with } K \approx \frac{a}{\pi R}. \quad \text{(V.137)}$$

Due to $R \ll a$, K is a very large number, so that the first new energy level $E_{1,1}$ is far beyond the low-lying energy levels $E_{N,0}$. In other words, an extra dimension cannot be seen at lower energies in experiments, if it is rolled up tightly enough. These effects can be seen only at sufficiently high energies.

11. Given the free one-dimensional SEq (5.36) and the function $\Phi(x)$, show that

$$\Psi(x,t) = A \frac{1}{\sqrt{t}} \int_{-\infty}^{\infty} e^{\frac{im}{2\hbar} \frac{(x-y)^2}{t}} \Phi(y)\, dy \qquad (V.138)$$

is a solution (A is a normalizing factor).

Solution: We compute the partial derivatives. We find

$$\partial_t \Psi(x,t) = -A \frac{1}{2t\sqrt{t}} \int_{-\infty}^{\infty} e^{\frac{im}{2\hbar} \frac{(x-y)^2}{2t}} \Phi(y)\, dy$$

$$+ A \frac{1}{\sqrt{t}} \int_{-\infty}^{\infty} \frac{im}{2\hbar} \left(-\frac{(x-y)^2}{t^2} \right) e^{\frac{im}{2\hbar} \frac{(x-y)^2}{t}} \Phi(y)\, dy$$

$$\partial_x \Psi(x,t) = A \frac{1}{\sqrt{t}} \int_{-\infty}^{\infty} \frac{im}{\hbar} \frac{(x-y)}{t} e^{\frac{im}{2\hbar} \frac{(x-y)^2}{t}} \Phi(y)\, dy$$

$$\partial_x^2 \Psi(x,t) = A \frac{1}{\sqrt{t}} \int_{-\infty}^{\infty} \left[\frac{im}{\hbar} \frac{1}{t} e^{\frac{im}{2\hbar} \frac{(x-y)^2}{t}} + \left(\frac{im}{\hbar} \frac{(x-y)}{t} \right)^2 \right] e^{\frac{im}{2\hbar} \frac{(x-y)^2}{t}} \Phi(y)\, dy.$$

$$(V.139)$$

It follows that

$$i\hbar\partial_t \Psi(x,t) = A \int_{-\infty}^{\infty} \left[-i\hbar \frac{1}{2t\sqrt{t}} + i\hbar \frac{1}{\sqrt{t}} \frac{im}{2\hbar} \left(-\frac{(x-y)^2}{t^2} \right) \right] e^{\frac{im}{2\hbar} \frac{(x-y)^2}{t}} \Phi(y)\, dy$$

$$-\frac{\hbar^2}{2m} \partial_x^2 \Psi(x,t) = A \int_{-\infty}^{\infty} \left[-\frac{\hbar^2}{2m} \frac{im}{\hbar} \frac{1}{t\sqrt{t}} - \frac{\hbar^2}{2m} \frac{1}{\sqrt{t}} \left(\frac{im}{\hbar} \frac{(x-y)}{t} \right)^2 \right] e^{\frac{im}{2\hbar} \frac{(x-y)^2}{t}} \Phi(y)\, dy.$$

$$(V.140)$$

Comparison of the right-hand sides immediately verifies the assertion.

Remark: One can show that:

$$\lim_{t\to 0} \Psi(x,t) = \Phi(x). \qquad (V.141)$$

Hence, (V.138) is another representation of the free one-dimensional SEq with the given initial condition $\Psi(x,0)$.

V.6   Exercises, Chap. 6

1. Show that for all $|z_i\rangle$ in (6.5), $\||z_i\rangle\|^2 = 1$ holds.
2. Given a MZI with symmetrical beam splitters, calculate the final state with and without a blocker if the initial state is given by $\alpha\,|H\rangle + \beta\,|V\rangle$.
3. Given an operator A with

$$A\,|H\rangle = a\,|H\rangle\,;\ \ A\,|V\rangle = b\,|V\rangle\,, \tag{V.142}$$

determine the explicit form of A.
Solution:

$$A = a\,|H\rangle\,\langle H| + b\,|V\rangle\,\langle V| \cong \begin{pmatrix} a & 0 \\ 0 & b \end{pmatrix}. \tag{V.143}$$

For the example $a = 1$ and $b = -1$, it follows that $A = |H\rangle\,\langle H| - |V\rangle\,\langle V| \cong \begin{pmatrix} 1 & 0 \\ 0 & -1 \end{pmatrix}$.

4. Which eigenvalues can a unitary operator have?
Solution: We start from

$$U\,|\varphi\rangle = \lambda\,|\varphi\rangle\,;\ \ U^\dagger = U^{-1};\ \ \langle\varphi\,|\varphi\rangle = 1. \tag{V.144}$$

It follows that

$$\langle\varphi|\,U^\dagger = \lambda^*\,\langle\varphi|\ \text{ or }\ \langle\varphi|\,U^\dagger U\,|\varphi\rangle = \lambda^*\lambda\,\langle\varphi\,|\varphi\rangle. \tag{V.145}$$

Due to $U^\dagger U = 1$ and $\langle\varphi\,|\varphi\rangle = 1$, we obtain immediately

$$|\lambda|^2 = 1. \tag{V.146}$$

Hence, the eigenvalues of unitary operators are on the unit circle and have the form $\lambda = e^{i\alpha}$ (and not just $\lambda = \pm 1$, as is often inferred incorrectly from $|\lambda|^2 = 1$).

5. Circularly- and linearly-polarized states are connected by $|r\rangle = \frac{1}{\sqrt{2}}\,|h\rangle + \frac{i}{\sqrt{2}}\,|v\rangle$ and $|l\rangle = \frac{1}{\sqrt{2}}\,|h\rangle - \frac{i}{\sqrt{2}}\,|v\rangle$. Show that this basis transformation is unitary (or that the transformation matrix is unitary).
Solution: The transformation between linearly- and circularly-polarized light can be described by the matrix $\frac{1}{\sqrt{2}}\begin{pmatrix} 1 & i \\ 1 & -i \end{pmatrix}$, which is unitary.

6. Give the matrix representation of the operators T, S and S' from (6.11), (6.12) and (6.13) and their combinations TST and $TS'T$.
Solution: We take into account that we are in a two-dimensional space. Therefore we can represent the basis states $|H\rangle$ and $|V\rangle$ e.g. by the vectors $\begin{pmatrix} 1 \\ 0 \end{pmatrix}$ and $\begin{pmatrix} 0 \\ 1 \end{pmatrix}$; the product $|V\rangle\,\langle H|$ is then

$$|V\rangle \langle H| \cong \begin{pmatrix} 0 \\ 1 \end{pmatrix} \begin{pmatrix} 1 & 0 \end{pmatrix} = \begin{pmatrix} 0 & 0 \\ 1 & 0 \end{pmatrix}. \qquad \text{(V.147)}$$

We can read the action of the beam splitter from (6.11); it is

$$T \cong \frac{1+i}{2} \begin{pmatrix} 1 & i \\ i & 1 \end{pmatrix}. \qquad \text{(V.148)}$$

For the mirror, it follows analogously from (6.12):

$$S \cong \begin{pmatrix} 0 & -1 \\ -1 & 0 \end{pmatrix} \qquad \text{(V.149)}$$

i.e. the well-known phase jump π. Then it follows for the case without a blocker

$$TST \cong \begin{pmatrix} 1 & 0 \\ 0 & 1 \end{pmatrix}. \qquad \text{(V.150)}$$

For the case with a blocker, we have to replace S by S':

$$S' \cong \begin{pmatrix} 0 & 0 \\ -1 & 0 \end{pmatrix} \qquad \text{(V.151)}$$

and obtain

$$TS'T \cong \frac{1}{2} \begin{pmatrix} 1 & i \\ -i & 1 \end{pmatrix}. \qquad \text{(V.152)}$$

7. Given the operator

$$U = a |H\rangle \langle H| + b |H\rangle \langle V| + c |V\rangle \langle H| + d |V\rangle \langle V| \cong \begin{pmatrix} a & b \\ c & d \end{pmatrix}; \quad \text{(V.153)}$$

for which values of the coefficients is U is a unitary operator? In other words: How is the general two-dimensional unitary transformation formulated? Solution: The equations $UU^\dagger = U^\dagger U = 1$ have to be satisfied, which reads in matrix representation:

$$\begin{pmatrix} a & b \\ c & d \end{pmatrix} \begin{pmatrix} a^* & c^* \\ b^* & d^* \end{pmatrix} = \begin{pmatrix} 1 & 0 \\ 0 & 1 \end{pmatrix} \text{ and } \begin{pmatrix} a^* & c^* \\ b^* & d^* \end{pmatrix} \begin{pmatrix} a & b \\ c & d \end{pmatrix} = \begin{pmatrix} 1 & 0 \\ 0 & 1 \end{pmatrix}. \quad \text{(V.154)}$$

This gives the equations

$$|a|^2 + |b|^2 = 1; \quad ac^* + bd^* = 0$$
$$ca^* + db^* = 0; \quad |c|^2 + |d|^2 = 1 \qquad \text{(V.155)}$$

and

$$|a|^2 + |c|^2 = 1; \quad a^*b + c^*d = 0$$
$$b^*a + d^*c = 0; \quad |b|^2 + |d|^2 = 1. \tag{V.156}$$

From the equations with the square values, it follows immediately that

$$|b|^2 = |c|^2 \text{ and } |a|^2 = |d|^2 \tag{V.157}$$

and we can use the *ansatz*

$$a = Ae^{i\alpha}; \quad b = Be^{i\beta}; \quad c = Be^{i\gamma}; \quad d = Ae^{i\delta}; \quad A^2 + B^2 = 1. \tag{V.158}$$

Thus, the remaining two equations $ac^* + bd^* = 0$ and $b^*a + d^*c = 0$ give

$$e^{i\alpha}e^{-i\gamma} + e^{i\beta}e^{-i\delta} = 0 \text{ and } e^{-i\beta}e^{i\alpha} + e^{-i\delta}e^{i\gamma} = 0. \tag{V.159}$$

A closer look reveals that these two equations are identical; as the result, we have for example:

$$e^{i\delta} = -e^{i(\beta-\alpha+\gamma)} \text{ or } \delta = \beta - \alpha + \gamma + \pi. \tag{V.160}$$

Thus, we have in matrix representation initially

$$U \cong \begin{pmatrix} Ae^{i\alpha} & Be^{i\beta} \\ Be^{i\gamma} & -Ae^{-i(\alpha-\beta-\gamma)} \end{pmatrix}; \quad A^2 + B^2 = 1. \tag{V.161}$$

This result may be written in a structurally simpler manner. To this end, we put $e^{i(\beta+\gamma+\pi)/2}$ outside the brackets:

$$U \cong e^{i(\beta+\gamma+\pi)/2} \begin{pmatrix} Ae^{i\left(\alpha-\frac{\beta+\gamma+\pi}{2}\right)} & Be^{i\frac{\beta-\gamma-\pi}{2}} \\ -Be^{-i\frac{\beta-\gamma-\pi}{2}} & Ae^{-i\left(\alpha-\frac{\beta+\gamma+\pi}{2}\right)} \end{pmatrix}; \quad A^2 + B^2 = 1, \tag{V.162}$$

or, more compactly (with $p = Ae^{i\left(\alpha-\frac{\beta+\gamma+\pi}{2}\right)}$ etc.):

$$U \cong e^{i\mu} \begin{pmatrix} p & q \\ -q^* & p^* \end{pmatrix}; \quad |p|^2 + |q|^2 = 1; \quad p, q \in \mathbb{C}; \quad \mu \in \mathbb{R} \tag{V.163}$$

as a general form of a two-dimensional unitary transformation.[159]
As an important special case, we obtain the real rotation

$$U_{\text{rotation}} \cong \begin{pmatrix} \cos\vartheta & \sin\vartheta \\ -\sin\vartheta & \cos\vartheta \end{pmatrix}; \quad p = \cos\vartheta; \quad q = \sin\vartheta; \quad \mu = 0. \tag{V.164}$$

---

[159] On extension with $e^{i(\beta+\gamma)/2}$, we find the equivalent representation $e^{i\mu} \begin{pmatrix} p & q \\ q^* & -p^* \end{pmatrix}$.

The symmetrical beam splitter follows with $\frac{1+i}{2} = \frac{1}{\sqrt{2}} e^{i\frac{\pi}{4}}$,

$$U_{\text{beam splitter}} \cong \frac{1+i}{2} \begin{pmatrix} 1 & i \\ -i & 1 \end{pmatrix}; \quad p = \frac{1}{\sqrt{2}}; \quad q = \frac{i}{\sqrt{2}}; \quad \mu = \frac{\pi}{4}. \quad \text{(V.165)}$$

The *Hadamard matrix* is found as

$$U_{\text{Hadamard}} \cong \frac{1}{\sqrt{2}} \begin{pmatrix} 1 & 1 \\ 1 & -1 \end{pmatrix}; \quad p = \frac{1}{i\sqrt{2}}; \quad q = \frac{1}{i\sqrt{2}}; \quad \mu = \frac{\pi}{2}. \quad \text{(V.166)}$$

8. Given a MZI without a blocker and with asymmetrical beam splitters (transmittance $\neq$ reflectance), determine the properties required of the beam splitters in order that a beam entering horizontally activates only detector 1, while detector 2 remains dark.
   Solution: We can represent an asymmetrical beam splitter as

$$T \cong (\alpha + i\beta) \begin{pmatrix} \alpha & i\beta \\ i\beta & \alpha \end{pmatrix}; \quad \alpha, \beta \in \mathbb{R}, > 0; \quad \alpha^2 + \beta^2 = 1 \quad \text{(V.167)}$$

where α is the amplitude transmission coefficient and β the reflection coefficient of the beam splitter. The factor i in front of β denotes the relative phase shift between transmitted and reflected beams. T is unitary; cf. (V.163). The action of the whole Mach–Zehnder interferometer can be described by

$$T_2 S T_1 = (\alpha_2 + i\beta_2) \begin{pmatrix} \alpha_2 & i\beta_2 \\ i\beta_2 & \alpha_2 \end{pmatrix} \begin{pmatrix} 0 & -1 \\ -1 & 0 \end{pmatrix} (\alpha_1 + i\beta_1) \begin{pmatrix} \alpha_1 & i\beta_1 \\ i\beta_1 & \alpha_1 \end{pmatrix}. \quad \text{(V.168)}$$

This term can be evaluated to give

$$T_2 S T_1 = -(\alpha_2 + i\beta_2)(\alpha_1 + i\beta_1) \begin{pmatrix} i\alpha_1\beta_2 + i\alpha_2\beta_1 & \alpha_1\alpha_2 - \beta_1\beta_2 \\ \alpha_1\alpha_2 - \beta_1\beta_2 & i\alpha_1\beta_2 + i\alpha_2\beta_1 \end{pmatrix}. \quad \text{(V.169)}$$

Hence, if we want detector 1 to always respond and detector 2 never, we must set $\alpha_1\alpha_2 = \beta_1\beta_2$. It follows that

$$\beta_2 = \frac{\alpha_1\alpha_2}{\beta_1}; \quad \alpha_2 = \frac{\beta_1\beta_2}{\alpha_1}. \quad \text{(V.170)}$$

With $\alpha_i^2 + \beta_i^2 = 1$, this yields

$$\alpha_2^2 + \beta_2^2 = \alpha_2^2 + \frac{\alpha_1^2\alpha_2^2}{\beta_1^2} = \frac{\alpha_2^2}{\beta_1^2}(\beta_1^2 + \alpha_1^2) = \frac{\alpha_2^2}{\beta_1^2} = 1$$
$$\alpha_2^2 + \beta_2^2 = \frac{\beta_1^2\beta_2^2}{\alpha_1^2} + \beta_2^2 = \frac{\beta_2^2}{\alpha_1^2}(\beta_1^2 + \alpha_1^2) = \frac{\beta_2^2}{\alpha_1^2} = 1, \quad \text{(V.171)}$$

or, due to $\alpha_i, \beta_i > 0$, finally

$$\alpha_2 = \beta_1; \quad \beta_2 = \alpha_1. \tag{V.172}$$

In other words, the transmission and reflection factors of the second beam splitter must be reversed relative to the first beam splitter:

$$T_1 \cong (\alpha_1 + i\beta_1) \begin{pmatrix} \alpha_1 & i\beta_1 \\ i\beta_1 & \alpha_1 \end{pmatrix}; \quad T_2 \cong (\beta_1 + i\alpha_1) \begin{pmatrix} \beta_1 & i\alpha_1 \\ i\alpha_1 & \beta_1 \end{pmatrix}. \tag{V.173}$$

For the total action of the Mach–Zehnder interferometer, it then follows:

$$T_2 S T_1 \cong -i\,(\beta_1 + i\alpha_1)\,(\alpha_1 + i\beta_1) \begin{pmatrix} 1 & 0 \\ 0 & 1 \end{pmatrix}$$

$$= (\alpha_1 + i\beta_1)\,(\alpha_1 - i\beta_1) \begin{pmatrix} 1 & 0 \\ 0 & 1 \end{pmatrix} = \begin{pmatrix} 1 & 0 \\ 0 & 1 \end{pmatrix}, \tag{V.174}$$

as expected. See also J. Pade and L. Polley, 'Wechselwirkungsfreie Quantenmessung' (Interaction-free quantum measurement, in German), Physik in der Schule 38/5 (2000) 343.

V.7 Exercises, Chap. 7

1. Show for $\rho = |\psi\,(x, t)|^2$ that:

$$\int_{-\infty}^{\infty} \rho\,(x, t)\,\mathrm{d}x = 1 \;\forall\, t. \tag{V.175}$$

Here we assume that (i) the potential is real, and (ii) $\Psi \underset{x \to \infty}{\sim} x^a$, with $a < -\frac{1}{2}$.

2. Infinite potential well: Given the wavefunctions

   (a) $\Psi\,(x, t) = e^{-i\omega_n t} \sqrt{\frac{2}{a}} \sin \frac{n\pi}{a} x$ and

   (b) $\Psi\,(x, t) = c_n e^{-i\omega_n t} \sqrt{\frac{2}{a}} \sin \frac{n\pi}{a} x + c_m e^{-i\omega_m t} \sqrt{\frac{2}{a}} \sin \frac{m\pi}{a} x$,

   calculate for both cases the probability of finding the quantum object in the interval (x_1, x_2)

$$w_{x_1, x_2}^{\text{quantum mechanics}} = \int_{x_1}^{x_2} \Psi^*\,(x, t)\,\Psi\,(x, t)\,\mathrm{d}x. \tag{V.176}$$

3. Given the SEq $i\hbar\dot{\psi} = H\psi$ with a real potential, derive from the continuity equation constructively (i.e. not just proving by insertion) that **j** is given by

$$\mathbf{j} = \frac{\hbar}{2mi} \left(\psi^* \nabla \psi - \psi \nabla \psi^* \right). \tag{V.177}$$

Solution: Since the potential is real, we have $H = H^*$ and therefore

$$i\hbar\dot{\psi} = H\psi; \quad -i\hbar\dot{\psi}^* = H\psi^*. \tag{V.178}$$

Using the continuity equation, we can write

$$\nabla \mathbf{j} = -\dot{\rho} = -\partial_t \psi^* \psi = -\dot{\psi}^* \psi - \psi^* \dot{\psi}. \tag{V.179}$$

On the right-hand side, we insert (V.178):

$$\nabla \mathbf{j} = -\left(-\frac{H\psi^*}{i\hbar} \right) \psi - \psi^* \left(\frac{H\psi}{i\hbar} \right) = \frac{i}{\hbar} \left(\psi^* H\psi - \psi H\psi^* \right). \tag{V.180}$$

Since the potential is real, the potential terms cancel and we have:

$$\nabla \mathbf{j} = \frac{\hbar}{2mi} \left(\psi^* \nabla^2 \psi - \psi \nabla^2 \psi^* \right). \tag{V.181}$$

On the right-hand side, we insert $\pm \nabla \psi^* \cdot \nabla \psi$ and obtain

$$
\begin{aligned}
\nabla \mathbf{j} &= \frac{\hbar}{2mi} \left(\psi^* \nabla^2 \psi + \nabla \psi^* \cdot \nabla \psi - \psi \nabla^2 \psi^* - \nabla \psi \cdot \nabla \psi^* \right) \\
&= \frac{\hbar}{2mi} \left(\nabla \left(\psi^* \nabla \psi \right) - \nabla \left(\psi \nabla \psi^* \right) \right) = \frac{\hbar}{2mi} \nabla \left(\psi^* \nabla \psi - \psi \nabla \psi^* \right).
\end{aligned}
\tag{V.182}
$$

Comparing the right and left sides yields

$$\nabla \left(\mathbf{j} - \frac{\hbar}{2mi} \left(\psi^* \nabla \psi - \psi \nabla \psi^* \right) \right) = 0, \tag{V.183}$$

and this is the desired result. (Strictly, it follows from the last equation due to $\nabla (\nabla \times \mathbf{A}) = 0$, however, that $\mathbf{j} = \frac{\hbar}{2mi} \left(\psi^* \nabla \psi - \psi \nabla \psi^* \right) + \nabla \times \mathbf{A}$, where $\mathbf{A}$ is an arbitrary field.)

4. Calculate j (one-dimensional) for $\psi = Ae^{\gamma x}$ and $\psi = Ae^{i\gamma x}$, with $\gamma \in \mathbb{R}$ and $A \in \mathbb{C}$.
5. Calculate $\mathbf{j}(\mathbf{r}, t)$ for $\Psi(\mathbf{r}, t) = Ae^{i(\mathbf{kr} - \omega t)}$.
6. Given a modification of the infinite potential well, namely the potential

$$V(x) = \begin{cases} iW & \text{for } 0 < x < a \\ \infty & \text{otherwise} \end{cases}; \ W \in \mathbb{R}, \tag{V.184}$$

calculate the energy spectrum and show that the norm of the (time-dependent) total wavefunction is independent of time only for $W = 0$.
Solution: The stationary SEq including boundary conditions reads

$$E\varphi = -\frac{\hbar^2}{2m}\varphi'' + iW\varphi; \quad \varphi(0) = \varphi(a) = 0; \tag{V.185}$$

here, we have to formulate a complex energy, i.e.

$$E = E_R + iE_I. \tag{V.186}$$

With

$$\gamma^2 = \frac{2m}{\hbar^2}(E - iW), \tag{V.187}$$

we find as solution

$$\varphi = Ae^{i\gamma x} + Be^{-i\gamma x}. \tag{V.188}$$

The boundary condition at $x = 0$ yields $B = -A$, the one at $x = a$ leads due to $A \neq 0$ to $0 = e^{i\gamma a} - e^{-i\gamma a}$ or $e^{2i\gamma a} = 1$. With the *ansatz* $\gamma = \gamma_R + i\gamma_I$, it follows that

$$e^{2i(\gamma_R + i\gamma_I)a} = e^{2i\gamma_R a}e^{-2\gamma_I a} = 1. \tag{V.189}$$

This gives immediately $\sin 2\gamma_R a = 0$, and therefore

$$\gamma_R a = n\pi \text{ and } \gamma_I = 0. \tag{V.190}$$

From (V.187), we can conclude that $E_I = W$ (due to $\gamma_I = 0$). For the real part of the energy, we obtain the well-known relation

$$E_{R,n} = \frac{\hbar^2\gamma_R^2}{2m} = \frac{\hbar^2}{2m}\left(\frac{n\pi}{a}\right)^2; \quad n = 1, 2, \ldots \tag{V.191}$$

so that the energies are given by

$$E_n = E_{R,n} + iW. \tag{V.192}$$

We insert this into the total wavefunction

$$\psi(x, t) = \sum_n c_n\varphi_n(x)e^{-\frac{it}{\hbar}E_n} \tag{V.193}$$

and obtain

$$\psi(x, t) = e^{\frac{t}{\hbar}W}\sum_n c_n\varphi_n(x)e^{-\frac{it}{\hbar}E_{R,n}}. \tag{V.194}$$

Depending on the sign of $W \neq 0$, $\psi(x, t)$ tends for $t \to \pm\infty$ to 0 or to ∞. Explicitly, it holds that:

$$\int |\psi(x,t)|^2 \, dx = e^{\frac{2t}{\hbar} W} \int \sum_n c_n^* \varphi_n^*(x) \, e^{\frac{it}{\hbar} E_{R,n}} \sum_m c_m \varphi_m(x) \, e^{-\frac{it}{\hbar} E_{R,m}} dx$$

$$= e^{\frac{2t}{\hbar} W} \sum_{n,m} c_n^* c_m e^{\frac{it}{\hbar} E_{R,n}} e^{-\frac{it}{\hbar} E_{R,m}} \int \varphi_n^*(x) \, \varphi_m(x) \, dx.$$

(V.195)

Due to the orthonormality of the eigenfunctions $\varphi_n(x)$, it follows that

$$\int |\psi(x,t)|^2 \, dx = e^{\frac{2t}{\hbar} W} \sum_n |c_n|^2 . \tag{V.196}$$

As expected, we cannot obtain $\int |\psi(x,t)|^2 \, dx = 1 \ \forall t$.

V.8   Exercises, Chap. 8

1. Given that $|\nu_1\rangle \langle \nu_1| + |\nu_2\rangle \langle \nu_2| = 1$. Show: $|\nu_e\rangle \langle \nu_e| + |\nu_\mu\rangle \langle \nu_\mu| = 1$.

2. Show that the matrices $\begin{pmatrix} c & 0 & se^{-i\delta} \\ 0 & 1 & 0 \\ -se^{i\delta} & 0 & c \end{pmatrix}$ and $\begin{pmatrix} 1 & 0 & 0 \\ 0 & c & s \\ 0 & -s & c \end{pmatrix}$ with $\delta \in \mathbb{R}$ are
   unitary. The abbreviations s and c stand for $\sin \alpha$ and $\cos \alpha$.

3. Show that the product of two unitary matrices is also unitary.

4. Is the beam splitter operator T from Chap. 6,

$$T = \frac{1+i}{2} [1 + i \, |H\rangle \langle V| + i \, |V\rangle \langle H|], \tag{V.197}$$

   a Hermitian, a unitary or a projection operator? $\{|H\rangle, |V\rangle\}$ is a CONS.

5. Given $A = \begin{pmatrix} 1 & i \\ -i & 1 \end{pmatrix}$:

   (a) Show that A is Hermitian, but not unitary.
       Solution:

$$A^\dagger = \begin{pmatrix} 1 & i \\ -i & 1 \end{pmatrix} = A; \ A^\dagger A = A^2 = 2A \neq \begin{pmatrix} 1 & 0 \\ 0 & 1 \end{pmatrix}. \tag{V.198}$$

   (b) Calculate e^{cA}.
       Solution: Due to $A^2 = 2A$, it follows that $A^n = 2^{n-1} A$ and therefore,

$$e^{cA} = 1 + \sum_{n=1}^\infty \frac{c^n}{n!} A^n = 1 + \sum_{n=1}^\infty \frac{c^n}{n!} 2^{n-1} A. \tag{V.199}$$

       This means that

$$e^{cA} = 1 + \frac{1}{2} \sum_{n=1}^{\infty} \frac{c^n}{n!} 2^n A = 1 + \left(\frac{e^{2c}}{2} - 1 \right) A. \qquad (V.200)$$

6. Given the operators

$$L_1 = \frac{|v\rangle \left(\langle u| + \langle w| \right) + \left(|u\rangle + |w\rangle \right) \langle v|}{\sqrt{2}}$$

$$L_2 = \frac{-|v\rangle \left(\langle u| - \langle w| \right) + \left(|u\rangle - |w\rangle \right) \langle v|}{i\sqrt{2}} \qquad (V.201)$$

$$L_3 = |u\rangle \langle u| - |w\rangle \langle w|.$$

(a) Are these Hermitian, unitary or projection operators?
   Solution: We can see directly that the operators are Hermitian, e.g.

$$L_2^{\dagger} = \frac{-\left(|u\rangle - |w\rangle \right) \langle v| + |v\rangle \left(\langle u| - \langle w| \right)}{-i\sqrt{2}} = L_2. \qquad (V.202)$$

But they are neither unitary nor projective; we have e.g. for L_3:

$$L_3^{\dagger} L_3 = L_3^2 = |u\rangle \langle u| + |w\rangle \langle w| \qquad (V.203)$$

and this term is neither L_3 nor the unity operator.
(b) Calculate $[L_1, L_2]$.
   Solution: We calculate first the individual terms, i.e.

$$\begin{aligned} 2i\, L_1 L_2 &= -\left(|u\rangle + |w\rangle \right) \left(\langle u| - \langle w| \right) \\ 2i\, L_2 L_1 &= \left(|u\rangle - |w\rangle \right) \left(\langle u| + \langle w| \right). \end{aligned} \qquad (V.204)$$

It follows that

$$\begin{aligned} [L_1, L_2] &= \frac{-|u\rangle \langle u| + |u\rangle \langle w| - |w\rangle \langle u| + |w\rangle \langle w|}{2i} \\ &\quad - \frac{|u\rangle \langle u| + |u\rangle \langle w| - |w\rangle \langle u| - |w\rangle \langle w|}{2i} \\ &= \frac{-2\,|u\rangle \langle u| + 2\,|w\rangle \langle w|}{2i} = i\left(|u\rangle \langle u| - |w\rangle \langle w| \right) = i L_3. \end{aligned} \qquad (V.205)$$

7. Show that the time evolution

$$|v(t)\rangle = -\sin \vartheta\, |v_1\rangle\, e^{-i\omega_1 t} + \cos \vartheta\, |v_2\rangle\, e^{-i\omega_2 t} \qquad (V.206)$$

is unitary.
8. Determine explicitly $\langle v_e | v(t) \rangle$ in (8.8), and $\langle v_\mu | v(t) \rangle$.
   Solution: With (8.1) or (8.2), it holds that

$$\langle \nu_e \,|\nu(t)\rangle = -\sin\vartheta \,\langle \nu_e \,|\nu_1\rangle\, e^{-i\omega_1 t} + \cos\vartheta\, \langle \nu_e\, |\nu_2\rangle\, e^{-i\omega_2 t}$$
$$= -\sin\vartheta \cos\vartheta e^{-i\omega_1 t} + \cos\vartheta \sin\vartheta e^{-i\omega_2 t} \tag{V.207}$$

as well as

$$\langle \nu_\mu \,|\nu(t)\rangle = -\sin\vartheta \,\langle \nu_\mu \,|\nu_1\rangle\, e^{-i\omega_1 t} + \cos\vartheta\, \langle \nu_\mu\, |\nu_2\rangle\, e^{-i\omega_2 t}$$
$$= \sin^2\vartheta e^{-i\omega_1 t} + \cos^2\vartheta e^{-i\omega_2 t}. \tag{V.208}$$

9. Determine explicitly p_e in (8.9), and p_μ.
   Solution: For p_e, we consider first

$$e^{-i\omega_1 t} - e^{-i\omega_2 t} = e^{-i\frac{\omega_1+\omega_2}{2}t}\left[e^{-i\frac{\omega_1-\omega_2}{2}t} - e^{i\frac{\omega_1-\omega_2}{2}t}\right] = 2i e^{-i\frac{\omega_1+\omega_2}{2}t}\sin\left(\frac{\Delta\omega}{2}t\right).$$
$$\tag{V.209}$$

It follows that

$$p_e = |\langle \nu_e \,|\nu(t)\rangle|^2 = \sin^2\vartheta\cos^2\vartheta \cdot 4\sin^2\left(\frac{\Delta\omega}{2}t\right) = \sin^2 2\vartheta \cdot \sin^2\left(\frac{\Delta\omega}{2}t\right).$$
$$\tag{V.210}$$

For p_μ, we use $p_e + p_\mu = 1$ and obtain

$$p_\mu = 1 - \sin^2 2\vartheta \cdot \sin^2\left(\frac{\Delta\omega}{2}t\right). \tag{V.211}$$

If we want to calculate p_μ explicitly, we start from

$$p_\mu = |\langle \nu_\mu\, |\nu(t)\rangle|^2 = |\sin^2\vartheta e^{-i\omega_1 t} + \cos^2\vartheta e^{-i\omega_2 t}|^2. \tag{V.212}$$

Due to $\left(\sin^2\vartheta + \cos^2\vartheta\right)^2 = 1$, this gives

$$p_\mu = \sin^4\vartheta + 2\sin^2\vartheta\cos^2\vartheta\cos(\Delta\omega t) + \cos^4\vartheta$$
$$= 1 + 2\sin^2\vartheta\cos^2\vartheta\,[\cos(\Delta\omega t) - 1]. \tag{V.213}$$

We transform the square brackets by means of $\cos 2x = \cos^2 x - \sin^2 x = 1 - 2\sin^2 x$ and obtain

$$p_\mu = 1 - 4\sin^2\vartheta\cos^2\vartheta\sin^2\left(\frac{\Delta\omega}{2}t\right) = 1 - \sin^2 2\vartheta \cdot \sin^2\left(\frac{\Delta\omega}{2}t\right). \quad (V.214)$$

10. Prove (8.10); find an approximation for ΔE in the case of very small rest masses.
    Solution:

$$\hbar \Delta \omega = \Delta E = E_1 - E_2 = \sqrt{p^2 c^2 + m_1^2 c^4} - \sqrt{p^2 c^2 + m_2^2 c^4}$$

$$= pc \left[\sqrt{1 + \frac{m_1^2 c^2}{p^2}} - \sqrt{1 + \frac{m_2^2 c^2}{p^2}} \right] \approx pc \left[1 + \frac{m_1^2 c^2}{2p^2} - 1 - \frac{m_2^2 c^2}{2p^2} \right]$$

$$= \frac{c^4}{2pc} (m_1^2 - m_2^2) := \frac{c^4 \Delta m^2}{2pc}.$$

$$(V.215)$$

11. Given the state

$$|\psi (t)\rangle = \sum_n c_n |\varphi_n\rangle \, e^{-i E_n t / \hbar} \qquad (V.216)$$

with the initial condition $|\psi (0)\rangle$. $\{|\varphi_n\rangle\}$ is a CONS. How are the constants c_n related to the initial conditions?
Solution: Since $\{|\varphi_n\rangle\}$ is a CONS, we have

$$\langle \varphi_m |\psi (0)\rangle = \sum_n c_n \langle \varphi_m |\varphi_n\rangle = \sum_n c_n \delta_{mn} = c_m \qquad (V.217)$$

and the state reads

$$|\psi (t)\rangle = \sum_n \langle \varphi_n |\psi (0)\rangle \, |\varphi_n\rangle \, e^{-i E_n t / \hbar} = \sum_n |\varphi_n\rangle \langle \varphi_n| e^{-i E_n t / \hbar} \, |\psi (0)\rangle .$$

$$(V.218)$$

12. Given two CONS $\{|\varphi_i\rangle\}$ and $\{|\psi_i\rangle\}$. A quantum system is in the superposition $|z\rangle = \sum_i d_i |\psi_i\rangle$.

(a) Calculate the probability of measuring the quantum system in the state $|\varphi_k\rangle$.
Solution:

$$p_k = |\langle \varphi_k |z\rangle|^2 = \left| \sum_i d_i \langle \varphi_k |\psi_i\rangle \right|^2 . \qquad (V.219)$$

(b) Show that $\sum_k p_k = 1$.
Solution:

$$\sum_k p_k = \sum_k \langle z |\varphi_k\rangle \langle \varphi_k |z\rangle = \langle z |z\rangle = 1. \qquad (V.220)$$

13. Given the model system

$$i\hbar \frac{d}{dt} |\psi (t)\rangle = H |\psi (t)\rangle \text{ with } H = 1 + A\sigma_y; \ A > 0, \qquad (V.221)$$

where σ_y is the y-Pauli matrix. (For the sake of simplicity, we do not distinguish between $=$ and $\cong$.)

(a) Determine the eigenvalues and eigenvectors of H.
Solution: The eigenvalue problem reads

$$H \, |\varphi\rangle = E \, |\varphi\rangle \ \text{or} \ \begin{pmatrix} 1 & -i A \\ i A & 1 \end{pmatrix} \begin{pmatrix} a \\ b \end{pmatrix} = E \begin{pmatrix} a \\ b \end{pmatrix}, \tag{V.222}$$

or explicitly

$$\begin{matrix} a - i Ab = Ea \\ i Aa + b = Eb \end{matrix} \ \text{or} \ \begin{matrix} a\,(1 - E) = i\, Ab \\ b\,(1 - E) = -i\, Aa. \end{matrix} \tag{V.223}$$

It follows that

$$b = -a \frac{E - 1}{i A}. \tag{V.224}$$

By inserting, we obtain $(a \neq 0)$

$$a \frac{1 - E}{i A} \, (1 - E) = -i\, Aa \ \text{or} \ (E - 1)^2 = A^2. \tag{V.225}$$

Hence, the eigenvalues read

$$E_1 = 1 + A; \ E_2 = 1 - A. \tag{V.226}$$

with (V.224), we obtain first for the associated eigenvectors

$$|\varphi_1\rangle = \begin{pmatrix} a \\ ia \end{pmatrix}; \ |\varphi_2\rangle = \begin{pmatrix} a \\ -ia \end{pmatrix}, \tag{V.227}$$

where a is arbitrary (due to the fact that the eigenvalue problem (V.222) is linear and a multiple of a solution is also a solution). We can fix a by an additional requirement; usually this is the normalization. If we require that $|\varphi_1\rangle$ be normalized, it follows that

$$\langle \varphi_1 \, | \varphi_1 \rangle = \begin{pmatrix} a^* & -ia^* \end{pmatrix} \begin{pmatrix} a \\ ia \end{pmatrix} = 2aa^* \overset{!}{=} 1. \tag{V.228}$$

The simplest choice for a is $1/\sqrt{2}$. Analogous statements hold for $|\varphi_2\rangle$, and we obtain finally the normalized eigenvectors (which are clearly orthogonal, as required):

$$|\varphi_1\rangle = \frac{1}{\sqrt{2}} \begin{pmatrix} 1 \\ i \end{pmatrix}; \ |\varphi_2\rangle = \frac{1}{\sqrt{2}} \begin{pmatrix} 1 \\ -i \end{pmatrix}. \tag{V.229}$$

(b) How does the general expression $|\psi\,(t)\rangle$ read for a time-dependent state? Solution: (1) Long version: The general expression $|\psi\,(t)\rangle$ for a time-dependent state follows as a solution of the system

$$i\hbar \frac{d}{dt} \, |\psi\rangle = H \, |\psi\rangle \ \text{or} \ i\hbar \frac{d}{dt} \begin{pmatrix} f \\ g \end{pmatrix} = \begin{pmatrix} 1 & -i A \\ i A & 1 \end{pmatrix} \begin{pmatrix} f \\ g \end{pmatrix}. \tag{V.230}$$

The exponential *ansatz* (it is a differential equation with constant coefficients):

$$\begin{pmatrix} f \\ g \end{pmatrix} = \begin{pmatrix} F \\ G \end{pmatrix} e^{\gamma t}; \quad F, G \text{ constant} \tag{V.231}$$

leads to

$$i\hbar\gamma \begin{pmatrix} F \\ G \end{pmatrix} = \begin{pmatrix} 1 & -iA \\ iA & 1 \end{pmatrix} \begin{pmatrix} F \\ G \end{pmatrix}, \tag{V.232}$$

and by comparison with the eigenvalue problem just treated, for γ the two solutions follow immediately:

$$\gamma_{1,2} = \frac{E_{1,2}}{i\hbar}.$$

Hence, the general state is

$$|\psi(t)\rangle = c_1 |\varphi_1\rangle e^{-iE_1 t/\hbar} + c_2 |\varphi_2\rangle e^{-iE_2 t/\hbar} \tag{V.233}$$

where the c_i are integration constants which are determined by the initial conditions. In explicit form, this reads

$$|\psi(t)\rangle = \frac{e^{-it/\hbar}}{\sqrt{2}} \left[c_1 \begin{pmatrix} 1 \\ i \end{pmatrix} e^{-iAt/\hbar} + c_2 \begin{pmatrix} 1 \\ -i \end{pmatrix} e^{iAt/\hbar} \right]. \tag{V.234}$$

(2) Short version: Since $\{|\varphi_1\rangle, |\varphi_2\rangle\}$ is a CONS, each state can be represented at time $t = 0$ as a linear combination:

$$|\psi(0)\rangle = c_1 |\varphi_1\rangle + c_2 |\varphi_2\rangle. \tag{V.235}$$

Since $\{|\varphi_1\rangle, |\varphi_2\rangle\}$ are states with sharp energies, the time evolution is given by

$$|\psi(t)\rangle = c_1 |\varphi_1\rangle e^{-iE_1 t/\hbar} + c_2 |\varphi_2\rangle e^{-iE_2 t/\hbar}. \tag{V.236}$$

(c) How is $|\psi(t)\rangle$ expressed for the initial state $|\psi(t=0)\rangle = \begin{pmatrix} 1 \\ 0 \end{pmatrix}$?

Solution: With the given initial condition, we have

$$\begin{pmatrix} 1 \\ 0 \end{pmatrix} = \frac{1}{\sqrt{2}} \left[c_1 \begin{pmatrix} 1 \\ i \end{pmatrix} + c_2 \begin{pmatrix} 1 \\ -i \end{pmatrix} \right], \tag{V.237}$$

and this leads immediately to

$$c_2 = c_1; \quad c_1 = \frac{1}{\sqrt{2}}. \tag{V.238}$$

Hence, the total state for this initial state reads

$$|\psi(t)\rangle = \frac{e^{-it/\hbar}}{\sqrt{2}}\left[\frac{1}{\sqrt{2}}\begin{pmatrix}1\\i\end{pmatrix}e^{-iAt/\hbar} + \frac{1}{\sqrt{2}}\begin{pmatrix}1\\-i\end{pmatrix}e^{iAt/\hbar}\right]$$

$$= e^{-it/\hbar}\begin{pmatrix}\cos\dfrac{At}{\hbar}\\[2mm]\sin\dfrac{At}{\hbar}\end{pmatrix}. \tag{V.239}$$

(d) Assume that we measure $|\psi(t)\rangle$ from part c. With which probability will we find the state $|\chi\rangle = \begin{pmatrix}1\\0\end{pmatrix}$ (i.e. the initial state)?

Solution: The probability follows as $|\langle\chi|\psi\rangle|^2$, i.e.

$$|\langle\chi|\psi\rangle|^2 = \left|(1\ 0)\begin{pmatrix}\cos\frac{At}{\hbar}\\\sin\frac{At}{\hbar}\end{pmatrix}\right|^2 = \cos^2\frac{At}{\hbar}. \tag{V.240}$$

After the measurement, the state is $|\chi\rangle$.

V.9 Exercises, Chap. 9

1. Given a Hermitian operator A and the eigenvalue problem $A\varphi_n = a_n\varphi_n$, $n = 1, 2, \ldots$, show that:

   (a) The eigenvalues are real.
   (b) The eigenfunctions are pairwise orthogonal. Here, it is assumed that the eigenvalues are nondegenerate.

2. Show that the expectation value of a Hermitian operator is real.
   Solution: Due to the Hermiticity, $\int \psi^* A\psi = \int (A\psi)^* \psi$ holds:

$$\langle A\rangle^* = \left(\int \psi^* A\psi\right)^* = \int \psi (A\psi)^* = \int (A\psi)^* \psi = \int \psi^* A\psi = \langle A\rangle. \tag{V.241}$$

3. Show that

$$\int \Psi_1^* A\Psi_2 dV = \int (A\Psi_1)^* \Psi_2 dV \tag{V.242}$$

holds for the operators $\mathbf{r}$, $\mathbf{p}$, H. Restrict the discussion to the one-dimensional case. Which conditions must the wavefunctions satisfy?
Solution: In one dimension, it holds for a Hermitian operator A that:

$$\int f^*(x)Ag(x)dx = \int (Af(x))^* g(x)dx. \tag{V.243}$$

Evidently, this equation is satisfied for $A = x$.

For $A = p = \frac{\hbar}{i}\frac{d}{dx}$, it follows with partial integration:

$$\int f^*(x)\frac{\hbar}{i}\frac{d}{dx}g(x)dx = \frac{\hbar}{i}\int f^* g'dx$$
$$= \frac{\hbar}{i}\left[(f^*g)_{-\infty}^{\infty} - \int f^{*\prime}gdx\right] = \int \left(\frac{\hbar}{i}\frac{d}{dx}f\right)^* gdx \tag{V.244}$$

where $(f^*g)_{-\infty}^{\infty} = 0$ must be fulfilled.

For $A = H$, we need to worry only about the space derivatives (assuming a real potential). With partial integration, we find:

$$\int f^*(x)g''(x)dx = (f^*g')_{-\infty}^{\infty} - \int f^{*\prime}g'dx$$
$$= (f^*g')_{-\infty}^{\infty} - \left[(f^{*\prime}g)_{-\infty}^{\infty} - \int f^{*\prime\prime}gdx\right] = \int f^{*\prime\prime}gdx \tag{V.245}$$

where $(f^*g')_{-\infty}^{\infty} - (f^{*\prime}g)_{-\infty}^{\infty} = 0$ must be fulfilled.

4. Show that for the infinite potential well (between 0 and a), $\langle x\rangle = \frac{a}{2}$.

Solution: With the eigenfunctions $\varphi_n = \sqrt{\frac{2}{a}}\sin\left(\frac{n\pi}{a}x\right)$; $n = 1, 2, ...$, we have:

$$\langle x\rangle = \frac{2}{a}\int_0^a x\sin^2\left(\frac{n\pi}{a}x\right)dx = \frac{2}{a}\int_0^a x\frac{1-\cos\left(\frac{2n\pi}{a}x\right)}{2}dx. \tag{V.246}$$

Because of

$$\int_0^a x\cos\left(\frac{2n\pi x}{a}\right)dx = \frac{a}{2n\pi}\frac{\partial}{\partial n}\int_0^a \sin\left(\frac{2n\pi x}{a}\right)dx$$
$$= -\frac{a}{2n\pi}\frac{\partial}{\partial n}\left(\cos\left(\frac{2n\pi a}{a}\right) - 1\right) = 0, \tag{V.247}$$

it follows that

$$\langle x\rangle = \frac{2}{a}\int_0^a \frac{x}{2}dx = \frac{2}{a}\frac{a^2}{4} = \frac{a}{2}. \tag{V.248}$$

5. Given the infinite potential well with walls at $x = 0$ and $x = a$; we consider the state

$$\Psi(x, t) = \sqrt{\frac{2}{a}}\sin\left(\frac{n\pi}{a}x\right)e^{-i\omega_n t}. \tag{V.249}$$

(a) Determine the position uncertainty Δx.

Solution: We have

$$(\Delta x)^2 = \langle x^2\rangle - \langle x\rangle^2 \tag{V.250}$$

and thus it follows that

$$(\Delta x)^2 = \frac{2}{a} \int_0^a \sin^2\left(\frac{n\pi}{a}x\right) x^2 dx - \left(\frac{2}{a} \int_0^a \sin^2\left(\frac{n\pi}{a}x\right) x dx\right)^2. \quad \text{(V.251)}$$

With

$$\int_0^a \sin^2\left(\frac{n\pi}{a}x\right) x^2 dx = \frac{a^3}{12}\frac{2n^2\pi^2 - 3}{n^2\pi^2}; \quad \int_0^a \sin^2\left(\frac{n\pi}{a}x\right) x dx = \frac{a^2}{4}$$

$$\text{(V.252)}$$

we obtain

$$(\Delta x)^2 = \frac{a^2}{6}\frac{2n^2\pi^2 - 3}{n^2\pi^2} - \frac{a^2}{4} = \frac{a^2}{4}\left[\frac{n^2\pi^2 - 6}{3n^2\pi^2}\right] \quad \text{(V.253)}$$

or

$$\Delta x = \frac{a}{2}\sqrt{\frac{n^2\pi^2 - 6}{3n^2\pi^2}} \xrightarrow[n\to\infty]{} \frac{a}{2\sqrt{3}} \approx 0.289a. \quad \text{(V.254)}$$

(b) Determine the momentum uncertainty Δp.
   Solution: We have
$$(\Delta p)^2 = \langle p^2 \rangle - \langle p \rangle^2. \quad \text{(V.255)}$$

We first calculate $\langle p \rangle$:

$$\langle p \rangle = \frac{2\hbar}{ai} \int_0^a \sin\left(\frac{n\pi}{a}x\right) \frac{d}{dx} \sin\left(\frac{n\pi}{a}x\right) dx = \frac{\hbar}{ai} \int_0^a \frac{d}{dx} \sin^2\left(\frac{n\pi}{a}x\right) dx \quad \text{(V.256)}$$
$$= \frac{\hbar}{ai}\left[\sin^2\left(\frac{n\pi}{a}x\right)\right]_0^a = 0.$$

For $\langle p^2 \rangle$, we have

$$\langle p^2 \rangle = -\frac{2\hbar^2}{a} \int_0^a \sin\left(\frac{n\pi}{a}x\right) \frac{d^2}{dx^2} \sin\left(\frac{n\pi}{a}x\right) dx = \frac{2\hbar^2}{a}\left(\frac{n\pi}{a}\right)^2 \int_0^a \sin^2\left(\frac{n\pi}{a}x\right) dx$$
$$= \frac{2\hbar^2}{a}\left(\frac{n\pi}{a}\right)^2 \frac{a}{2n\pi}\left[\frac{n\pi}{a}a - \cos\frac{n\pi}{a}a \sin\frac{n\pi}{a}a\right] = \left(\frac{n\pi\hbar}{a}\right)^2. \quad \text{(V.257)}$$

It follows that

$$\Delta p = \frac{n\pi\hbar}{a}. \quad \text{(V.258)}$$

This gives for the product of the uncertainties:

$$\Delta x \cdot \Delta p = \frac{a}{2}\sqrt{\frac{n^2\pi^2-6}{3n^2\pi^2}} \cdot \frac{n\pi\hbar}{a} = \frac{\hbar}{2}\sqrt{\frac{n^2\pi^2-6}{3}} > \frac{\hbar}{2}. \qquad (V.259)$$

The last inequality holds due to $\pi^2 > 9 \to \frac{\pi^2-6}{3} > 1$.

Occasionally, one encounters the fallacy that in the infinite potential well, due to $E = \frac{p^2}{2m}$, a sharp energy leads to a sharp momentum. But that is true only for the absolute value of the momentum; the momentum itself is not sharp. The physical reason is that the states are standing waves, which correspond to a back-and-forth movement with correspondingly different momenta (hence $\langle p \rangle = 0$); in other words, for a given energy, p is not uniquely defined by $p = \pm\sqrt{2mE}$.

6. In the infinite potential well, a normalized state is given by

$$\Psi(x,t) = c_n\varphi_n(x)e^{-i\omega_n t} + c_m\varphi_m(x)e^{-i\omega_m t}; \quad c_n, c_m \in \mathbb{C}; \quad n \neq m. \quad (V.260)$$

Calculate $\langle x \rangle$.

Solution: We note first that due to the normalization, we have

$$\int_0^a \Psi^*\Psi dx = |c_n|^2 + |c_m|^2 = 1. \qquad (V.261)$$

with

$$\langle x \rangle = \int_0^a \Psi^* x \Psi dx, \qquad (V.262)$$

it follows that

$$\langle x \rangle = |c_n|^2 \int_0^a x\varphi_n^2 dx + |c_m|^2 \int_0^a x\varphi_m^2 dx + \left\{ c_n^* c_m e^{i(\omega_n-\omega_m)t} \int_0^a \varphi_n x \varphi_m dx + c.c \right\}$$

$$= \frac{2}{a}|c_n|^2 \int_0^a x\sin^2 k_n x dx + \frac{2}{a}|c_m|^2 \int_0^a x\sin^2 k_m x dx$$

$$+ \left\{ \frac{2}{a}c_n^* c_m e^{i(\omega_n-\omega_m)t} \int_0^a x\sin k_n x \cdot \sin k_m x dx + c.c \right\}. \qquad (V.263)$$

Because of $k_n = \frac{n\pi}{a}$, this gives

$$\int_0^a x\sin^2 k_n x dx = \frac{a^2}{4} \qquad (V.264)$$

and

$$\int_0^a x \sin k_n x \cdot \sin k_m x \, dx = \frac{2nma^2}{\pi^2 \left(n^2 - m^2\right)^2} \left[(-1)^{n+m} - 1\right]; \quad n \neq m. \quad (V.265)$$

This means

$$\langle x \rangle = \frac{2}{a} |c_n|^2 \frac{a^2}{4} + \frac{2}{a} |c_m|^2 \frac{a^2}{4}$$

$$+ \left\{ \frac{2}{a} c_n^* c_m e^{i(\omega_n - \omega_m)t} \frac{2nma^2}{\pi^2(n^2 - m^2)^2} \left[(-1)^{n+m} - 1\right] + c.c \right\} \quad (V.266)$$

$$= \frac{a}{2} + \frac{4nma}{\pi^2(n^2 - m^2)^2} \left[(-1)^{n+m} - 1\right] \left\{ c_n^* c_m e^{i(\omega_n - \omega_m)t} + c.c \right\}.$$

with $c_n = |c_n| e^{i\varphi_n}$, it follows that

$$c_n^* c_m e^{i(\omega_n - \omega_m)t} = |c_n| |c_m| e^{i(\omega_n - \omega_m)t + i(\varphi_m - \varphi_n)}. \quad (V.267)$$

Hence, we can compensate the phases by choosing a new zero of time; thus, we can set $\varphi_m - \varphi_n = 0$. In addition, we use the shorthand notation $\Delta\omega_{nm} = \omega_n - \omega_m$. It follows then that:

$$\langle x \rangle = \frac{a}{2} + \frac{8nma |c_n| |c_m|}{\pi^2 \left(n^2 - m^2\right)^2} \left[(-1)^{n+m} - 1\right] \cos \Delta\omega_{nm} t \quad (V.268)$$

or

$$\langle x \rangle = \frac{a}{2} \cdot \left\{ 1 - \frac{32nm|c_n||c_m|}{\pi^2(n^2 - m^2)^2} \cos \Delta\omega_{nm} t \right. \quad \begin{array}{l} \text{for } n + m = \left\{ \begin{array}{l} \text{even} \\ \text{odd.} \end{array} \right. \end{array} \quad (V.269)$$

Calculation exercise: Show that

$$\frac{32nm |c_n| |c_m|}{\pi^2 \left(n^2 - m^2\right)^2} < 1. \quad (V.270)$$

Solution: Because of $|c_n|^2 + |c_m|^2 = 1$ and $(|c_n| - |c_m|)^2 \geq 0$, we have $|c_n| |c_m| \leq \frac{1}{2}$, and therefore

$$\frac{32nm |c_n| |c_m|}{\pi^2 \left(n^2 - m^2\right)^2} \leq \frac{16nm}{\pi^2 \left(n^2 - m^2\right)^2} \leq \frac{16n(n+1)}{\pi^2 (2n+1)^2} \leq \frac{16n^2}{\pi^2 4n^2} = \frac{4}{\pi^2} < 1. \quad (V.271)$$

7. Consider an infinite square well with potential limits at $x = 0$ and $x = a$. The initial value of the wavefunction is $\Psi(x, 0) = \Phi \in \mathbb{R}$ for $b - \varepsilon \leq x \leq b + \varepsilon$

and $\Psi(x, 0) = 0$ otherwise (of course, $0 \leq b - \varepsilon$ and $b + \varepsilon \leq a$). Remember that the eigenfunctions $\varphi_n(x) = \sqrt{\frac{2}{a}} \sin k_n x$ with $k_n = \frac{n\pi}{a}$ form a CONS.

(a) Normalize the initial state.
Solution:

$$\int_0^a |\Psi(x, 0)|^2 \, dx = \int_{b-\varepsilon}^{b+\varepsilon} \Phi^2 \, dx = \Phi^2 \cdot 2\varepsilon = 1 \quad \rightarrow \quad \Phi = \frac{1}{\sqrt{2\varepsilon}}. \quad \text{(V.272)}$$

(b) Calculate $\Psi(x, t)$.
Solution: Start with

$$\Psi(x, t) = \sum c_n \varphi_n(x) e^{-i\frac{E_n t}{\hbar}} \quad \text{with} \quad c_n = \int \varphi_n^*(x) \Psi(x, 0) \, dx. \quad \text{(V.273)}$$

It follows then that

$$c_n = \sqrt{\frac{2}{a}} \int_{b-\varepsilon}^{b+\varepsilon} \sin k_n x \cdot \Phi \, dx = \sqrt{\frac{2}{a}} \Phi \left(-\frac{\cos k_n (b + \varepsilon) - \cos k_n (b - \varepsilon)}{k_n} \right),$$

$$\text{(V.274)}$$

and with this,

$$c_n = -\sqrt{\frac{2}{a}} \Phi \frac{-2 \sin bk_n \cdot \sin \varepsilon k_n}{k_n} = \frac{2}{\sqrt{a}} \sin bk_n \frac{\sin \varepsilon k_n}{\sqrt{\varepsilon} k_n}. \quad \text{(V.275)}$$

For the total wavefunction, we find:

$$\Psi(x, t) = \frac{2}{\sqrt{a}} \sum \sin bk_n \frac{\sin \varepsilon k_n}{\sqrt{\varepsilon} k_n} \varphi_n(x) e^{-i\frac{E_n t}{\hbar}}. \quad \text{(V.276)}$$

(c) Find the probability of measuring the system in the state n.
Solution: It is

$$|c_n|^2 = \frac{4}{a} \sin^2 bk_n \frac{\sin^2 \varepsilon k_n}{\varepsilon k_n^2} = \frac{4a \sin^2 \left(n\pi \frac{b}{a} \right) \sin^2 \left(n\pi \frac{\varepsilon}{a} \right)}{\varepsilon} \frac{1}{n^2 \pi^2}. \quad \text{(V.277)}$$

8. Show that for the expectation value of a physical quantity A,

$$i\hbar \frac{d}{dt} \langle A \rangle = \langle [A, H] \rangle + i\hbar \left\langle \frac{\partial}{\partial t} A \right\rangle \quad \text{(V.278)}$$

holds. Show that for time-independent operators, the expectation value of the corresponding physical quantity is conserved, if A commutes with H.

9. Show that

$$\frac{d}{dt} \langle \mathbf{r} \rangle = \frac{1}{m} \langle \mathbf{p} \rangle \text{ and } \frac{d}{dt} \langle \mathbf{p} \rangle = -\langle \nabla V \rangle. \tag{V.279}$$

10. Under which conditions is the orbital angular momentum $\mathbf{l} = \mathbf{r} \times \mathbf{p}$ a conserved quantity?

Solution: To check if the angular momentum $\mathbf{l} = \mathbf{r} \times \frac{\hbar}{i} \nabla$ is a conserved quantity, due to the relation

$$i\hbar \frac{d}{dt} \langle \mathbf{l} \rangle = \langle [\mathbf{l}, H] \rangle, \tag{V.280}$$

we need to calculate its commutator with H. Since the angular momentum is a vector, we have three equations; we limit ourselves to $[l_x, H]$ and transfer the result to the two other components. We use

$$l_x = (\mathbf{r} \times \mathbf{p})_x = y p_z - z p_y = \frac{\hbar}{i} \left(y \frac{\partial}{\partial z} - z \frac{\partial}{\partial y} \right) \tag{V.281}$$

and

$$H = -\frac{\hbar^2}{2m} \nabla^2 + V = H_0 + V. \tag{V.282}$$

(a) First we show that $[l_x, H_0] = 0$ using $H_0 = \frac{\mathbf{p}^2}{2m}$ and the relation $[x, p_x] = i\hbar$ plus the analogues for y, z. We split the expression $\left[y p_z - z p_y, \mathbf{p}^2 \right]$ and consider only $\left[y p_z, \mathbf{p}^2 \right]$; the other term follows by interchanging y and z. In this way (as always we assume that the order of the partial derivatives is irrelevant), we obtain:

$$\left[y p_z, \mathbf{p}^2 \right] = \left[y p_z, p_x^2 + p_y^2 + p_z^2 \right] = \left[y p_z, p_y^2 \right] \tag{V.283}$$

since p_x^2 and p_z^2 commute with $y p_z$. The remaining term is rearranged:

$$\left[y p_z, p_y^2 \right] = y p_z p_y^2 - p_y^2 y p_z = y p_z p_y^2 - p_y \left(y p_y - i\hbar \right) p_z$$
$$= y p_z p_y^2 - p_y y p_y p_z + p_y i\hbar p_z = y p_z p_y^2 - \left(y p_y - i\hbar \right) p_y p_z + p_y i\hbar p_z$$
$$= y p_z p_y^2 - y p_y^2 p_z + 2 i\hbar p_y p_z = 2 i\hbar p_y p_z. \tag{V.284}$$

It follows then that

$$\left[y p_z - z p_y, \mathbf{p}^2 \right] = 2 i\hbar p_y p_z - 2 i\hbar p_z p_y = 0, \tag{V.285}$$

or $[l_x, H_0] = 0$, and analogously for l_y, l_z.

(b) It remains to calculate

$$[l_x, H] = [l_x, V]. \tag{V.286}$$

This expression (operator equation!) is evaluated as

$$
\begin{aligned}
[l_x, V] &= \frac{\hbar}{i}\left(y\frac{\partial}{\partial z} - z\frac{\partial}{\partial y}\right)V - V\frac{\hbar}{i}\left(y\frac{\partial}{\partial z} - z\frac{\partial}{\partial y}\right) \\
&= \frac{\hbar}{i}\left(y\frac{\partial V}{\partial z} - z\frac{\partial V}{\partial y}\right) + V\frac{\hbar}{i}\left(y\frac{\partial}{\partial z} - z\frac{\partial}{\partial y}\right) - V\frac{\hbar}{i}\left(y\frac{\partial}{\partial z} - z\frac{\partial}{\partial y}\right) \\
&= \frac{\hbar}{i}\left(y\frac{\partial V}{\partial z} - z\frac{\partial V}{\partial y}\right) = \frac{\hbar}{i}(\mathbf{r} \times \nabla V)_x,
\end{aligned}
$$

$$(V.287)$$

or, for all three components in compact form:

$$
[\mathbf{l}, H] = \frac{\hbar}{i}(\mathbf{r} \times \nabla V). \tag{V.288}
$$

In sum, this means that

$$
\frac{d}{dt}\langle\mathbf{l}\rangle = -\langle\mathbf{r} \times \nabla V\rangle. \tag{V.289}
$$

The right-hand side is zero for $V = V(r)$ with $r = |\mathbf{r}|$, for then we have $\nabla V(r) = \frac{\mathbf{r}}{r}\frac{\partial V(r)}{\partial r}$. Hence, in general, i.e. excepting the radially-symmetric case, the angular momentum $\mathbf{l}$ is a not conserved quantity in an external potential.

11. Given the Hamiltonian H with a discrete and non-degenerate spectrum E_n and eigenstates $\varphi_n(\mathbf{r})$, show that the energy uncertainty ΔH vanishes, iff the quantum object is in an eigenstate of the energy.
Solution: Time-dependent and stationary SEq are:

$$
i\hbar\frac{\partial}{\partial t}\Psi(\mathbf{r}, t) = H\Psi(\mathbf{r}, t); \quad H\varphi_n(\mathbf{r}) = E_n\varphi_n(\mathbf{r}). \tag{V.290}
$$

The general solution, given by ($\{\varphi_n(x)\}$, is a CONS)

$$
\Psi(\mathbf{r}, t) = \sum_n c_n\varphi_n(\mathbf{r})e^{-i\omega_n t}; \quad \sum_n |c_n|^2 = 1. \tag{V.291}
$$

We have to calculate

$$
\langle H\rangle = \int \Psi^* H\Psi dV \text{ and } \langle H^2\rangle = \int \Psi^* H^2\Psi dV. \tag{V.292}
$$

Inserting gives

$$\langle H \rangle = \sum_{nm} \int c_n^* \varphi_n^* e^{i\omega_n t} c_m E_m \varphi_m e^{-i\omega_m t} dV = \sum_{nm} c_n^* e^{i\omega_n t} c_m E_m e^{-i\omega_m t} \int \varphi_n^* \varphi_m dV$$

$$\langle H^2 \rangle = \sum_{nm} \int c_n^* \varphi_n^* e^{i\omega_n t} c_m E_m^2 \varphi_m e^{-i\omega_m t} dV = \sum_{nm} c_n^* e^{i\omega_n t} c_m E_m^2 e^{-i\omega_m t} \int \varphi_n^* \varphi_m dV.$$

(V.293)

Because of the orthonormality of the eigenfunctions ($\int \varphi_n^* \varphi_m dV = \delta_{nm}$), it follows that

$$\langle H \rangle = \sum_{n} |c_n|^2 E_n ; \quad \langle H^2 \rangle = \sum_{n} |c_n|^2 E_n^2 \qquad (V.294)$$

and therefore

$$(\Delta H)^2 = \langle H^2 \rangle - \langle H \rangle^2 = \sum_{n} |c_n|^2 E_n^2 - \left(\sum_{n} |c_n|^2 E_n \right)^2 ; \quad \sum_{n} |c_n|^2 = 1.$$

(V.295)

We rewrite this as

$$(\Delta H)^2 = \sum_{n} |c_n|^2 E_n^2 \cdot \sum_{m} |c_m|^2 - \sum_{n} |c_n|^2 E_n \cdot \sum_{m} |c_m|^2 E_m$$

$$= \sum_{nm} |c_n|^2 |c_m|^2 E_n^2 - \sum_{nm} |c_n|^2 |c_m|^2 E_n E_m = \sum_{nm} |c_n|^2 |c_m|^2 E_n (E_n - E_m) .$$

(V.296)

The last double sum of course yields the same result if we interchange n and m. Using this fact, we write

$$(\Delta H)^2 = \tfrac{1}{2} \sum_{nm} |c_n|^2 |c_m|^2 E_n (E_n - E_m) + \tfrac{1}{2} \sum_{nm} |c_m|^2 |c_n|^2 E_m (E_m - E_n)$$

$$= \tfrac{1}{2} \sum_{nm} |c_n|^2 |c_m|^2 (E_n - E_m)^2 .$$

(V.297)

We see that all terms in the sum are non-negative. Hence, ΔH is zero iff each term $|c_n|^2 |c_m|^2 (E_n - E_m)^2$ vanishes. Since the terms for $n = m$ or $E_n = E_m$ are zero anyway, each of the terms $|c_n|^2 |c_m|^2$ with $n \neq m$ has to vanish separately in order to arrive at $\Delta H = 0$. This is the case iff (a) all c_n are zero (trivial solution, physically uninteresting); or (b) all c_n are zero except one, say c_N. But in this case, the state Ψ is an eigenstate of H with the energy E_N:

$$\Psi(x, t) = c_N \varphi_N(x) e^{-i\omega_N t} . \qquad (V.298)$$

This property of the variance exists for all Hermitian operators; it disappears iff the state is an eigenstate of the operator for which the mean value is evaluated (see Appendix O, Vol. 1).

V.10   Exercises, Chap. 11

1. Show that the equation

$$\sum_i c_i A_{ji} = ac_j \qquad \text{(V.299)}$$

may be written in the matrix representation as

$$\mathbb{A}\mathbf{c} = a\mathbf{c} \qquad \text{(V.300)}$$

with the matrix $\{A_{ji}\} \equiv \mathbb{A}$ and the column vector $\mathbf{c}$. Is the equation also valid for non-square matrices?

2. Do the functions of one variable which are continuous in the interval $[0, 1]$ form a Hilbert space?
Solution: No, since sequences of continuous functions may lead to discontinuous functions; example $\lim_{n\to\infty} x^n$ in the interval $[0, 1]$. Thus, the criterion of completeness is not satisfied.

3. The space $l^{(2)}$ consists of all vectors $|\varphi\rangle$ with infinitely many components (coordinates) $c_1, c_2, \ldots$, such that

$$\||\varphi\rangle\|^2 = \sum_n |c_n|^2 < \infty. \qquad \text{(V.301)}$$

Show that also the linear combination of two vectors $|\varphi\rangle$ and $|\chi\rangle$ belongs to this space, and that the scalar product $\langle \varphi | \chi \rangle$ is defined.
Solution: $|\chi\rangle$ has the coordinates $d_1, d_2, \ldots$. Then we have

$$\|\alpha |\varphi\rangle + \beta |\chi\rangle\|^2 = \sum_n |\alpha c_n + \beta d_n|^2 \le 2 \sum_n \left(|\alpha|^2 |c_n|^2 + |\beta|^2 |d_n|^2\right) < \infty. \qquad \text{(V.302)}$$

The inequality is satisfied due to

$$|a + b|^2 = 2\left(|a|^2 + |b|^2\right) - |a - b|^2. \qquad \text{(V.303)}$$

For the scalar product, we obtain using the Schwarz inequality $|\langle \varphi | \chi \rangle| \le \||\varphi\rangle\| \cdot \||\chi\rangle\|$ (see Appendix G, Vol. 1):

$$|\langle \varphi | \chi \rangle| = \left|\sum_n c_n^* d_n\right| \le \sqrt{\sum_n |c_n|^2} \cdot \sqrt{\sum_n |d_n|^2} < \infty. \qquad \text{(V.304)}$$

4. Given the operator A and the equation

$$i \frac{d}{dt} |\psi\rangle = A |\psi\rangle, \qquad \text{(V.305)}$$

which condition must A fulfill, so that the norm of $|\psi\rangle$ is conserved?
Solution: With $\frac{d}{dt}|\psi\rangle = |\dot\psi\rangle$ and $i\langle\dot\psi| = -\langle\psi|A^\dagger$, it holds that

$$i\frac{d}{dt}\langle\psi\,|\psi\rangle = i\langle\dot\psi\,|\psi\rangle + i\langle\psi\,|\dot\psi\rangle = -\langle\psi|A^\dagger|\psi\rangle + \langle\psi|A|\psi\rangle. \quad (V.306)$$

Since this equation must hold for all allowed $|\psi\rangle$, the conservation of the norm implies $-A^\dagger + A = 0$; hence, the operator A has to be Hermitian. In other words: a linear differential equation of the first order *must* have the structure $i\frac{d}{dt}|\psi\rangle = A|\psi\rangle$ with $A^\dagger = A$ in order for the norm to be conserved.

5. Given the operator A. Derive the equation

$$i\hbar\frac{d}{dt}\langle A\rangle = \langle[A,H]\rangle + i\hbar\langle\dot A\rangle \quad (V.307)$$

in the bra-ket formalism.
Solution: Start from

$$i\hbar\frac{d}{dt}\langle A\rangle = i\hbar\frac{d}{dt}\langle\psi|A|\psi\rangle = i\hbar\langle\dot\psi|A|\psi\rangle + i\hbar\langle\psi|\dot A|\psi\rangle + i\hbar\langle\psi|A|\dot\psi\rangle. \quad (V.308)$$

This leads with $i\hbar|\dot\psi\rangle = H|\psi\rangle$ to

$$i\hbar\frac{d}{dt}\langle A\rangle = -\langle\psi|H^\dagger A|\psi\rangle + \langle\psi|AH|\psi\rangle + i\hbar\langle\psi|\dot A|\psi\rangle, \quad (V.309)$$

and, with $H = H^\dagger$, it follows finally

$$i\hbar\frac{d}{dt}\langle A\rangle = \langle\psi|AH - HA|\psi\rangle + i\hbar\langle\psi|\dot A|\psi\rangle = \langle[A,H]\rangle + i\hbar\langle\dot A\rangle. \quad (V.310)$$

6. Given the Hamiltonian H with discrete and non-degenerate spectrum, (a) in the formulation with space variables and (b) as abstract operator; what is in each case the matrix representation of the time-dependent SEq?

   (a) Solution: The eigenvalue equation or stationary SEq reads

$$H\varphi_n(x) = E_n\varphi_n(x), \quad (V.311)$$

   and the time-dependent SEq is

$$i\hbar\partial_t\psi(x,t) = H\psi(x,t). \quad (V.312)$$

   Since $\{\varphi_n(x)\}$ is a CONS, we can write $\psi(x,t)$ as

$$\psi(x,t) = \sum_n c_n(t)\varphi_n(x). \quad (V.313)$$

We insert this expression in the SEq and obtain

$$i\hbar \sum_n \partial_t c_n(t)\,\varphi_n(x) = \sum_n c_n(t)\,H\varphi_n(x). \tag{V.314}$$

Multiplication by $\varphi_m^*(x)$ and integration leads, due to the orthonormality of $\{\varphi_n(x)\}$, to

$$i\hbar \sum_n \partial_t c_n(t) \int \varphi_m^*(x)\,\varphi_n(x)\,dx = \sum_n c_n(t) \int \varphi_m^*(x)\,H\varphi_n(x)\,dx$$

$$i\hbar\partial_t c_m(t) = \sum_n c_n(t) \int \varphi_m^*(x)\,H\varphi_n(x)\,dx.$$

$$\tag{V.315}$$

Above, we have seen that $\int \varphi_m^*(x)\,H\varphi_n(x)\,dx = E_m\delta_{nm}$. With this, we obtain for the time-dependent SEq:

$$i\hbar\partial_t c_m(t) = E_m c_m(t), \tag{V.316}$$

or, in matrix form

$$i\hbar\partial_t \mathbf{c}(t) = H_{\text{Matrix}}\mathbf{c}(t) \tag{V.317}$$

with the column vector and the matrix

$$\mathbf{c} = \begin{pmatrix} c_1 \\ c_2 \\ \vdots \end{pmatrix}; \quad H_{\text{Matrix}} = \begin{pmatrix} E_1 & 0 & \cdots \\ 0 & E_2 & \cdots \\ \vdots & \vdots & \ddots \end{pmatrix}. \tag{V.318}$$

The solution of the ordinary differential equation (V.316) reads

$$c_m(t) = c_m(0)\,e^{-iE_m t/\hbar}, \tag{V.319}$$

and the solution $\psi(x,t)$ obtains its familiar form:

$$\psi(x,t) = \sum_n c_m(0)\,\varphi_n(x)\,e^{-iE_m t/\hbar}. \tag{V.320}$$

(b) Solution: For a change, we calculate by a slightly different route for the abstract case. We start from

$$i\hbar\partial_t |\psi\rangle = H|\psi\rangle \tag{V.321}$$

where the stationary SEq is given by

$$H|\varphi_n\rangle = E_n|\varphi_n\rangle. \tag{V.322}$$

Due to $\partial_t |\varphi_n\rangle = 0$, it follows that

$$i\hbar\partial_t \langle\varphi_n|\psi\rangle = \langle\varphi_n| H |\psi\rangle = \sum_m \langle\varphi_n| H |\varphi_m\rangle \langle\varphi_m|\psi\rangle, \qquad (V.323)$$

and with

$$c_n = \langle\varphi_n|\psi\rangle \text{ and } H_{nm} = \langle\varphi_n| H |\varphi_m\rangle = E_m\delta_{nm} \qquad (V.324)$$

we find

$$i\hbar\partial_t c_n = \sum_m H_{nm}c_m, \qquad (V.325)$$

or, written compactly using the column vector $\mathbf{c}$ and the matrix H_{Matrix}:

$$i\hbar\partial_t\mathbf{c} = H_{\text{Matrix}}\mathbf{c}. \qquad (V.326)$$

V.11 Exercises, Chap. 12

1. Given an eigenstate $|k\rangle$ of the momentum operator; how is this state described in the position representation?
2. Show by using $\langle x| k\rangle = \frac{1}{\sqrt{2\pi}}e^{ikx}$ that the improper vectors $|k\rangle$ form a CONS.
3. Given an improper vector $|\varphi_\lambda\rangle$, what is the associated eigendifferential $\left|\varphi_{\lambda,\Delta\lambda}\right\rangle$?
4. Given the state $|k\rangle$ with the sharply-defined momentum k; we have $\langle x| k\rangle = \frac{1}{\sqrt{2\pi}}e^{ikx}$.

   (a) What is the (abstract) eigendifferential?
       Solution: With $k_n = n\Delta k$ (fixed screening), it follows that

$$|k_n, \Delta k\rangle = \frac{1}{\sqrt{\Delta k}} \int_{k_n}^{k_n+\Delta k} |k'\rangle\, dk'. \qquad (V.327)$$

   (b) How is the eigendifferential expressed in the position representation?
       Solution: Multiplication by $\langle x|$ gives

$$\langle x| k_n, \Delta k\rangle = \frac{1}{\sqrt{2\pi}}\frac{1}{\sqrt{\Delta k}} \int_{k_n}^{k_n+\Delta k} e^{ik'x}dk'$$
$$= e^{ik_n x}\frac{e^{i\Delta k\cdot x} - 1}{\sqrt{2\pi\Delta k}ix} = e^{i(k_n+\Delta k/2)x}\frac{2\sin(\Delta k\cdot x/2)}{\sqrt{2\pi\Delta k}\cdot x}. \qquad (V.328)$$

(c) Show that the eigendifferentials of (b) are orthonormal.
Solution: For the scalar product, we have

$$\langle k_m, \Delta k | k_n, \Delta k \rangle$$

$$= \int_{-\infty}^{\infty} e^{-ik_m x - i\Delta k \cdot x/2} \frac{2 \sin(\Delta k \cdot x/2)}{\sqrt{2\pi \Delta k x}} \cdot e^{ik_n x + i\Delta k \cdot x/2} \frac{2 \sin(\Delta k \cdot x/2)}{\sqrt{2\pi \Delta k x}} dx$$

$$= \frac{2}{\pi \Delta k} \int_{-\infty}^{\infty} \frac{\sin^2(\Delta k \cdot x/2)}{x^2} \cdot e^{i(k_n - k_m)x} dx.$$

$$(V.329)$$

Insertion and substitution of $y = x \Delta k$ leads to

$$\langle k_m, \Delta k | k_n, \Delta k \rangle = \frac{2}{\pi} \int_{-\infty}^{\infty} \frac{\sin^2(y/2)}{y^2} \cdot e^{i(n-m)y} dy = \delta_{k_n, k_m} = \delta_{n,m}.$$

$$(V.330)$$

On the last integral: The term with $\sin(n - m)y$ vanishes due to the point
symmetry of the sine function. Regarding the cosine, a formula tabulation
or your own calculation gives:

$$\int_0^{\infty} \frac{\sin^2 az}{z^2} \cos bz \, dz = \begin{cases} \frac{\pi}{2}(a - \frac{b}{2}) & b < 2a \\ 0 & b \geq 2a. \end{cases}$$

$$(V.331)$$

5. Given the SEq in the abstract formulation

$$i\hbar \frac{d}{dt} |\psi\rangle = H |\psi\rangle,$$

$$(V.332)$$

(a) Formulate the equation in the position representation and in the momentum
representation.
Solution: We have

$$i\hbar \frac{d}{dt} \langle x | \psi \rangle = \int \langle x | H | x' \rangle \langle x' | \psi \rangle \, dx'$$

$$i\hbar \frac{d}{dt} \langle k | \psi \rangle = \int \langle k | H | k' \rangle \langle k' | \psi \rangle \, dk'.$$

$$(V.333)$$

Since H is diagonal in the position representation, we obtain $i\hbar \frac{d}{dt} \langle x | \psi \rangle = \langle x | H | x \rangle \langle x' | \psi \rangle$ or $i\hbar \frac{d}{dt} \psi(x) = H(x)\psi(x)$.
(b) How can one calculate the matrix element $\langle k | H | k' \rangle$, if H is known in the
position representation?
Solution: We insert the one (identity operator) two times:

$$\langle k | H | k' \rangle = \int \langle k | x \rangle \langle x | H | x' \rangle \langle x' | k' \rangle \, dx \, dx'.$$

$$(V.334)$$

Since H is diagonal in the position representation, we obtain with $\langle x\,|k\rangle = \frac{1}{\sqrt{2\pi}}e^{ikx}$:

$$\langle k|\,H\,|k'\rangle = \frac{1}{2\pi}\int e^{-ikx}\,H\,(x)\,e^{ik'x}\,\mathrm{d}x. \qquad (V.335)$$

6. Given a CONS $\{|\varphi_n\rangle\}$; formulate the projection operator

$$P_1 = |\varphi_1\rangle\,\langle\varphi_1| \qquad (V.336)$$

in the position representation.

Solution: For a state $|\Psi\rangle$, in the bra-ket notation we have:

$$P_1\,|\Psi\rangle = |\varphi_1\rangle\,\langle\varphi_1\,|\Psi\rangle = c\,|\varphi_1\rangle\,;\ \ c = \langle\varphi_1\,|\Psi\rangle\,. \qquad (V.337)$$

In the position representation, it follows that

$$(P_1\Psi)\,(r) = c\varphi_1\,(r) = \varphi_1\,(r)\cdot\int \mathrm{d}^3r'\ \varphi_1^*\left(r'\right)\Psi\left(r'\right). \qquad (V.338)$$

We see explicitly that the operator is not diagonal in the position representation. In detail:

$$P_1|\Psi\rangle = |\varphi_1\rangle\langle\varphi_1|\Psi\rangle \leftrightarrow \langle r|P_1|\Psi\rangle = \langle r|\varphi_1\rangle\cdot\langle\varphi_1|1|\Psi\rangle$$

$$= \langle r|\varphi_1\rangle\cdot\int \mathrm{d}^3r'\ \langle\varphi_1|r'\rangle\langle r'|\Psi\rangle \qquad (V.339)$$

$$\text{or}\ \int \mathrm{d}^3r'\ \langle r|P_1|r'\rangle\langle r'|\Psi\rangle = \varphi_1(r)\cdot\int \mathrm{d}^3r'\ \varphi_1^*(r')\Psi(r'). \qquad (V.340)$$

7. A and B are self-adjoint operators with $[A, B] = i\hbar$, and $|a\rangle$ is an eigenvector of A for the eigenvalue a. Then we have

$$\langle a\,|[A, B]|\,a\rangle = \langle a\,|AB - BA|\,a\rangle = (a - a)\,\langle a\,|B|\,a\rangle = 0. \qquad (V.341)$$

On the other hand, we also have:

$$\langle a\,|[A, B]|\,a\rangle = \langle a\,|i\hbar|\,a\rangle = i\hbar \neq 0. \qquad (V.342)$$

Question: where is the flaw in this argument?

Solution: One can show (see Appendix I, Vol. 1) that at least one of the two operators must be *unbounded* if we have $[A, B] = i\hbar$ (which is not apparent at first glance, and therefore—and because this term is introduced only in Chap. 13— the exercise is a bit unfair). This means that the eigenvectors are not normalizable and the corresponding scalar products do not exist. This is an example of how one has to be somewhat more cautious when dealing with unbounded operators and continuous spectra.

V.12   Exercises, Chap. 13

1. Let A be a linear and B an anti-linear operator; $|\varphi\rangle$ is a state. Compute or simplify $A\,(i\,|\varphi\rangle)$ and $B\,(i\,|\varphi\rangle)$.

2. Show that the complex conjugation $\mathcal{K}$ is an anti-linear operator.
   Solution:
   $$\mathcal{K}i\,|\varphi\rangle = -i\mathcal{K}\,|\varphi\rangle \text{ or } \mathcal{K}i = -i\mathcal{K}. \tag{V.343}$$

3. Show that the commutator $C = [A,\,B]$ of two Hermitian operators A and B is anti-Hermitian.
   Solution: We have

   $$C^{\dagger} = (AB - BA)^{\dagger} = BA - AB = -C. \tag{V.344}$$

4. The Hermitian operators A and B fulfill $[A,\,B] \neq 0$. Consider the operator $Q = c\,[A,\,B]$. For which c is Q a Hermitian operator?

5. Consider the operator $Q = AB$, where A and B are Hermitian matrices. Under what conditions is Q a Hermitian operator?

6. Show in the bra-ket representation that:

   (a) Hermitian operators have real eigenvalues.
   (b) The eigenfunctions of Hermitian operators are pairwise orthogonal (assuming the spectrum is not degenerate).

7. Show that the mean value of a Hermitian operator A is real, and the mean value of an anti-Hermitian operator B is imaginary.
   Solution:

   $$\langle A \rangle^{\dagger} = \langle\psi|\,A\,|\psi\rangle^{\dagger} = \langle\psi|\,A^{\dagger}\,|\psi\rangle = \langle\psi|\,A\,|\psi\rangle = \langle A \rangle \rightarrow \langle A \rangle \in \mathbb{R}$$
   $$\langle B \rangle^{\dagger} = \langle\psi|\,B\,|\psi\rangle^{\dagger} = \langle\psi|\,B^{\dagger}\,|\psi\rangle = -\langle\psi|\,B\,|\psi\rangle = -\langle B \rangle \rightarrow \langle B \rangle \in \mathbb{I}.$$
   $$\tag{V.345}$$

8. What is the quantum-mechanical operator for the classical term $\mathbf{p} \times \mathbf{l}$?
   Solution: We have here a product of operators and have to check first whether the translation of the classically identical terms $\mathbf{p} \times \mathbf{l}$ and $-\mathbf{l} \times \mathbf{p}$ into quantum mechanics yields the same result. If not, we have to symmetrize, as shown in Chap. 3. We consider the x components. We have

   $$(\mathbf{p} \times \mathbf{l})_x = p_y l_z - p_z l_y = p_y \left(x p_y - y p_x\right) - p_z \left(z p_x - x p_z\right)$$
   $$-(\mathbf{l} \times \mathbf{p})_x = -l_y p_z + l_z p_y = -\left(z p_x - x p_z\right) p_z + \left(x p_y - y p_x\right) p_y. \tag{V.346}$$

With $[x, p_x] = i\hbar$ and $\left[x, p_y\right] = \left[x, p_z\right] = 0$ etc. in mind, evaluation of the brackets leads to

$$(\mathbf{p} \times \mathbf{l})_x = p_y x p_y - p_y y p_x - p_z z p_x + p_z x p_z = x p_y^2 + x p_z^2 - p_y y p_x - p_z z p_x$$

$$= x \left(p_y^2 + p_z^2\right) + (i\hbar - y p_y) p_x + (i\hbar - z p_z) p_x$$

$$-(\mathbf{l} \times \mathbf{p})_x = -z p_x p_z + x p_z p_z + x p_y p_y - y p_x p_y = x \left(p_y^2 + p_z^2\right) - p_x z p_z - p_x y p_y.$$

$$\text{(V.347)}$$

We see that $\mathbf{p} \times \mathbf{l} \neq (\mathbf{p} \times \mathbf{l})^\dagger = -\mathbf{l} \times \mathbf{p}$ holds true; consequently, we have to symmetrize and obtain for the quantum-mechanical operator

$$(\mathbf{p} \times \mathbf{l})_{\text{classical},x} \rightarrow (\mathbf{p} \times \mathbf{l})_{\text{quantum},x} = \frac{(\mathbf{p} \times \mathbf{l})_x - (\mathbf{l} \times \mathbf{p})_x}{2}$$
$$= x \left(p_y^2 + p_z^2\right) + (i\hbar - z p_z - y p_y) p_x \qquad \text{(V.348)}$$

plus cyclic permutations for the two other components. This operator is obviously Hermitian.

One can transform the result into a more pleasing expression, see Appendix G, Vol. 2, 'Lenz vector'. Moreover, we see that

$$\mathbf{p} \times \mathbf{l} + \mathbf{l} \times \mathbf{p} = 2i\hbar\mathbf{p} \qquad \text{(V.349)}$$

so that we can write

$$(\mathbf{p} \times \mathbf{l})_{\text{classical}} \rightarrow \frac{\mathbf{p} \times \mathbf{l} + \mathbf{p} \times \mathbf{l} - 2i\hbar\mathbf{p}}{2} = \mathbf{p} \times \mathbf{l} - i\hbar\mathbf{p}. \qquad \text{(V.350)}$$

9. Calculate the mean value of σ_z for the normalized state $\begin{pmatrix} a \\ b \end{pmatrix}$.

Solution: It is

$$\langle \sigma_z \rangle = \left(a^* \; b^*\right) \begin{pmatrix} 1 & 0 \\ 0 & -1 \end{pmatrix} \begin{pmatrix} a \\ b \end{pmatrix} = |a|^2 - |b|^2 = 2|a|^2 - 1. \qquad \text{(V.351)}$$

10. Given the time-independent Hamiltonian H; what is the associated time evolution operator $U(t)$?

11. Let U be the operator $U = e^{iA}$, where A is a Hermitian operator. Show that U is unitary.

12. What are the eigenvalues that a unitary operator can have?

13. Show that the time evolution operator $e^{-i\frac{Ht}{\hbar}}$ is unitary.

14. Show that scalar products, matrix elements, eigenvalues and expectation values are invariant under unitary transformations.

Solution:

Scalar products and matrix elements:

$$\langle \Psi' | \Phi' \rangle = \langle \Psi | U^\dagger U | \Phi \rangle = \langle \Psi | \Phi \rangle$$
$$\langle \Psi' | A' | \Phi' \rangle = \langle \Psi | U^\dagger U A U^\dagger U | \Phi \rangle = \langle \Psi | A | \Phi \rangle. \tag{V.352}$$

Expectation value (see matrix element):

$$\langle A' \rangle = \langle \Psi' | A' | \Psi' \rangle = \langle \Psi | U^\dagger U A U^\dagger U | \Psi \rangle = \langle \Psi | A | \Psi \rangle = \langle A \rangle. \tag{V.353}$$

Eigenvalue: With

$$A | a_n \rangle = a_n | a_n \rangle \tag{V.354}$$

it follows that:

$$U A | a_n \rangle = \begin{cases} U A U^\dagger U | a_n \rangle = A' | A_n' \rangle \\ U a_n | a_n \rangle = a_n | A_n' \rangle, \end{cases} \tag{V.355}$$

or, compactly:

$$A' | A_n' \rangle = a_n | A_n' \rangle. \tag{V.356}$$

15. P_1 and P_2 are projection operators. Under which conditions are $P = P_1 + P_2$ and $P = P_1 P_2$ projection operators?

16. Formulate the matrix representation of the operator $P = |e_1\rangle \langle e_1|$ in $\mathbb{R}^3$.
    Solution:

$$P = |e_1\rangle \langle e_1| \cong \begin{pmatrix} 1 \\ 0 \\ 0 \end{pmatrix} (1\ 0\ 0) = \begin{pmatrix} 1\ 0\ 0 \\ 0\ 0\ 0 \\ 0\ 0\ 0 \end{pmatrix}. \tag{V.357}$$

17. What is the general definition of a projection operator?

18. Given the CONS $\{|\varphi_n\rangle\}$; for which c_n is the operator $A = \sum c_n |\varphi_n\rangle \langle \varphi_n|$ a projection operator?

19. Which eigenvalues can a projection operator have?

20. Given the CONS $\{|\varphi_n\rangle\}$ in a Hilbert space of dimension N. Consider the operator

$$P = \sum_{n \le N'} |\varphi_n\rangle \langle \varphi_n| \tag{V.358}$$

with $N' \le N$. Show that P is a projection operator.

21 Given the Operator A with a degenerate spectrum:

$$A |\varphi_{n,r}\rangle = a_n |\varphi_{n,r}\rangle;\ r = 1, \ldots g_n. \tag{V.359}$$

(a) Formulate the projection operator onto the states with subscript n?
    Solution: It is

$$P_n = \sum_{r=1}^{g_n} |\varphi_{n,r}\rangle \langle \varphi_{n,r}|;\ \sum_n P_n = 1. \tag{V.360}$$

(b) Formulate the spectral representation of A.

Solution: It is

$$A = A \cdot 1 = A \sum_n P_n = A \sum_n \sum_r |\varphi_{n,r}\rangle\langle\varphi_{n,r}| = \sum_{n,r} |\varphi_{n,r}\rangle a_n \langle\varphi_{n,r}|.$$
(V.361)

22. Given the operators $A = |\varphi\rangle\langle\varphi|$ and $B = |\psi\rangle\langle\psi|$. Let $\langle\varphi|\psi\rangle = \alpha \in \mathbb{C}, \alpha \neq 0$. For which α is the operator $C = AB$ a projection operator?

Solution: C must be idempotent, i.e. $C^2 = C$. This means

$$ABAB = AB \text{ or } |\varphi\rangle\langle\varphi|\psi\rangle\langle\psi|\varphi\rangle\langle\varphi|\psi\rangle\langle\psi| = |\varphi\rangle\langle\varphi|\psi\rangle\langle\psi|$$
$$\rightarrow |\varphi\rangle \alpha\alpha^*\alpha \langle\psi| = |\varphi\rangle \alpha \langle\psi|$$
(V.362)

or in short form,

$$\alpha\alpha^* = 1 \rightarrow \alpha = e^{i\delta}, \delta \in \mathbb{R}.$$
(V.363)

The Hermiticity of C means that

$$C = AB = C^\dagger = B^\dagger A^\dagger = BA$$
(V.364)

and this leads to

$$|\varphi\rangle\langle\varphi|\psi\rangle\langle\psi| = |\psi\rangle\langle\psi|\varphi\rangle\langle\varphi| \text{ or } |\varphi\rangle\alpha\langle\psi| = |\psi\rangle\alpha^*\langle\varphi|.$$
(V.365)

Multiplication from the left by $\langle\varphi|$ yields

$$|\varphi\rangle\alpha\alpha^* = |\psi\rangle\alpha^*\langle\varphi|\varphi\rangle$$
(V.366)

or

$$|\psi\rangle = \frac{e^{i\delta}}{\langle\varphi|\varphi\rangle}|\varphi\rangle.$$
(V.367)

Hence, $|\psi\rangle$ and $|\varphi\rangle$ must be collinear, and $\langle\psi|\psi\rangle\langle\varphi|\varphi\rangle = 1$.

23. Given the operator $Q = B^\dagger B$, where B is unitary. How can Q be more simply written?

24. Given the operator $Q = B^\dagger B$, where B is not unitary. Show that the eigenvalues of Q are real and that they are not negative.

25. Given the operator $A = \beta|\varphi\rangle\langle\psi|$. Let $\langle\psi|\varphi\rangle = \alpha \neq 0$; α and β are complex constants. The states $|\varphi\rangle$ and $|\psi\rangle$ are normalized. Which conditions must $|\varphi\rangle$, $|\psi\rangle$, α and β fulfill to ensure that A is a Hermitian, a unitary, or a projection operator?

Solution:

(a) If A is a Hermitian operator, it must hold that

$$A = \beta \, |\varphi\rangle \, \langle\psi| = A^\dagger = \beta^* \, |\psi\rangle \, \langle\varphi| . \tag{V.368}$$

We multiply from the right by $|\varphi\rangle$ and obtain

$$\beta \, |\varphi\rangle \, \langle\psi| \, \varphi\rangle = \beta^* \, |\psi\rangle \, \langle\varphi| \, \varphi\rangle \Rightarrow \beta \, |\varphi\rangle \, \alpha = \beta^* \, |\psi\rangle \Rightarrow |\psi\rangle = \frac{\alpha\beta}{\beta^*} \, |\varphi\rangle . \tag{V.369}$$

Since the states are normalized, it follows that $|\alpha| = 1$. $|\psi\rangle$ and $|\varphi\rangle$ differ from each other only by a phase factor.

(b) If A is a unitary operator, it must hold that $A^\dagger A = 1$, i.e.

$$A^\dagger A = \beta^* \, |\psi\rangle \, \langle\varphi| \, \beta \, |\varphi\rangle \, \langle\psi| = |\beta|^2 \, |\psi\rangle \, \langle\psi| = 1. \tag{V.370}$$

This is satisfied for $|\beta| = 1$ and $|\psi\rangle \, \langle\psi| = 1$. The requirement $AA^\dagger = 1$ leads analogously to $|\varphi\rangle \, \langle\varphi| = 1$. This means that $|\alpha| = 1$; also here, $|\psi\rangle$ and $|\varphi\rangle$ must agree up to a phase factor and $|\varphi\rangle \, \langle\varphi| = 1$.

(c) If A is a projector, it must hold that $A^2 = A$, i.e.

$$A^2 = \beta^2 \, |\varphi\rangle \, \alpha \, \langle\psi| = A = \beta \, |\varphi\rangle \, \langle\psi| . \tag{V.371}$$

This is satisfied for $\alpha\beta = 1$ and $|\psi\rangle = \frac{\alpha\beta}{\beta^*} \, |\varphi\rangle$. Hence, with the result of part (a), it follows that $|\alpha| = 1$ and $|\beta| = 1$.

26. Given a CONS $\{|\varphi_n\rangle\}$ and an operator

$$A = \sum_{n,m} c_{nm} \, |\varphi_n\rangle \, \langle\varphi_m| ; \; c_{nm} \in \mathbb{C}. \tag{V.372}$$

How must the coefficients c_{nm} be chosen in order that A be a Hermitian, a unitary, or a projection operator?

Solution:

(a) If A is Hermitian, it must apply that

$$A = \sum_{n,m} c_{nm} \, |\varphi_n\rangle \, \langle\varphi_m| = A^\dagger = \sum_{n,m} c^*_{nm} \, |\varphi_m\rangle \, \langle\varphi_n| = \sum_{n,m} c^*_{mn} \, |\varphi_n\rangle \, \langle\varphi_m| . \tag{V.373}$$

Here, we interchanged the summation indices in the last step. The comparison shows immediately that

$$c_{nm} = c^*_{mn}. \tag{V.374}$$

If we represent (c_{nm}) as a matrix C, the last equation means none other than the familiar adjoint: commutation of columns and rows plus complex conjugation.

(b) If A is unitary, it must apply that:

$$AA^\dagger = \sum_{n,m} c_{nm} |\varphi_n\rangle \langle\varphi_m| \sum_{n',m'} c^*_{n'm'} |\varphi_{m'}\rangle \langle\varphi_{n'}|$$
$$= \sum_{n,m,n',m'} c_{nm} c^*_{n'm'} |\varphi_n\rangle \delta_{m\dot{m}'} \langle\varphi_{n'}| = 1. \tag{V.375}$$

It follows that

$$1 = \sum_{n,m,n'} c_{nm} c^*_{n'm} |\varphi_n\rangle \langle\varphi_{n'}| = \sum_{n,n'} |\varphi_n\rangle \langle\varphi_{n'}| \sum_m c_{nm} c^*_{n'm}. \tag{V.376}$$

Since $\{|\varphi_n\rangle\}$ is a CONS, we have $1 = \sum_n |\varphi_n\rangle \langle\varphi_n|$. Hence, we must have

$$\sum_m c^*_{n'm} c_{nm} = \delta_{nn'}. \tag{V.377}$$

Thus, the different rows of the matrix $C = (c_{nm})$ have to be normalized and pairwise orthogonal.

(c) If A is a projector, it must apply that

$$A^2 = \sum_{n,m} c_{nm} |\varphi_n\rangle \langle\varphi_m| \sum_{n',m'} c_{n'm'} |\varphi_{n'}\rangle \langle\varphi_{m'}| = A = \sum_{n,m} c_{nm} |\varphi_n\rangle \langle\varphi_m|, \tag{V.378}$$

and from part (a), $c_{nm} = c^*_{mn}$. It follows that

$$\sum_{n,m} c_{nm} |\varphi_n\rangle \langle\varphi_m| = \sum_{n,m,n',m'} c_{nm} c_{n'm'} |\varphi_n\rangle \delta_{mn'} \langle\varphi_{m'}|$$
$$= \sum_{n,m,m'} c_{nm} c_{mm'} |\varphi_n\rangle \langle\varphi_{m'}| \tag{V.379}$$

or

$$\sum_{n,m} c_{nm} |\varphi_n\rangle \langle\varphi_m| = \sum_{n,m} |\varphi_n\rangle \langle\varphi_m| \sum_l c_{nl} c_{lm}. \tag{V.380}$$

Hence, we must have

$$\sum_l c_{nl} c_{lm} = c_{nm}. \tag{V.381}$$

For the matrix $C = (c_{nm})$, this means $C^2 = C$.

27. A CONS $\{|\varphi_n\rangle, n = 1, 2, \ldots, N\}$ spans a vector space $\mathcal{V}$.

(a) Show that each operator A acting in $\mathcal{V}$ can be represented as

$$A = \sum_{n,m} c_{nm} |\varphi_n\rangle \langle\varphi_m|. \tag{V.382}$$

Solution: If we let act A on a state of the CONS, the result has to be representable as a superposition of the $|\varphi_n\rangle$ (due to the completeness of the system). Thus we have

$$A \left| \varphi_m \right\rangle = \sum_n c_{nm} \left| \varphi_n \right\rangle .$$ (V.383)

Multiplication from the right by $\langle \varphi_m |$ and summation over m gives (again due to completeness) the desired result.

(b) Consider the special case ($N = 3$):

$$A \left| \varphi_1 \right\rangle = - \left| \varphi_2 \right\rangle ; \quad A \left| \varphi_2 \right\rangle = - \left| \varphi_3 \right\rangle ; \quad A \left| \varphi_3 \right\rangle = - \left| \varphi_1 \right\rangle + \left| \varphi_2 \right\rangle .$$ (V.384)

What is the operator A? (Determine the coefficients c_{nm}, i.e. formulate A as a linear combination of products $\left| \varphi_i \right\rangle \left\langle \varphi_j \right|$).
Solution: It follows that

$$A = - \left| \varphi_2 \right\rangle \left\langle \varphi_1 \right| - \left| \varphi_3 \right\rangle \left\langle \varphi_2 \right| - \left(\left| \varphi_1 \right\rangle - \left| \varphi_2 \right\rangle \right) \left\langle \varphi_3 \right| .$$ (V.385)

28. How is the generalized Heisenberg uncertainty relation formulated for each of the pairs $(x, l_x) , \left(x, l_y \right) , (x, l_z)$?
Solution: We have

$$x l_x - l_x x = \frac{\hbar}{i} \left[x \left(y \partial_z - z \partial_y \right) - \left(y \partial_z - z \partial_y \right) x \right] = 0$$ (V.386)

as well as

$$x l_y - l_y x = \frac{\hbar}{i} \left[x \left(z \partial_x - x \partial_z \right) - \left(z \partial_x - x \partial_z \right) x \right] = \frac{\hbar}{i} \left[x z \partial_x - z \partial_x x \right]$$
$$= \frac{\hbar}{i} \left[x z \partial_x - z x \partial_x - z \right] = i \hbar z ,$$ (V.387)

and analogously

$$x l_z - l_z x = i \hbar y .$$ (V.388)

Hence, it follows due to

$$\Delta A \cdot \Delta B \geq \frac{1}{2} \left| \langle [A, B] \rangle \right|$$ (V.389)

immediately that

$$\Delta x \cdot \Delta l_y = 0$$
$$\Delta x \cdot \Delta l_y \geq \frac{1}{2} \left| \langle [x, l_y] \rangle \right| = \frac{\hbar}{2} \left| \langle z \rangle \right|$$
$$\Delta x \cdot \Delta l_z \geq \frac{1}{2} \left| \langle [x, l_z] \rangle \right| = \frac{\hbar}{2} \left| \langle y \rangle \right| .$$ (V.390)

29. For the Pauli matrices, the following uncertainty relation holds:

$$\Delta \sigma_x \Delta \sigma_y \geq \left| \langle \sigma_z \rangle \right| .$$ (V.391)

For which normalized states $\psi = \begin{pmatrix} a \\ b \end{pmatrix}$ is the right-hand side a minimum/maximum?

Solution: We have (see also (V.351))

$$\langle \sigma_z \rangle = \begin{pmatrix} a^* & b^* \end{pmatrix} \begin{pmatrix} 1 & 0 \\ 0 & -1 \end{pmatrix} \begin{pmatrix} a \\ b \end{pmatrix} = |a|^2 - |b|^2 = 2|a|^2 - 1 \qquad (V.392)$$

(because of the normalization $|a|^2 + |b|^2 = 1$). $|\langle \sigma_z \rangle|$ is maximal ($|\langle \sigma_z \rangle| = 1$) for $|a| = 0, 1$; and minimal ($\langle \sigma_z \rangle = 0$) for $|a| = \pm \frac{1}{\sqrt{2}}$.

30. What is the generalized uncertainty relation for H and $\mathbf{p}$?

Solution: It is

$$\Delta H \cdot \Delta p_i \geq \frac{1}{2} |\langle [H, p_i] \rangle|. \qquad (V.393)$$

With

$$H p_i - p_i H = V p_i - p_i V = -(p_i V) \qquad (V.394)$$

it follows:

$$\Delta H \cdot \Delta p_i \geq \frac{\hbar}{2} \left| \frac{\partial V}{\partial x_i} \right|. \qquad (V.395)$$

31. The position operator in the Heisenberg picture, x_H, is given by

$$x_H = e^{i \frac{tH}{\hbar}} x e^{-i \frac{tH}{\hbar}}. \qquad (V.396)$$

How does this operator depend explicitly on time? The potential is assumed to be constant, $\frac{dV}{dx} = 0$. Hint: Use the equation

$$e^{iA} B e^{-iA} = B + i[A, B] + \frac{i^2}{2!} [A, [A, B]] + \frac{i^3}{3!} [A, [A, [A, B]]] + \cdots \quad (V.397)$$

or

$$i\hbar \frac{d}{dt} x_H = [x_H, H] \qquad (V.398)$$

(or both for practice).

32. A Hamiltonian H depends on a parameter q, $H = H(q)$. In addition, $E(q)$ is a nondegenerate eigenvalue and $|\varphi(q)\rangle$ the corresponding eigenvector:

$$H(q) |\varphi(q)\rangle = E(q) |\varphi(q)\rangle. \qquad (V.399)$$

Show that

$$\frac{\partial E(q)}{\partial q} = \langle \varphi(q)| \frac{\partial H(q)}{\partial q} |\varphi(q)\rangle. \qquad (V.400)$$

(This equation is also called the Feynman–Hellmann theorem.)
Solution: It holds that

$$\langle \varphi (q)| H (q) |\varphi (q)\rangle = E (q) . \qquad \text{(V.401)}$$

We differentiate both sides with respect to q and obtain

$$\frac{\partial E (q)}{\partial q} = \left\langle \frac{\partial}{\partial q} \varphi (q) \middle| H (q) |\varphi (q)\rangle + \langle \varphi (q)| \frac{\partial H (q)}{\partial q} |\varphi (q)\rangle + \langle \varphi (q)| H (q) \middle| \frac{\partial}{\partial q} \varphi (q) \right\rangle$$

$$= \langle \varphi (q)| \frac{\partial H (q)}{\partial q} |\varphi (q)\rangle + E (q) \left(\left\langle \frac{\partial}{\partial q} \varphi (q) \middle| \varphi (q)\rangle + E (q) \langle \varphi (q) \middle| \frac{\partial}{\partial q} \varphi (q) \right\rangle \right)$$

$$= \langle \varphi (q)| \frac{\partial H (q)}{\partial q} |\varphi (q)\rangle + E (q) \frac{\partial}{\partial q} \langle \varphi (q) |\varphi (q)\rangle = \langle \varphi (q)| \frac{\partial H (q)}{\partial q} |\varphi (q)\rangle ,$$

$$\text{(V.402)}$$

due to

$$\frac{\partial}{\partial q} \langle \varphi (q) |\varphi (q)\rangle = \frac{\partial}{\partial q} 1 = 0. \qquad \text{(V.403)}$$

33. $\{|n\rangle\}$ is a CONS. Every solution of the SEq may be written as

$$|\psi\rangle = \sum_l a_l |l\rangle \qquad \text{(V.404)}$$

and every operator A as

$$A = \sum_{mn} c_{mn} |n\rangle \langle m| . \qquad \text{(V.405)}$$

Can the non-Hermitian operator A (i.e. $c_{mn} \neq c_{nm}^*$ for at least one pair n, m) have a real expectation value (for arbitrary states $|\psi\rangle$) under these conditions?
Solution: Let

$$A^\dagger = \sum_{mn} c_{mn}^* |m\rangle \langle n| = \sum_{mn} c_{nm}^* |n\rangle \langle m| \neq A \qquad \text{(V.406)}$$

or

$$c_{mn} \neq c_{nm}^*. \qquad \text{(V.407)}$$

For the expectation value, it holds that

$$\langle A \rangle = \sum_{mnll'} a_l^* \langle l| c_{mn} |n\rangle \langle m| a_{l'} |l'\rangle = \sum_{mnll'} a_l^* c_{mn} \delta_{nl} \delta_{ml'} a_{l'} = \sum_{mn} a_n^* c_{mn} a_m.$$

$$\text{(V.408)}$$

This expectation value must be real, i.e.

$$\langle A \rangle = \sum_{mn} a_n^* c_{mn} a_m = \langle A \rangle^* = \sum_{mn} a_n c_{mn}^* a_m^* = \sum_{mn} a_m c_{nm}^* a_n^*. \qquad \text{(V.409)}$$

If the last equation has to hold for all possible values of $\{a_n\}$ (i.e. for all solutions of the SEq), we must have

$$c_{mn} = c^*_{nm}. \tag{V.410}$$

Thus, we have a contradiction.

34. We consider the Hamiltonian $H = 1 + a\sigma_y$, already introduced in the exercises for Chap. 8.

(a) What is the expected result of the measurement of the x-component of the spin in the state $|\psi_t\rangle$ with $|\psi_0\rangle = \begin{pmatrix} 1 \\ 0 \end{pmatrix}$?

Solution: The x-component of the spin is represented by the operator $s_x = \frac{\hbar}{2}\sigma_x$. Since $|\psi_t\rangle$ is normalized, the expectation value is given by $\langle s_x \rangle = \langle \psi_t | s_x | \psi_t \rangle$ with $|\psi_t\rangle = e^{-it/\hbar} \begin{pmatrix} \cos\frac{at}{\hbar} \\ \sin\frac{at}{\hbar} \end{pmatrix}$ (see Chap. 8), i.e.

$$\langle s_x \rangle = \frac{\hbar}{2} \left(\cos\frac{at}{\hbar} \ \ \sin\frac{at}{\hbar} \right) \begin{pmatrix} 0 & 1 \\ 1 & 0 \end{pmatrix} \begin{pmatrix} \cos\frac{at}{\hbar} \\ \sin\frac{at}{\hbar} \end{pmatrix}$$

$$= \frac{\hbar}{2} 2 \cos\frac{at}{\hbar} \sin\frac{at}{\hbar} = \frac{\hbar}{2} \sin\frac{2at}{\hbar}. \tag{V.411}$$

(b) What is the uncertainty Δs_x in this state?

Solution: It is $\Delta^2 s_x = \langle s_x^2 \rangle - \langle s_x \rangle^2$. Because of $s_x^2 = \frac{\hbar^2}{4} \begin{pmatrix} 1 & 0 \\ 0 & 1 \end{pmatrix}$, we have $\langle s_x^2 \rangle = \frac{\hbar^2}{4}$ and it follows that

$$\Delta s_x = \sqrt{\langle s_x^2 \rangle - \langle s_x \rangle^2} = \frac{\hbar}{2}\sqrt{1 - \sin^2\frac{2at}{\hbar}} = \frac{\hbar}{2}\left| \cos\frac{2at}{\hbar} \right|. \tag{V.412}$$

(c) Calculate the commutator $[s_x, s_y]$ and formulate the uncertainty relation for the observables s_x and s_y for arbitrary times t.

Solution: start with $[s_x, s_y] = i\hbar s_z$. With this, the generalized uncertainty relation reads

$$\Delta s_x \Delta s_y \geq \frac{\hbar}{2} |\langle s_z \rangle|. \tag{V.413}$$

We calculate the two sides separately. First the left side: We know Δs_x. For Δs_y, we calculate first $\langle s_y \rangle$. It is

$$\langle s_y \rangle = \frac{\hbar}{2} \left(\cos\frac{at}{\hbar} \ \ \sin\frac{at}{\hbar} \right) \begin{pmatrix} 0 & -i \\ i & 0 \end{pmatrix} \begin{pmatrix} \cos\frac{at}{\hbar} \\ \sin\frac{at}{\hbar} \end{pmatrix} = 0 \tag{V.414}$$

and therefore $\Delta^2 s_y = \langle s_y^2 \rangle = \left\langle \frac{\hbar^2}{4} \begin{pmatrix} 1 & 0 \\ 0 & 1 \end{pmatrix} \right\rangle = \frac{\hbar^2}{4}$. All together, it follows that

$$\Delta s_x \, \Delta s_y = \frac{\hbar^2}{4} \left| \cos \frac{2at}{\hbar} \right|. \tag{V.415}$$

Now the right side: we have:

$$\frac{\hbar}{2} |\langle s_z \rangle| = \frac{\hbar}{2} \left| \left(\cos \frac{at}{\hbar} \quad \sin \frac{at}{\hbar} \right) \frac{\hbar}{2} \begin{pmatrix} 1 & 0 \\ 0 & -1 \end{pmatrix} \begin{pmatrix} \cos \dfrac{at}{\hbar} \\ \sin \dfrac{at}{\hbar} \end{pmatrix} \right| \tag{V.416}$$

$$= \frac{\hbar^2}{4} \left| \cos^2 \frac{at}{\hbar} - \sin^2 \frac{at}{\hbar} \right| = \frac{\hbar^2}{4} \left| \cos \frac{2at}{\hbar} \right|.$$

Hence, both sides are equal (so to speak the closest realization of the uncertainty relation).

35. Given an eigenvalue problem $A \, |a_m\rangle = a_m \, |a_m\rangle$ ($\{|a_m\rangle\}$ is a CONS); we can define a function of the operator by

$$F(A) \, |a_m\rangle := F(a_m) \, |a_m\rangle. \tag{V.417}$$

(a) Show that:

$$F(A) = \sum_m F(a_m) \, P_m \tag{V.418}$$

with $P_m = |a_m\rangle \, \langle a_m|$.
Solution: We have

$$F(A) \, |a_m\rangle = F(a_m) \, |a_m\rangle \rightarrow F(A) \, |a_m\rangle \langle a_m| = F(a_m) \, |a_m\rangle \langle a_m|, \tag{V.419}$$

and due to the completeness of the eigenvectors, it follows that

$$F(A) = \sum_m F(a_m) \, |a_m\rangle \langle a_m| = \sum_m F(a_m) \, P_m. \tag{V.420}$$

(b) Show that if $F(a)$ is real for all eigenvalues a_m, then $F(A)$ is self-adjoint.
Solution: Let $F^*(a_m) = F(a_m)$. Then it follows that

$$[F(A)]^\dagger = \sum_m F^*(a_m) \, P_m = \sum_m F(a_m) \, P_m = F(A). \tag{V.421}$$

36. What are the conditions which the elements of a two-dimensional normal matrix have to fulfill?
Solution: With

$$A = \begin{pmatrix} a & b \\ c & d \end{pmatrix}; \quad A A^\dagger = A^\dagger A, \tag{V.422}$$

it follows that

$$\begin{pmatrix} a & b \\ c & d \end{pmatrix} \begin{pmatrix} a^* & c^* \\ b^* & d^* \end{pmatrix} = \begin{pmatrix} a^* & c^* \\ b^* & d^* \end{pmatrix} \begin{pmatrix} a & b \\ c & d \end{pmatrix}, \tag{V.423}$$

and thus

$$\begin{pmatrix} aa^* + bb^* & ac^* + bd^* \\ ca^* + db^* & cc^* + dd^* \end{pmatrix} = \begin{pmatrix} a^*a + c^*c & a^*b + c^*d \\ b^*a + d^*c & b^*b + d^*d \end{pmatrix}. \tag{V.424}$$

This gives the two equations

$$bb^* = c^*c; \ ac^* + bd^* = a^*b + c^*d \tag{V.425}$$

or

$$bb^* = c^*c; \ (a - d) c^* = \left(a^* - d^*\right) b. \tag{V.426}$$

These two equations have the particular solution $a = d$ and $bb^* = c^*c$. For $a \neq d$, the solution reads

$$A = \begin{pmatrix} a & b \\ \frac{a-d}{a^*-d^*}b^* & d \end{pmatrix}. \tag{V.427}$$

37. Given the matrix

$$A = \begin{pmatrix} 0 & \gamma^2 \\ 1 & 0 \end{pmatrix}; \ \gamma \neq 0. \tag{V.428}$$

(a) Is A normal?
    Solution: We have

$$A^\dagger = \begin{pmatrix} 0 & 1 \\ \gamma^{*2} & 0 \end{pmatrix} \tag{V.429}$$

and therefore

$$AA^\dagger = \begin{pmatrix} 0 & \gamma^2 \\ 1 & 0 \end{pmatrix} \begin{pmatrix} 0 & 1 \\ \gamma^{*2} & 0 \end{pmatrix} = \begin{pmatrix} |\gamma|^4 & 0 \\ 0 & 1 \end{pmatrix} \tag{V.430}$$

and

$$A^\dagger A = \begin{pmatrix} 0 & 1 \\ \gamma^{*2} & 0 \end{pmatrix} \begin{pmatrix} 0 & \gamma^2 \\ 1 & 0 \end{pmatrix} = \begin{pmatrix} 1 & 0 \\ 0 & |\gamma|^4 \end{pmatrix}. \tag{V.431}$$

Hence, the matrix is not normal for $|\gamma| \neq 1$.

(b) Show that A is diagonalizable for almost all γ, but not by a unitary transformation.
    Solution: A has the eigenvalues $\pm\gamma$. A is diagonalizable, if there is a matrix $B = \begin{pmatrix} a & b \\ c & d \end{pmatrix}$ such that

$$\begin{pmatrix} a & b \\ c & d \end{pmatrix}\begin{pmatrix} 0 & \gamma^2 \\ 1 & 0 \end{pmatrix} = \begin{pmatrix} \gamma & 0 \\ 0 & -\gamma \end{pmatrix}\begin{pmatrix} a & b \\ c & d \end{pmatrix} \quad \text{with } \det B = ad - bc \neq 0$$
(V.432)

holds. From the last equation, it follows that

$$\begin{pmatrix} b & a\gamma^2 \\ d & c\gamma^2 \end{pmatrix} = \begin{pmatrix} \gamma a & \gamma b \\ -\gamma c & -\gamma d \end{pmatrix}.$$
(V.433)

This gives the two equations $b = \gamma a$ and $d = -\gamma c$. From them, it follows that

$$B = \begin{pmatrix} a & \gamma a \\ c & -\gamma c \end{pmatrix} \quad \text{with } \det B = -2ac\gamma \neq 0.$$
(V.434)

Hence, neither a nor c must vanish.

Now we have to determine whether B is unitary. We have

$$\begin{aligned} BB^\dagger &= \begin{pmatrix} a & \gamma a \\ c & -\gamma c \end{pmatrix}\begin{pmatrix} a^* & c^* \\ \gamma^* a^* & -\gamma^* c^* \end{pmatrix} \\ &= \begin{pmatrix} |a|^2\left(1 + |\gamma|^2\right) & ac^*\left(1 - |\gamma|^2\right) \\ a^*c\left(1 - |\gamma|^2\right) & |c|^2\left(1 + |\gamma|^2\right) \end{pmatrix}; \quad ac \neq 0. \end{aligned}$$
(V.435)

One sees directly that B is unitary only for $|\gamma| = 1$.

38. In the derivation of the uncertainty relation, the functions must be in the domains of definition of the operators and of the operator products involved. If they are not, we do not obtain meaningful statements. As an example we consider the function:

$$f(x) = \frac{\sin x^2}{x}.$$
(V.436)

(a) Is $f(x)$ square-integrable?
   Solution: We have

$$\int_{-\infty}^{\infty} f^2(x)dx = \int_{-\infty}^{\infty} \frac{\sin^2 x^2}{x^2}dx = 2\sqrt{\pi}.$$
(V.437)

   Hence, the function is square-integrable.
(b) Is $f(x)$ in the domain of definition of the operator x?
   Solution: No, $xf(x) = \sin x^2$ is not square-integrable. In other words: $f(x)$ is not in the domain of definition of x.
(c) Can a meaningful uncertainty relation be derived for $f(x)$?
   Solution: For Δx, we have

$$(\Delta x)^2 = \langle x^2 \rangle - \langle x \rangle^2 = \int_{-\infty}^{\infty} \sin^2 x^2 dx - \left(\int_{-\infty}^{\infty} \frac{\sin^2 x^2}{x} dx \right)^2 \qquad \text{(V.438)}$$

and a similar formulation holds for Δp. Even if we accepted the value ∞ for $\int \sin^2 x^2 dx$, the statement $\Delta x \Delta p \geq \frac{\hbar}{2}$ is trivially satisfied, i.e. meaningless, similar to $\infty \geq \frac{\hbar}{2}$.

(d) Can similar statements be made for the function $g(x) = \frac{\sin x}{x}$?

39. Given two operators A and B which commute with their commutator, $[A, [A, B]] = [B, [A, B]] = 0$. Show that:

$$\left[B, A^n \right] = n \, [B, A] \, A^{n-1}. \qquad \text{(V.439)}$$

Solution: We use mathematical induction. The equation clearly is valid for $n = 1$. If it applies for n, it follows for $n + 1$:

$$\left[B, A^{n+1} \right] = B A^{n+1} - A^{n+1} B = B A^{n+1} - ABA^n + ABA^n - A^{n+1} B$$

$$= [B, A] \, A^n + A \, [B, A^n] = [B, A] \, A^n + n A \, [B, A] \, A^{n-1} = (n + 1) \, [B, A] \, A^n. \qquad \text{(V.440)}$$

40. Show that the momentum operator is given in the coordinate representation by $p = \frac{\hbar}{i} \frac{d}{dx}$. Make use only of the commutator $[x, p] = i\hbar$ and derive, making use of the previous exercise, the result:

$$[p, f(x)] = \frac{\hbar}{i} \frac{df(x)}{dx}. \qquad \text{(V.441)}$$

Solution: We expand the function $f(x)$ in a Taylor series e.g. around zero: $f(x) = \sum_n f^{(n)}(0) \frac{x^n}{n!}$. Then it holds that

$$[p, f(x)] = \sum_n f^{(n)}(0) \frac{1}{n!} \left[p, x^n \right]. \qquad \text{(V.442)}$$

We transform the commutator on the right side by means of (V.439) and obtain with $[p, x^n] = n \, [p, x] \, x^{n-1}$:

$$[p, f(x)] = \sum_n f^{(n)}(0) \frac{1}{n!} n \, [p, x] \, x^{n-1} = \frac{\hbar}{i} \sum_n f^{(n)}(0) \frac{x^{n-1}}{(n-1)!} = \frac{\hbar}{i} \frac{df(x)}{dx}$$

$$\text{(V.443)}$$

i.e. (V.441). The form of this equation suggests the *ansatz* $p = \alpha\frac{d}{dx}$; we insert and obtain with the auxiliary function $g(x)$, because of

$$[p, f(x)]g(x) = \frac{\hbar}{i}\frac{df(x)}{dx} \cdot g(x) \tag{V.444}$$

initially:

$$[p, f(x)]g(x) = \alpha\frac{d}{dx}f(x)g(x) - \alpha f(x)\frac{d}{dx}g(x) = \alpha g(x)\frac{d}{dx}f(x). \tag{V.445}$$

Finally, the constant α is determined by the comparison with (V.444), giving $\frac{\hbar}{i}$, and we obtain the desired result.

41. Given two operators A and B which commute with their commutator, $[A, [A, B]]$ $= [B, [A, B]] = 0$. Show that:

$$e^{A+B} = e^A e^B e^{-\frac{1}{2}[A,B]}. \tag{V.446}$$

This is a special case of the *Baker-Campbell-Hausdorff formula* (relation, theorem). The general case considers e^{A+B} for two operators, which do not have to commute with their commutator (this is used e.g. in (V.397)). By the way, these authors published their work in 1900, well before the birth of quantum mechanics.

(a) First, prove the equation

$$[B, e^{xA}] = e^{xA}[B, A]x. \tag{V.447}$$

Solution: With the power series expansion of the e-function, we have

$$[B, e^{xA}] = \sum \frac{1}{n!}x^n[B, A^n]$$
$$e^{xA}[B, A]x = \sum \frac{1}{n!}x^{n+1}A^n[B, A]. \tag{V.448}$$

We compare same powers of x:

$$\frac{1}{(n+1)!}x^{n+1}[B, A^{n+1}] = \frac{1}{n!}x^{n+1}A^n[B, A]. \tag{V.449}$$

This equation is already proved, cf. (V.439).

(b) Define

$$G(x) = e^{xA}e^{xB} \tag{V.450}$$

and show the following equation holds:

$$\frac{dG}{dx} = (A + B + [A, B] x) G. \tag{V.451}$$

Integrate this equation.

Solution: Taking the derivative of G gives

$$\frac{dG}{dx} = \frac{d}{dx} e^{xA} e^{xB} = Ae^{xA} e^{xB} + e^{xA} e^{xB} B. \tag{V.452}$$

The second term on the right side is transformed by means of the result of part (a):

$$e^{xA} e^{xB} B = e^{xA} Be^{xB} = \left(Be^{xA} + e^{xA} [A, B] x \right) e^{xB}$$
$$= (B + [A, B] x) e^{xA} e^{xB} - \tag{V.453}$$

It follows that

$$\frac{dG}{dx} = (A + B + [A, B] x) G. \tag{V.454}$$

We can integrate this equation directly, since the operator $(A + B)$ commutes with $[A, B]$:

$$G(x) = G_0 e^{(A+B)x + \frac{1}{2}[A,B]x^2}. \tag{V.455}$$

Due to $G(x = 0) = 1$, the integration constant G_0 is 1, and the final result follows for $x = 1$:

$$e^A e^B = e^{A+B+\frac{1}{2}[A,B]}. \tag{V.456}$$

V.13   Exercises, Chap. 14

1. Given an observable A and a state $|\varphi\rangle$. Show by means of Postulates (2.1) and
(2.2) that the expected result of a measurement of A is given by $\langle A \rangle = \langle \varphi| A |\varphi\rangle$.
To simplify the discussion, we consider an observable A whose eigenvalues are
discrete and non-degenerate and whose eigenvectors form a CONS, $A |n\rangle = a_n |n\rangle$.

Solution: Let $|\varphi\rangle$ have the form $|\varphi\rangle = \sum_n c_n |n\rangle$, where at least one coefficient
is not zero. Then we know from Postulate 2.1 that the probability of finding $|\varphi\rangle$
in the state $|n\rangle$ (i.e. of measuring a_n) is given by

$$p_n = |\langle n |\varphi\rangle|^2 = \langle \varphi| P_n |\varphi\rangle \tag{V.457}$$

where $P_n = |\varphi_n\rangle \langle \varphi_n|$ is the projection operator onto the subspace n. As always,
we assume that $|\varphi\rangle$ is normalized.

As we have already seen in Chap. 9, this is where the expectation value comes into play. If one has measured the quantity a_n on a single system, one cannot draw any definite conclusions about the state prior to the measurement, since we are dealing with probabilities. To get more information, one can—at least in principle—proceed by preparing an ensemble, i.e. by preparing many individual systems so that they are all in the same state $|\varphi\rangle$. Now one determines which of the states $|n\rangle$ is occupied by each member of the ensemble. If the number of measurements N is very large, we have an experimental statement about the expectation value of A in the state $|\varphi\rangle$, namely

$$\langle A \rangle = \lim_{N \to \infty} \frac{1}{N} \sum_{m=1}^{N} A_m \tag{V.458}$$

where A_m is the result of the mth measurement. These results vary from one measurement to the next, but always show one of the values a_n. The theoretical value (expectation value) is (we assume no degeneracy):

$$\langle A \rangle = \sum_n p_n a_n = \sum_n \langle \varphi | n \rangle a_n \langle n | \varphi \rangle = \langle \varphi | A | \varphi \rangle \tag{V.459}$$

where we have used the spectral decomposition $A = \sum_n |n\rangle a_n \langle n|$ (see Chap. 13).

2. Show that the operator $s_x + s_z$ is Hermitian, but does not represent a measurable physical quantity if understood literally, i.e., as the instruction to measure the x-component plus (and) the z-component of the spin. The spin matrices s_i are related to the Pauli matrices σ_i by $s_i = \frac{\hbar}{2}\sigma_i$.

Solution: The operator $s_x + s_z = \frac{\hbar}{2} \begin{pmatrix} 1 & 1 \\ 1 & -1 \end{pmatrix}$ is obviously Hermitian. Its eigenvalues are $\pm\frac{\hbar}{\sqrt{2}}$ (check yourself); and, according to our postulates, one obtains one of theses values as the result of a measurement. If, on the other hand, we measure s_x and s_z separately and then add the results, the measurement gives for each of the two operators $\pm\frac{\hbar}{2}$, and thus in the sum one of the three results $\hbar$, 0 or $-\hbar$; i.e. values which clearly do not match up with $\pm\frac{\hbar}{\sqrt{2}}$.
Of course, the core of the problem is that the two operators s_x and s_z do not commute.

3. (An example concerning projections, probabilities and expectation values.) The angular momentum operator **L** for angular momentum 1 can be represented in the vector space $\mathbb{C}^3$ by the following matrices (see Chap. 16, Vol. 2):

$$L_x = \frac{\hbar}{\sqrt{2}} \begin{pmatrix} 0 & 1 & 0 \\ 1 & 0 & 1 \\ 0 & 1 & 0 \end{pmatrix}; \; L_y = \frac{\hbar}{\sqrt{2}} \begin{pmatrix} 0 & -i & 0 \\ i & 0 & -i \\ 0 & i & 0 \end{pmatrix}; \; L_z = \hbar \begin{pmatrix} 1 & 0 & 0 \\ 0 & 0 & 0 \\ 0 & 0 & -1 \end{pmatrix}.$$

$$\tag{V.460}$$

(a) Which measured results are possible in a measurement of L_i ($i = x, y, z$)?
    Solution: The measured result must be one of the eigenvalues of L_i. For L_z,
    one sees directly that the eigenvalues are $+\hbar$, 0, $-\hbar$ (diagonal elements). The
    calculation shows (check yourself) that L_x and L_y have these eigenvalues,
    also.

(b) What are the corresponding eigenvectors for L_z?
    Solution: The calculation shows (check yourself) that the eigenvectors asso-
    ciated with the eigenvalues $+\hbar$, 0, $-\hbar$ are

$$|+\hbar\rangle = \begin{pmatrix} 1 \\ 0 \\ 0 \end{pmatrix}; \quad |0\rangle = \begin{pmatrix} 0 \\ 1 \\ 0 \end{pmatrix}; \quad |-\hbar\rangle = \begin{pmatrix} 0 \\ 0 \\ 1 \end{pmatrix}. \tag{V.461}$$

(c) What are the probabilities of measuring the results $+\hbar$, 0, $-\hbar$ on the state

$$|\psi\rangle = \begin{pmatrix} 1 \\ i \\ -2 \end{pmatrix} ? \tag{V.462}$$

Solution: First, we have to normalize the state:

$$|\psi\rangle_{\text{norm}} = \frac{1}{\sqrt{6}} \begin{pmatrix} 1 \\ i \\ -2 \end{pmatrix}. \tag{V.463}$$

We see directly that the probabilities for obtaining the measured result $+\hbar$
or 0 or $-\hbar$ are given by $\frac{1}{6}$ or $\frac{1}{6}$ or $\frac{2}{3}$. The probabilities sum up to 1, as indeed
they must.

4. Given the state

$$|\psi\rangle_v = \frac{|x_1\rangle e^{-i\omega t} + |x_2\rangle e^{-2i\omega t}}{\sqrt{2}} \tag{V.464}$$

with normalized and mutually orthogonal states $|x_i\rangle$: We measure the x_1 com-
ponent of $|\psi\rangle_v$. After the measurement, we have

$$|\psi\rangle_n = |x_1\rangle e^{-i\omega t}. \tag{V.465}$$

Illustrate this state reduction by considering the change in the real or imaginary
part of $|\psi\rangle$.
Solution: By measuring at time T, we cause the state $|\psi\rangle_v$ to collapse into $|x_1\rangle$
(apart from a possible phase). This is the initial value of the time evolution
after the measurement. Since the energy is sharp, the time behavior is given
by $|x_1\rangle e^{-i\omega t}$. The Hilbert space is two-dimensional (spanned by $|x_1\rangle$ and $|x_2\rangle$).
Because of the complex prefactors, we thus have a four-dimensional space. To

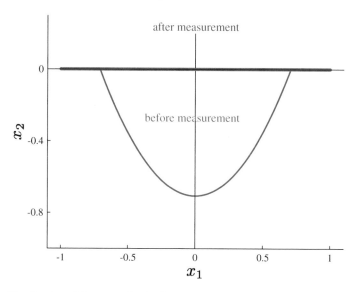

Fig. V.4 Visualization of the state reduction by a measurement

enable a visual presentation, we restrict ourselves to considering the real part. Before the measurement, we have

$$\text{Re}\,|\psi\rangle_v = \frac{|x_1\rangle \cos \omega t + |x_2\rangle \cos 2\omega t}{\sqrt{2}} \cong \frac{1}{\sqrt{2}} \begin{pmatrix} \cos \omega t \\ \cos 2\omega t \end{pmatrix}; \qquad \text{(V.466)}$$

and afterwards,

$$\text{Re}\,|\psi\rangle_n = |x_1\rangle \cos \omega t \cong \begin{pmatrix} \cos \omega t \\ 0 \end{pmatrix}. \qquad \text{(V.467)}$$

For the visualization we use $\cos 2\omega t = \cos^2 \omega t - \sin^2 \omega t = 2\cos^2 \omega t - 1$. Before the measurement, $\text{Re}\,|\psi\rangle_v$ moves on the parabola $x_2 = 2x_1^2 - 1$; afterwards, back and forth on the x_1 axis, as indicated in Fig. V.4.

Further Reading

1. J. Audretsch (ed.), *Verschränkte Welt-Faszination der Quanten (Entangled World-Fascination of Quantum, in German)* (Wiley-VCH, Weinheim, 2002)
2. J. Audretsch, *Entangled Systems* (Wiley-VCH, Weinheim, 2007)
3. J.-L. Basdevant, J. Dalibard, *Quantum Mechanics* (Springer, Berlin, 2002)
4. D.R. Bes, *Quantum Mechanics* (Springer, Berlin, 2004)
5. B.H. Bransden, C.J. Joachain, *Quantum Mechanics* (Pearson Education Limited, Harlow, 2000)
6. C. Cohen-Tannoudji, B. Diu, F. Laloë, *Quantum Mechanics*, vols. 1 & 2 (Hermann Paris/Wiley, New York, 1977)
7. F. Embacher, Homepage with much material about quantum theory (also for school), University Vienna (2012), http://homepage.univie.ac.at/franz.embacher/
8. R.P. Feynman, R.B. Leighton, M. Sand, *Quantum Mechanics*. The Feynman Lectures on Physics, vol. 3 (Addison-Wesley Reading, Massachusetts, 1965)
9. T. Fließbach, *Quantenmechanik (Quantum Mechanics, in German)*, 3rd edn. (Spektrum Akademischer Verlag, Heidelberg, 2000)
10. K. Gottfried, T.-M. Yan, *Quantum Mechanics: Fundamentals* (Springer, New York, 2006)
11. K.T. Hecht, *Quantum Mechanics* (Springer, New York, 2000)
12. C.J. Isham, *Quantum Theory-Mathematical and Structural Foundations* (Imperial College Press, London, 2008)
13. T. Lancaster, S.J. Blundell, *Quantum Field Theory for the Gifted Amateur* (Oxford University Press, Oxford, 2014)
14. R.D. Klauber, *Student Friendly Quantum Field Theory*, 2nd edn. (Sandrove Press, Fairfield, 2015)
15. M. Le Bellac, *Quantum Physics* (Cambridge University Press, Cambridge, 2006)
16. H. Lüth, *Quantenphysik in der Nanowelt (Quantum Physics in the Nanoworld, in German)* (Springer, Berlin, 2009)
17. E. Merzbacher, *Quantum Mechanics*, 3rd edn. (Wiley, New York, 1998)
18. A. Messiah, *Quantum Mechanics*, vols. 1 & 2 (North-Holland Publishing Company, Amsterdam, 1964)
19. Münchener Internetprojekt zur Lehrerfortbildung in Quantenmechanik (Munich internet project for teacher training in quantum mechanics, in German) (2012), http://homepages. physik.uni-muenchen.de/milq/
20. G. Münster, *Quantentheorie (Quantum Theory, in German)* (Walter de Gruyter, Berlin, 2006)

© Springer Nature Switzerland AG 2018
J. Pade, *Quantum Mechanics for Pedestrians 1*, Undergraduate Lecture
Notes in Physics, https://doi.org/10.1007/978-3-030-00464-4

21. W. Nolting, *Grundkurs Theoretische Physik 5, Quantenmechanik, Teil 1: Grundlagen und Quantenmechanik Teil 2: Methoden und Anwendungen (Quantum Mechanics, Part 1: Basics and Part 2: Methods and Applications, in German)* (Verlag Zimmermann-Neufang, Ulmen, 1992)
22. A. Peres, *Quantum Theory-Concepts and Methods* (Kluwer Academic Publishers, Doordrecht, 1995)
23. A.I.M. Rae, *Quantum Mechanics*, 5th edn. (Taylor and Francis, New York, 2008)
24. H. Rollnik, *Quantentheorie 1 & 2 (Quantum Theory 1 & 2, in German)*, 2nd edn. (Springer, Berlin, 2003)
25. H. Schulz, *Physik mit Bleistift (Physics with a Pencil, in German)*, 4th edn. (Verlag Harri Deutsch, Frankfurt am Main, 2001)
26. F. Schwabl, *Quantum Mechanics*, 3rd edn. (Springer, Berlin, 2002)
27. N. Zettili, *Quantum Mechanics, Concepts and Applications*, 2nd edn. (Wiley, New York, 2009)

Index of Volume 1

Heisenberg, 345
interaction, 345
Schrödinger, 345
Poisson bracket, 389, 392
Polarization
circular, 18
elliptical, 18
linear, 17
Position representation, 158
Postulates of quantum mechanics, 351
Potential
real, 94
time-independent, 31
Potential well
infinite, 55
Preparation, 180, 370
Probability, 195
Probability amplitude, 190
Probability current density, 95
Probability density, 87
Projection, 47
Propagator, 175, 193
Property, 179, 192

Q
Quantization, canonical , 393
Quantum cryptography, 130, 339
Quantum eraser, 330
Quantum hopping, 311
Quantum Zeno effect, 82, 319
anti-, 319

R
Relativistic quantum mechanics, 405
Rotation, active and passive, 22

S
Scalar product, 44, 141, 143, 265
Schrödinger equation, 192
properties, 29
stationary, 32
time-dependent, 10
time-independent, 31
Self interference, 77
Separation of variables, 6
Special relativity, 375
Spectral representation, 178
Spectrum, 33
degenerate, 173
Spinor, 414, 424
Square-integrable, 61

Standard deviation, 115, 336
State, 126
flavor, 100
improper, 153
mass, 100
proper, 153
unstable, 320
State function, 11
State reduction, 92, 191
State space, 20, 41, 188
String theory, 70
Superposition principle, 6, 29, 313
System
closed, 367
isolated, 192, 367
open, 200, 367
Systems of units, 205

T
Tensor, metric, 381
Theorem
of Feynman-Hellmann, 184
Stone's, 298
Time evolution, 197
irreversible, 191
reversible, 193

U
Uncertainty
s. Standardabweichung, 115
Uncertainty principle, 300
Uncertainty relations, 170
Unitary space, 44, 142, 144

V
Variance, 115, 336
Vector space, 41, 264

W
Wave
plane, 8
Wave equation, 6
Wave function, 11
collapse, 92, 128, 191, 196
Wave-particle duality, 77
Which-way information, 77, 327

Z
Zero vector, 44, 268

Index of Volume 2

© Springer Nature Switzerland AG 2018
J. Pade, *Quantum Mechanics for Pedestrians 1*, Undergraduate Lecture
Notes in Physics, https://doi.org/10.1007/978-3-030-00464-4

Printed in the United States
By Bookmasters